Climate Change, Climate Science and Economics

G. Cornelis van Kooten

Climate Change, Climate Science and Economics

Prospects for an Alternative Energy Future

Springer

G. Cornelis van Kooten
Professor of Economics and Geography
Canada Research Chair in Environmental Studies
Department of Economics
University of Victoria
Victoria, BC, Canada

ISBN 978-94-007-4987-0 ISBN 978-94-007-4988-7 (eBook)
DOI 10.1007/978-94-007-4988-7
Springer Dordrecht Heidelberg New York London

Library of Congress Control Number: 2012947807

Printed on acid-free paper

Springer is part of Springer Science+Business Media (www.springer.com)

Preface

My exposure to climate related research actually began as a doctoral student in Agricultural and Resource Economics at Oregon State University. In a research methods course, one of my advisors, Dr. John ('Jack') Edwards, commented upon his efforts to complete his Ph.D. dissertation in Economics from the University of Chicago. He was examining the effects of climate on agriculture and, after running numerous regressions that confounded cause and effect, he finally completed the research by applying a large dose of economic theory and a simple regression model. But it was not until I had taken a position in Agricultural Economics at the University of Saskatchewan that, in the mid-1980s, the question of climate change and global warming came to my attention.

Elaine Wheaton of the Saskatchewan Research Council in Saskatoon put together a team of researchers to examine the effect of anthropogenic climate change on Canada's boreal forests, potential strategies for forest-sector adaptation, and the role that Canada's forests might play in mitigating climate change. We completed Phase I of the research for Environment Canada by 1987 and put together a larger team of researchers for the promised Phase II, but then the Canadian government backed away from climate research and there was no Phase II. During this period, I also worked with Louise Arthur who was at the University of Manitoba, where her research focused on the impacts of climate change on prairie agriculture. We combined our talents and published several papers on climate change related to agriculture and forestry in western Canada. The research was unfunded for the most part and, by the early 1990s, we were overtaken by well-funded American researchers, who had the ability to develop large-scale, national-level models of the forest and agricultural sectors – our research was relegated to the periphery.

During the 1990s and until recently, I focused primarily on the economics of mitigating climate change through carbon sequestration in terrestrial ecosystems. I discuss this line of research in Chap. 9. It led to my appointment in 2002 as the Canada Research Chair in Environmental Studies and Climate in the Department of Economics at the University of Victoria (UVic), and to my involvement as a contributing author to the chapter on forestry in the 2007 report of the United Nations' Intergovernmental Panel on Climate Change (IPCC). At UVic, I was surprised at the politicization of climate research and the lack of focus, despite ongoing efforts in a number of areas. I found that the best venue for participating in climate research on campus was through the Institute for Integrated Energy Systems (or IESVic), which was an offshoot of the Department of Mechanical Engineering – essentially a think tank founded by David Scott. Although IESVic's primary focus was on fuel cell research, there was a spirit of openness and questioning regarding climate change and the means to address it. In this regard, I am grateful to IESVic's second director, Ged McLean (who to my dismay left academic life for industry), and subsequent directors Ned Djilali and Peter Wild, for encouraging me to engage with IESVic. I benefitted as well from many discussions with Lawrence Pitt (who has some of the best insights into climate change and renewable energy), Andrew Rowe, and many graduate students at IESVic, all of whom opened a door to the wonderful world of energy systems analysis in the context of reducing carbon dioxide emissions and enhancing energy efficiency. I am also grateful to my own graduate students during this period, including, among many, Jesse Maddaloni, Ryan Prescott, Julia Zhu, Geerte Cotteleer, Alison Eagle and Linda Wong.

But I was really dragged into the climate change morass (for that is what it truly is) when I provided pre-publication comments for the book *Taken by Storm*, written by Christopher Essex and Ross McKitrick. I had been asked to provide comments by Ross McKitrick of the University of Guelph, whom I had known at the University of British Columbia where, as a member of faculty, I had been a member of his Ph.D. supervisory committee. It was my first exposure to some of the issues concerning the theory of catastrophic anthropogenic global warming. My comments on the rear jacket of the book read as follows: "Any politician who failed to read this book and yet is willing to commit society's resources to avert global warming has been derelict in his or her duty to the public. Professors Essex and McKitrick present a powerful case." At that point, I still had reservations about the view taken by the book, but my dust jacket rendition stoked up the ire of the environmentalists, who subsequently sent me numerous emails requesting intimate details about my research funding (particularly if any research, regardless of the subject area, had ever been funded by an oil company), my employment history and so on, much of which was readily available on the internet.

I also had lunch with Dr. Timothy Ball, a retired professor of climate science at the University of Winnipeg now living in Victoria. Professor Ball indicated that, as a result of research that questioned whether carbon dioxide was a principal driver of climate change, he had been the object of vitriolic attacks on his person and felt he had been denied promotion on several occasions because of his views. As a leading thorn in the side of climate scientists, Ross also confided to me that he too had been the object of malicious attacks. McKitrick is often dismissed as an economist, unfit to comment on climate science, except that he is expert in statistical analysis and has extensive experience constructing models that predict economic outcomes – models of the kind that form the foundation of the emission scenarios used to drive climate model projections.

In July 2009, just as my study leave began, Dr. Calvin Beisner telephoned to ask if I would write a synopsis of the current state of economics related to climate change for his Cornwall Alliance for the Stewardship of Creation. A previous version had been written by Ross, who had probably suggested my name to Calvin, so I agreed that this would be a good way to begin my sabbatical, given that I had already agreed to teach a new course on the topic for a minor in Climate Studies at my UVic. The underlying document and associated research marked the beginning of this book. During the rest of 2009 and during 2010 and 2011, I discovered that economists were working on aspects related to almost all of the topics of climate science presented in this book, particularly climate reconstructions, paleoclimatology and climate modeling, as well as topics normally considered the purview of economists.

In this book, I attempt to address the science of climate change as objectively as possible. However, based solely on my choice of topics and the scientific literature I cite, it is unavoidable that I will be accused of bias. Therefore, it behooves me to be clear about my own stance. Until my encounter with the environmental movement (attributable to what I had written about *Taken by Storm*), a meeting with David Anderson (then Canada's Minister of the Environment under Jean Chretien's Liberal government) around the same time, and subsequent discussions with climate scientists at the University of Victoria, I had no specific view on what might be causing climate change. I was content to confine my research to economic issues relating to the uptake of carbon in forest ecosystems. It was only after much reflection, particularly during my 2009–2010 sabbatical leave that coincided, in late 2009, with the release of the so-called 'climategate' material (see Chap. 5), that I determined the case being made by climate modelers to be weak one, and not entirely supported by the empirical evidence. I simply could not understand how climate change could be an irreversible catastrophe that would put an end to the human race, how CEOs of energy companies had committed crimes against humanity (equivalent to Hitler, Pol Pot and other mass murders), or that rising temperatures were the greatest security threat to the United States – all claims made by climate scientists. Scientists had, in my mind, oversold their case.

The main obstacle that I could not get around, which potentially could prove to have involved one of the greatest science cover-ups ever (although the verdict is probably still out), concerned the Medieval Warm Period. Historical writings and anthropological evidence gathered and/or reported by Ian Plimer, Brian Fagan, Jared Diamond, Bjørn Lomborg and many others indicate that there was a period between about 900 and 1300 AD when the earth was warmer than today. Until recently, no one questioned the existence of the Medieval Warm Period, but climate scientists did just that in making the case that present temperatures are the warmest humans have ever experienced. The attempt to exorcise the Medieval Warm Period from the scientific record makes fascinating reading, and is discussed in Chap. 3. As carbon dioxide was not the culprit, what could explain this warming?

Another issue that stands out in my mind concerns the role of observation. Science has always been observation based: scientific theories are tested against the empirical evidence (deductive reasoning) or observations lead to 'generalizations' or theories (inductive reasoning). In contrast, 'predictions' of climate change are based on climate models that are not tested against observation, except through back-casting exercises as a form of validation. Back-casting as a means of validating models has not had much success in terms of predicting the future, as McKitrick well understood, and this is just as true for climate models as economic and other models. Knowing this, climate scientists validate their models by comparing them with other climate models that essentially share the same biases. A particular bias is that human activities are presumed to be the main drivers responsible for climate change. It is not surprising that not all scientists accept the outcomes of climate models, and it is not surprising to find scientists who propose alternative theories of how climate changes. It is important, even for the purposes of this book, to examine some of the questions surrounding the science of climate change. After all, economic prescriptions for addressing climate change can only be as good as the underlying science, and where the underlying science casts doubt, such doubt must be taken into account in developing policy.

Finally, I am disappointed by the climate debate. Very few people appear to know much about the nuances of climate change, yet they are willing to make all sorts of pronouncements regarding what governments must do to prevent global catastrophe. For example, in 2011 I presented a seminar in Geography at UVic focusing on weather index insurance. The seminar was highjacked when so-called climate experts in the audience took me to task for not understanding the science, issues of causality, and how science works. One questioned my knowledge of the 'urban heat island' (Chap. 2) suggesting that it only referred to the increase in nighttime temperature caused by the radiation of heat from pavement and other 'black bodies' that had been absorbed during the day. He was wrong, of course, because black bodies radiate heat at all times (see Chap. 4). Another claimed it is impossible to determine whether economic activity as measured by GDP causes temperature to rise or vice versa (see Chap. 2) – I thought causality in this case is rather obvious! A third person claimed that the research of some scientists was suspect because of their religious beliefs and/or political leanings. And these are individuals who are influential in teaching students and/or bringing UVic's scientific climate expertise to the general public. Rather shocking in my opinion. I can only hope that this book will enlighten them and others.

Finally, some acknowledgements are in order. I want to acknowledge many national and international non-governmental organizations and government agencies, publishers, and individuals who granted permission to use their material in whole or part. Permission to use historical surface temperature data was kindly granted by: (1) the United Kingdom's Meteorological Office at the Hadley Centre/Climate Research Centre at the University of East Anglia, whose website (www.metoffice.gov.uk/hadobs) contains public sector information licensed under the Open Government License v1.0; (2) the U.S. National Atmospheric Space Administration's Goddard Institute for Space Studies (http://data.giss.nasa.gov/); (3) the U.S. National Oceanic and Atmospheric Administration, which also provided information on weather stations (http://www.ncdc.noaa.gov/ghcnm/); (4) the Berkeley Earth Surface Temperature project at the University of California, Berkeley (http://www.berkeleyearth.org); and (5) Environment Canada for Canadian data. Permission to use temperature data derived from NASA satellites was provided by Dr. John Christy, who indicated that such data were in the public domain.

Paleoclimatic proxy data (e.g., tree ring data) and temperature reconstructions based on ice core, tree ring and other proxy data are stored electronically with the World Data Center for Paleoclimatology in Boulder, Colorado, and are available from their website http://www.ncdc.noaa.gov/paleo/paleo.html. These are open access data and a statement to that effect, "NOAA/National Climatic Data Center Open Access to Physical Climate Data Policy, December 2009," appears on their website.

Stephen McIntyre granted permission to use any material found on his website http://ClimateAudit.org. Likewise, Anthony Watts granted permission to use material from his website http://wattsupwiththat.com/. Watts' material included information about the quality of U.S. weather stations and the hockey stick http://wattsupwiththat.com/2009/10/05/united-nations-pulls-hockey-stick-from-climate-report/. Indur Goklany also granted permission to use any data found in his many reports and papers. Data on atmospheric carbon dioxide content are found at http://cdiac.ornl.gov/trends/co2/contents.htm and are used with permission. Arctic sea ice extent data used in this book were derived from AMSR-E sensor and provided by Japan Aerospace Exploration Agency (JAXA) through the IARC-JAXA Information System (IJIS).

Some material found in Chaps. 10, 11 and 12 are published as Chapter 2.20, entitled "Wind Energy Policy," in *Comprehensive Renewable Energy*, edited by Ali Sayigh, Copyright Elsevier 2012, and are used with permission. Some figures and tables in Chap. 10 were created using data from Key World Energy Statistics 2009 published in 2010 by the International Energy Agency of the OECD and used with permission.

I have benefitted from research funding from the Canada Research Chairs program, the Landbouw Economisch Instituut (LEI) in The Hague, Agriculture and Agri-Food Canada's Linking Environment and Agriculture Network, and various Social Science and Humanities Research Council Grants.

I am also grateful to various individuals who have contributed in one way or another to the research. Parts of Chap. 6 were written in consultation with Roel Jongeneel and Martijn vander Heide of the LEI and Brian Scarfe of my own Department. I am grateful for support from and discussions with Lawrence Pitt, Ross McKitrick, my colleague David Giles, Calvin Beisner, Tom Pedersen, Brad Stennes, Kurt Niquidet, James Nolan, Richard Tol, James Wanliss and many others. I owe a particular debt of gratitude to Linda Voss for research support and her many hours going through various drafts and preparing the document for publication. Finally, I want to thank my wife Mary for her support of my research, and providing an atmosphere conducive to completing a project of this magnitude.

Contents

List of Figures

List of Tables

Chapter 1
Introduction

I'm offended that science is being perverted in the name of global warming – today's environmental cause célèbre. ... [T]he world seems to have lost its collective mind and substituted political belief for the spirit of scientific inquiry.

– From the Preface of *Global Warming: False Alarm*
by Ralph Alexander

All interesting and important policy questions involve choosing among competing values. In this regard, human-caused climate change poses a most challenging policy conundrum for politicians. Does anthropogenic climate change demand drastic and immediate action? If so, what is the cost? Are the consequences of global warming catastrophic and, if so, what form do such consequences take? How willing are we to give up inexpensive fossil fuel energy and accept the consequences of mitigating climate change? Should citizens of rich countries deny citizens of poor countries access to cheap fossil-fuel energy? However well intended, it is naïve and irresponsible to ignore the unavoidable tradeoffs that efforts to reduce carbon dioxide emissions will entail.

Along with the tradeoffs come opportunity costs. The best measure of cost is the opportunities forgone – the value of alternatives sacrificed. Money spent to combat climate change cannot be spent to eradicate malaria (which kills 2 million people per year, mostly children under 5 years), to improve female literacy (a key requirement for social progress), to fight hunger, malnutrition and communicable diseases, or to build roads, electric power plants and transmission grids, and water and sewage treatment plants, that help reduce human misery.

The world is discovering that combating climate change will be extremely difficult and could be expensive. It is especially vexing because:

- The atmosphere is a commons with unrestricted access. The benefits of burning fossil fuels accrue to individuals, but the costs of emissions are borne by all. This makes climate change the greatest of all collective action problems. It requires the cooperation of all countries, each with different interests and incentives.

G.C. van Kooten, *Climate Change, Climate Science and Economics: Prospects for an Alternative Energy Future*, DOI 10.1007/978-94-007-4988-7_1,

- The costs and benefits of climate change and of its mitigation will be unequally distributed. This means different countries will bargain strategically to advance their individual perceived interests.
- Carbon dioxide is considered to be a persistent resident in the atmosphere. If overnight we eliminated every source of human CO_2 emissions, the atmosphere could continue warming for 100 years or more, assuming climate models are correct in their predictions.
- If current trends continue, developing countries will quite soon become the largest emitters of greenhouse gases; China has already become number one. Their leaders understand that increasing energy consumption is a prerequisite for continued economic development. Because carbon based fuels are cheap and ubiquitous, they will remain the fuel of choice, even where countries make concerted efforts to increase reliance on non-carbon sources of energy.
- Reducing emissions fast enough and by enough to avoid allegedly dangerous human interference with the climate system requires an unprecedented transformation of energy systems. For example, to cut global emissions in half by 2050 requires that, on average, the world economy will then have the same carbon intensity as Switzerland had in 2004 – an immense and unprecedented challenge to national and international institutions.

It's clear: Whether anthropogenic or natural, whether dangerous or benign, climate change is inevitable. Our challenge is to deal with it responsibly. This book offers some perspectives on the issues and how we might begin to think about policy.

1.1 The Political Side of Climate

It is a warm Sunday morning in August and the Trans-Canada Highway between Victoria and Nanaimo on Vancouver Island in British Columbia is clogged with vacationers heading to the ferries that will take them to the mainland, to Vancouver and beyond. The highway is congested because of the sheer size of the recreational vehicles on the road, including the oversize trucks required to haul the large travel trailers. A truck-trailer unit is probably worth at least $60,000, with most units costing substantially more; for every 5 km, each unit burns about a liter of gasoline or diesel, with its attendant CO_2 emissions. The roadway itself is of little help despite being a four-lane divided highway, because it is not a freeway and traffic lights abound. Vacationers are on the road early because the Provincial Parks require them to vacate their spots by 11:00 a.m., and there is a mad rush to get to the ferry terminal as soon as possible to queue up for the next available ship.

Clearly, Canadians and citizens of rich countries live in a world where materialism is the prevailing god. However, a significant proportion of the population opposes this opulence, perhaps because they are envious or because they feel guilty about their own wealth and spending habitats in relation to the poverty in which many of those with whom they share the planet find themselves. Perhaps, there is a real concern about the impact of their actions on nature, the environment and/or future generations – that this consumption of the Earth's bounty cannot continue at this level, although some of the most vocal environmentalists have been found to drive large, gas-guzzling vehicles and own (and heat) palatial homes with few inhabitants.

During much of history, people have pointed to the inevitable limits to natural resources as a constraint on what can be consumed – the unsustainability of economic growth and ever increasing material wellbeing (e.g., see van Kooten and Bulte 2000, pp. 256–260). And yet, the limits have never been reached, with real prices of almost all commodities falling over time rather than rising as would be required if there were true scarcity. The past several decades have witnessed a shift to what might be perceived as a greater threat than even resource scarcity. There is a possibility that human activities, especially those leading to emissions of greenhouse gases in affluent rich countries, are having a negative impact on the Earth's climate, causing unprecedented warming.

Influential scientists and public figures have argued, often quite vehemently, that global warming is the greatest threat to civilization. Sir John Houghton, a former chief executive of the UK Meteorological Office and co-chair of the Scientific Assessment Working Group of the Intergovernmental Panel on Climate Change (IPCC), was quoted by *The Guardian* newspaper on July 28, 2003 as saying: "… the impacts of global warming are such that I have no hesitation in describing it as a 'weapon of mass destruction'." In 2008 testimony before the U.S. House Select Committee on Energy Independence & Global Warming, James E. Hansen, director of NASA's Goddard Institute for Space Studies, opined that the "CEOs of fossil energy companies … should be tried for high crimes against humanity and nature," naming ExxonMobil and Peabody Coal in particular.[1]

Among politicians, former Democratic vice-president Al Gore shared the 2008 Nobel Peace Prize with contributors to the 2007 Fourth Assessment Report of the IPCC for his film, *An Inconvenient Truth*. Gore points out that:

> Two thousand scientists, in a hundred countries, engaged in the most elaborate, well organized scientific collaboration in the history of humankind, have produced long-since a consensus that we will face a string of terrible catastrophes unless we act to prepare ourselves and deal with the underlying causes of global warming. [As a result of global warming,] the relationship between our civilization and the Earth has been radically transformed. … Adopting a central organizing principle means embarking on an all-out effort to use every policy and program, every law and institution, to halt the destruction of the environment.[2]

---

[1] See http://www.columbia.edu/~jeh1/2008/TwentyYearsLater_20080623.pdf (viewed October 13, 2009).

[2] All quotes here and elsewhere that are not otherwise cited can be found at: http://www.c3headlines.com/global-warming-quotes-climate-change-quotes.html and/or http://www.laurentian.ca/Laurentian/Home/Research/Special+Projects/Climate+Change+Case+Study/Quotes/Quotes.htm?Laurentian_Lang=en-CA (viewed July 20, 2009).

Gore's former boss, Democratic President Bill Clinton, has also gotten on the bandwagon: "I worry about climate change. It's the only thing that I believe has the power to fundamentally end the march of civilization as we know it, and make a lot of the other efforts that we're making irrelevant and impossible." And President Barack Obama, who was awarded the 2009 Nobel Peace Prize (but not for anything related to climate), has referred to climate change as an "irreversible catastrophe" (*The Economist*, September 26, 2009, p. 36). U.S. Senator John Kerry, as Chairman of the Senate's Foreign Relations Committee, views climate change as the greatest threat to U.S. security in the twenty-first Century.[3]

Climate pundits now exploit any published piece of certified (i.e., 'peer-reviewed') research that supports one side of the debate or the other, regardless of the caveats attached by the researchers themselves. The debate is as much about propaganda as fact, with citizens needing to decide in their own minds how to respond (whether at the voting booth or in choosing a new automobile) – at least citizens in rich democratic countries can decide how to respond. The vast majority of humanity is likely unconcerned and ignorant about climate change, because they struggle simply to survive.

Various groups and bloggers now exist solely to challenge the notion that human actions are bringing about catastrophic climate change. Skeptics are often desultorily characterized as 'right wing' simply because they do not adhere to a politically correct environmentalist agenda, and members of such groups are referred to as 'climate deniers' because they oppose the so-called 'scientific consensus' noted by Al Gore in the above quote – as if scientific knowledge and progress occur by consensus. In juxtaposition to Al Gore's *An Inconvenient Truth* are the 2007 film *The Great Global Warming* Swindle by British filmmaker Martin Durkin and the 2009 film *Not Evil Just Wrong: The True Cost of Global Warming Hysteria*,[4] by Irish filmmakers Phelim McAleer and Ann McElhinney. The latter film dramatizes the harm to poor people from higher energy prices. Likewise, there now exist groups and bloggers that have explicitly been created to oppose those that challenge the IPCC view on global warming.

Other than their opposing standpoints, the two sides of the debate are unequal in both monetary and political terms. President Obama illustrated this divide in an October 23, 2009 address to MIT: "The naysayers, the folks who would pretend that this is not an issue, they are being marginalized. But I think it's important to understand that the closer we get [to a global climate agreement], the harder the opposition will fight and the more we'll hear from those whose interest or ideology run counter to the much needed action that we're engaged in."[5] President Obama echoes a powerful battle cry: if you are not with us, you are against us, and therefore you are the enemy. Just like the evidence coming from files and emails posted on the internet by a hacker or whistleblower seeking to expose scientists at the Climate

[3] See Bender (2009). Kerry also co-sponsored the Senate climate bill in Fall 2009.

[4] See www.noteviljustwrong.com (viewed April 14, 2010).

[5] See www.nytimes.com/2009/10/24/us/politics/24obama.text.html (viewed December 4, 2009).

Research Unit at the University of East Anglia in the UK (see *The Economist*, November 28, 2009), this does not bode well for encouraging debate and the ability to address climate change in a sensible and effective way.

The climate science that lies behind claims of catastrophic global warming receives research funding of some $2 billion annually.[6] The U.S. Climate Change Program is a multi-agency climate research program that also engages research organizations in other countries; it received $1.15 billion in 2007, $1.20 billion in 2008, and requested $1.30 billion for 2009 (McMullen and Jabbour 2009, Appendix B). The U.S. Climate Change Technology Program receives about the same amount to conduct research on alternative energy. In comparison, the opposing side of the debate receives funding that is at best measured in terms of a few million dollars per year; yet, every dollar spent comes under close scrutiny by media and bloggers – the very notion of critiquing or challenging the 'scientific consensus' is often mired in allegations of bias and allegiance to international oil corporations.

Blogs such as DeSmogBlog.com (supported by various renewable energy interests including E-boom Finance), ExxposeExxon.com (supported by environmental lobby groups), and exxonsecrets.org (a Greenpeace website) only serve to illustrate the intensity to which skeptics are scrutinized.[7] Indeed, Greenpeace castigated Exxon for providing $76,100 in funding to the Smithsonian Astrophysics Observatory because it is the home of Willie Soon and Sallie Baliunas (two so-called 'deniers'), who were assumed to be the recipients "unless Exxon explains itself."[8]

Given this imbalance in funding, it is no wonder that more research has found evidence of global warming and its ill effects than research to the contrary. Yet, what is often left unsaid is that the science of climate change is pursued under the pretense that human activities that emit greenhouse gases are directly responsible

[6] Information on spending is available in the annual report to Congress, entitled *Our Changing Planet*, with the latest available at http://www.usgcrp.gov/usgcrp/Library/ocp2009/ocp2009.pdf (viewed October 9, 2009).

[7] In 2000, the ExxonMobil Foundation provided a grant of $15,000 to the Harvard-Smithsonian Center, known to have links with Willie Soon, Sallie Baliunas and Craig Idso, well-known 'deniers'; the Foundation and ExxonMobil Corp also contributed $160,000 over 3 years to the George T. Marshall Institute (headed by George O'Keefe, formerly of the American Petroleum Institute), and more than $900,000 to the Competitive Enterprise Institute (Nesmith, 2003). Paul Krugman even wrote: "A leaked memo from a 1998 meeting at the American Petroleum Institute, in which Exxon … was a participant, describes a strategy of providing 'logistical and moral support' to climate change dissenters, 'thereby raising questions about and undercutting the 'prevailing scientific wisdom'.' And that's just what Exxon Mobil has done: lavish grants have supported a sort of alternative intellectual universe of global warming skeptics" (*New York Times*, April 17, 2006). http://query.nytimes.com/gst/fullpage.html?res=9407EEDD173FF934A25757C0A9609C8B63&sec=&spon=&&scp=3&sq=Exxon%20skeptic%20climate&st=cse (viewed October 9, 2009). Thus, Krugman and the environmental lobby pursued Exxon into providing $100 million in funding to Stanford's Global Climate and Energy Project, which conducts research into alternative fuels … [and funds] carbon-capture research with the EU (see Colvin 2007 at (viewed October 9, 2009): http://money.cnn.com/magazines/fortune/fortune_archive/2007/04/30/8405398/index2.htm)

[8] At http://members.greenpeace.org/blog/exxonsecrets/2009/05/26/exxon_admits_ 2008_ funding_ of_global_warm (viewed October 13, 2009).

for global warming. Indeed, the IPCC was set up to provide the scientific basis for anthropogenically-driven climate change. This bias alone justifies a strong critique of what the science is trying to achieve – the critics say the research is inherently oriented towards proving a particular hypothesis whose conclusion has already been accepted as a belief, rather than collecting data, analyzing it critically, and providing outsiders the opportunity to determine if the analysis and conclusions are truly scientific. The reasons for this view are examined more fully in Chap. 2, but revelations coming from the leaked emails from the Climate Research Unit suggest that climate scientists may indeed have circumvented the scientific process. This too is considered in more detail in Chaps. 2 and 5.

What is surprising is that poorly funded researchers have made such a strong case against the 'consensus' that they have become the targets of environmental activists who seek to discredit their work and sources of funding. Indeed, the attacks have become so bitter that some scientists opposed to the consensus have even come under attack for their religious beliefs.

Increasingly, the debate will be resolved in the courts. For example, in 2007 the New Party took legal action to prevent showing of *An Inconvenient Truth* as an educational tool in the United Kingdom.[9] The Court found that the film falsely claimed that (1) melting snows on Mount Kilimanjaro constituted evidence of global warming; (2) evidence from ice cores proved that rising CO_2 causes temperature increases over 650,000 years (the Court found that over this period, increases in CO_2 actually lagged temperature by 800–2000 years); (3) the drying up of Lake Chad was caused by global warming; (4) polar bears had drowned due to disappearing arctic ice (the study on which this was based showed that four polar bears drowned because of a particularly violent storm); (5) global warming could stop the Gulf Stream throwing Europe into an ice age, while this is a scientific impossibility; (6) global warming was responsible for species losses, including coral reef bleaching, when there is no evidence to support this assertion; (7) sea levels could rise by 7 m, thereby displacing millions of people (rather sea levels are expected to rise by about 40 cm over the next 100 years and that there is no threat of massive migration); (8) rising sea levels has resulted in the evacuation of people on certain Pacific islands to New Zealand, which could not be substantiated; and (9) hurricane Katrina was the result of global warming, when it is not possible to attribute one-off events to climate change.

Many oppose resolving scientific issues in the courts, but that is sometimes the only way to resolve deep conflicts in society, thus enabling public policy to proceed. Rightly or wrongly, there is considerable precedence for doing so, including the famous trial regarding the teaching of evolution versus creation as a scientific explanation for origins. This may not resolve the scientific debate, but there always remains the option to overturn a verdict at a future date as more becomes known about the science. Therefore, lawsuits have their place in determining scientific

---

[9] The case involved Stewart Andrew Dimmock (Claimant/Respondent) versus Secretary of State for Education & Skills (Defendant/Appellant), case number CO/3615/2007, Royal Courts of Justice, Strand, London, September 27, 2007. For more information and a full transcript of the Court's ruling, see (viewed October 6, 2009): http://www.newparty.co.uk/articles/inaccuracies-gore.html

issues, and they should not be ruled out. Indeed, they are preferable to the mantra that a consensus exists about some scientific controversy where a majority of scientists favor one side of the issue over another.

Given that the Climate Research Unit (CRU) at the University of East Anglia apparently destroyed raw station-level, historical weather data, the collection and storage of which was partially funded by American taxpayers, the prospect for future lawsuits looks increasingly likely. And with revelations about the quality of reporting in the latest Intergovernmental Panel on Climate Change report on climate change (IPCC WGI 2007),[10] it would be unsurprising if climate legislation did not come under greater scrutiny by opposition parties in democratically elected parliaments; nor would it be surprising to find that regulations to limit greenhouse gas emissions issued by agencies, such as the U.S. Environmental Protection Agency (EPA), became the object of lawsuits.[11]

Science has always been observation based. Scientific theories are always tested against the empirical evidence (deductive reasoning), or observations are used to formulate 'generalizations' or theories (inductive reasoning). In contrast, 'predictions' of climate change are based on climate models that are nothing more than mathematical representations of complex and chaotic systems. It is not surprising that not all scientists accept the outcomes of climate models, and it is not surprising to find scientists who propose alternative theories of how climate changes.[12] It is important, even for the purposes of this book, to examine some of the questions surrounding the science of climate change. After all, economic prescriptions for addressing climate change can only be as good as the underlying science, and where the underlying science casts doubt, such doubt must be taken into account in developing policy. Already in the second chapter, we deal with doubt and uncertainty – it is simply unavoidable in a discussion of future climate.

1.2 Engaging Climate Change Research

This book focuses on the economics of climate change. But it also focuses on areas where economists have been found engaging the climate change research agenda – economists are contributing to research in a variety of climate-related fields, whether

[10] See *The Economist* (February 6, 2010, p. 85), which is rather tolerant of the problems within the IPCC and somewhat harsher on its critics.

[11] Texas and other parties took the EPA to court arguing that it has exceeded its mandate in attempting to regulate greenhouse gas emissions. See Bakst (2010).

[12] Foremost are astrophysicists who attribute global changes in climate to solar forcing (Parker 1999; Wu et al. 2009), cosmic rays and their impact on cloud formation (Svensmark and Calder 2007), or other astronomical factors. Climate scientists are vigorous in attacking these theories, particularly those related to sunspots (e.g., Weaver 2008), although some non-anthropogenic explanations of warming, such as the cosmic ray-cloud formation theory, are beginning to be tested (http://cdsweb.cern.ch/record/1181073/, viewed April 23, 2010). These and other theories, and critiques of the science, are discussed in Chap. 5.

reconstructing past climates or contributing to climate models that forecast future climate. In this regard, they are no different than physicists, geologists, political scientists, engineers, ecologists, statisticians, and many others who are busy conducting research in areas that climate scientists consider their bailiwick. Indeed, the most frequent complaint one hears about statements on climate made by ‘outsiders’ is that they lack the expertise to comment on the science. This is an interesting means of suppressing debate and stifling scientific progress in understanding climate change, especially in the face of evidence that people outside a scientific discipline are more likely to solve a challenge facing the discipline than those in it. So-called outsiders often have a better chance of solving a problem bedeviling a science than the experts in the field.[13]

In addition to the work of economists in the fields of climate reconstruction and climate modeling, there is another reason why it is important for economists to be familiar with climate science. Economic analysis of costs and benefits, and policy recommendations, cannot proceed without some knowledge about the underlying climate science. Clearly, if recent temperatures and projected future global warming are within historical experience, there is likely little to be done to mitigate climate change and the best policy is to do nothing more than facilitate adaptation; after all, adaptation appears to have been successful in the past and, given today’s much better technologies, will be appropriate now.

The question of whether current and projected temperatures are within historical experience deals with two aspects of climate science – past climate records and the modeling of future climates. With respect to the climate record, the research of scientists at the CRU at East Anglia and their counterparts at other institutions is invaluable, as is the integrity of that work. It is important to know how previous periods of warming and cooling compare with more recent temperatures. Likewise, it is important to understand the credibility of future climate projections, and the extent to which such projections are outside or within the realm of human experience. Thus, the ability of climate models to replicate the complex workings of the oceans, atmosphere, terrestrial ecosystems, and other systems and factors that determine temperature and precipitation is important. For example, what are the principal drivers in climate models? How well do the mathematical equations in climate models represent the real world? How many parameters in climate models are fixed by the principles of physics and how many can be varied by the modeler? Is the interaction between oceans, atmosphere and terrestrial systems modeled accurately? Are explanations of solar activity and cloud formation adequately addressed in the models? Are there alternative theories that better explain historical temperatures and come to other conclusions about future trends in temperatures and precipitation? Without precise answers to these questions, answers that are unlikely to be forthcoming without further research, which might take a decade or more, it is difficult to set policy and make decisions.

[13] See *The Economist* (August 7, 2010, pp. 79–80), Lakhani and Jeppesen (2007), Lakhani et al., (2007), and http://blog.innocentive.com/2009/05/07/the-innocentive-insider-surprising-but-true/ (viewed August 20, 2010).

If anthropogenic carbon dioxide emissions play little or no role as a driver of climate change, there is little that can be done to mitigate climate change from an economic policy perspective, except to retain or increase institutional flexibility so economies can better adapt, even if climate change and its worst outcomes are to become reality. As an analogy, there is nothing one can do to stop hurricane force winds from blowing, but you can get out of the way and/or invest in infrastructures that minimize their consequences. If human emissions of greenhouse gases do result in global warming, policies to mitigate such emissions take on a larger role, although such knowledge is insufficient by itself to take action, let alone drastic action. It depends on factors related to past experience and on projections from climate models as to how bad it might get. So a first step in determining an optimal economic response depends on getting the science right, or at least enabling the debate to continue and further our understanding of the science. It is necessary to know something about the potential ability of science and technology to mitigate (e.g., remove greenhouse gases from the atmosphere or introduce particulates into it to cool the globe) and/or adapt to any adverse impacts from global warming.

It is also important to know something about the damages from climate change – their potential magnitudes. What damages can be expected under any and all possible future climate regimes? Not only does one need monetary estimates of potential damages, but one needs to know the degree of certainty associated with such estimates because each depends on a number of uncertain factors.

And there lies the problem: Nothing is known with certainty – neither the climate science nor the potential harm that future climate might cause. Everything associated with climate change is characterized by wicked uncertainty. While there are methods for making decisions under extreme uncertainty (Ben-Haim 2001; Ben-Tal and Nemirovski 2002; Lempert et al. 2006), these are difficult to apply in the context of global warming, and fraught with controversy. In the end, much of the policy debate is based on speculation and belief, often pitting one ideology against another – one worldview against another (Nelson 2010). And the unfortunate loser in all of this is science – objectivity and rationality.

Efforts to mitigate climate change must balance the costs of doing so with benefits in the form of the damages that are avoided. In his February 11, 2009 John Locke Foundation debate with John R. Christy of the University of Alabama at Huntsville, William Schlesinger of the Cary Institute of Ecosystem Studies (and previously Duke University) identified at least five major concerns related to global warming[14]:

1. Climate models generally agree that most of polar and temperate regions of the north and south will be much warmer a decade from now, although there will be little warming along the equator. There will be large breakup and melting of the Antarctic ice sheet similar to what happened to the Larsen Ice Shelf, which broke off and entered the sea; therefore, sea levels will rise. By mid-century, there will

[14] See http://www.globalwarming.org/2009/02/12/john-christy-debates-william-schlesinger/ and video at http://www.johnlocke.org/lockerroom/lockerroom.html?id=18946 (October 13, 2009).

be great threat of land loss to rising sea levels in North Carolina and New York City, with very large costs to taxpayers.
2. Over the past century, some parts of United States got drier, some wetter; climate models suggest that, in the future, the southwestern part will experience greater drought while the eastern part will get wetter. South-central Europe will experience greater droughts as will the U.S. central plains. The result will be an adverse impact on food prices
3. The ranges of various insects will change, and that of the corn borer, for example, will increase considerably. The result will be to reduce crop yields.
4. Malaria will increase its range, and malaria will be found along the Gulf coast and Eastern coast of United States and will become commonplace elsewhere as well.
5. The ranges of various plants and trees will change. The current prevalence of beech and maple forest in northern states of the U.S. will disappear, being replaced by oak, hickory and savannah. Owners of forest companies will be concerned about this because trees that are planted today might not grow at their expected high levels of productivity late in this century.

Schlesinger's conclusion is that we permit global warming at our peril, as climate change potentially threatens our livelihood – governments need to act to stop warming.

To these potential threats, we might well add the following as these have been widely reported:

6. Loss of biodiversity as most evidenced by the threat to mega fauna, such as polar bears that lose their ability to hunt seals as more open water replaces the ice platforms from which they hunt.
7. Massive migrations of peoples as a result of climate change, which in turn leads to global unrest (e.g., see Fagan).

These and other issues are the subject of this book.

1.3 Plan of the Book

The focus of this book is on the economics of climate change and economic policy, and the great uncertainty involved. The book begins in Chap. 2 with a discussion of the historical climate data collected at weather stations since the mid 1800s. There we consider issues related to the construction of average temperatures, and how extraneous factors have affected data collection over time. We also consider how we might construct financial weather derivatives that rely on data from weather stations spread across a landscape, a problem similar to that of finding an average global temperature.

Then, in Chap. 3, we discuss the science of reconstructing historical climate records. This is important if we are to determine whether current temperatures are outside human experience. That is, we wish to examine the record going back some two millennia, say, as this is well within recorded history. If temperatures in the past were as warm as or warmer than those of today, then scientists need to explain the mechanism by which this occurred since CO_2 is not the trigger – to validate their models, climate modelers should replicate past warming without the atmospheric CO_2 levels experienced today. However, whether temperatures were higher 500, 1,000 or more years ago does not necessarily imply that predictions of anthropogenic global warming are invalid. The picture is certainly more complicated than that.

Climate models are the subject of Chap. 4. Here economists are involved because economic models are needed to determine the emission scenarios that form the basis of the predictions from climate models (some of which even include an economic feedback). Thus, in Chap. 4, there is a description of the greenhouse gas emission scenarios that the United Nations' Intergovernmental Panel on Climate Change (IPCC) uses to drive climate model outcomes. This is followed by a discussion of energy balances and climate models, and a simple energy balance model is used to illustrate the sensitivity of climate models to different assumptions regarding feedbacks.

In Chap. 5, we provide alternative explanations for global warming that can be found in the peer-reviewed literature. Included in this chapter, for example, is a discussion of the role that clouds play in causing global temperatures to increase. One view suggests that this role may have been overstated – that the feedbacks from cloud formation are negative (lowering temperature) as well as positive (raising temperatures). Another view that still needs to be fully tested places a greater emphasis on the role of the sun in blocking cosmic rays from outside the solar system. When the sun is less active (i.e., less sunspots), more cosmic rays strike the earth and this leads to overall cooling. Yet another view attributes temperature changes to natural disturbances related to ocean oscillations, and less to anthropogenic emissions of greenhouse gases.

Although economists have played a role in climate modeling and in debates about temperature measurement and the validity and interpretation of the historic record, and indeed are now becoming increasingly interesting in studying temperature and other data,[15] economic methods and economic policy are systematically addressed in Chaps. 6, 7 and 8. The theory of measuring economic surpluses (including non-market ones) is described in Chap. 6. In particular, the method of cost-benefit analysis is outlined, including the use of discounting. The economics of climate change are discussed in Chap. 7. The discussion includes, among other things, the question of an appropriate instrument for curbing greenhouse gas emissions; regulatory regimes, carbon taxes and various forms of emissions trading are contrasted. The need to focus on alternatives other than reducing greenhouse gas emissions is

---

[15] At the World Congress of Environmental and Resource Economists held in Montreal, June 28–July 2, 2010, a number of papers examined, for example, the statistical methods underlying reconstructions of past temperatures, statistical effects of local dimming caused by particulates, and multivariate statistical methods for investigating climate data series (see www.wcere2010.org).

also considered. More specific analysis of proposed or actual policies that have been implemented is left to Chap. 8. Several bills to address climate change that were considered by U.S. legislators are examined, as are policies in Europe and elsewhere. Policy implementation is where theory meets reality, and to get this wrong could adversely impact costs to such an extent that the best alternative might be to ignore global warming and concentrate on adaptation.

Many countries have decided to avoid painful emission reductions, instead focusing on policies that attempt to slow global warming by other means. Carbon offset credits are earned by a variety of means outlined by the UN's Kyoto Process, including the sequestration of carbon in terrestrial sinks. The challenges of relying on biological carbon sinks and biomass fuels are the topic of Chap. 9.

If global warming is the result of anthropogenic emissions of greenhouse gases, principally carbon dioxide from fossil fuel burning, then it is clear that any solution to the issue of climate change must address the use of fossil fuel energy. The role of fossil fuel energy and clean energy alternatives are examined in Chaps. 10 and 11. The former chapter examines the prospects for renewable energy, while the latter considers the potential obstacles facing wind energy because of its intermittent nature. These same obstacles face most other renewable forms of energy.

The book concludes in Chap. 12 with some thought-provoking observations. The intent of that chapter and the entire book is to challenge the reader to think about climate change and potential global warming from a scientific rather than ideological perspective.

References

Bakst, D. (2010). *Texas fight! What other states can learn from Texas v. U.S. EPA*. Viewed August 18, 2010, from http://www.masterresource.org/2010/08/texas-fight-us-epa/

Bender B. (2009, September 10) *The Boston Globe*. Viewed October 9, 2009, from http://www.boston.com/news/politics/politicalintelligence/2009/09/kerry_to_kick_o.html

Ben-Haim, Y. (2001). *Information-gap decision theory*. San Diego, CA: Academic Press.

Ben-Tal, A., & Nemirovski, A. (2002). Robust organization – Methodology and applications. *Mathematical Programming, Series B, 92*(3), 453–480.

Fagan, B. M. (2008). *The great warming: Climate change and the rise and fall of civilizations*. New York: Bloomsbury Press.

Lakhani, K. R., & Jeppesen, L. B. (2007). Getting unusual suspects to solve R&D puzzles. *Harvard Business Review, 85*(5), 30–32.

Lakhani, K. R., Jeppesen, L. B., Lohse, P. A., & Panetta, J. A. (2007). *The value of openness in scientific problem solving* (Harvard Business School Working Paper 07-050, 57 pp.). Boston. Viewed July 18, 2012. Retrieved from http://www.hbs.edu/research/pdf/07-050.pdf.

Lempert, R. J., Groves, D. G., Popper, S. W., & Bankes, S. C. (2006). A general, analytic method for generating robust strategies and narrative scenarios. *Management Science, 52*(4), 514–528.

McMullen, C. P., & Jabbour, J. (Eds.). (2009). *Climate change 2009: Science compendium*. Nairobi: United Nations Environment Programme.

Nelson, R. H. (2010). *The new holy wars: Economic religion versus environmental religion in contemporary America*. University Park: Pennsylvania State University Press.

Nesmith, J. (2003, June 2). *Seattle Post-Intelligencer*. Viewed October 13, 2009, from http://www.seattlepi.com/national/124642_warming02.html

Parker, E. N. (1999, June 3). Solar physics: Sunny side of global warming. *Nature, 399*(6735), 416–417.

Svensmark, H., & Calder, N. (2007). *The chilling stars: A new theory of climate change*. Cambridge: Icon Books Ltd.

van Kooten, G. C., & Bulte, E. H. (2000). *The economics of nature: Managing biological assets*. Oxford: Blackwell Publishers.

Weaver, A. (2008). *Keeping our cool: Canada in a warming world*. Toronto: Viking Canada/Penguin Group.

Wu, J., Yu, Z., Zeng, H. A., & Wang, N. (2009). Possible solar forcing of 400-year wet-dry climate cycles in Northwestern China. *Climatic Change, 96*(4), 473–482.

Chapter 2
Weather and the Instrumental Record

To kill an error is as good a service as, and sometimes better than, the establishing of a new truth or fact

– Charles Darwin

Weather is a fascinating topic. Whenever people from different regions of Canada happen to be talking on the telephone, whether for personal or business reasons, the discussion at one point or another turns to the weather – a comparison of the weather situation at that time in the regions in which the discussants find themselves. During winter months, those living on the west coast of the country generally own the bragging rights. January temperatures are inevitably significantly warmer in places like Vancouver and Victoria than in interior continental regions, such as Edmonton or Regina. The Pacific Ocean moderates temperatures so that they rarely fall below freezing, but nearness to such a huge sink can also lead to heavy rainfall and many days without sunshine, something those living on the coast rarely tell their colleagues or friends living elsewhere in the frozen 'white north.'

Before examining the scientific aspects of climate change, we want to focus on observed weather data, often referred to as instrumental data because instruments are used to collect such things as temperature, precipitation, wind speed, and so on. Much of this data is now collected and recorded automatically as opposed to having someone go to the place where the instruments are located – the 'weather station' – and read the values on the instruments and record them. The mere activity of reading and recording weather data is a source of error: readings are not always taken at the same times of day, and the human involved may forget to take a reading, read the instruments incorrectly, and/or make an error recording the information.

In addition to measurement error, there are errors associated with the aggregation of data from weather stations located varying distances from each other and from the point at which we wish to have information, and at various elevations. Aggregation of weather data is an important problem: for example, it is important for determining

G.C. van Kooten, *Climate Change, Climate Science and Economics: Prospects for an Alternative Energy Future*, DOI 10.1007/978-94-007-4988-7_2,

average regional or global temperatures, and it is particularly important in constructing weather indexes for insurance purposes.

The historical record is thus beset by both measurement and aggregation problems (although there are likely others as well). As a result, one needs to be very careful in working with weather records. In this chapter, we examine the meaning of average temperature, the aggregation of weather data and even the more pedestrian issue of what constitutes a valid weather station. We begin in the next section with the simple problem of aggregating data from two weather stations, and then consider the problem at a global scale. We end the chapter be examining the use of aggregated weather data for insurance purposes and the construction of financial derivatives based on weather data.

2.1 Collecting and Gridding Weather Data

Located at the southern tip of Vancouver Island off the west coast of Canada, British Columbia's capital city of Victoria enjoys perhaps the best climate in Canada. Because of the surrounding natural beauty, it is also a place where citizens have a strong environmental conscious and sensitivity, in particular, to policies for averting climate change. The provincial government has already implemented a carbon tax and intends to address future expansion of its electrical generating capacity in a carbon neutral way, relying on wind, run-of-river, biomass and other renewable generation technologies. The Capital Regional District (CRD), which includes Victoria, has an extensive recycling program, restricts the use of chemicals of all kinds, is looking into ways to reduce its carbon footprint, and demonstrates a concern for the environment in many other ways. Recently, the University of Victoria established a network of school-based weather stations throughout the CRD that could help climate scientists understand local weather patterns and, thereby, enable them to do a better job of scaling down climate projections from global climate models to the regional level. The weather network is used below to describe a particularly difficult problem associated with temperature comparisons over space and time, namely, the effect of non-climatic factors.

Obtaining a surface air temperature is an elusive exercise and there are strict guidelines as to how and where weather stations should be sited (Watts 2009, p.8; Williams et al. 2008). For example, as discussed later in the chapter, they are not to be located on roofs where heat from the building could lead to erroneous readings. Further, surface air temperatures vary even at a single location, depending on how close to the ground the reading is taken. Thus, temperatures at 2 m above the ground will differ from those at 20 m; the ideal is to obtain a temperature reading that takes into account the temperatures at various heights, which implies that a weather station would need to measure continually the temperatures at points in a stack from next to ground level to 20 or more meters above ground. As James Hansen points out: The temperature reported at a weather station "is truly meaningful only to a person who happens to visit the weather station at the precise moment when the reported temperature is measured, in other words, to nobody."[1]

[1] See http://data.giss.nasa.gov/gistemp/abs_temp.html (viewed February 18, 2010).

2.1.1 Working with Instrumental Data

Given the elusive nature of temperature at a single location, how does one aggregate temperatures across a city or region? Is the temperature reading from a single weather station in a city representative of the city's surface air temperature at that time? No doubt, the temperature reading provides some indication as to how cold or warm it is, but does it provide the accuracy needed to say that a city's temperatures are increasing or decreasing by extremely small, barely measureable amounts over a decade?

Some further questions might include: How are data from various weather stations aggregated? How are missing data treated? How are missing weather stations handled? And how does one deal with the fact that weather stations are predominately found in the northern hemisphere with few in the southern hemisphere, that for many weather stations data are unreliable for much of the record, and that temperature measurements across vast ocean expanses are unavailable or spotty at best?

The Climate Research Unit (CRU) at the University of East Anglia in the United Kingdom keeps records of weather station data from around the globe. It receives funding for this from various sources that have included the U.S. Department of Energy and the U.S. Environmental Protection Agency. Recently, the CRU announced that, in the 1980s, it had destroyed the historical weather station data that it had employed in its analyses, because it had insufficient storage space to maintain it.[2] As a result, we are left with reconstructed historical data with no way to verify whether the reconstruction was done correctly. Further, the individual primarily responsible for the reconstruction of the historical record to obtain gridded temperatures (specifically the 5° latitude × 5° longitude gridded data) cannot recall how he did this because "he had deleted his notes on how he performed the homogenization [reconstruction]. This means that it is not possible to reconstruct how the raw data turned into his temperature curve."[3] Of course, the historical weather station data are not lost; only it is not possible to know exactly how the CRU constructed its temperature series.

---

[2] On August 13, 2009, Andrew Orlowski of *The Register* reported that the CRU had destroyed weather data. According to the CRU, "data storage availability in the 1980s meant that we were not able to keep the multiple sources for some sites, only the station series after adjustment for homogeneity issues. We, therefore, do not hold the original raw data but only the value added (i.e. quality controlled and homogenized) data" (viewed February 18, 2010 at: http://www.theregister.co.uk/2009/08/13/cru_missing/). Despite allegations, it is not entirely clear what has been lost – the actual raw data, the notes (and importantly the computer code used to generate the homogenized data), or both (e.g., see McKitrick 2010c). For example, in response to data requests, the CRU has claimed it downloaded data to the U.S. Department of Energy. Even if the data are available, McKitrick indicates that without knowing which weather stations are included in the reconstruction, it is impossible to check the CRU's results. See also next note.

[3] See "A Superstorm for Global Warming Research" by Marco Evers, Olaf Stampf and Gerald Traufetter, Spiegel Online, http://www.spiegel.de/international/world/0,1518,druck-686697,00.html (viewed April 6, 2010). Jones et al. (2010) claim that a detailed description of the weighting method used to construct the $5^{\circ} \times 5^{\circ}$ latitude-longitude gridded boxes of temperatures, and thus their 'temperature curve,' can be found in an earlier paper (Jones et al. 2001). However, the current author could not determine from Jones et al. (2001) how it was done. This is an issue discussed in more detail in the next section.

It is possible to illustrate the challenge of measuring and aggregating temperature data using two simple examples. Although some think that measurement issues are the sole purview of climate scientists, the examples indicate that anyone with some ability to analyze numbers should be able to understand the nature of the problem. If one understands these examples, it should be quite easy to understand the controversies related to historical reconstructions of temperature and other climate data as found in the remaining sections of this chapter and especially in Chap. 3.

2.1.2 A Mini Heat Island Effect

Researchers at the University of Victoria in British Columbia, Canada recently established a school-based system of 131 weather stations across southern Vancouver Island – the greater Victoria region – and communities farther north on the Island, including the city of Nanaimo some 100 km to the north.[4] The system has been a media relations coup, with television stations able to provide viewers with real-time temperature readings from a school located close to where they live. Information from two schools can be used to illustrate the measurement issues that plague climate scientists in 'constructing' regional or global temperature averages.

A major problem facing scientists is that of removing the effects of socioeconomic activities (non-climatic factors), because a failure to do so confounds the impact that CO_2 emissions have on temperatures. For example, if one weather station is impacted by a heat source and another is not, this will cause the average of the two stations' temperatures to be higher than it should be. Suppose that a weather station is increasingly surrounded by developments such as buildings, parking lots, industrial or commercial outlets. The temperature readings at the station will exhibit an overall upward trend over time, but this is due to increasing economic activity around the weather monitoring facility and not a result of factors that cause global warming. This phenomenon is sometimes referred to as an urban heat island (UHI) effect.

How do scientists remove the non-climatic influence from their temperature data series? How do they 'homogenize' the temperature data? Clearly, the adjustments are ad hoc at best and, as discussed further in Sect. 2.3, attempts to remove the non-climatic signal have not been successful. The difficulty facing climate scientists in their efforts to homogenize temperature data can be illustrated using information on temperatures for Lambrick Park High School and Gordon Head Middle School in the municipality of Saanich, adjacent to Victoria. Both schools are included in the Victoria Weather Network system.

On February 18, 2010, at 1:22 pm, the temperature at Lambrick Park High School was 10.7 °C, while it was 9.5 °C at Gordon Head Middle School; on July 9, 2010,

[4] The project was funded primarily by Canada's Natural Sciences and Engineering Research Council (NSERC). Information on the schools (and some non-school 'hosts' of weather monitoring stations), along with weather data and graphs, are available from http://www.victoriaweather.ca/.

the respective temperatures were 25.9 and 24.4 °C at 11:09 am. The two schools are separated only by a road and a field, and are only some 250–300 m apart; yet, on those 2 days, the temperature differed by 1.2–1.5 °C, or 2.2–2.7 °F. This temperature difference falls in the range of the IPCC's (2007) projected rise in average global temperatures over the next century. However, two observations of average daily temperature are insufficient to make the case that the differences are somehow systemic, although it does highlight the problem of determining a meaningful measure of regional average (or mean) temperature.

Given these two schools, one expects their daily average temperature readings to vary by only a small amount, with the difference negative on some occasions and positive on others. However, an inspection of the weather stations at the two schools was revealing. Ideally, weather stations should be located in open fields away from anything that might affect temperatures, whether an air conditioning unit that emits heat to the outside as it cools the inside of a building, an exhaust pipe that emits heat, or a structure (e.g., building, parking lot) that absorbs the sun's energy and then radiates longwave energy back into the atmosphere thereby affecting the temperature at the monitoring station.

The weather station at Gordon Head School is located on the edge of a newly (circa 2007) constructed structure on top of the original one-storey roof. Although not ideally located as heat absorbed by the tar roof and other surroundings clearly affect temperature readings, it is likely the best one can do given there is no open space that might be free of interference and potential vandalism. This is not the case for the weather station at Lambrick Park School. An inspection of this weather station clearly indicates that it is located near several heat sources – several exhaust pipes related to the school's heating system and machinery from the wood working shop (which is perhaps the largest of its kind in the region and includes a kiln). The problem with interference from these heat sources might have been avoided by placing the monitoring facility at the front rather than the back of the school, but this would have placed the weather station even closer to the one at Gordon Head School.

Based on the locations of the weather stations at the two schools, one would expect a regular difference between the readings at Lambrick Park School and Gordon Head School, with the former consistently higher than the latter. The reason for this relates to the effect that extraneous factors, such as exhaust vents, have on temperature readings at Lambrick Park School. Consider observations for 2010 and 2011.[5] During summer 2010, observations on average temperature were missing for 16 consecutive days for Gordon Head School and 5 days for Lambrick Park School; in 2011, there were 9 days of missing observations, 1 in March at both schools and 8 in April at Gordon Head School.

---

[5] Data were obtained from the website indicted in the previous footnote (and were available March 6, 2012). A request for permission to display a chart illustrating the differences in average, maximum and minimum daily temperatures between the two schools was denied by the University of Victoria professor overseeing the school-based weather network.

The average daily temperature difference is always non-negative, varying from a low of 0–1 °C, with a mean difference in average daily temperature of 0.38 °C. These differences are not insignificant given temperature changes of 0.3 °C are projected by climate models to require as many as two decades to realize under some business-as-usual scenarios.

The same pattern emerges when one subtracts the daily maximum (minimum) temperature at Gordon Head School from that at Lambrick Park School. Over the 2-year period, the average difference in daily maximum temperatures was 0.66 °C, while the average difference in daily minimum temperatures was 0.29 °C.[6] Further, the largest difference in daily maximum temperatures was 3.3 °C, while it was 1.4 °C for daily minimum temperatures. There were 24 days out of 730 when the maximum temperature was highest at Gordon Head School (at most by only 0.4 °C) and 28 days when the minimum temperature was higher (at most by 0.7 °C). Surprisingly, this never occurred on the same days.

Clearly, the temperatures at Lambrick Park School are affected by the woodworking shop below the weather station and the nearby heating vents, while the weather station at Gordon Head School is not similarly impacted. The fact that average daily temperatures at one school consistently exceed those at the other is an example that, on a larger scale, is referred to as an urban heat island effect. Further support of the UHI effect is provided by the fact that differences between daytime (maximum) temperatures are greater than the differences between nighttime temperatures, and that, at certain times of year, there is little difference in the temperatures between the two weather stations.

Given that climate scientists construct average temperature series that they claim have removed the influence of the socioeconomic factors (the urban heat island), one might ask in the current context the following question: How does one go about constructing an average of the two schools' temperatures? One approach is to ignore the persistent presence of the heat source and take the simple average of the two temperatures – the average of the two daily averages, maxima and/or minima, or the hour-by-hour readings. The averaged results inevitably introduce non-climatic factors into the aggregated temperature.

A second option is to somehow adjust or 'correct' the Lambrick Park readings for the presence of the heat sources. During periods when school is out of session, Lambrick Park temperatures are much closer to those of Gordon Head; for example, on December 30, 2010 when the former school was used only for basketball practice, the difference in daily average temperature was only 0.1 °C, while there was only a 0.4 °C difference in the maximum temperature for that day and no difference in minimum temperatures (although minimum temperatures differed by 0.5 °C the following day). One could adjust the Lambrick Park weather station temperatures by using the proportional difference between the schools' temperatures during periods when students are not in school (presumably days when the schools are not used

[6] The differences in average, maximum and minimum daily temperatures between the two schools are highly statistically significant – the chance that the Gordon Head temperatures might actually be higher is less than 0.005.

or little used) to adjust (i.e., lower) the temperature readings on days when students are present. If the temperature readings at Gordon Head School also turn out to be influenced by a heat source, then both schools' temperatures will need to be adjusted relative to the nearest weather station that is considered to provide 'unpolluted' readings.

Another approach might be to look at the changes in temperatures at the two schools. If changes in temperature track closely and if the Gordon Head School readings are somehow considered to be free of external influence, then the readings as Lambrick Park could be adjusted downwards to a time, say, when the difference in temperatures was greatest (or some average of the greatest differences in temperature). The problem is that any such approach is arbitrary and based on the assumed purity of the Gordon Head temperatures.

Finally, if the temperatures at Gordon Head School are considered free of any external human influence, and given the closeness of the two schools, the best option is to ignore temperatures from Lambrick Park School and rely only on those from Gordon Head School. Including temperature readings from Lambrick Park School must introduce non-climatic factors into the final homogenized temperature construction as there is no fool proof way of removing the distorting influences present at the Lambrick Park weather monitoring station.

Clearly, the temperature readings at Gordon Head School are also impacted by non-climatic factors, as the thermometer is located above an asphalt roof top. To realize what is going on, consider a hot and sunny summer day. You can quite easily stand on bare feet on a grass field, but, if you try to do this on an asphalt parking lot, you could well burn your feet. A thermometer held above the parking lot will yield a much higher temperature reading than the same thermometer held above the grass field. As discussed in Chap. 4, the reason has to do with black body radiation – the parking lot gives off much more heat (even on a hot summer day) than does the grassland. This is an example of the urban heat island effect.

Now suppose that one wished to use the information from the weather stations in the school network to create a single, homogenized temperature series that best represents Greater Victoria's true temperature (as is done on the weather network site). How does one address the socioeconomic and other extraneous influences on temperature? Practically, there is no way that this can be done because it is impossible to account for exhaust outlets, tar roofs, buildings, et cetera, that radiate heat. Any adjustment is bound to be ad hoc. One might weight the temperature readings at each school by the inverse of the number of students registered at the school (or the inverse squared), by the size of the school's footprint (again inverted so that larger schools with a supposedly greater non-climatic influence are weighted less), by the energy consumed at each school, or by some combination of these (or other) weighting factors. But none of these adjustments will eliminate the confounding influence of socioeconomic factors on temperatures. This is a problem not only when it comes to the reconstruction from weather station data of historical average temperatures for a region, but also for the development of weather products that can be used in guiding decisions in the primary sectors as discussed in Sect. 2.4.

Given that scientists do not and cannot know to what extent thermometers at various weather stations across a region, country or the globe are impacted by extraneous factors, any attempt to create an undefiled temperature summary is doomed to fail. Therefore, the question is not whether temperature reconstructions still include non-climatic factors, but to what extent? Before considering this further, there are other problems related to the aggregating of weather data that need to be considered.

2.1.3 *Aggregating Temperature Data*

Consider the example in Table 2.1 where we have 11 weather stations located randomly in various directions from a grid point, and we want to determine the representative temperature at that grid point. Stations J and K are not included initially, as these are assumed to be added at a future date.

Assume that the temperature data at each weather station in Table 2.1 was recorded at exactly the same time, which is unlikely to be the case. Then there are several ways to determine the temperature at the grid point (Table 2.2). First, we can take a simple average of all the stations, which gives a grid point temperature of 8.99 °C at the time the readings were taken.[7] Second, it makes some sense to weight the readings by the distances that stations are from the grid point. If weather station data are weighted by the ratio of the distances monitoring stations are from the grid point, then we obtain a grid-point temperature of 8.55 °C, almost half a degree lower than if a simple average were used. To avoid including stations that are considered 'too far away,' it might be necessary to exclude those that lie beyond some arbitrary distance from the grid point. Suppose we exclude all weather stations beyond a 100 km radius, which would mean that data from stations D and E are excluded. In that case, the simple average temperature would fall to 8.01 °C and the weighted grid-point temperature would be 8.25 °C. Notice that the simple average is now nearly 1 °C higher.

Since the choice of an arbitrary distance for determining which weather stations to include in the construction of gridded temperatures results in loss of information, an alternative is to include all stations and use the inverse of distance squared as the means for weighting observations. Then nearby stations count much more than stations farther away, even more than was the case with 1/distance. With the inverse of distance-squared, the grid-point temperature based on all nine stations becomes 8.48 °C, a full 0.5 °C below the simple average.

The method used to construct gridded temperature data is important because global averages of temperature begin by determining an average temperature for

[7] Given that temperature readings are taken to one decimal place, it might be more appropriate to consider only one rather than two significant decimals. However, climate scientists regularly provide summary measures 'accurate' to the thousandth degree Celsius or even higher (e.g., the CRU reports average temperature anomalies to the third significant decimal) (Jones et al. 2010).

Table 2.1 Calculating grid point temperature: weather station information for hypothetical example

	Current stations									Added stations	
	A	B	C	D	E	F	G	H	I	J	K
Temperature (°C)	10.7	8.2	6.5	13.8	11.0	9.2	3.3	9.8	8.4	–0.4	15.1
Distance from grid point (km)	28	15	62	140	154	95	53	61	78	120	18

Table 2.2 Calculated grid-point temperature (°C) for hypothetical example

Scenario	Simple average	Weighted by 1/distance	Weighted by 1/distance^2
Original nine stations	8.99	8.55	8.48
100 km radius limit	8.01	8.25	8.43
Exclude A, C, and E	8.78	8.10	8.04
Exclude B, D, and I	8.42	8.43	8.92
Adding J and K	8.69	9.69	10.44

each grid. In our simple hypothetical example, five different measures of the gridded temperature can logically be justified, with the difference between them amounting to as much as 0.98 °C. What happens if one or more weather stations are lost, or one or more are added?

Consider first the case where three weather stations are arbitrarily removed from the data set, and then two new stations are added. Suppose that stations A, C and E are arbitrarily removed, and then stations B, D and I are removed. Finally, assume two new stations are added, J located 120 km away from the grid point and recording a temperature of –0.4 °C, and K located only 18 km away with a temperature reading of 15.1 °C. A summary of results is provided in Table 2.2. In our example, when stations are added the simple average at the grid point falls, but the weighted average can rise by as much as 1.45 °C, not an insignificant amount.

The example provided in Tables 2.1 and 2.2 is contrived and there exist other methods for obtaining gridded values. Nonetheless, it serves to illustrate some important points. First, weather stations need to be properly sited. Meteorologist Anthony Watts initiated a project to photograph every weather station in the U.S. to determine if they were reliable. He found that 89% of weather stations violated the National Weather Service's own standards for locating monitoring stations, with too many "stations located next to the exhaust fans of air conditioning units, surrounded by asphalt parking lots and roads, on blistering-hot rooftops, and near sidewalks and buildings that absorb and radiate heat" (Watts 2009). This issue is discussed in greater detail Sect. 2.2.

Second, spatial statistical methods are needed to determine the nearby stations required to fill-in missing raw data when a reading at a particular station is 'lost'. Williams et al. (2008) indicate that the U.S. Historical Climatology Network (USHCN) (see Sect. 2.2) uses information from "the best correlated nearby stations:" Suspect readings at one weather station are determined, for example, by comparing them to readings at other sites (presumably sites that are more trustworthy).

The best way to determine the "best correlated nearby stations" is to use spatial statistical methods. Then, to measure distances between weather stations, and between stations and a grid point, one should employ the methods available using a geographical information system (GIS) tool, namely, altitude above sea level as well as geographic distance. GIS is a vast improvement over measurements taken from photos or other means, while spatial statistical methods account for spatial autocorrelation that could otherwise result in misleading records substituting for data gaps. At the time in the 1980s when the CRU apparently destroyed much of its raw weather data in favor of the constructed gridded data, spatial statistics and GIS were in their infancy. It is quite possible, therefore, that more modern constructs of gridded temperature data would differ from those of the past.[8]

Raw temperature data are also adjusted to take into account time-of-observation bias to insure that data from different weather stations are adjusted to the same time of day to remove a potential non-climatic source of error (e.g., temperatures taken at midnight are generally colder than those taken during daylight hours). It is also necessary to take into account the history of a weather station, as even the highest-quality weather stations have been relocated at some point in their history, including USHCN weather stations. Weather monitoring sites were relocated from cities to airports or from roof tops to grassy areas, often resulting in cooler readings than were observed at the previous sites. Depending on how the adjustments were applied to correct for these artificial changes, average temperatures may have been reduced – earlier readings were lowered after the gauges were moved because, at the new location, there was a reduced non-climate influence. That is, previous readings were considered to have been affected by an urban heat island effect, so they were lowered. Thus, if stations were moved in the first half of the twentieth century, the temperatures for this era were also reduced, perhaps making the first half of the century appear cooler than warranted and, thereby, the second half warmer in comparison.

Missing data at one or more weather stations are also 'filled in,' but it is not clear how this is done as there are a number of different techniques for doing so. Choice of any method is rather ad hoc. Finally, an urban warming bias is removed from the data, but, as noted in Chap. 3, the methods used to do so appear not to have been very successful.

Yet, as one commentator noted with respect to the CRU data, it did not really matter that the CRU destroyed the raw observational data as these were unreliable in any event.[9] Until weather data were automatically recorded, most weather stations recorded only a daily maximum and a daily minimum temperature, with the two temperatures sometimes recorded for different calendar days and with no consistency

---

[8] The Berkeley Earth Surface Temperature project, which is discussed in the next section, claims to have accomplished this; see, e.g., Rohde et al. (2011).

[9] "Climategate: So Jones Lost the Data? It Was Worthless, Anyway" by Vincent Gray, February 15, 2010: http://pajamasmedia.com/blog/climategate-so-jones-lost-the-data-it-was-worthless-anyway/ (viewed February 22, 2010). See also previous notes 2 and 3.

across weather stations as to the time of day that a maximum or minimum temperature was recorded. Thus, one station might record a maximum at 3:00 pm, while a nearby station recorded its maximum temperature at 1:00 or 2:00 pm. A minimum might be recorded at 11:30 pm at one station, while the nearby station recorded its minimum for that calendar day at just past midnight on the following day. The maximum and minimum temperatures for the calendar day were averaged to provide the mean daily temperature at that weather station. This is hardly a scientific method of determining a mean daily temperature. Given that the raw data were flawed, there is no reason that the constructed gridded data are not likewise flawed.

It should be noted that the historic temperature data provided by USHCN weather stations consist primarily of daily maximum and minimum temperatures (Williams et al. 2008). Again, if these data are used to construct historical temperature records to be compared against current records and paleoclimatic reconstructions (see Chap. 3), a lot of qualifiers need to be inserted when making conclusions about climate trends, especially if these trends are used to guide policy.

Finally, it seems important to know how gridded data are constructed from raw data. If weather stations drop out of the record at any time, or are simply ignored, this has implications for the derived temperature (and precipitation and wind speed) values. As noted, however, the methods used to create the gridded data (specifically the computer codes) are also lost to researchers.

2.2 Available Instrumental Data

There are two main types of global temperature data. One is based on records of surface temperatures collected primarily from weather stations and sea surface temperature (SST) records collected by ships. SSTs are preferred to marine air temperatures because they are considered more reliable, although air temperatures are more in the spirit of how data are collected on land. The second type of data is satellite data, which are much more recent and are considered a better proxy of temperatures around the globe because the coverage is more uniform.

2.2.1 Surface Level Data Collection

There are three major global indices of temperatures that incorporate weather station data: (1) the Climate Research Unit (CRU) at East Anglia University, (2) the National Aeronautics and Space Administration's Goddard Institute of Space Studies (NASA-GISS), and (3) the National Oceanic and Atmospheric Administration (NOAA). Each of these groups primarily relies on the Global Historical Climatology Network (GHCN) for their input data, which contain the daily maximum, minimum, mean and adjusted mean temperatures. A large proportion of the GHCN network is composed of the U.S. Historical Climatology Network (USHCN). There are 7,280

Table 2.3 GHCN monthly temperature sources

Data source	# of mean temp stations	# of max/min temp stations
NCAR's (National Center for Atmospheric Research) world monthly surface station climatology	3,563	0
NCDC's max/min temperature data set	3,179	3,179
Deutscher Wetterdienst's global monthly surface summaries data set	2,559	0
Monthly climatic data for the world	2,176	0
World weather records (1971–1980)	1,912	0
World weather records (1961–1970)	1,858	0
U.S. summary of the day data set	1,463	1,463
U.S. historical climatology network	1,221	1,221
Climatological database for N hemisphere land areas	920	0
Australian national climate center's data set for Australia	785	785
North American climate data, NCDC	764	764
Bo-Min's data set for the People's Republic of China	378	0
USSR network of CLIMAT stations	243	0
Daily temperature & precipitation data for 223 USSR stations (NDP-040)	223	223
Two long-term databases for People's Republic of China (NDP-039)	205	60
ASEAN climatic atlas	162	162
Pakistan's meteorological and climatological data set	132	132
Diaz's data set for high-elevation areas	100	0
Douglas' data set for Mexico	92	0
Ku-nil's data set for Korea	71	71
Jacka's data set for Antarctic locales	70	0
Monthly data for the Pacific Ocean/Western Americas	60	0
U.S. historical climatology network (Alaska)	47	47
Muthurajah's data set for Malaysia	18	18
Hardjawinata's data set for Indonesia	13	13
Fitzgerald's data set for Ireland	11	11
Sala's data set for Spain	3	0
Al-kubaisi's data set for Qatar	1	1
Al-sane's data set for Kuwait	1	1
Stekl's data set for Ireland	1	1

Source: http://www.ncdc.noaa.gov/oa/climate/ghcn-monthly/source-table1.html (viewed April 24, 2010)

weather stations in the GHCN database, and 1,221 in the USHCN database (Table 2.3). However, it is nearly impossible to determine which weather stations are included in each of the three reconstructions.[10]

[10] An excellent source of climate data is the KNMI website: http://climexp.knmi.nl/. Available information on weather station data and where it can be found is also provided by Steve McIntyre at http://climateaudit.org/station-data/ (viewed April 24, 2010).

The GHCN uses a 12-digit number to identify weather stations: the first 3 digits are the country code, digits 4 through 8 refer to the nearby World Meteorological Organisation (WMO) station, and digits 9 through 11 to the particular station. All 11 digits are needed to identify a station, while the 12th digit identifies the 'duplicate number' since GHCN archives versions that are scribally distinct.[11] A list of the GHCN sources of monthly weather station data is provided in Table 2.3. This gives some notion of the variability of data sources and thereby the potential quality of the data that are provided. Interestingly, 22,231 stations provide mean monthly temperature data and 8,152 provide monthly maximum and minimum temperatures; yet, only 7,280 weather stations provide data that can be used to provide global temperature summaries.[12]

For the United States, weather data are available from the USHCN of the U.S. National Climatic Data Center and the Automated Surface Observation System (ASOS). As noted, the USHCN consists of 1,221 official, government weather-monitoring stations that have data going back to at least 1880 and sometimes earlier. The method used to collect USHCN data has essentially remained unchanged, with accurate visual observations of temperatures taken twice daily to record minimum and maximum daily temperature; a clever device marks the high and low reading each day, thereby eliminating any guesswork (Sussman 2010).

The ASOS system consists of some 1,000 automated weather monitoring sites that 'came on line' in the 1980s, with more than 400 added since 1998. These are located at airports, downtown areas, and so on, hardly places that are not affected by heat from nearby buildings, parked cars, pavement, and so on – an 'urban heat island.' The University of Victoria's network of school-based weather stations (discussed in Sect. 2.1) is similar in nature to ASOS weather monitoring sites – a delightful means of entertaining folks and establishing trends, but more difficult to justify as a serious scientific record of true temperatures. The temperature averages from both networks will likely provide a record of higher temperatures than seen previously, but it would be an error to 'splice' such data onto other historical data to say something about past trends in temperatures compared to current ones.

The NASA-GISS temperature series constitutes a reconstruction by NASA's James Hansen and his colleagues (Hansen et al. 1999, 2001, 2010). (Despite working as a scientist for the U.S. government, Hansen is a prominent advocate of catastrophic anthropogenic global warming.[13]) The NASA-GISS data employ information from 7,364 stations, which includes all of the 7,280 GHCN stations and two Southern Ocean stations, with the remainder nearly all from Antarctica, including (it appears) 14 ships. However, the GISS-NASA website indicates that only 6,257 are used to create global temperatures (and no ships).[14]

---

[11] See McIntyre, http://climateaudit.org/station-data/ (viewed April 24, 2010).

[12] A description of the GHCN data can be found in Peterson and Vose (1997) and Peterson et al. (1998).

[13] See Goddard (2011). As several commentators have already observed, this would appear to be a conflict of interest. How impartial can a climate-data gatekeeper be if that same person is a vociferous proponent of human driven global warming?

[14] A list of weather stations and numbers is available from (viewed February 18, 2010): http://data.giss.nasa.gov/gistemp/station_data/.

NOAA's National Climatic Data Center (NCDC) maintains the U.S. Global Climate Observing System (GCOS).[15] GCOS was established in 1992 and "builds upon, and works in partnership with, other existing and developing observing systems such as the Global Ocean Observing System, the Global Terrestrial Observing System, and the Global Observing System and Global Atmospheric Watch of the World Meteorological Organization."[16] The GCOS Surface Network consists of 1,025 land-based weather stations, but reconstructions of past temperatures are based on GHCN-monthly data that

> … contain mean temperature data for 7,280 stations and maximum/minimum temperature data for 4,966 stations. All have at least 10 years of data. The archive also contains homogeneity-adjusted data for a subset of this network (5,206 mean temperature stations and 3,647 maximum/minimum temperature stations). The homogeneity-adjusted network is somewhat smaller because at least 20 years of data were required to compute reliable discontinuity adjustments and the homogeneity of some isolated stations could not be adequately assessed. … In general, the best spatial coverage is evident in North America, Europe, Australia, and parts of Asia. Likewise, coverage in the Northern Hemisphere is better than the Southern Hemisphere.[17]

It is clear that the NOAA reconstruction is based on weather station data that rely primarily on the GHCN, just as was the case with NASA-GISS. Note that the number of weather stations included in the reconstructions necessarily changes because not all are in existence over the time period of the record. The NOAA reconstructions are not discussed further because of their similarity to the other surface temperature reconstructions.

The Hadley Centre of the UK Meteorology Office is associated with the CRU and is one place that retains historical weather data and creates 'homogenized' reconstructions of global temperatures on a gridded basis (fine 2° latitude × 2° longitude and coarser 5° latitude × 5° longitude grids). The homogenized reconstructions not only provide gridded temperature averages, but they also homogenize or sanitize the data to eliminate non-climate factors resulting from socioeconomic activities that directly affect temperature measurements. Socioeconomic activities include such things as the encroachment of urban and other development on weather monitoring stations; adjustments are made, for example, to eliminate the effect on temperature data of the heat generated by pavement from a parking lot that now surrounds a monitoring station, where previously the station was located in an open field (discussed in more detail below).

Collection and analysis of the data fell under the leadership of Phil Jones, who was until 2010 the Director of the CRU. The CRU provides several data products on its website (http://www.cru.uea.ac.uk/cru/data/hrg/). The CRU TS 1.x, 2.x and 3.x series contain historical temperature data that have not been adjusted for non-climate factors, while the HadCRUT data (with CRUTEM containing only the land temperature data) have been corrected to remove the influence of non-climate factors

---

[15] See http://www.ncdc.noaa.gov/oa/usgcos/index.htm (viewed April 24, 2010).

[16] From http://www.ncdc.noaa.gov/oa/usgcos/programdescription.htm (April 24, 2010).

[17] From http://www.ncdc.noaa.gov/oa/climate/ghcn-monthly/index.php (April 24, 2010).

(Jones et al. 1985, 1986b, 2001). The HadCRUT3 average global temperature series is the reconstruction that is most often reported, although questions remain concerning the degree to which non-climatic factors have been removed (see Sect. 2.3).

The Hadley Centre-CRU's data might well be considered identical to that available from NASA, particularly given that the CRU constructed data on behalf of the U.S. government. One difference between the GISS and Hadley/CRU data is that Hadley reports temperature anomalies from the 1961–1990 average global monthly temperatures, while GISS reports anomalies from the period 1951–1980 (with average annual global temperature over this period said to be 14 °C).[18] The similarity between the NASA-GISS and Hadley-CRU reconstructions of historical global temperatures is indicated by the 98% correlation that we find between the HadCRUT3 and GISS adjusted annual temperature series, although the monthly correlation has fallen to just over 90% for the period after 1987 due to adjustments in the GISS data described in Chap. 3 (Sect. 3.2).

When we look at the three surface temperature reconstructions of historical global temperatures, it is impossible to duplicate what was done.[19] In essence, the reconstructions and homogenization of the raw weather station data have to be accepted on faith. As Steve McIntyre has shown, it is difficult to determine where one might find the raw data, it is next to impossible to figure out which weather stations were included in which reconstruction and for what years, and it is simply impossible to determine the methods used to aggregate and summarize data into gridded boxes and correct the data for non-climate factors. How the temperature data are constructed is discussed by Jones (1988, 1994) and Jones et al. (Jones et al. 1986a, c, 1999), although it is not clear from these studies how this is done as they reference reports completed for the U.S. Department of Energy (Jones et al. 1985, 1986b).[20] These reports were completed in the 1980s and may not be relevant to how the data are handled today. A report provided to the U.S. Department of Energy by Jones et al. (2010) is not helpful regarding methods.

By late 2011, a global average temperature series had been independently constructed by researchers from the University of California at Berkeley under the auspices of physicist Richard Muller. The Berkeley Earth Surface Temperature (BEST) project took a different approach from those of NASA, NOAA and the CRU. Over the past 150 years, some 39,000 land-based sites had recorded temperatures for various lengths of time. The data from these weather stations are generally of poor

[18] See http://data.giss.nasa.gov/gistemp/ (viewed March 9, 2010).

[19] Indeed, Jones and Moberg (2003, p.208) admit that it is difficult to say what homogeneity adjustments have been applied since the original data sources do not always include this information.

[20] See http://climateaudit.org/station-data/ (viewed April 24, 2010). As McKitrick (2010c), points out: The 1985 technical reports to the U.S. Department of Energy are indeed exhaustive, but they refer to data sets that have since been superseded, and thus are not adequate for understanding the post-1980 CRUTEM series (para 48, pp.26–27). "Following the publication of the CRUTEM3 data series (Brohan et al. 2006), it was not possible to discern from information on the CRU website, or in accompanying publications, which locations and weather stations had been used to produce the gridcell anomalies" (para 54, p.30).

quality (see below), so NASA, NOAA and the CRU employed various subsets of what might be considered the better quality stations. However, the BEST group used sophisticated statistical algorithms to sort through and make sense of the massive amount of temperature data available from all of the weather stations. Unsurprisingly, the results are not terribly different from those of NASA, NOAA and the CRU; after all, the data sources are the same in each case – instrumental records from land-based weather stations. The BEST results indicate that average global temperatures have generally risen by about 0.9 °C since the 1950s, while the other reconstructions indicate a rise of about 0.6 °C (Rohde et al. 2011).[21] These issues are discussed further below.

2.2.2 *Satellite Data*

Since 1978, microwave data from satellites have provided the most accurate temperature measures from anywhere around the globe, including at different layers in the atmosphere. Satellites can better measure average temperatures at various places in the atmosphere. Further, "the geographic coverage of the Earth is so complete that we can now calculate global average temperature variations with high precision – to about one or two hundredths of a degree C per month" (Spencer 2010, p.4).

Two groups have employed microwave data from satellites to construct temperature series. The best known is a group of scientists at the University of Alabama at Huntsville (UAH); John Christy, Director of the Earth System Science Center at UAH, and Roy W. Spencer, a senior research scientist at UAH and former senior climate scientist at NASA's Marshall Space Flight Center in Huntsville, have developed a global temperature data set from satellite data beginning in 1979.[22] The second group is Remote Sensing Systems (RSS) of Santa Rosa, California. We focus on the UAH data, but the two data series are nearly identical, differing only in how the temperature data are presented. Initially, there had been some disagreement on the method used to construct temperatures from satellite microwave data, but this had been quickly resolved. Temperature series based on satellite data have indicated a slower rate of warming than the surface temperature data, with no warming evident since 1998 (see below).

2.2.3 *Weather Stations and Global Temperature*

Temperature records are available for some 39,000 weather stations, but sufficiently long time series are available for a much smaller subset of monitoring sites. As of 2010, there were 7,350 weather stations in the world plus 14 ships, although only

[21] Four papers have been submitted for potential publication to the *Journal of Geophysical Research* (Muller et al. 2011a, b; Rohde et al. 2011; Wickham et al. 2011).

[22] Both researchers have questioned the role of humans in driving climate change, with Spencer recently arguing that three-quarters of the observed increase in temperatures is due to changes in natural cloud formation (i.e., of non-human origin) (Spencer 2010). This is discussed further in Chap. 5.

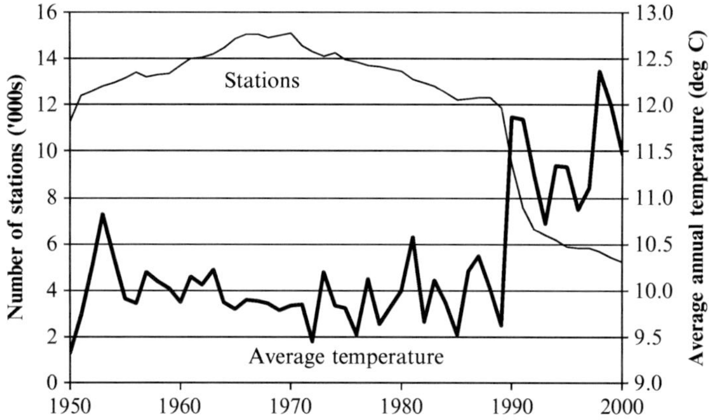

Fig. 2.1 Effect of weather station numbers on global temperature, 1950–2000

6,257 stations were used to provide global climate information.[23] There are 1,221 high-quality weather stations in the United States alone (Williams et al. 2008), about one-fifth of the global total. The three groups (NASA, NOAA and CRU) employ various subsets of the weather station data. For example, amidst claims that some historic raw data have been lost, the CRU at East Anglia University maintains monthly temperatures for slightly more than 3,000 weather stations that is uses to reconstruct historic monthly average global temperatures, and average temperature at a gridded level.[24]

The coverage of past and current weather stations is denser for the United States, southern Canada, Europe and Japan than it is for other regions, particularly central South America, much of Africa and Antarctica (where coverage is nearly non-existent), as well as Australia, parts of Asia and the ex-Soviet Union where coverage is nothing like it is in the United States and Europe. Temperatures over oceans are much more difficult to obtain and water temperatures are used as opposed to air temperatures, as in the case of land-based weather stations. Sea temperature readings in the northern hemisphere are more abundant than those in the southern oceans; temperature data for oceans are sparse, while oceans account for around 70% of the Earth's surface. When account is taken of these factors, it is clear there are huge gaps in global coverage of surface temperature readings.

The number of weather stations and the quality of the data they provide increased from 1,850 to 1,950, although coverage during periods of war was poorer than at other times. The number of weather stations from which data were available peaked during the period 1950 through 1990 (particularly in the late 1960s), but fell dramatically after 1990 (see Fig. 2.1). As noted in Sect. 2.1, a change in the number of sites used to construct summary temperatures creates measurement challenges that

[23] http://data.giss.nasa.gov/gistemp/station_data/ (viewed February 18, 2010).

[24] Information found at http://www.cru.uea.ac.uk/cru/data/temperature/#datdow (viewed March 5, 2010). See also Brohan et al. (2006).

are not easily or consistently overcome as the numbers of observations change from year to year, or month to month. It also creates an enormous challenge to derive a measure of global average temperature, especially one that can easily be understood and recreated by other researchers.

As Richard Muller of the BEST project notes[25]:

> The temperature-station quality is largely awful. The most important stations in the U.S. are included in the Department of Energy's Historical Climatology Network. … 70% of these stations have such poor siting that, by the U.S. government's own measure, they result in temperature uncertainties of between two and five degrees Celsius or more. We do not know how much worse are the stations in the developing world. Using data from all these poor stations, the U.N.'s Intergovernmental Panel on Climate Change estimates an average global 0.64°C temperature rise in the past 50 years, 'most' of which the IPCC says is due to humans. Yet the margin of error for the stations is at least three times larger than the estimated warming. … Moreover, the three major temperature analysis groups … analyze only a small fraction of the available data, primarily from stations that have long records. … On top of that, stations have moved, instruments have changed and local environments have evolved. Analysis groups try to compensate for all this by homogenizing the data, though there are plenty of arguments to be had over how best to homogenize long-running data taken from around the world in varying conditions. These adjustments often result in corrections of several tenths of one degree Celsius, significant fractions of the warming attributed to humans.

Jones et al. (2001) claim to have devised a means "to adjust the grid box temperature series for changes in the number of contributing stations through time and to reduce all series to a consistent level of variance" – also referred to as 'homogenization.' The homogenization appears to rely on an adjustment to the temperature anomalies using the square root of the ratio of the effective number of reporting weather stations at a particular time to the number required to leave variance unchanged as more stations are added. What is most disconcerting, however, is that the temperature anomalies for any given grid box (whether 2° latitude × 2° longitude or 5° × 5°) are based only on the climate monitoring facilities in that grid box, and not those in other grid boxes. In terms of the analysis of hypothetical station monitoring data in the previous section, this is akin to arbitrarily choosing a distance for determining whether to include a weather station, regardless of its quality. In essence, some information is discarded. Further, the constructed temperatures in any grid box are assumed to be independent of those in neighbouring grid boxes, which is unlikely to be the case.

One factor contributing to the reduction in weather stations was the collapse of the Soviet Union, but it was not the only factor. Joe D'Aleo, founder of the Weather Channel, and E. Michael Smith, a computer analyst, report that the National Climatic Data Center (NOAA-NCDC) and the National Aeronautics and Space Administration's Goddard Institute of Space Studies (NASA-GISS) dropped many meteorological stations even though many continued to make appropriate reports; the ones dropped were generally located in colder climates (D'Aleo and Watts 2010).[26] The commentators argue that these actions make the reported temperature

[25] Quote by R. Muller, Wall Street Journal, October 21, 2011 (viewed November4, 2011): http://online.wsj.com/article/SB10001424052970204422404576594872796327348.html.

[26] See http://www.spaceref.com/news/viewpr.html?pid=30000 (viewed March 9, 2010).

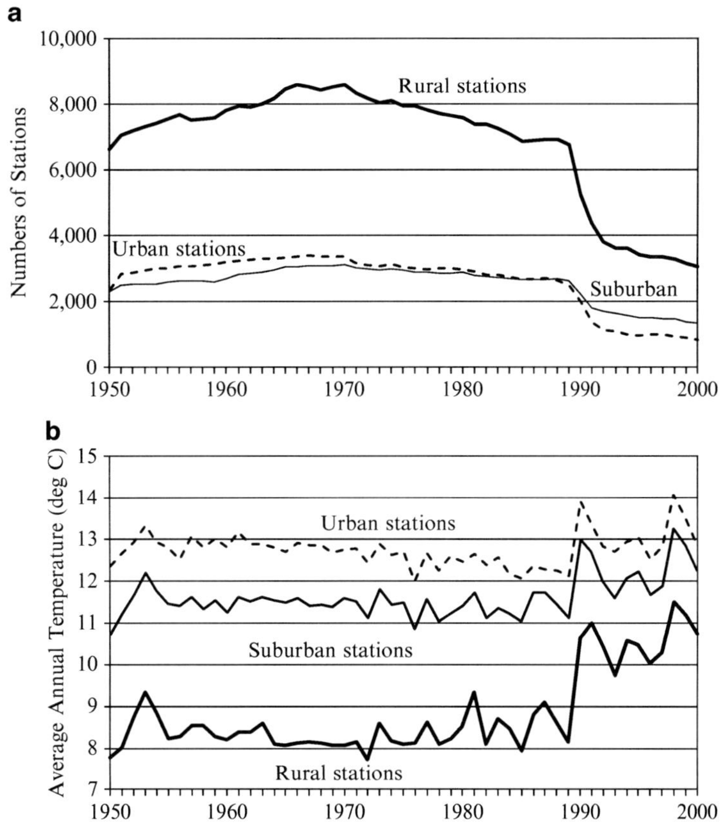

Fig. 2.2 (**a**) Numbers of weather stations by location, 1950–2000, (**b**) Average annual temperature by location of weather stations, 1950–2000

trends unreliable and likely lead to an unknown warming bias. This is seen in Fig. 2.1, where the recorded rise in average global temperatures appears to be directly related to the reduction in the number of weather stations used to construct the average global temperature.[27]

It is interesting to note that urban, suburban and rural weather stations have been dropped since the late 1960s, and precipitously since 1990, although the majority of those dropped by NASA-GISS are found in rural areas (Fig. 2.2a). In each case, a reduction in the number of stations led to an increase in average global temperatures, but this was more pronounced for rural stations, as indicated in Fig. 2.2b. It would appear, therefore, that the observed rise in temperatures after 1990 was due largely to a reduction in the number of weather stations used to construct the average temperatures.

---

[27] See http://www.uoguelph.ca/~rmckitri/research/nvst.html (as viewed March 9, 2010). Data provided by R. McKitrick, University of Guelph. Also see D'Aleo and Watts (2010).

2.2.4 *Quality of Data Collection*

Despite the difficulties inherent in the raw data available to researchers and the fact that the number of weather stations for which temperature data are available changes from one year to the next, the CRU claims that its measures of global average monthly temperature are quite accurate.

> Annual values are approximately accurate to +/–0.05°C (two standard errors) for the period since 1951. They are about four times as uncertain during the 1850s, with the accuracy improving gradually between 1860 and 1950 except for temporary deteriorations during data-sparse, wartime intervals. Estimating accuracy is a far from a trivial task as the individual grid-boxes are not independent of each other and the accuracy of each grid-box time series varies through time (although the variance adjustment has reduced this influence to a large extent).[28]

Without knowing something about what the true or actual average global temperature might be, it is impossible to assess whether this claim is accurate or not.

If one examines the HadCRUT3 data, one is struck by the revisions that are made. For example, the average temperature anomaly for 1850 was –0.44675 in the data reconstruction of March 2009, while it had been adjusted to –0.44283 by March 2010, an increase of 0.00392 °C. For 1900, the anomaly was –0.22308 in the March 2009 but it had been adjusted to –0.22483 by March 2010, a reduction of 0.00175 °C. Because measurement accuracy is to three significant digits, the absolute differences are 0.002–0.004 °C. Although one understands that these adjustments are made because of differences in the methods used to construct grid point data from raw data, there remains a question concerning how this can be done if, as pointed out above, some or all of the raw data for those years are no longer available. Further, such adjustments are not insignificant and, for 1850 at least, appear to fall outside the range of statistical error (although that is difficult to determine without the raw data).

As discussed further in Chap. 3, the NASA-GISS temperature reconstruction has been altered to such an extent that early readings were reduced while later ones were increased, so that 2010 appears as the warmest year in the 1880–2010 record. The main reason for this appears to be the unexplained removal of records from some weather stations. The Berkeley Earth surface temperature project tracks the temperature reconstructions of NOAA and the CRU quite well, but less so that of NASA.

2.2.5 *Are the U.S. Temperature Data Reliable?*

Anthony Watts is a California meteorologist with some 25 years of experience. In 2007, he undertook to investigate whether Stevenson Screen thermometer shelters, which are used in U.S. weather stations, painted with traditional whitewash (slaked lime in water

[28] See 'Answers to Frequently-asked Questions' at the CRU website (viewed March 10, 2010): http://www.cru.uea.ac.uk/cru/data/temperature/#datdow. The HadCRUT3 data and other data products are also available from this website. See also http://climexp.knmi.nl/.

Table 2.4 Analysis of U.S. historical climatology network weather stations

Station rating	#	%	Description
CRN1	19	2	Flat, horizontal ground surrounded by a clear surface with a slope less than 19°; vegetation ground cover less than 10 cm high; sensors >100 m from artificial heating or reflecting surfaces (e.g., buildings, concrete surfaces, parking lots); far from large bodies of water, unless representative of area, and located >100 m away: no shading for a sun elevation greater than 30
CRN2	76	8	Same as CRN1, but surrounding vegetation less than 25 cm high; no artificial heat sources within 30 m; no shading for a sun elevation greater than 5°
CRN3	142	15	(Error 1 °C) same as CRN2, except no artificial heating sources within 10 m
CRN4	578	61	(Error >2 °C) same as CRN2, except artificial heating sources allowed within 10 m
CRN5	133	14	(Error >5 °C) Sensor located next to or above artificial heating source, such as a building, roof top, parking lot, concrete surface

Source: Watts (2009), Williams et al. (2008)

that leaves a white calcium residue) gave different temperature readings than those painted with semi-gloss latex that had different infrared properties (Watts 2009). Whitewash had been used up until 1979, after which time there was a change to latex paint. In a traditional experiment comparing screens that were whitewashed, unpainted (the control) and painted with latex, Watts discovered that average maximum temperatures differed by 0.3 °F (0.17 °C) while average minimum temperatures differed by 0.8 °F (0.44 °C). This is a significant difference as it amounts to perhaps as much as 35% of the projected low-end, 1.2 °C increase in global temperatures over the next century, or some 17% or more of the temperature increase over the past 130 years.

More importantly, however, Watts set about to investigate actual weather stations in the Chico, California area to determine if he could find field evidence of this difference. Instead, he discovered that the weather stations were impacted by other factors, including exhaust air from a cooling device in a cell phone tower at one location and improper siting of the thermometer relative to the screen at another. Clearly, the temperature readings were impacted by non-climatic factors. As a result, Watts then decided to check out other weather stations to determine how many were improperly set up or located. Finally, the project was extended to the entire U.S., with the objective of at least getting photos of all 1,221 USHCN climate-monitoring stations.

Some 650 volunteers provided photographs of 1,003, of which 994 have so far been rated according to the Climate Reference Network's (CRN) five-point rating system.[29] Results as of May 31, 2009 are provided in Table 2.4. Although statistical

[29] See www.surfacestations.org (viewed February 1, 2011). The rating system is provided in a manual by the National Oceanic and Atmospheric Administration (NOAA) and National Climatic

analysis of the data is planned, Watts is also hoping to include a few more CRN1 and CRN2 stations as only 95 of the 994 stations investigated were unaffected by a nearby heating source (within a 30 m radius).

While recognizing that many of the U.S. Historical Climatology Network's (USHCN) stations are improperly sited and thus unreliable, a NOAA study by Menne et al. (2010) attempted to determine whether this mattered. The authors wanted to know if the lower-rated stations actually provided higher readings than the higher-rated stations. They found that the weather stations in the lower CRN categories (the ones considered 'unreliable') actually displayed a somewhat slower upward trend in temperature than the ones considered 'better-sited' (and thus are somehow more 'reliable'). The NOAA study constitutes an important piece of evidence and has deflected some of the criticism concerning the appropriateness of many weather stations.

It is clear that, if one only demonstrates that properly sited (CRN1 and CRN2) weather stations lead to a higher average increase in temperature compared to other stations, it does not constitute proof that there is no problem. Indeed, the NOAA study might actually confirm the notion that temperatures are artificially high because many temperature readings come from poor-quality, improperly-sited weather stations. Suppose that temperature readings at poor-quality stations are consistently higher than those at high-quality stations because of the extraneous influence of heat sources. That is, temperature is a function of declining weather station quality or, more accurately, a function of economic development. Menne et al. simply found that the slope (or first derivative) of this function is higher for the better-rated weather stations than the lower-rated ones. Contrary to their conclusion, this finding may well support the charge that data from poor-quality stations bias average temperatures upwards – that economic development biases temperature readings upwards. This is the case if the (unknown) relation describing the impact of economic development on temperature is convex so that the temperature and rate at which temperature rises (slope of the development-temperature relation) are lower at more reliable (rural) than at less reliable (urban) weather stations. Thus Menne et al.'s results simply confirm that the effect of development leads not only to higher temperature readings but also to higher rates of temperature increase.

What might be more troublesome is the admission that weather stations are not properly sited and that so few climate-monitoring places meet the CRN1 or CRN2 standard in a country such as the United States.[30] This does not bode well for other countries. Further, it may be irresponsible to use temperature data from weather stations

Data Center (NCDC), entitled *Climate Reference Network (CRN) Site Information Handbook* and dated December 10, 2002. At (viewed 14 April 2010): www1.ncdc.noaa.gov/pub/data/uscrn/documentation/program/X030FullDocumentD0.pdf. It is worthwhile noting that the U.S. is the only country that attempts to rank the quality of its weather stations.

[30] A network of 'super' stations that meet all of the proper siting criteria has now been established in the U.S., but data from this network are only available for about 2 years.

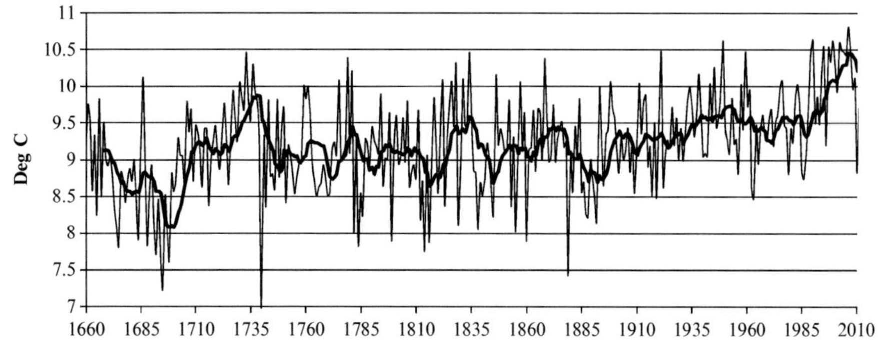

Fig. 2.3 Average annual and 10-year moving average of temperatures in Central England, HadCET series, 1659–2010

that one knows are influenced by extraneous factors to make the case that society must spend untold resources to reduce human emissions of CO_2 and other greenhouse gases because today's warming is unprecedented by historical standards. At the very least, non-climatic factors have to be taken into account, a topic to which we turn our attention in the next section.

2.3 The Instrumental Record

The oldest continuous monthly average temperature series comes from readings taken at various locations in central England. Data are available since 1659.[31] The annual and 10-year moving averages are provided in Fig. 2.3, but the data have been homogenized to take out the urban warming influence at urban stations using nearby rural station data (Manley 1953, 1974; Parker et al. 1992). The issue of homogenizing data to remove the influence of non-climate factors was discussed earlier in conjunction with the school-based weather data in Victoria, British Columbia, and is considered in more detail below. Here we only note that, considering the Little Ice Age affected the record during the period from 1650 until nearly 1900, it is perhaps surprising that Central England temperatures have not risen to a greater extent. Climate change does not deal with local temperatures, so we must turn our attention to available global temperature data.

2.3.1 Global Temperature Reconstructions

The global instrumental temperature data that are provided to, and employed by, researchers are reconstructions based on daily average and maxima and minima temperature records from various weather stations around the globe. As already noted,

[31] Data to construct Figure 2.3 are from http://www.hadobs.org/ (viewed August 26, 2010).

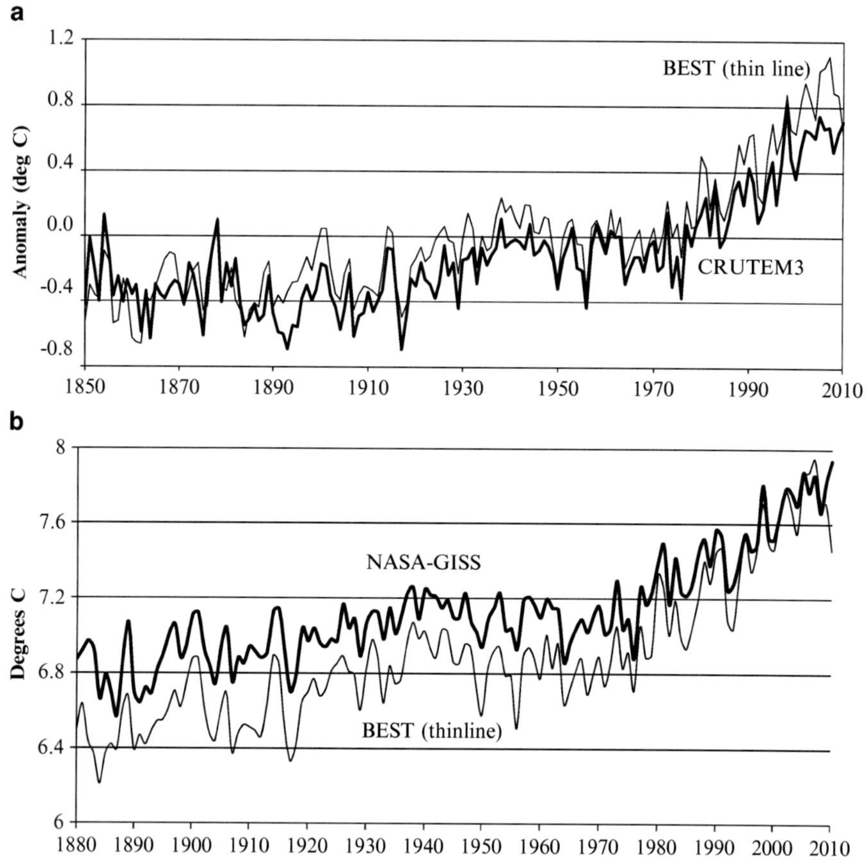

Fig. 2.4 (**a**) Average annual global land surface temperature anomaly, BEST and Hadley CRUTEM3 temperature series, 1850–2010, (**b**) Absolute annual average global surface temperatures, NASA-GISS land surface temperature product and BEST, 1880–2010

duplicating the data manipulations that have gone into those reconstructions is difficult to impossible; in the case of the CRU, for example, it is no longer possible to identify all of the raw data that were employed in the reconstructions. The data products available from the CRU consist of records of (reconstructed) temperatures that are uncorrected for non-climatic factors, and homogenized records that corrected for these extraneous factors. The unadjusted data are referred to as CRU TS products, while the adjusted data are referred to as HadCRUT or CRUTEM products, with the latter only based on land surface temperatures and the former both land and ocean temperatures. (The central England temperature product is referred to as HadCET.)

The CRUTEM3 and HadCRUT3 historical temperature record reconstructions have been the most trusted and commonly used despite the problems noted above. The homogenized HadCRUT3 temperature series goes back to 1850. The GISS reconstruction employs much of the same raw weather station data as that employed by the CRU and begins in 1880. Both data series are found in Fig. 2.4. In Fig. 2.4a,

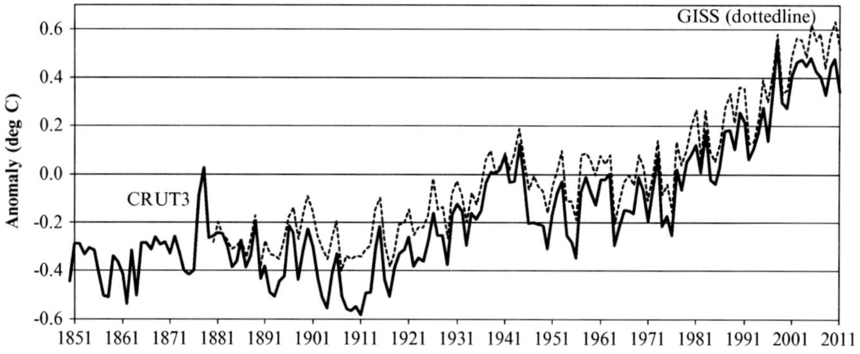

Fig. 2.5 HadCRUT3 and NASA-GISS global temperature anomaly, annual, 1851–2011

global surface temperature anomalies constructed by the CRU and the Berkeley group are compared. The CRUTEM3 and BEST temperature anomalies clearly indicate that global temperatures had been relatively flat for the first 125 years, perhaps rising some 0.05–0.20 °C. An increase in temperature in the period since 1850 would not be surprising given that the Little Ice Age ended sometime around 1850–1870.

In Fig. 2.4b we compare the NASA-GISS absolute global average land-surface temperatures with those of BEST. The Berkeley group finds that the average temperature for the period 1951–1980 is 7.11 °C (GISS provides a value of 14 °C, but then for the average global as opposed to land temperatures only.) In Fig. 2.4b, the BEST average is added to both sets of temperature anomalies, and then the BEST series is adjusted (by subtracting 0.28 °C) so that they have the same maximum temperature (although these occur in different years). Not surprisingly, the record over the past approximately 130 years indicates that temperatures have generally risen. If we look at actual temperatures for the NASA-GISS and BEST land-surface series in Fig. 2.4b, we find that the 10-year average global surface temperature has increased by roughly 0.9 °C in the NASA-GISS record and 1.2 °C in the BEST record. This translates into an increase of about 0.07–0.1 °C per decade. The greatest increase in temperatures has come since the mid 1950s, however; by 0.6 °C in the NASA-GISS and CRU reconstructions and 0.9 °C in the Berkeley reconstruction (Rohde et al. 2011).[32]

The rise in temperatures has not been persistent throughout the 1850–2011 record. Consider Fig. 2.5, where we plot the annual CRUT3 and NASA-GISS temperature series.[33] Temperatures appear to have been relatively flat for the period from the beginning of the record to about the early 1930s, although the quality of the data is admittedly poor. After this there is a steep decade-long rise in global temperatures to 1941, after which they decline somewhat and then level off until the late 1970s. The 1941–1978 period is one of rapid industrial expansion and rising

[32] BEST data are available at http://www.berkeleyearth.org/ (viewed January 6, 2012).

[33] These data differ slightly from that in Figure 2.4(a); they are from Jones et al. (2010), who prepared the data for the U.S. government.

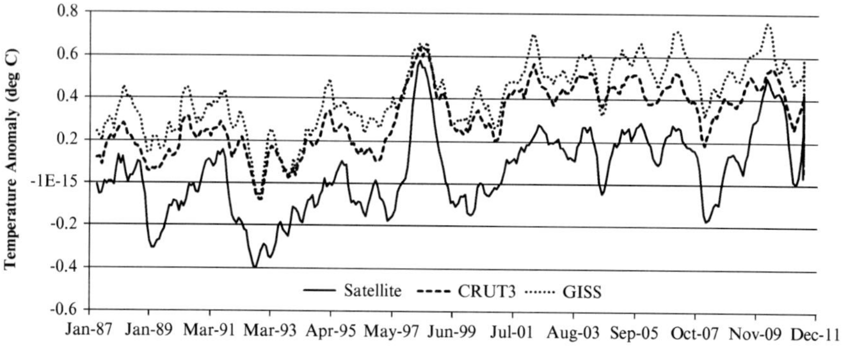

Fig. 2.6 Average monthly global temperature anomaly, 4-month moving average, GISS, Hadley and satellite data series, January 1987–December 2011

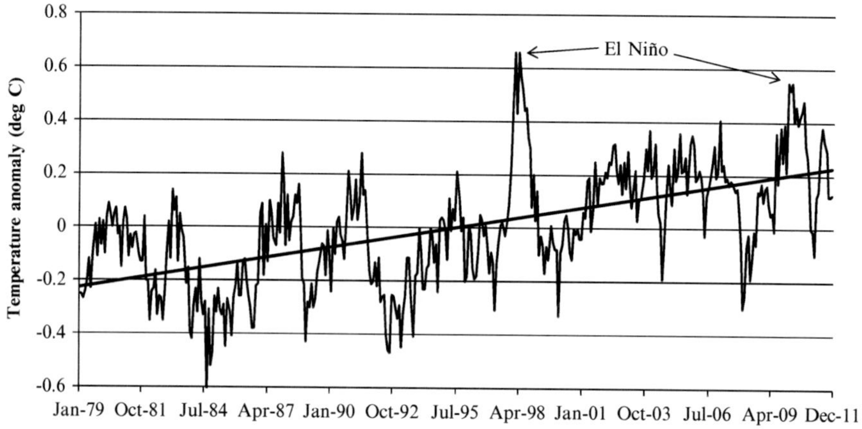

Fig. 2.7 Average monthly global temperature anomalies, satellite data, December 1978–July 2011

atmospheric CO_2, a period during which one would have expected temperatures to have risen if there is a link between human emissions of CO_2 and temperatures. However, lack of a direct correlation between rising CO_2 emissions and temperatures does not negate such a relationship as other factors are also in play, and there may even be a delay of several years between a rise in atmospheric CO_2 content and temperature increase.

After 1978, temperatures again rise rapidly, but stabilize beginning in the late 1990s. Average global temperatures have remained constant or even fallen somewhat since 1998 (Figs. 2.5, 2.6, and 2.7); that year was characterized by a particularly strong El Niño event, although there is not total agreement on this. In the GISS reconstruction, temperatures rise after 1999, actually peaking in mid 2010 (Fig. 2.5) in sharp contrast to the satellite and HadCRUT3 temperature products. GISS uses nighttime radiance observations to correct for non-climatic effects, but also employs observations from fewer weather stations (as discussed in the previous section). A 4-month moving average of global temperatures in Fig. 2.6 indicates sharp drop in temperatures in the last period of 2011.

If we consider only the monthly satellite data (Fig. 2.7), we see a quite pronounced upward blip in temperatures during 1997–1998 that corresponds to an El Niño event (see Chap. 3). Another less intense El Niño event is also evident in Fig. 2.7, occurring near the end of 2009 and first few months of 2010. Notice also the sharp decline in global average temperature towards the end of 2010 and through 2011.

Consider again the monthly satellite data in Fig. 2.7. The data are for the period from December 1978 through December 2011. They clearly indicate an upward trend in global mean temperatures. A linear ordinary least squares (OLS) regression provides the following statistical trend:

$$\underset{(-2.51)}{T = -0.0312} + \underset{(11.17)}{0.5408} \times T_{L1} + \underset{(5.94)}{0.2878} \times T_{L2} + \underset{(2.85)}{0.00016} \times \textit{time}, \; R^2 = 0.727, \; n=395,$$

where T is the monthly temperature anomaly (or de-meaned temperature), T_{L1} and T_{L2} are monthly temperatures lagged 1 and 2 months, respectively, *time* is the monthly time trend, n is the number of observations, and t-statistics are provided in parentheses.

The lags of the dependent variable are used as control variables because temperatures in 1 month tend to affect those in the next several months (e.g., if a particular month is colder than normal, one expects the following months to be colder as well). Non-climatic factors such as a measure of economic or volcanic activity are not included in the regression. The *time* variable then represents rising atmospheric CO_2 due to fossil fuel emissions; that is, we expect temperatures to rise over time due to human forcing. Notice that the regression model explains more than three-quarters of the variation in global average monthly temperature, and all estimated coefficients, including on the trend of interest, are highly statistically significant.[34] Thus, temperature has been increasing by some 0.0023 °C per year or 0.023 °C per decade. If this trend continues, then we can expect an increase in temperature of only 0.2 °C by 2100.

Monthly ocean and land temperature anomalies are available from the satellite data and shown in Fig. 2.8. Not unexpectedly, land temperatures show greater variation than ocean temperatures. Consider ocean temperatures.

Oceans constitute a huge heat sink that can affect climate for many years. Little is known about the non-radiative heat exchanges between the oceans and the atmosphere, but these can warm or cool the atmosphere by as much as 1 °C. That is, heat exchanges between the oceans and the atmosphere take place independent of forcings that originate with anthropogenic activities. One study estimated the heat

[34] When other lags were included in the regression, they turned out to be statistically insignificant, while their inclusion did not change the coefficients on the two lags of the dependent variable that were included.

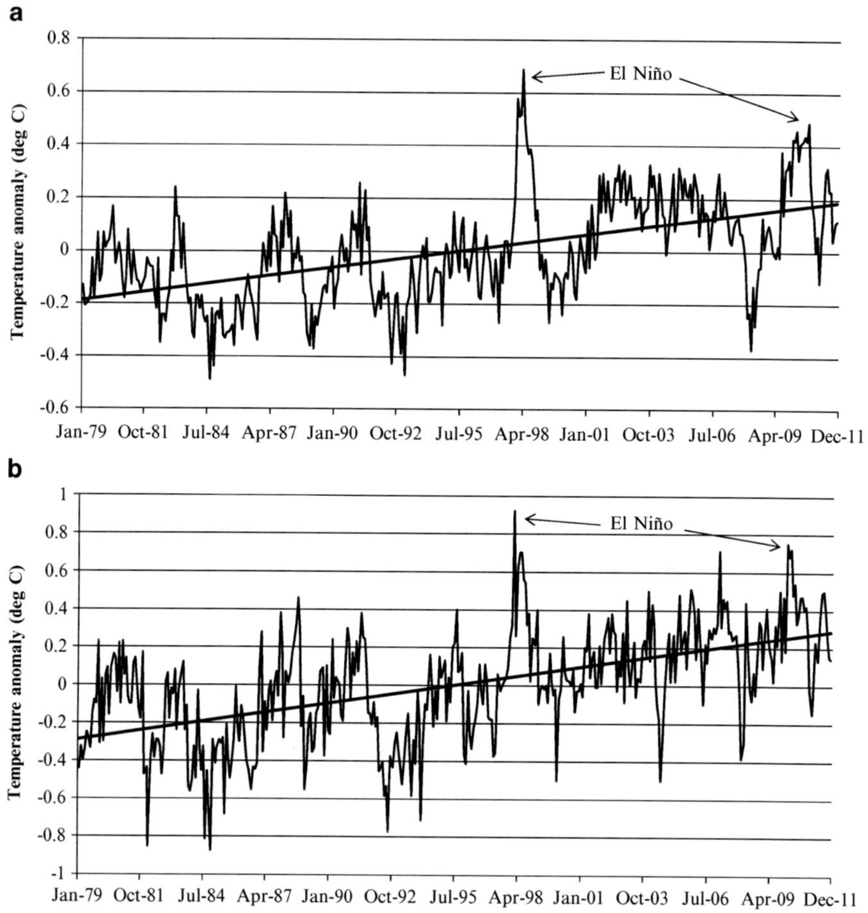

Fig. 2.8 (**a**) Average monthly ocean temperature anomalies, satellite data, December 1978 through December 2011, (**b**) Average monthly global land surface temperature anomalies, satellite data, December 1978–December 2011

content of the upper 750 m of the Earth's oceans from 1993 to 2005 using data from satellites, ocean moorings, floats and shipboard sensors.[35] It found the average temperature of the upper ocean increased by 0.09 °C from 1993 to 2003, and then fell 0.03 °C from 2003 to 2005. Using results from oceanographic cruises in the North Atlantic Ocean at 24.5°N, Vélez-Belchí et al. (2010) found that there was a warming of 0.27 °C from 1957 to 1998, but a significant cooling of −0.15 °C in the upper ocean from 1998 to 2004. This was corroborated using data from the Argo ocean network (http://www.argo.net), which indicated a strong −0.13 °C cooling between

[35] Source: http://news.mongabay.com/2006/0926-oceans.html (viewed April 28, 2010).

1998 and 2006. Laidre et al. (2010) find to the contrary that oceans are warming, perhaps more than originally thought, but this conclusion is based on data from narwhals in Baffin Bay. Using monitoring devices connected to narwhals constitutes a rather crude method for measuring changes in ocean temperatures, and need to be corroborated with other data.

The satellite data indicate that ocean temperatures have not changed much since 2005, and are influenced by El Niño events. Thus, as shown in Fig. 2.8, ocean temperatures have fallen since the last such event; this is confirmed by Vélez-Belchí et al. (2010) who, during the period 1957–2006, found the ocean to be at its warmest and, surprisingly, also at its saltiest in 1998. The latter result is surprising because more glacier melting might be expected in warm years, which would reduce ocean salinity as more fresh water enters the sea. Ocean temperatures clearly fluctuate over time, despite suggestions that ocean temperatures have generally risen. While absorption and release of energy by oceans are extremely important to potential future warming, the processes whereby oceans affect atmospheric temperatures are ill understood and not at all modeled (see Chap. 4).

Longer term ocean temperatures from satellite data are plotted in Fig. 2.8a, along with the trend in ocean temperatures since 1978. The linear trend is

$$\underset{(-2.51)}{\mathrm{T} = -0.0312} + \underset{(11.17)}{0.5408} \times \mathrm{T}_{\mathrm{L}1} + \underset{(5.94)}{0.2878} \times \mathrm{T}_{\mathrm{L}2} + \underset{(2.85)}{0.00016} \times \mathit{time}, \mathrm{R}^2 = 0.727,\ \mathrm{n} = 395,$$

where again t-statistics are provided in parentheses. The results indicate that global ocean temperatures have advanced by about 0.020 °C per decade. If this rate continues throughout the twenty-first Century, the oceans would warm by some 0.17 °C by 2100. If one considers only global land temperatures, a similar regression analysis indicates a warming trend of 0.05 °C per decade, or 0.45 °C by 2100.

Statistical evidence of trends in temperature data is not ubiquitous. For example, (McKitrick and Vogelsang 2012) examine whether there is an upward trend in temperatures in tropics (20°N to 20°S latitude). Using monthly temperature data from the Hadley Center for 1958–2010, they first discover a statistical shift in the data occurring in December 1977. If a level-shift term for this date is included in the regression, no statistically significant upward trend in temperature can be detected – the trend is not statistically different from zero.

Since the 1998 El Niño event, average monthly global ocean temperatures have hardly changed at all, effectively increasing only during 2010 in response to another somewhat weaker El Niño event; data for 2011 indicate that a significant drop in ocean temperatures is underway. It would appear, therefore, that ocean warming will not become a driver of rising sea levels any time soon, although it is possible that ocean temperatures are lower than global warming would indicate due to melting ice.

The satellite data indicate that, while global temperatures have risen since the late 1970s, the increase has been small. The slight rise in global temperatures in the satellite record is the result of the lack of increase since 1998. This was the case even though mean global land temperatures for the first 4 months of 2010 were the highest of any 4-month period recorded in the satellite data (Fig. 2.8b). However, if ocean and land temperatures are aggregated to obtain an average global temperature, then the warmest 4-month period was February through May 1998, and this was the case also for the Hadley CRUT3 record. After April 2010, there was a significant drop in land and global temperatures in the satellite record, and especially ocean temperatures. Interestingly and perhaps somewhat strangely, the NASA-GISS record indicates a continuous upward trend, with temperatures in 2010 considered to be higher than those in 1998 (see Figs. 2.4b and 2.5). This is discussed further in Chap. 3.

If one were to examine only the trend between the end of 1978 and 1998, the temperature increase would appear to be significant. The message that one needs to take from this is that the increase in temperature that one finds depends on the period chosen. For some periods, temperatures have risen rapidly; other periods temperatures have remained flat or even declined. While this is true for the past three decades and for the twentieth century in its entirety, it is also true for the much longer term. In order to determine whether current temperatures are outside the realm of past human experience, it is necessary to go back several millennia, and not just 100 or 150 years.

2.3.2 *Effect of Non-climatic Factors on Temperature Measures*

If temperature data are to be an accurate gauge of the impact that human emissions of CO_2 and other greenhouse gases have on global temperatures, it is necessary to correct the raw temperature data for non-climatic factors – correcting for the urban heat island effect (see Sect. 2.1) and other socioeconomic factors (e.g., population growth). This is exactly the sort of thing which the Climate Research Unit at East Anglia University claimed it had done in moving from the CRU TS temperature series to the CRUT series; apparently this contamination has also been removed from the NASA-GISS temperature data. Indeed, scientists are convinced that there is no evidence of a non-climatic bias in the instrumental, non-satellite record (see, e.g., Parker 2010). Yet, as noted above, it is not clear how climate scientists actually removed the non-climatic influence, but there is strong empirical evidence indicating that they were not successful, and that the increases in temperatures seen in Figs. 2.4 and 2.5, for example, are partly the result of socioeconomic factors unrelated to CO_2 emissions.

What do we mean by the term non-climatic factors? Suppose a weather station is initially located in a field with no nearby heat source, or at least none within a 100 m or more, and is rated as a CRN1 climate-monitoring facility. Clearly, as time passes, the field in which the weather station is located might be developed and, consequently, a heat source located nearby. Indeed, the field need not be developed. All that might be required is the construction of a cell phone tower within a few meters of the monitoring station, with heat from the cooling mechanism affecting temperature data. This is what Anthony Watts found at one of three weather stations located in the Chico, California area. Cell phones are a very recent phenomenon and it should come as no surprise that a cell phone tower might be located near a weather station, perhaps because the property is owned by a non-profit organization that might need a little extra revenue to meet expenses.

But it is not just the 'heat island effect' or encroaching developments that systematically affect temperature readings. Rather, any "socioeconomic activity can lead to purely local atmospheric modifications (such as changes in water vapor and fine particle levels), which, along with other land-surface modifications and data inhomogeneities, can cause apparent trends in temperature data that are not attributable to general climatic changes" (McKitrick 2010c). (We saw this in Sect. 2.1 for two schools in Victoria's school weather network, where a monitoring station was inappropriately located.) Although it is impossible to consider all the separate socioeconomic activities that might affect each and every weather monitoring facility, it is nonetheless possible to determine whether such effects impact temperatures in a systematic way. If general measures of local socioeconomic activity affect summary measures of local temperatures, this constitutes evidence that non-climatic factors show up in the temperature record. More succinctly, this constitutes evidence that rising temperatures are not the result only of increasing atmospheric CO_2.

This is precisely what de Laat and Maurellis (2004, 2006), McKitrick and Michaels (2004, 2007) found: population growth, changes in GDP per capita, GDP density per unit area, coal consumption (which increased particulate matter in the atmosphere as well as emissions of CO_2), and other factors explained rising temperatures over the period 1980–2010. McKitrick and Michaels (2004, 2007), for example, regressed temperature observations for each of 440 land-based (2.5° latitude × 2.5° longitude) grid cells for the period 1979–2002 on socioeconomic variables. They found that these explained about half of the increase in global temperatures over the past two decades. While one expects evidence that socioeconomic (non-climatic) factors affect temperatures in the unadjusted CRU TS data, one should not have found it in the adjusted CRUTEM (land-only) data.

Needless to say, McKitrick and Michaels, and de Laat and Maurellis, were berated by the climate scientists who cannot believe that, after their adjustments to homogenize the surface data, the data remain contaminated.[36] The IPCC authors responsible for evaluating this contradictory research dismissed it as follows:

> McKitrick and Michaels (2004) and De Laat and Maurellis (2006) attempted to demonstrate that geographical patterns of warming trends over land are strongly correlated with

[36] McKitrick (2010b) provides an interesting and entertaining commentary on the attempts to prove his results and those of de Latt and Maurellis false. Some of this is discussed in the next several paragraphs. See also McKitrick (2010a), which addresses an error in the IPCC WGI (2007) report that pertains to his research. This paper was sent to seven journals – three journals would not even send it out for review, while a fourth journal would not even correspond with the author. The paper was examined by seven reviewers, six of whom agreed with the methods and results, and recommended publication. The editors of two journals turned down publication because, in one instance, the paper did not really address the journal's aims and, in the other, the editor agreed with the one dissenting reviewer (despite evidence that the reviewer was not familiar with the statistical methods employed).

> geographical patterns of industrial and socioeconomic development, implying that urbanisation and related land surface changes have caused much of the observed warming. However, the locations of greatest socioeconomic development are also those that have been most warmed by atmospheric circulation changes …, which exhibit large-scale coherence. Hence, the correlation of warming with industrial and socioeconomic development ceases to be statistically significant. In addition, observed warming has been, and transient greenhouse-induced warming is expected to be, greater over land than over the oceans …, owing to the smaller thermal capacity of the land (IPCC WGI 2007, Chapter 3, p.244).

On what grounds does the IPCC dismiss the research?

First, in the last sentence of the quote, the IPCC authors argue that only activities on land matter and that taking into account ocean data leads to the finding of contamination. In essence, the IPCC argues that the contrary findings are spurious. However, McKitrick and Michaels use only the CRUTEM data set, which includes no sea measures.

Second, the statement that "the correlation of warming with industrial and socioeconomic development ceases to be statistically significant" when geographic patterns are taken into account is not backed up. The IPCC authors provide no evidence that, if spatial aspects are included, the effect of industrial and socioeconomic developments on warming is no longer statistically significant.[37]

In response to the IPCC objections, McKitrick (2010a) investigated claims that spatial considerations would change the statistical significance of the results. He was able to control for spatial effects in two ways. First, he included atmosphere circulation measures as control variables, namely, those related to the Arctic Oscillation (AO), North Atlantic Oscillation (NAO),[38] Pacific Decadal Oscillation (PDO), and the El Niño-Southern Oscillation (ENSO). Second, he controlled for spatial autocorrelation, which is a statistical method that corrects bias arising when what happens at one location is correlated with what happens at another (usually nearby) location. This type of correlation could make the effect on temperatures at one site look bigger (or smaller) than they really are because the influence on temperature from socioeconomic activities in grid A includes the effect on temperatures of activities in grid B. Upon re-estimating his results to take these factors into account, and using both the CRU and satellite data, McKitrick found that his and Patrick Michaels' earlier results continued to hold.

Finally, McKitrick and Nierenberg (2011) re-examined all of the evidence and data. They took into account spatial autocorrelation and known atmospheric circulation patterns that might result in spurious correspondence with socioeconomic variables. Regardless of the statistical tests they performed or explanatory variables that they added to address concerns by climate scientists, the authors found the effect of socioeconomic variables on temperature trends to be extremely robust for the GISS

---

[37] References in the above quote to other sections in the same chapter of the IPCC report were removed as they provide no evidence whatsoever on this matter (see also McKitrick 2010b).

[38] The NAO is not to be confused with Atlantic Multi-decadal Oscillation (AMO); the former is caused by surface-atmospheric pressure changes, whereas the latter is the result of changes in ocean temperatures and currents and other factors that are not entirely known.

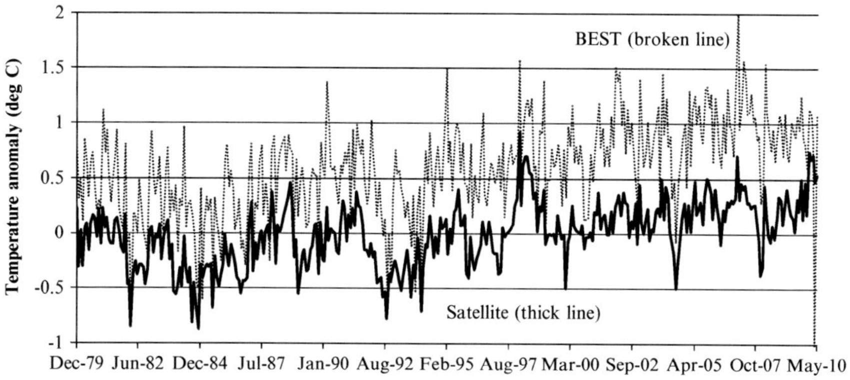

Fig. 2.9 Comparison of monthly land surface temperatures, satellite and BEST constructions, December 1978–May 2010

and CRU land surface temperature products. When satellite data or data from an ensemble of climate models for the same period were substituted for the GISS and CRU temperature data, the statistical significance of the socioeconomic, non-climatic factors disappeared! The satellite and climate model data did not exhibit the same bias. This research clearly indicates that the temperature products provided by the CRU, NOAA and NASA-GISS have not removed non-climate factors and, therefore, can attribute the rise in temperatures observed during the twentieth century to be the result primarily of human emissions of greenhouse gases.

Lastly, the Berkeley group argues that it has removed any non-climatic factors in its construct of global land temperatures. However, the real test will occur when the BEST climate data are substituted into the McKitrick model, and, like the satellite data, the statistical significance between temperatures and the socioeconomic, non-climatic factors disappears. An examination of Fig. 2.9 suggests that this might not be the case. Monthly satellite land-based temperature anomalies are compared to the monthly anomalies in the BEST data for the period December 1978 through May 2010. Although the simple correlation between the two series is only 0.79, the BEST data exhibit much greater variance.[39]

One can only conclude that scientists should not rely on data from surface-based weather stations that have subsequently been reconstructed to remove industrial and socioeconomic (non-climatic) factors. Certainly, the instrumental data that are defended by most climate scientists are contaminated and cannot in good conscience be used as a basis for making policy.[40] Only satellite data are unaffected by non-climatic factors and are the only uncontaminated data for determining temperature trends.

[39] BEST data are available at http://www.berkeleyearth.org/ (viewed January 6, 2012).

[40] Climate scientists continue to insist that the temperature reconstructions from surface-based observations are free of non-climate factors. What is perplexing is that, in making such claims, no statistical evidence is provided and there are no citations to peer-reviewed studies that do provide statistical evidence of contamination (see, e.g., Parker 2010).

2.4 Weather Derivatives

One place where weather station data play an important role is in the emerging market for weather insurance. Weather derivatives refer to financial instruments that are traded in markets and are purchased to reduce economic agents' exposure to weather risks. They began to trade on the Chicago Mercantile Exchange (CME) in 1999 and now constitute one of the fastest growing financial instruments. The two weather derivatives that currently trade on the CME are heating degree days (HDDs) and cooling degree days (CDDs), which are related. A HDD refers to the number of degrees that a day's average temperature is below 65 °F (18 °C), which is the temperature below which buildings need to be heated. In juxtaposition, a CDD is the number of days multiplied by the degrees in each day that the average temperature exceeds 65 °F. For example, if in a given week, the daily temperature averages 60 °F, there are 35 heating degree days (= 5 °F×7 days); if during a week, the daily temperature averages 75 °F, there are 70 CDDs. The price of weather derivatives traded on the CME in the winter is based on an index made up of monthly HDD values. By purchasing a contract to buy or sell HDDs or CDDs at a future date, a business can protect itself against losses caused by unexpected shifts in weather conditions. As of 2010, future CDDs and HDDs are available on the CME for Atlanta, Chicago, Cincinnati, New York, Dallas, Philadelphia, Portland, Des Moines, Las Vegas, Tucson, Minneapolis, Kansas City, Boston, Houston and Sacramento.

Suppose an energy company is in the business of selling fuel oil for heating purposes. Further, suppose that it expects to sell a great deal of fuel oil the next month because temperatures have historically been low during this month. However, if temperatures are higher than expected, the company could lose substantial revenues because it will have too large a fuel oil inventory, say. To protect against this risk, it could rely on forecasts from the national weather service and buy fuel in accordance with those forecasts. Alternatively, it could enter the futures market and purchase HDDs; the price of an HDD is set at $20, so the market 'speculates' on the number of heating degree days that will be realized in a month. Assume the market forecasts that there will be 300 HDDs in the coming month. A trader (speculator) can purchase 300 heating degree days for $6,000 and, if the realized number of HDDs for the month is higher (say 320), sell them for $6,400, yielding a profit of $400. The energy company wants to avoid the risk associated with an unusually warm month, so it will contract to sell HDDs today (which are purchased by the speculator) with the proviso that it will purchase the realized amount back from the speculator at the end of the month. If the realized outcome is 320 HDDs, the company will lose $400 but benefit from sales of fuel oil; if the realized outcome is 250 HDDs, the company will fulfill the contract by paying the speculator $5,000 (= 250×$20) thereby gaining $1,000 to be set against the loss in fuel oil sales.

The futures market in financial weather derivatives is perfect because there is no asymmetry – no trader can have insider information as no one can predict the weather. Indeed, empirical evidence indicates that the futures market in these weather derivatives does a better job at predicting the numbers of HDDs and CDDs than the weather forecasting services (Chincarini 2011).

2.4.1 Weather Index Insurance

Weather index insurance has been promoted by agricultural economists in lieu of crop insurance (Skees 2008; Turvey 2001, 2005; Turvey et al. 2006; Vedenov and Barnett 2004; Woodward and Garcia 2008; Xu et al. 2008). In Canada and the U.S., crop insurance schemes are government run and are not actuarially sound – crop insurance subsidizes farmers. Crop insurance is the business of government partly because of adverse selection. This occurs when some farmers would not voluntarily participate in crop insurance because they are able to adapt to drought by diversifying their farm operations; they raise livestock, plant a range of crops (including drought tolerant varieties), employ summerfallow (leaving fields fallow and controlling weeds to conserve moisture for next year's crop), and/or own fields spread across a landscape. They are affected only by large-scale droughts, in which case they would more likely be eligible for disaster relief rather than insurance per se. For these farmers, the costs of private crop insurance exceed expected benefits. Only farmers who are likely to claim benefits would participate in private crop insurance, giving rise to adverse selection. Requiring all farmers to participate eliminates this problem, but at a cost to society.

No crop insurance program can eliminate the problem of moral hazard, which occurs because, once farmers are in a crop insurance program, they take no steps to reduce their risks of exposure to drought. For example, farmers no longer diversify their operations, but specialize in only one activity, generally grain production as this enables them to spend only part of the year actually engaged in farming (say, 3 weeks during spring planting and 3 weeks during fall harvest). Farmers also do not reduce risk by planting drought-tolerant varieties with lower potential yield. The farmers' decisions are contrary to the desires of the insurer – farmers take on risks they would otherwise avoid.

The advantages of weather index insurance are several. First, weather index insurance eliminates the problem of adverse selection (or hidden information), because participation by one farmer does not affect the premium that other farmers would pay. Rainfall and temperature outcomes are uncorrelated with participation rates. Second, it eliminates the problem associated with moral hazard (or hidden action); because the farmer cannot influence the weather, she cannot affect the payout. The payoff structure is independent of actual crop yields. Finally, weather indexes can be quite simple – they can be easy to understand. What is needed is a record of temperature and/or precipitation that is sufficiently long to enable the insurer to develop an actuarially-sound premium for whatever index is chosen.

Consider the following example. A farmer wishes to insure herself against weather that is too hot during 10 crucial days in the growing season. Suppose she purchases an insurance contract that provides a payout of $2,000 for every degree Celsius that the temperature exceeds 38 °C, and that this is the case for each day during this crucial period in the growing season. Further, suppose there are three weather stations considered to be sufficiently 'close' to the farm. These are located at distances of 10, 20 and 30 km from the farm, and, on a given day during the crucial growing period, record temperatures of 35.0, 40.0 and 43.5 °C, respectively.

The simplest weather index to use would be the average temperature. For this day, the average temperature is 39.5 °C, so the farmer would get a payout of $3,000 (=$2,000 × 1.5 °C) for that day.

One might reasonably argue that, because the weather stations are located at unequal distances from the farm, the farmer is overpaid for this particular day since the nearest weather station has the lower temperature reading. Indeed, if weather station data are weighted by the inverse of the distance from the farm to the weather station, the average is reduced to 38.3 °C, so the farmer would only receive a payout of $600. It would be difficult to argue that one payout is somehow fairer than another, or that one weather index is fairer than another. For example, the nearest weather station might be located at a much higher altitude than the farm and the other two weather stations. What is important here is that the insurance company and farmer agree upon the basis or method used to construct the weather index for insurance purposes, whether a simple average or a weighted one; they might even need to agree on which weather stations to include in the index, and how temperature readings are affected by extraneous factors (e.g., newly-installed heating vents the affect temperature readings; see Sect. 2.1).

There are several criteria that are important in the choice of a weather index, however. From the farmer's perspective, it is important that there exists a relationship between the weather index and the potential loss from too high temperatures. The premium that the farmer pays depends on the link between the weather risk and income. From the perspective of the insurance company, the premium must cover its risk plus a return to investment. To determine the premium, it is necessary to know something about the probability that temperatures during the 10 crucial days exceed 38 °C. This requires the insurer to have reliable data on past temperatures to establish its exposure to risk. Since both parties (agents) to the insurance contract employ the same weather station data, which are presumed to be fair in that they do not favor either agent, any index using this data, any trigger point and/or payout scheme (e.g., one that pays more as the index rises) can be decided upon by the parties. It does not matter since neither agent can affect the outcome of the weather index. The premium that the farmer pays is only to be tied to the actual weather index that is to be used and to the magnitude of the potential payouts.

While weather index insurance eliminates problems of adverse selection and moral hazard associated with standard crop insurance, the only drawback of weather index insurance relates to *basis risk*, which is defined as "the risk that payoffs of a hedging instrument do not correspond to the underlying exposures" (Norton et al. 2010). Continuing with our example, basis risk occurs when the farmer receives a payout and none is warranted; the weather index indicates temperatures are higher than the threshold but temperatures at the farm are below the threshold and there is no crop damage. Alternatively, it is possible that no payout is provided when one is warranted; this occurs when the farmer experiences crop damage because on-farm temperatures exceed the threshold, but the weather index indicates the opposite. Thus, there is a tradeoff between a crop insurance program that relies on crop yields to determine payouts and one that employs a weather index.

2.4.2 Some Simple Examples of Weather Index Insurance

Although weather index insurance is not in common use, proponents argue that it is preferable to standard approaches of dealing with weather risk. Consider some examples. Turvey et al. (2006) examine insurance for ice wine, which requires that grapes are harvested only on days when temperatures are precisely in the range −8 to −12 °C. The weather index would be the number of days within a certain period of the year when temperatures remain between −8 and −12 °C during daylight hours. Choice of a weather station nearest the vineyard or an agreed-upon automated weather monitoring station installed for this purpose (e.g., similar to those in the Victoria school weather network).[41] Farmers would insure against the risk of leaving grapes on the vine to be harvested for the more lucrative ice wine market by paying a premium that would provide a payout if there are insufficient days available to harvest all of the grapes. The payout would fall as a function of the number of days that the vineyard owner was able to harvest grapes for ice wine.

Growing degree days (GDDs) refers to the number of days multiplied by the number of degrees Celsius that average daily temperature exceeds 5 °C. If there are insufficient GDDs, grain crops will be of a lower quality (lower grade) and sell at a lower price. A weather index based on GDDs would enable farmers to ensure against the risk of insufficient warm days to permit the grain to achieve the expected desired quality. Farmers could be provided a payout that increases with the extent to which GDDs fall below a threshold value. Likewise, if weather indexes were available, farmers could insure against too much rain during the growing season, which reduces grain's protein content, or too much precipitation during harvest, which could require expensive grain drying. Farmers could also insure against early frosts if such financial weather derivatives were available.

A particular type of weather insurance risk that does not involve a weather index per se is described by Chantarat et al. (2008). When a famine develops in the Horn of Africa, for example, a humanitarian response is generally not forthcoming until it is too late and people are experiencing the ravages of starvation including death. Once they recognize the need for aid, relief agencies need to mobilize resources – they need to obtain funds to purchase food and contract to have it delivered. The authors examine weather-related insurance for Africa that would pay relief agencies in timely fashion rather than relying on fund raising once a famine is identified. Aid agencies would purchase 'famine relief insurance.' The weather index that is recommended is the mid-upper arm circumference (MUAC) of children aged 6–59 months. The insurer would pay out when "the proportion of children aged 6–59 months in a community who suffer a mid-upper arm circumference z-score

[41] The problem with a newly-installed, site specific monitoring station is the lack of a historical record of temperatures that the insurance company can use for calculating the insurance premium. The insurer will need to rely on information from nearby weather stations, which militates against the need for a site specific station.

≤−2.” For a large region, like the Horn of Africa, it would require the services of a very large insurance companies, such as Lloyd’s of London which is the world’s largest insurance market.

The MUAC indicator was been adopted by the United Nations’ World Food Program after the 1985 famine in Ethiopia as an early warning of famine. MUAC measurements indicated that a famine was imminent in Somalia and other parts of the Horn of Africa as early as November 2010, but political factors related to control of much of Somalia by an Islamic militia, the unwillingness of the Ethiopian government to admit to a failure of their anti-famine policies and other factors meant that action to alleviate the famine was delayed. However, experience indicates that the cost of political delay raises the cost of helping those in need from $7 per head to $23 (*The Economist*, July 30, 2011, p.46).

2.4.3 Adapting to Climate Change

One would think that highly resource dependent economies would have developed a variety of triggers that promote mitigation of and adaptation to severe weather events that can result in large economic damages from wildfires (including costs of fire suppression), reduced power generating capacity, crop loss, and/or flooding. Yet, resource owners do not appear to insure against such events or even anticipate them. Timber companies do not insure against reduced timber harvests (perhaps because the government compensates for such losses), while governments make no effort to account for higher risks of disturbance in forestry (e.g., allocate more budget to fire fighting, train fire-fighting crews or reduce fuel loads). Agricultural producers in rich countries especially rely on government sponsored crop insurance schemes, or wait upon government subsidies in cases of drought. Electrical system operators make no allowance for potential shortfalls in power generation. This situation could be ameliorated by relying to a greater extent on weather indexes for insurance purposes, whether this involves actual financial instruments sold by insurance companies or the development of thresholds that policymakers can use to trigger action to increase fire preparedness in forestry, or take action to store additional water behind reservoirs while relying on power from other sources (including imports).

Because the primary sectors in rich countries generally account for a small but important proportion of GDP, governments are able to cope with weather risk on an ongoing basis. There is no urgency to develop financial instruments that the private sector can use to protect against weather risk. This is not true in poor countries, but in those countries financial markets are not sufficiently developed to facilitate protection against weather risks. Yet, even rich countries could benefit by explicitly accounting for weather and other climate risks.

There is evidence to suggest that the Pacific Decadal Oscillation (PDO), the North Atlantic Oscillation (NAO), the El Niño Southern Oscillation (ENSO), and the Pacific North American (PNA) tele-connectivity index impact agriculture and forestry in western Canada, and elsewhere across the globe. These events have similar impacts to what might be expected under climate change. An El Niño results in warmer and drier winters in western Canada, while the La Niña phase of the ENSO is associated with cold and wet conditions (Shabbar et al. 1997). In the summer, there is a greater risk of drought with a La Niña event, with losses in the U.S. corn belt estimated to have been in the range of $10 billion for each of the 1983 and 1988 La Niña events; an El Niño, on the other hand, is associated with a warm growing season with above normal June-July precipitation that favors higher Canadian wheat yields (Garnett 2007; Hsieh et al. 1999; Khandekar 2004). There is also evidence that these climatic events influence wildfires.[42]

Much like weather index insurance, there are potential social and private benefits by including knowledge about climate events in the decision calculus. For example, if the risk of a cold winter is greatly reduced because an El Niño event is taking place, farmers might take advantage by planting more area to winter wheat. This increases their profits as it reduces field operations, reduces soil erosion and provides temporary nesting habitat for migrating waterfowl. Indeed, using a global agricultural model, Chen et al. (2002) indicate that agriculture could significantly benefit by taking just ENSO events into account. Likewise, governments might plan an increased allocation of resources to fight wildfire if they know that the summer is likely drier and warmer than usual; or that there is greater risk of flooding from snowmelt.

It is clear that financial instruments can be developed to enable private individuals and governments to better cope with weather risks. These could also facilitate adaptation to longer term climate change. While financial weather derivatives are a recent innovation, it is expected that they will play a larger role in the future, particularly if climate change is indeed resulting in warming and greater variability in temperatures and precipitation.

2.5 Discussion

Environmental groups are concerned about global warming because it is considered a threat to the Earth's ecosystems and thereby human welfare. Their solution is the drastic reduction of fossil fuel use. But this should only be done if it can be shown that increasing levels of CO_2 in the atmosphere are the result of human emissions and that atmospheric CO_2 does indeed lead to warming on a global scale. Otherwise, any policies we implement will have no effect on climate but could dramatically reduce the wellbeing of the poorest people on the planet, often the same people who are thought to be most affected by climate change.

A first step in understanding whether policy to mitigate or avert climate change will be worth undertaking, or whether it will simply make us poorer while having little

[42] Preliminary research by University of Victoria PhD student, Zhen Zhu, finds that the PNA, PDO and El Niño indexes predict wildfire intensity in British Columbia's interior. With some indexes, however, the more important predictor is a lag of nearly 1½ years as opposed to the closer lag of 4–6 months. Perhaps it requires a longer period of warm dry weather before forests are susceptible to fire.

or no affect on global temperatures, is to understand the climate record. Thus far, we have examined only the instrumental record, which begins in the mid to late 1800s, about the time that the so-called Little Ice Age was coming to an end (see Chap. 3). As a result, it should not be surprising that we find global temperatures have risen by some 1 °C, or 0.07 °C per decade, over the past 130 years. At the same time, atmospheric concentrations of CO_2 have also risen rapidly, and especially so since the end of World War II. Has there been a more rapid increase in temperatures later in the record, when atmospheric CO_2 was higher, or is there no evidence to this effect?

It is not an easy task to reconstruct and make sense of the historical temperature record. The quality of recorded temperature and other weather data remains an issue, especially records that exist earlier in the period for which data are available. The pattern of coverage and the number of weather stations are problematic. Changes in the number of weather stations and their locations constitute a challenge to the construction of consistent data on a global scale, as does the lack of coverage over oceans. As suggested in this chapter, the only truly reliable global data are likely available only since 1979, when satellite data became available. If that is the case, then it is only the record of the past three decades that can provide us with any indication as to the human contribution to climate change. The fact that satellite data have only been around for this short a period makes it nearly impossible to say anything definitive.

Nonetheless, weather station data are important. They are important, for example, to the development of financial weather derivatives that economic agents can use to reduce their exposure to weather risks. They may also be important as a tool that facilitates adaptation to climate change.

In this chapter, the problem of averaging temperatures, whether to obtain a regional or global average temperature or establishing a basis for temperatures at a particular location, was also examined. We concluded that, perhaps, it is not even possible to make sense of a global or regional average temperature. Certainly, the mathematician Christopher Essex does not think such a thing makes sense because temperatures differ continuously over space, as we demonstrated with the two Victoria schools; an average temperature depends on the number of readings one takes and such readings cannot be taken at precisely the same instant in time (Essex and McKitrick 2002; Essex et al. 2007). Overall, the challenge of creating temperature records is a daunting one.

One thing is clear, however: the length of the extant instrumental temperature record is too short to tell us whether humans might be culpable in causing observed warming. It behooves us, therefore, to determine whether current temperatures are warmer than they have been in human history. If they are, this might perhaps constitute some evidence of anthropogenic global warming. We turn to this issue in the next chapter.

References

Brohan, P., Kennedy, J. J., Harris, I., Tett, S. F. B., & Jones, P. D. (2006). Uncertainty estimates in regional and global observed temperature changes: A new dataset from 1850. *Journal of Geophysical Research-Atmospheres, 111*, D12106.

Chantarat, S., Turvey, C. G., Mude, A. G., & Barrett, C. B. (2008). Improving humanitarian response to slow-onset disasters using famine-indexed weather derivatives. *Agricultural Finance Review, 68*(1), 169–195.

Chen, C.-C., McCarl, B., & Hill, H. (2002). Agricultural value of ENSO information under alternative phase definition. *Climatic Change, 54*(3), 305–325.

Chincarini, L. (2011). No chills or burns from temperature surprises: An empirical analysis of the weather derivatives market. *Journal of Futures Markets, 31*(1), 1–33.

D'Aleo, J., & Watts, A. (2010, January 27). *Surface temperature records: Policy driven deception?* (SPPI Original Paper, 111 pp). Retrieved January 29, 2010, from http://scienceandpublicpolicy.org/images/stories/papers/originals/surface_temp.pdf

De Laat, A. T. J., & Maurellis, A. N. (2004). Industrial CO_2 emissions as a proxy for anthropogenic influence on lower tropospheric temperature trends. *Geophysical Research Letters, 31*, L05204. doi:10.1029/2003GL019024.

De Laat, A. T. J., & Maurellis, A. N. (2006). Evidence for influence of anthropogenic surface processes on lower tropospheric and surface temperature trends. *International Journal of Climatology, 26*(7), 897–913.

Essex, C., & McKitrick, R. (2002). *Taken by storm: The troubled science, policy and politics of global warming*. Toronto: Key Porter Books.

Essex, C., McKitrick, R., & Andresen, B. (2007). Does a global temperature exist? *Journal of Non-Equilibrium Thermodynamics, 32*(1), 1–28.

Garnett, E. R. (2007). *The use of El Niño information in forecasting grain yields in the Canadian prairie provinces*. Winnipeg: Canadian Wheat Board.

Goddard, S. (2011, February 15). The hottest year on record? *SPPI Reprint Series* (20 pp). Retrieved from http://scienceandpublicpolicy.org/images/stories/papers/reprint/the_hottest_year_ever.pdf

Hansen, J. E., Ruedy, R., Glascoe, J., & Sato, M. (1999). GISS analysis of surface temperature change. *Journal of Geophysical Research, 104*(D24), 30997–31022.

Hansen, J. E., Ruedy, R., Sato, M., Imhoff, M., Lawrence, W., Easterling, D., Peterson, T., & Karl, T. (2001). A closer look at United States and global surface temperature change. *Journal of Geophysical Research, 106*(D20), 23947–23963.

Hansen, J. E., Ruedy, R., Sato, M., & Lo, K. (2010). Global surface temperature change. *Reviews of Geophysics, 48*, RG4004. doi:10.1029/2010RG000345.

Hsieh, W. W., Tang, B., & Garnett, E. R. (1999). Teleconnections between Pacific sea surface temperatures and Canadian prairie wheat yield. *Agriculture and Forest Meteorology, 96*, 209–217.

IPCC WGI. (2007). *Climate change 2007. The physical science basis.* Working Group I contribution to the fourth assessment report of the Intergovernmental Panel on Climate Change. Cambridge: Cambridge University Press.

Jones, P. D. (1988). Hemispheric surface air temperature variations: Recent trends and an update to 1987. *Journal of Climate, 1*(6), 654–660.

Jones, P. D. (1994). Hemispheric surface air temperature variations: A reanalysis and an update to 1993. *Journal of Climate, 7*(11), 1794–1802.

Jones, P. D., & Moberg, A. (2003). Hemispheric and large-scale surface air temperature va-riations: An extensive revision and update to 2001. *Journal of Climate, 16*(2), 206–223.

Jones, P. D., Raper, S. C. B., Santor, B., Cherry, B. S. G., Goodess, C., Kelly, P. M., Wigley, T. M. L., Bradley, R. S., & Diaz, H. F. (1985). *A grid point surface air temperature data set for the Northern hemisphere, 1850–1984* (DoE Technical Report No. TR022). US. Department of Energy, Contract No. DE-AC02-79EV10098. DOE/EV/10098-2.

Jones, P. D., Raper, S. C. B., Bradley, R. S., Diaz, H. F., Kelly, P. M., & Wigley, T. M. L. (1986a). Northern hemisphere surface air temperature variations: 1851–1984. *Journal of Climate and Applied Meteorology, 25*(2), 161–179.

Jones, P. D., Raper, S. C. B., Goodess, C., Cherry, B. S. G., & Wigley, T. M. L. (1986b, July). *A grid point surface air temperature data set for the Southern hemisphere, 1850–1984* (DoE Technical Report No. TR027, p. 256). US. Dept. of Energy under Contract No. DE-AC02-79EV10098. DOE/EV/10098-2. Dist. Category UC-11.

Jones, P. D., Raper, S. C. B., & Wigley, T. M. L. (1986c). Southern hemisphere surface air temperature variations: 1851–1984. *Journal of Climate and Applied Meteorology, 25*(9), 1213–1230.

Jones, P. D., New, M., Parker, D. E., Martin, S., & Rigor, I. G. (1999). Surface air temperature and its changes over the past 150 years. *Reviews of Geophysics, 37*(2), 173–199.

Jones, P. D., Osborn, T. J., Briffa, K. R., Folland, C. K., Horton, E. B., Alexander, L. V., Parker, D. E., & Rayner, N. A. (2001). Adjusting for sampling density in grid box land and ocean surface temperature time series. *Journal of Geophysical Research-Atmospheres, 106*(D4), 3371–3380.

Jones, P. D., Parker, D. E., Osborn, T. J., & Briffa, K. R. (2010). Global and hemispheric temperature anomalies – Land and marine instrumental records. *Trends: A compendium of data on global change*. From http://cdiac.ornl.gov/trends/temp/jonescru/jones.html

Khandekar, M. L. (2004). *Canadian prairie drought: A climatological assessment* (Pub. No: T/787, pp. 37). Edmonton: Environmental Policy Branch, Alberta Environment.

Laidre, K. L., Heide-Jørgensen, M. P., Ermold, W., & Steele, M. (2010). Narwhals document continued warming of Southern Baffin Bay. *Journal of Geophysical Research, 115*, C10049. doi:10.1029/2009JC005820.

Manley, G. (1953). The mean temperature of Central England, 1698 to 1952. *Quarterly Journal of the Royal Meteorological Society, 79*(340), 242–261.

Manley, G. (1974). Central England temperatures: Monthly means 1659 to 1973. *Quarterly Journal of the Royal Meteorological Society, 100*(425), 389–405.

McKitrick, R. (2010a). Atmospheric circulations do not explain the temperature-industrialization correlation. *Statistics, Politics and Policy, 1*(1). doi:10.2202/2151-7509.1004 (Article 1).

McKitrick, R. (2010b, April 5). Circling the bandwagons: My adventures correcting the IPCC. *SPPI Reprint Series.* Retrieved January 19, 2011, from http://scienceandpublicpolicy.org/images/stories/papers/reprint/Circling_the_Bandwagons_Correcting_the_IPCC.pdf

McKitrick, R. (2010c, February 26). Evidence submitted to the independent climate change email review (ICCER), Sir M. Russell, Chairman (80 pp). Retrieved April 15, 2010, from http://sites.google.com/site/rossmckitrick/McKitrick-ICCER-Evidence.pdf

McKitrick, R., & Michaels, P. J. (2004). A test of corrections for extraneous signals in gridded surface temperature data. *Climate Research, 26*(2), 159–173.

McKitrick, R., & Michaels, P. J. (2007). Quantifying the influence of anthropogenic surface processes and inhomogeneities on gridded global climate data. *Journal of Geophysical Research, 112*, D24S09. doi:10.1029/2007JD008465.

McKitrick, R., & Nierenberg, N. (2011). Socioeconomic patterns in climate data. *Journal of Economic and Social Measurement, 35*(3–4), 149–175.

McKitrick, R., & Vogelsang, T. J. (2012). *Multivariate trend comparisons between autocorrelated climate series with possible intercept shifts* (Working Paper). Guelph: Department of Economic, University of Guelph.

Menne, M. J., Williams, C. N., Jr., & Palecki, M. A. (2010). On the reliability of the U.S. surface temperature record. *Journal of Geophysical Research-Atmospheres, 115*, D11108. doi:10.1029/2009JD01309.

Muller, R. A., Curry, J., Groom, D., Jacobsen, R., Perlmutter, S., Rohde, R., Rosenfeld, A., Wickham, C., & Wurtele, J. (2011a). *Decadal variations in the global atmospheric land temperatures* (Berkeley Earth Surface Temperature (BEST) Project working paper). Retrieved from http://www.berkeleyearth.org/

Muller, R. A., Curry, J., Groom, D., Jacobsen, R., Perlmutter, S., Rohde, R., Rosenfeld, A., Wickham, C., & Wurtele, J. (2011b). *Earth atmospheric land surface temperature and station quality in the United States* (Berkeley Earth Surface Temperature (BEST) project working paper). Retrieved from http://www.berkeleyearth.org/

Norton, M., Osgood, D., & Turvey, C. G. (2010). *Weather index insurance and the pricing of spatial basis risk*. Paper presented at the annual conference of the AAEA, CAES and WAEA, Denver.

Parker, D. E. (2010). Urban heat island effects on estimates of observed climate change. *WIREs Climate Change, 1*(1, January/February), 123–133.

Parker, D. E., Legg, T. P., & Folland, C. K. (1992). A new daily Central England temperature series, 1772–1991. *International Journal of Climatology, 12*(4), 317–342.

Peterson, T. C., & Vose, R. S. (1997). An overview of the global historical climatology network temperature database. *Bulletin of the American Meteorological Society, 78*(12), 2837–2849.

Peterson, T. C., Vose, R. S., Schmoyer, R., & Razuvaev, V. (1998). Global Historical Climatology Network (GHCN) quality control of monthly temperature data. *International Journal of Climatology, 18*(11), 1169–1179.

Rohde, R., Curry, J., Groom, D., Jacobsen, R., Muller, R. A., Perlmutter, S., Rosenfeld, A., Wickham, C., & Wurtele, J. (2011). *Berkeley earth temperature averaging process* (Berkeley Earth Surface Temperature (BEST) project working paper). Retrieved from http://www.berkeleyearth.org/

Shabbar, A., Bonsal, B., & Khandekar, M. (1997). Canadian precipitation patterns associated with the Southern Oscillation. *Journal of Climate, 10*(12), 3016–3027.

Skees, J. R. (2008). Innovations in index insurance for the poor in lower income countries. *Agricultural and Resource Economics Review, 37*(1), 1–15.

Spencer, R. W. (2010). *The great global warming blunder. How Mother Nature fooled the world's top climate scientists*. New York: Encounter Books.

Sussman, B. (2010). *Climategate. A veteran meteorologist exposes the global warming scam*. New York: WND Books.

Turvey, C. G. (2001). Weather derivatives for specific event risks in agriculture. *Review of Agricultural Economics, 23*(2), 333–351.

Turvey, C. G. (2005). The pricing of degree-day weather options. *Agricultural Finance Review, 65*(1), 59–85.

Turvey, C. G., Weersink, A., & Chiang, S.-H. C. (2006). Pricing weather insurance with a random strike price: The Ontario ice-wine harvest. *American Journal of Agricultural Economics, 88*(3), 696–709.

Vedenov, D. V., & Barnett, B. J. (2004). Efficiency of weather derivatives as primary crop insurance instruments. *Journal of Agricultural and Resource Economics, 29*(3), 387–403.

Vélez-Belchí, P., Hernández-Guerra, A., Fraile-Nuez, E., & Benítez-Barrios, V. (2010). Changes in temperature and salinity tendencies of the upper subtropical North Atlantic ocean at 24.5°N. *Journal of Physical Oceanography, 40*(11), 2546–2555. doi:10.1175/2010JPO4410.1.

Watts, A. (2009). *Is the U.S. surface temperature record reliable?* Chicago: The Heartland Institute and SurfaceStations.org.

Wickham, C., Curry, J., Groom, D., Jacobsen, R., Muller, R. A., Perlmutter, S., Rosenfeld, A., & Wurtele, J. (2011). *Influence of urban heating on the global temperature land average using rural sites identified from MODIS classifications* (Berkeley Earth Surface Temperature (BEST) project working paper). Retrieved from http://www.berkeleyearth.org/

Williams, C. N., Menne, M. J., Vose, R. S., & Easterling, D. R. (2008). *United States historical climatology network monthly temperature and precipitation data* (ORNL/CDIAC-87, NDP-019). Oak Ridge: Carbon Dioxide Information Analysis Center, Oak Ridge National Laboratory.

Woodward, J. D., & Garcia, P. (2008). Basis risk and weather hedging effectiveness. *Agricultural Finance Review, 68*(1), 99–118.

Xu, W., Odening, M., & Mußhoff, O. (2008). Indifference pricing of weather derivatives. *American Journal of Agricultural Economics, 90*(4), 979–993.

Chapter 3
Climate Science and Paleoclimatology

The burden of proof for destructive climate change firmly rests with those whose remedy requires an overturning of economic and political assumptions without precedent. We need to apply the best thinking of which we are capable. We haven't done that so far. In the postmodern dispensation that now beguiles us, this will be an uphill trudge. It is always more fun to damn the facts and embrace wishes. The great game of climate-change baseball is in the late innings, but Reality bats last.

William Anderson, Harvard University, in First Things, February 2010

By the time of the 2007 Report of the Intergovernmental Panel on Climate Change (IPCC 2007), many commentators were confidently asserting that the scientific debate about the causes and perhaps even the potential future impacts of global warming had been settled, and that there was evidence indicating that recent decades were the warmest humans had ever seen (see Chap. 2). It was assumed that the vast majority of scientists agreed that human activities were overwhelmingly responsible for already observed global warming. This claim of an 'overwhelming consensus' among scientists is itself a non-scientific statement that has never been tested. The truth, of course, is that a 'consensus' does not and never has existed, and, even if there were a consensus among scientists, this does not imply the truth of the matter. The validity of scientific statements is not resolved by consensus or by a popular vote, although there are occasions when courts are asked to weigh the evidence and decide in favor of one side or the other because the unresolved issue has immediate policy implications (e.g., when a British court was asked to rule on the showing of *An Inconvenient Truth* in public schools).

Every attempt to document a consensus regarding anthropogenic global warming failed. In 2004 *Science* published the results of a study by Naomi Oreskes (2004) of the University of California at San Diego claiming that "without substantial disagreement, scientists find human activities are heating the earth's surface." But an attempt by Benny Peiser of Liverpool John Moores University to replicate Oreskes' study found that she had made serious mistakes in handling data and, after re-examining

G.C. van Kooten, *Climate Change, Climate Science and Economics: Prospects for an Alternative Energy Future*, DOI 10.1007/978-94-007-4988-7_3,

the issue, he reached a contrary conclusion. Oreskes claimed that an analysis of 928 abstracts in the ISI database containing the phrase "climate change" proved the alleged consensus. It turned out that she had searched the database using three keywords ("global climate change") instead of the two ("climate change") she reported in the original article, thereby reducing the search results by an order of magnitude.[1] By searching just the keywords "climate change," Peiser found almost 12,000 articles in the same database for the relevant decade used by Oreskes. Excluded from Oreskes' list were "countless research papers that show that global temperatures were similar or even higher during the Holocene Climate Optimum and the Medieval Warm Period when atmospheric CO_2 levels were much lower than today; that solar variability is a key driver of recent climate change; and that climate modeling is highly uncertain".[2] Further, even using the three key words she actually used, "global climate change," brought up 1,247 documents, of which 1,117 included abstracts. An analysis of those abstracts shows

- only 1% explicitly endorsed what Oreskes called the 'consensus view';
- 29% implicitly accepted it "but mainly focus[ed] on impact assessments of envisaged global climate change;"
- 8% focused on 'mitigation;'
- 6% focused on methodological issues;
- 8% dealt "exclusively with paleo-climatological research unrelated to recent climate change;"
- 3% "reject[ed] or doubt[ed] the view that human activities are the main drivers of the 'observed warming over the last 50 years';"
- 4% focused 'on natural factors of global climate change;' and
- 42% did "not include any direct or indirect link or reference to human activities, CO_2 or greenhouse gas emissions, let alone anthropogenic forcing of recent climate change."

A more recent survey of the same database but covering more recent years showed that scientific opinion was shifting away from belief in catastrophic anthropogenic warming, and not toward it (Schulte 2008), while a survey of climate scientists showed that the matter remains very much debated among them (Bray and von

[1] The following erratum was printed in *Science* on January 21, 2005: The final sentence of the fifth paragraph should read "That hypothesis was tested by analyzing 928 abstracts, published in refereed scientific journals between 1993 and 2003, and listed in the ISI database with the keywords 'global climate change'."

[2] Benny J. Peiser, Letter to *Science*, January 4, 2005, submission ID: 56001. *Science* Letters Editor Etta Kavanagh eventually decided against publishing even a shortened version of the letter that she requested because "the basic points of your letter have already been widely dispersed over the internet" (e-mail from Etta Kavanagh to Benny Peiser, April 13, 2005). Peiser replied: "As far as I am aware, neither the details nor the results of my analysis have been cited anywhere. In any case, don't you feel that *Science* has an obligation to your readers to correct manifest errors? After all, these errors continue to be employed by activists, journalists and science organizations. ... Are you not aware that most observers know only too well that there is absolutely 'no' consensus within the scientific community about global warming science?" The correspondence between Peisner and the editors of *Science* is at www.staff.livjm.ac.uk/spsbpeis/Scienceletter.htm. (viewed April 11, 2011).

Storch 2007). Further, over 31,000 scientists, including over 9,000 with PhDs signed the Global Warming Petition stating, "There is no convincing scientific evidence that human release of carbon dioxide, methane, or other greenhouse gases is causing or will, in the foreseeable future, cause catastrophic heating of the Earth's atmosphere and disruption of the Earth's climate. Moreover, there is substantial scientific evidence that increases in atmospheric carbon dioxide produce many beneficial effects upon the natural plant and animal environments of the Earth." However, just as consensus does not resolve scientific disputes, neither does a petition.

During the latter part of November, 2009, hackers broke into the computers at East Anglia University in the United Kingdom, targeting in particular the University's Climate Research Unit (CRU); or perhaps a whistleblower released the information. The CRU specializes in the study of past climate. Its research reproducing historical temperatures from information based on tree rings, stalagmites in caves, ice cores, and lake sediment cores has come under increasing scrutiny, partly over controversy regarding the Medieval Warm Period (900–1300 AD) and, to a lesser extent, the Little Ice Age (1350–1850). The CRU supports the view that recent increases in temperatures are unprecedented by historical standards, that projected warming will be catastrophic, and that humans are responsible.[3]

The numerous documents and emails obtained from the CRU computers were posted anonymously on the internet, thereby providing a unique insight into the lengths to which scientists will go to protect their beliefs and data.[4] Overall, the emails and other information posted on the web paint a negative picture of how climate science is done, and raises questions concerning the view that recent and projected temperatures are outside historical norms. In particular, as one high-profile weekly news magazine noted: The scientists "believe in global warming too much, and that their commitment to the cause leads them to tolerate poor scientific practice, to close themselves off from criticism, and to deny reasonable requests for data" (*The Economist*, 28 November 2009, p. 93).[5] Several months after 'climategate'

---

[3] While supporting the view that current temperatures are unprecedented, the CRU now acknowledges that perhaps temperatures during the Medieval Warm Period were warmer than currently (see Vinther et al. 2010). This is discussed further below.

[4] Emails from East Anglia University can be searched at http://www.eastangliaemails.com/ (viewed April 15, 2010). Overviews of many of the key (controversial) emails are available in a United States Senate Report (U.S. Senate Committee on Environment and Public Works 2010) and from Australian science writer Joanne Nova (2010). A recent (2010) book by Steven Mosher and Thomas W. Fuller, *Climategate The CRUtape Letters* (ISBN 1450512437; self published but available from Amazon.com), provides a history of the climategate emails that ties them to the scientific issues as they evolved.

[5] *The Economist* is quite apologetic for the attitude of climate scientists, arguing that the scientific failings are typical practice. However, it fails to point out that more technical analyses of computer codes raise concerns about the crude methods used to link proxy temperature data from tree rings to observed (albeit also 'adjusted') data originating from weather stations; two of many interpretations are provided by Marc Sheppard (sinister) and John Graham-Cumming (apologetic) at (both viewed December 3, 2009): www.americanthinker.com/2009/11/crus_source_code_climategate_r.html and http://www.jgc.org/blog/2009/11/very-artificial-correction-flap-looks.html, respectively *The Economist*'s bias was revealed in a lengthy article in the March 20, 2010 issue entitled

broke, Phil Jones, Director of the CRU, admitted in a BBC interview on February 13, 2010, that he did not believe "the debate on climate change is over" or that "the vast majority of climate scientists think" it is resolved.[6]

In one study, Anderegg et al. (2010) constructed a list of 903 names of people who were convinced by the evidence that anthropogenic climate change was happening as described by the IPCC, and a list of 472 names of those who were deemed not to be convinced by the evidence. The former group included 619 IPCC Working Group I authors (IPCC WGI 2007), while the latter included many people who opposed government action to mitigate climate change (and may have even been convinced by the evidence).[7] Upon comparing the qualifications of the two groups, the authors found that the group of convinced scientists were more highly cited (and thus considered to have more climate expertise and prominence) than those in the unconvinced group. No statistical analysis was provided. What is most disconcerting about the analysis is the attempt to compare 'apples and oranges' – to compare experts on policy with those who essentially wrote the scientific case for anthropogenic warming. The policy experts deal with reality – what is politically feasible in mitigating greenhouse gas emissions without destroying the fabric of society – and what is not possible is the reduction in fossil fuel use that the convinced scientists propose as a solution (Gerondeau 2010; Levitt and Dubner 2009). The subject of reducing fossil fuel emissions is addressed in more detail in Chaps. 9, 10, 11, and 12.

The IPCC's Working Group I scientists only know what historical temperatures have done and what might happen in a world of higher temperatures, but have no comparative advantage in predicting the extent of damages and social unrest/upheaval that global warming or attempts to mitigate it might cause. Here we are in a fuzzy arena where theology, philosophy and social science trump climate science.

It would appear that climate change or global warming is no longer about climate science, but it is about beliefs and politics. As 'climategate' has shown, climate science has deteriorated into a conflict rather than a debate between global warming alarmists and skeptics. It has become a matter of winning the hearts and minds of ordinary citizens, convincing them that climate change is either the greatest disaster

---

"Spin, science and climate change" (pp. 83–86). Disconcertingly, the spin referred to detractors of catastrophic anthropogenic global warming, who the article suggests do not conduct peer reviewed research but only operate through blogs, in contrast to those real scientists who do believe in human-driven global warming. For another perspective, see http://www.youtube.com/watch?v=U5m6KzDnv7k (viewed April 9, 2011).

[6] See http://news.bbc.co.uk/2/hi/science/nature/8511670.stm (viewed February 16, 2010).

[7] The current author appears to have been included among those unconvinced by the evidence. However, his reason for signing one of the documents used by Anderegg et al. (2010) related to Canada's climate policies and not to the climate science (which he only began to investigate seriously in preparing the current book).

ever to face humankind or a benign change in weather patterns that is well within what humans have experienced in the past several millennia. Indeed, it has recently been characterized by some as a religious debate (Nelson 2010; Sussman 2010; Wanliss 2010).

In this chapter we demonstrate that there has been anything but a 'scientific consensus' regarding just one aspect of the climate science, namely, the historical record. Evidence of controversy was already presented in Chap. 2, where we saw that there is disagreement among scientists about whether the instrumental record even signals a general warming that can be attributed to human greenhouse gas emissions, or whether the warming is an artefact of socioeconomic activities that affect local climates and thus local temperature measurements. Here we extend our examination of the claim made by researchers at East Anglia University's CRU and their global collaborators that current temperatures are high by historical standards, that they have risen at a historically unprecedented rate, and that the rate of increase will be even faster in the future.

We begin in the next section by first examining those facts of global warming that might well be considered indisputable. To get our bearings, we then examine raw (unadjusted) temperature data from a number of weather stations, followed by a discussion of the paleoclimatic record. Finally, we combine the paleoclimatic data and the instrumental data in a discussion of the controversial 'hockey stick' diagram. In later chapters, we discuss projections of future temperature increases from climate models and some further controversies related to the science of climate change.

3.1 Indisputable 'Facts' of Global Warming?

What is everyone apparently agreed upon when it comes to climate change? That is not entirely clear, but there appears to be a consensus concerning the following:

1. Beyond dispute is the fact that the level of CO_2 in the atmosphere has increased since the beginning of the Industrial Revolution, and has risen from about 270 ppm by volume (denoted ppmv or more often simply ppm) to nearly 400 ppm today. This is indicated in Fig. 3.1. The upward trend in CO_2 has continued pretty much unabated and has in fact risen somewhat faster in recent years.
2. Temperatures have risen in the 150 or more years since the end of the Little Ice Age, which occurred around 1850 AD. This too is indisputable. The rise in temperatures is calculated to be slightly more than 0.05 °C per decade, or about 0.7 °C over the past 100 years, or, based on Chap. 2, about 1 °C over 130 years (0.07 °C per decade). This can be seen in Figs. 2.3 and 2.4.
3. Carbon dioxide is a greenhouse gas – it makes a contribution to global warming. This is not a point of disagreement. What is disputed is the extent of its contribu-

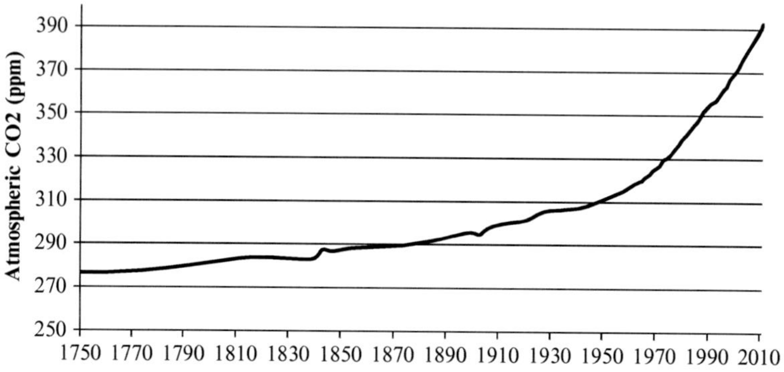

Fig. 3.1 Average atmospheric CO_2 concentrations based on ice core data (1750–1957) and instrumental data sampled at Mauna Loa, Hawaii (1958–2011) (See ftp://ftp.cmdl.noaa.gov/ccg/co2/trends/co2_mm_mlo.txt (as viewed August 25, 2011) for instrumental data; http://cdiac.ornl.gov/ftp/trends/co2/maunaloa.co2 for pre-1958 data. Data for 2011 are an average of January through July only

tion – the overall sum of the various positive (leading to further warming) and negative feedbacks caused by the initial CO_2 forcing.

4. Also indisputable is the fact that human activities have contributed to this rise in atmospheric CO_2 concentrations.

Everything else about global warming remains controversial, with peer-reviewed scientific papers providing evidence of the lack of consensus.

CO_2 is one of the most important greenhouse gases, with other greenhouse gases generally measured in terms of their CO_2 equivalence, denoted CO_{2e}. For convenience, we will simply use CO_2 to refer to carbon dioxide plus other greenhouse gases measured in terms of their CO_2 equivalence. However, CO_2 is a relatively minor greenhouse gas compared with water vapor.

In terms of its relative contribution to the greenhouse effect, water vapor accounts for 95.00%, followed by CO_2 (3.62%), nitrous oxide or N_2O (0.95%), methane or CH_4 (0.36%), and CFCs and miscellaneous gases (0.07%). However, once clouds are factored in, the contribution of water vapor to greenhouse warming may be less, varying between 66 and 85%, because clouds reflect the sun's rays. In climate models, it is the enhanced greenhouse effect – the 'forcing' effect of CO_2 in increasing water vapor in the atmosphere – that causes climate change of the magnitude found in the IPCC reports. Because warmer air causes more water to evaporate from the oceans, the initial CO_2-induced warming is thought to lead to a greater amount of water vapor which, in turn, increases temperatures even more – a climate feedback. It is this climate feedback and whether other factors (e.g., cosmic rays) affect water vapor and cloud formation that is a source of disagreement (which is discussed further in Chap. 5).

In the remainder of this chapter, we focus on the historical record to determine if global temperatures are already above those experienced in the past. Several important sources of disagreement are reviewed, the most important of which relates to the so-called 'hockey stick' – a graph showing temperatures to be flat for some 1,000 years and then rising sharply beginning around 1900. This controversy relates to paleoclimatic data and whether historical temperature reconstructions, or proxy data derived from tree ring data, ice core samples, lake-bed sediments and other sources, indicate that it was ever warmer than today, whether there is evidence of stable temperatures over the past two millennia (another version of the hockey stick), and whether CO_2 and temperature go hand-in-hand over time (as there is some suggestion that CO_2 lags temperature rise by 200–800 years as noted already in Chap. 1). Other controversies pertain to surface versus satellite temperature data and, more recently, whether ocean temperature data are more relevant. These issues relate to the means by which temperature and other weather data are collected, and the use of computer models to predict future climate change. The focus in this chapter, however, is only on the historical temperature record, primarily the paleoclimatic record.

It is important to recognize that *instrumental* records of precipitation, temperature and other weather data are available at a global level only after about 1850, and then not for most regions (see Chap. 2). People recorded temperature and/or precipitation at various times before the 1800s, with the best historical record available likely being the Central England temperature record. However, there are insufficient systematic records to construct large-scale regional or global temperature averages prior to the mid to late 1800s, just as the Little Ice Age was ending. As we already saw (Fig. 3.1), instrumental measurements of atmospheric CO_2 only began in 1958, although proxy measures are available from ice core samples for earlier years. The lack of instrumental records makes it difficult to say anything about current versus historical temperatures, for example, as the record of instrumental measurements is too short. Nonetheless, as discussed in Sect. 3.3, historical temperature proxy data can provide some indication regarding past climates.

3.2 Evidence from Individual Weather Stations

Consider two cities in western Canada, one located on the west coast and another in the interior. The city of Victoria at the southern end of Vancouver Island off the west coast of Canada enjoys a mild winter climate, while Edmonton is characterized by a continental climate as it is located east of the Rocky Mountains in central Alberta. Weather station data are available from Environment Canada for both cities for the period since 1880, but to construct the raw data sets it was necessary to combine data from two or more weather stations at each location since no single station had a continuous record over this period. Even then, several data points are missing for Victoria. Although not shown in the figure, average winter temperatures in Victoria

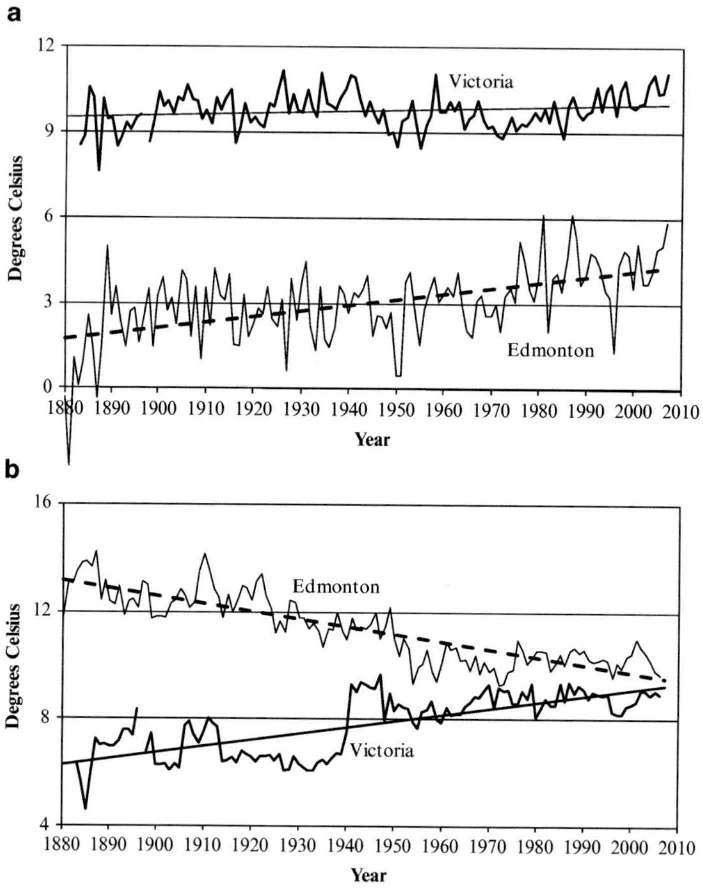

Fig. 3.2 (**a**) Average annual temperatures and trend, Victoria & Edmonton, 1880–2007, (**b**) Average maximum minus average minimum annual temperatures, Victoria & Edmonton, 1880–2007

are some 10–15 °C warmer than those in Edmonton, although Edmonton summer temperatures are generally a bit higher. Average annual temperatures are provided in Fig. 3.2a.

The record indicates that temperatures in both locations have risen over the past nearly 130 years, with those in Victoria having risen only slightly and those in Edmonton by some 2 °C. This is not unexpected given that the world was just coming out of the Little Ice Age and interior continental regions generally warm more than coastal regions, as these are affected by the ocean sink.

A more interesting story is told when one takes the differences between the average maximum and minimum temperatures, which are plotted in Fig. 3.2b. In Victoria, the maximum temperature appears to be rising relative to the minimum, suggesting that summers are getting warmer. This supports the view that temperatures have been rising, even in Victoria. A more telling result is that of Edmonton, where the difference between average maximum and minimum temperatures has fallen, even while average temperatures have risen. This suggests that there may be an 'urban heat island' effect, which occurs whenever a weather station is located in an area experiencing urban growth. Thus, a weather station that is located in an open field early in the historical record slowly gets surrounded by increasingly dense urban, commercial and/or industrial developments. The heat given off by the surrounding buildings at night in the winter prevents temperatures from falling to levels experienced earlier in the record.[8] Clearly, Edmonton weather station data exhibit a heat island phenomenon. This is to be expected for a location such as Edmonton – a rapidly growing industrial city at the heart of an oil and gas industry that began in the late 1940s.

Next, we examine the raw data for various weather stations in the United States because the majority of weather stations that have consistently recorded temperatures over a long period are found in the U.S. Plots of average recorded annual temperatures for seven locations in the U.S. are provided in Fig. 3.3a. There does not appear to be a definitive trend over the period 1926–2003 for these locations. In order to get a better notion of trend, for each location we standardized the data so that we obtain the variation from mean over the period in question. By standardizing the data, we obtain z-scores that are comparable across locations because the scores also adjust for fluctuations that may naturally be greater at one location than another. To construct a z-score that substitutes for a temperature, subtract the mean of the series from each observation and divide by the standard deviation of the temperature series:

$$z_{is} = \frac{x_{is} - \overline{x}_s}{\sigma_s},$$

where $z_{i,s}$ is the z-score for observation i at location s, $x_{i,s}$ is the original observation i at location s, and $\overline{x}_s$ and σ_s are the mean and standard deviation of the temperatures at location s, respectively. Note, however, that z-scores assume the underlying data have a normal distribution and that may not be correct. A plot of the z-scores for three locations and the average of the z-scores of 19 weather stations are found in Fig. 3.3b.

It is difficult to discern an overall upward trend in temperatures from Fig. 3.3. Although regressing the average z-score on time provides evidence of a very slight upward trend of nearly 0.05 °C per decade ($z = -0.1789 + 0.0045 \times \text{year}$, $R^2 = 0.0329$), the trend will vary by the number and locations of the cities included in the average. However, it is also important to note that temperatures peak in 1997–1998 when

[8] Since a body always gives off infrared radiation to its surroundings, buildings, pavement and so on contribute to higher temperatures during the daytime as well as nighttime (see Chap. 5). Thus, the heat island effect is not simply a nighttime phenomenon.

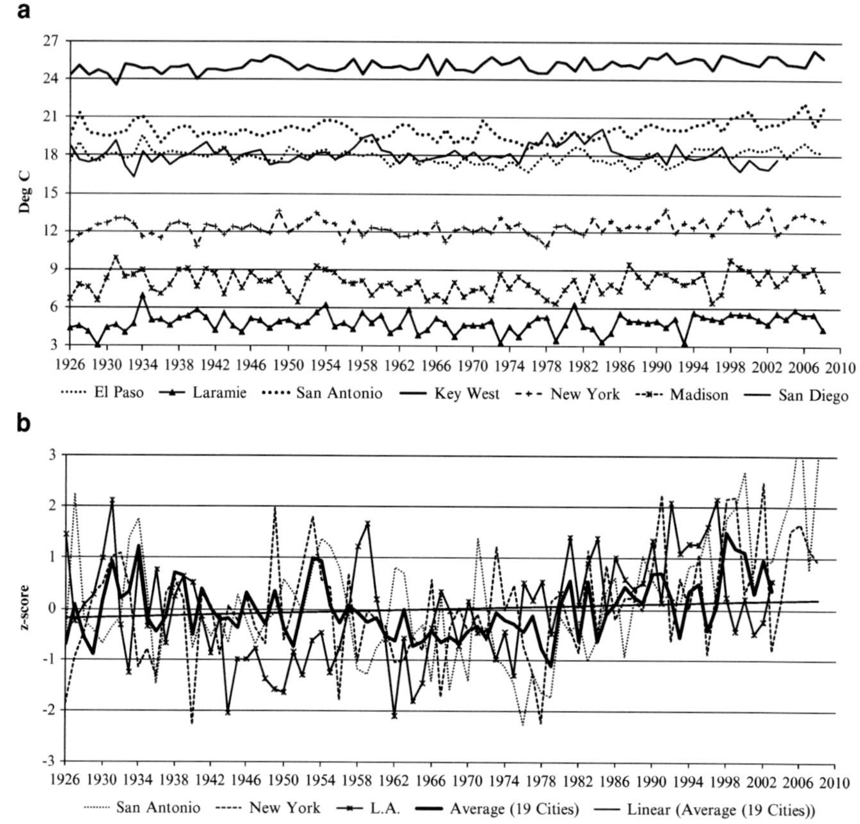

Fig. 3.3 (**a**) Average annual temperatures for selected U.S. cities, 1926–2008, (**b**) Average annual temperature anomaly for three U.S. cities, average of 19 U.S. weather stations and linear trend, 1926–2008

there was a particularly strong El Niño event, without which temperatures may have remained flat for the period in question.

The period since 1926, or even since 1850, is simply too short for us to determine whether current temperatures are higher than 'normal' – whether recent weather patterns and projected global warming are somehow outside the realm of human experience. To get a better feel for this, we need to investigate temperatures over several millennia.

As noted in Chap. 2, the most reliable temperature data probably come from the 1,221 high-quality weather stations that make up the U.S. Historical Climate Network.[9]

---

[9]Data for stations are available at http://cdiac.ornl.gov/epubs/ndp/ushcn/access.html (viewed April 26, 2010).

Table 3.1 Ten warmest years based on average contiguous 48 U.S. surface air temperature anomalies (°C)[a], 1880–2006

August 20, 2007		May 27, 2009	April 26, 2010	
Year	Anomaly	Year	Year	Anomaly
1934	1.25	1934	1998	1.32
1998	1.23	1998	2006	1.30
1921	1.15	1921	1934	1.20
2006	1.13	2006	1921	1.08
1931	1.08	1931	1999	1.07
1999	0.93	1999	1931	0.96
1953	0.90	1953	1990	0.92
1990	0.87	1990	2001	0.92
1938	0.86	1938	2005	0.92
1939	0.85	1954	2007	0.87

Source: http://data.giss.nasa.gov/gistemp/graphs/Fig.D.txt (dates viewed provided)
[a]Anomaly relative to 1951–1980 average

Quite some sleuthing is required to determine annual averages for the U.S. contiguous 48 states using the USHCN data, for example, which is why averages from other sources are generally used. NASA-GISS provides an average of U.S. surface temperature anomalies for the contiguous 48 states that have been cited by various commentators. These data have been homogenized in an attempt to remove non-climate influences: Measures of average U.S. temperatures rely on USHCN data, and then adjust the raw temperature data to remove non-climate influences. The adjusted data appear to change quite frequently, however (see below). One way to adjust the data is to employ population density, which has been a standard method. A more recent effort to remove non-climate sources of contamination adjusts observed surface temperatures using nighttime radiance (the amount of light emitted from various regions as measured by satellite data) for the period March 1996 through February 1997 (Hansen et al. 2010).

In Table 3.1, we provide NASA-GISS information regarding the 10 warmest years in the lower contiguous United States based on instrumental records. Rankings of the warmest years are provided for three different periods.[10] Since data for the August 2007 report are available only through 2006, 2007 is not included in the earliest ranking given in the table. Notice that scientists have adjusted the data in ways that make more recent years appear warmer. Thus, the number of years from the past two decades that appear in the top 20 warm years has increased from 7 to 8 and finally to 11. In the May 2009 listing, 2007 is the 14th warmest year in the historical record, but it has moved up to tenth by the April 2010 listing. Based on data

[10] In all three cases, the data are taken from http://data.giss.nasa.gov/gistemp/graphs/Fig.D.txt, as viewed August 20, 2007, May 27, 2009, and April 26, 2010. The middle observation is reported by Brian Sussman (2010, p. 58), with the others by the current author. Sussman does not report the temperature anomalies and there is no way to retrieve them from the internet location at which they are found.

Table 3.2 Ten coldest/warmest January-February-March-April seasons in the past 500 years, Stockholm, Sweden, temperature anomalies in °C from 1961 to 1990 average

Rank	Year	Value	Rank	Year	Value
Ten coldest years			Ten warmest years		
1	1569	–7.26	1	1863	5.68
2	1573	–6.47	2	1990	4.71
3	1557	–5.87	3	1743	4.58
4	1595	–5.83	4	1525	4.33
5	1572	–5.43	5	1989	4.11
6	1942	–5.32	6	1605	4.08
7	1614	–4.96	7	1822	4.04
8	1600	–4.81	8	1790	3.93
9	1574	–4.53	9	1762	3.81
10	1940	–4.24	10	2008	3.80

Source: Leijonhufvud et al. (2010)

released in early 2011, GISS data show 2010 to be the warmest year on record, by 0.01 °C over 1998 (see Goddard 2011). While it may be true that the latest adjustments based on nighttime radiance are scientifically better than earlier adjustments, it seems odd that the most recent years are now showing up as among the warmest in the temperature record, contrary to global evidence presented in Fig. 2.6 and, importantly, averages based on raw temperature data for the U.S. presented in Fig. 3.3b.

It should also be noted that, based on the raw USHCN weather station data, 22 out of 50 states recorded their highest temperature during the 1930s (Sussman 2010, pp. 55–56). Likewise, Vinther et al. (2006) find that 1941 was the warmest year experienced in Greenland between 1784 and 2005, while the 1930s and 1940s were the warmest decades; 1863 was the coldest year while the coldest decade was the 1810s (although it corresponded to two volcanic eruptions in 1809 and 1815, the latter Tambora). A reconstruction of winter and spring temperatures for Stockholm by Leijonhufvud et al. (2010) that goes back 500 years finds that 1863 is the warmest winter/spring, while 1569 is the coldest (Table 3.2). The Stockholm temperature reconstruction is based on documentary evidence combined with instrumental data, and it strongly suggests the existence of a Little Ice Age (LIA).

Why are recent U.S. temperatures considered to be so warm compared to other years in the record? As pointed out in Chap. 2 and in the discussion above, climate scientists appear not to have been able to eliminate the contamination due to non-climatic or socioeconomic influences from the temperature record. This is not to suggest that global temperatures have not increased since the late 1800s, but rather that the recent decade may not have been the warmest ever. Certainly, the use of nighttime radiance is fraught with problems, including the fact that some jurisdictions illuminate their skies to a much greater extent than others, not because they are somehow richer, but because of political and historical factors pertaining to public lighting, sprawl, and so on. Further, it is difficult to wrap one's head around the idea that radiance observations for 1996–1997 can be used to adjust temperature data going back several decades or more. Missing data at stations, changing station locations, spatial coverage, and varying record lengths across stations affect the temperature reconstructions. Given that these are problems for weather stations that are considered to be of the highest quality and that are located in the world's richest country, one is left to speculate about the quality of data from weather monitoring stations elsewhere on the globe.

The the main reason why recent temperatures appear to be the warmest on record in the NASA-GISS temperature reconstruction, however, concerns Arctic temperatures. Outside of satellite observations of temperatures (which are not used in the GISS reconstruction), there are very few weather stations in the north. Yet, Hansen and his colleagues extrapolate these limited observations to the entire Arctic. In 2010, therefore, the average temperature for the entire far north ranged from 4 to 6° C above normal on the basis of observed temperatures in Nuut, Greenland, and a couple of other northern stations. As Goddard (2011) points out, neither the satellite data nor the HadCRUT reconconstructions come to a similar conclusion; the recent warm years are the result solely of incorrect procedures for averaging temperatures over a vast area based on extremely limited observations. The pitfalls of this were discussed in Chap. 2.

As noted in Chap. 2, the Berkeley Earth Surface Temperature project seeks to shed light on questions regarding the instrumental temperature record, and thus that of the warmest year. Even so, these efforts concern the warmest year in the past 130 or so, and not that of the last two millennia. We now turn to this issue.

3.3 Eliminating the Medieval Warm Period and Little Ice Age

The Intergovernmental Panel on Climate Change (IPCC), and thus much of the climate science community, takes the view that, although there is some variability within the system, on balance the Earth's climate is generally in equilibrium and has been so for thousands of years. In this world, there are three things that can cause the climate to change. These forcings, as they are known, are volcanoes, solar cycles (sunspot cycles) and human activities. Volcanoes spew particulates into the atmosphere that reflect sunlight back into space, thereby resulting in global cooling. However, volcanic ash might fall on snow and ice, thereby reducing the reflectivity (or albedo) of the surface while absorbing heat, thus leading to warming. The overall impact depends on a variety of factors and the time frame considered. The 11-year sunspot cycles, on the other hand, are thought to have little impact on global temperatures (IPCC WGI 2007, pp. 476–479). Consequently, this leaves anthropogenic emissions of greenhouse gases as the IPCC's main explanation for climate change.

Not surprisingly, the Medieval Warm Period (MWP), which is dated from about 900 to 1300 AD, created a problem for this view of past climate, as evidenced by the climategate emails. The MWP makes it difficult to accept the view that fossil fuel consumption (CO_2), large-scale cattle rearing (methane), tropical deforestation

(CO_2) and other activities are responsible for climate change. The MWP stands in contrast to the notion that such human activities will cause temperatures to rise to levels never seen before. Clearly, the MWP was not the result of anthropogenic emissions of CO_2 and other gases. It was a natural event, but one that is not explained in the IPCC account. If temperatures during the MWP were as high as or higher than those experienced thus far, there has to be something other than an anthropogenic forcing that accounted for this warm period.

The MWP also creates a dilemma for climate modelers. If the MWP was real, it would then be incumbent upon climate modelers to duplicate the Medieval Warming to demonstrate the veracity of their models. After all, information on atmospheric CO_2 and other greenhouse gases, and aerosols and particulates, is available from such things as lake bed sediments and ice cores. Thus, climate modelers cannot claim that their models are to be trusted simply because they are based on scientific relationships (mathematical equations) when such models cannot reconstruct an event such as the MWP. If, on the other hand, the MWP was the result of extra-terrestrial forces (sunspots, cosmic rays, earth orbit, tilt of the earth, etc.) that cannot be taken into account by climate models, there is no reason why these forces cannot also explain current climate events (as discussed in Chaps. 4 and 5). The same is true if modelers find there is some non extra-terrestrial explanation previously not taken into account.

There is simply too much evidence for the Medieval Warm Period to ignore. It comes from historical writings – the Viking colonization of Greenland, grape growing in England, crop production at high elevations, and so on (e.g., see Diamond 2005; Fagan 2008; Ladurie 1971; Lomborg 2007; Plimer 2009, pp. 31–99). Yet, climate scientists and climate modelers have deflected criticism by arguing that the MWP was not a period of global warming, but, rather, a period of heterogeneous warming with some regions experiencing a burst of warming at the same time that others experienced a cool period.[11] Backcasts of temperatures from climate models appear to confirm this position as various climate models' simulated temperatures for the past millennium do not indicate extended periods where temperatures were 'out of equilibrium,' but, rather, confirm the notion of long-term equilibrium with average temperatures fluctuating slighlty about the shaft of a 'hockey stick' (discussed below).[12]

The Little Ice Age also poses a problem for climate scientists because it provides a possible explanation for the warming observed in the instrumental record – any upturn in global temperatures would be expected once this period of 'natural' cooling came to an end. Indeed, the rise in temperatures seen in the instrumental record is not at all unexpected given that the instrumental record begins about the same time that the LIA ended. Some climate scientists argue that the LIA was confined only to Northern Europe,[13] while the IPCC downplays the LIA, arguing

[11] See IPPC Working Group I(IPCC WGI 2007, pp. 466–474). The Working Group I (WGI) report is entitled 'Climate Change 2007. The Physical Science Basis.'

[12] See (IPCC WGI 2007, p. 479).

[13] For example, Tom Pedersen, Director of the Pacific Institute for Climate Studies (PICS) at the University of Victoria in British Columbia, claims that the low temperatures of the LIA were local occurrences as opposed to a wider, global trend (personal communication, April 25, 2010).

that temperatures during this period were at most 0.1–0.3 °C cooler than normal (IPCC WGI 2007, p. 108). Contrary evidence to these views of the LIA is provided below, particularly in Fig. 3.6c.

In his exhaustive study of historical climate that includes records not employed by climate scientists, Ladurie (1971) was certainly convinced that the LIA was a global phenomenon that affected areas beyond Europe.[14] In his study on the Little Ice Age, Brian Fagan (2000) relied on much anthropological evidence to indicate that the LIA impacted North America and places as far away as New Zealand.[15] Khim et al. (2002) find evidence for both the MWP and LIA in sediment cores from the northern end of the Antarctic Peninsula. Plimer (2009, p. 74) notes that the "cold climate and glacier expansion in the Little Ice Age are documented from all continents and on major islands from New Zealand in the Southern Pacific Ocean to Svalbard in the Arctic Sea." Likewise, in their historical temperature reconstructions for regions in China, Ge et al. (2010) find that recent warming has likely been exceeded in the past 1,000 or more years, the rate of recent warming was not unusual, and the observed warming of the twentieth Century comes after an exceptionally cold period in the 1800s. This is confirmed by Ran et al. (2011), for example, who indicate that temperatuires in the MWP exceeded those of the twentieth century and the first decade of the twenty-first century by at least 0.5 °C. These authors also conclude that solar radiation may have been an important forcing mechanism explaining past ocean temperatures.

The evidence that a global warming period, known as the Medieval Warm Period, was followed by a global cooling, or Little Ice Age, is simply overwhelming. It is impossible for climate science to ignore. Hence, it came as no surprise that, without observational evidence to the contrary, the First Assessment Report of the Intergovernmental Panel on Climate Change (IPCC 1990) included the MWP as a historic event, and simply argued that the continued increase in atmospheric CO_2 would soon lead to global warming that exceeded that of the MWP. The graph used in the 1990 IPCC report is duplicated here as Fig. 3.4.[16]

Still, the MWP remained a thorn in the side of the IPCC for the reason mentioned above – climate science could not explain the warming period using the climate models upon which predictions of catastrophic warming are based. Therefore,

---

[14] Ladurie (1971) points to the advance and retreat of glaciers in North America and Greenland (pp. 99–107), records of flowering dates for the cherry blossom and other plants in Japan (p. 270), lake freezing dates in Japan (p. 272), evidence from giant cacti in Arizona (p. 40), and many examples from other regions, as support for the existence of the LIA outside Europe.

[15] Fagan's study is particularly instructive when it is contrasted with his Medieval Warm Period (Fagan 2008). The only clear conclusion is that warm weather is greatly preferred to cold, which is why the MWP is sometimes referred to as an 'optimum.' Certainly there were droughts and plagues of locusts, but evidence from various sources indicates that droughts, crop failure and yields were much worse during cold periods than warm ones (Fagan 2000, 2008; Idso and Singer 2009; Ladurie 1971; Plimer 2009, pp. 63–86).

[16] Steve McIntyre discusses the origins of this figure (Fig. 7c in the IPCC report) in a May 9, 2008 blog at http://www.climateaudit.org

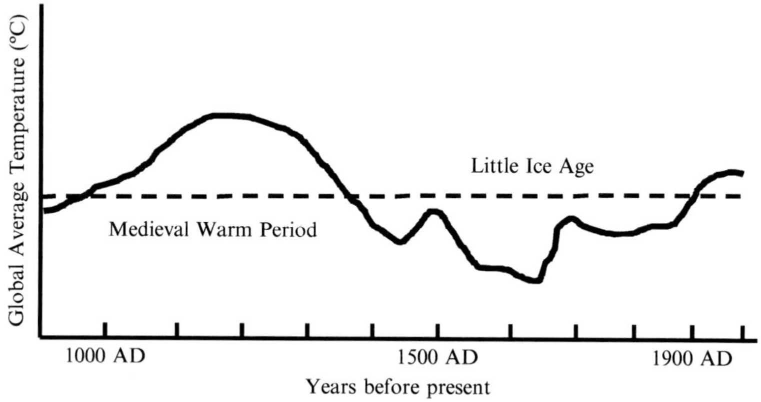

Fig. 3.4 Modern historical temperature trends as found in the 1990 IPCC report

climate researchers looked for a way to eliminate the Medieval Warm Period, and with it the Little Ice Age. In testimony before the U.S. Senate Committee on Environment & Public Works' Hearing on Climate Change and the Media, on Wednesday, December 6, 2006, David Deming of the University of Oklahoma stated: "In 1995, I published a short paper in the academic journal *Science*. … The week the article appeared, I was contacted by a reporter for National Public Radio. He offered to interview me, but only if I would state that the warming was due to human activity. When I refused to do so, he hung up on me. I had another interesting experience around [this] time… I received an astonishing email from a major researcher in the area of climate change. He said, 'We have to get rid of the Medieval Warm Period'."[17] Climate scientists found the evidence to eliminate the MWP and LIA in a Yale University PhD dissertation by Michael Mann (1998) and two follow-up papers by Mann et al. (1998, 1999), which are often referred to as MBH98 and MBH99.

Despite the work of Mann and his colleagues, the controversy about the MWP and LIA has not died down. As discussed below, McIntyre and McKitrick (2003, 2005a, b) and others found errors in MBH98 and MBH99, while climate scientists themselves were uneasy about the MBH conclusions. The following examples are documented in the climategate emails.[18] Michael Mann confirmed the truth of David Deming's Congressional testimony in a climategate email of June 4, 2003. In reference to an earlier statement or email by Jonathan Overpeck of the University of

[17] See http://epw.senate.gov/public/index.cfm?FuseAction=Hearings.Testimony&Hearing_ID=bfe4d91d-802a-23ad-4306-b4121bf7eced&Witness_ID=6b57de26-7884-47a3-83a9-5f3a85e8a07e (viewed October 12, 2009).

[18] Climategate emails are at http://www.eastangliaemails.com/index.php (viewed June 8, 2010). A searchable database for the 2009 and 2011 climategate emails is available at (viewed December 7, 2011): http://foia2011.org/

Arizona, who along with the CRU's Keith Briffa was a lead coordinating author of the paleo-climate section of the Fourth Assessment Report (IPCC WGI 2007), Mann wrote: "... a good earlier point that peck [Overpeck] made w/ regard to the memo, that it would be nice to try to 'contain' the putative 'MWP'." Clearly, the 'putative' MWP and also the LIA remained a major problem for paleoclimate scientists even after MBH98 and MBH99, and some felt uneasy about attempts to suppress these periods in the historical record.

Phil Jones of the CRU subsequently argued in 2004 that neither the MWP or LIA could be denied, but that the MWP was "no way as warm" as the last two decades of the twentieth Century and that no decade of the LIA averaged more than 1 °C below the 1961–1990 average global temperature, although this was based on "gut feeling, no science." Further unease was expressed in an email written February 16, 2006 by Briffa: "Let us not try to over-egg the pudding ... [as] there have been many different techniques used to aggregate and scale the data – but the efficacy of these [techniques] has not yet been established." Yet, no objections or qualifiers were included in the Fourth Assessment Report's conclusion that the most recent years were the warmest of the past 1,300.

Despite the statements in the previous paragraph, Jones later concluded on the basis of data from Greenland that the MWP was as warm as or warmer than anything seen recently. Yet, he and his coauthors concluded that current warming trends in Greenland "will result in temperature conditions that are warmer than anything seen in the past 1,400 years" (Vinther et al. 2010). This conclusion is unwarranted by the evidence and rooted solidly in the belief that temperature increases of the period from 1975 to 1998 (see Chap. 2) will continue indefinitely into the future. It also assumes that the temperature reconstructions of the past are accurate, something which we consider in the following discussions.

3.3.1 Analyzing Paleoclimatic Data

One way to compare current temperatures with past ones is through the use of proxy temperature data. Paleoclimatologists can infer past temperature records from ice cores, tree rings from long-living trees, lake bed sediments, stalagmites in caves, and coral reefs. Tree rings have a 1–3 year temporal resolution, speleothems (stalagmites, stalactites and similar rock formations) may also be resolved annually, ice cores resolve information on a decadal scale, historical documents (and anthropological evidence) have temporal resolutions of 10–30 years, and lake sediments have resolutions at the decadal to century time scales.

3.3.1.1 Constructing a Temperature Proxy

A temperature proxy refers to a measure, such as tree-ring width, that is sufficiently correlated with temperature to enable the reconstruction of temperature records where instrumental data are not available. Consider tree-ring width as a temperature proxy, with foresters having found that tree rings are wider when temperatures are warmer, *given sufficient precipitation so that it does not unduly constrain growth.* How do we obtain a temperature proxy based on a particular historical record of tree-ring widths?

First it is necessary to determine the effect that temperature has on the tree-ring widths of the tree or trees in the series.[19] This is done by regressing observed temperatures on tree-ring widths for a period where we have both observed temperatures and tree-ring widths. This is the *calibration period*. The *response function* to be estimated is as follows:

$$R_t = \beta_0 T_t + \beta_1 T_{t-1} + \beta_2 T_{t-2} + \cdots + \delta_i Z_{i,t} + \cdots + \varepsilon_t, \quad (3.1)$$

where R_t is the width of the tree ring measured at time t; T refers to temperature (with lags to indicate the effect of previous years' temperatures on tree growth); $Z_{i,t}$ refers to other potential factors, such as CO_2 levels that might affect tree-ring width; ε_t refers to the structure of the error terms (e.g., whether the model assumes errors are normally distributed or have some other distribution); and the βs and δs are parameters that need to be estimated. Given the statistical problems associated with lags (especially as this relates to the construction of a temperature proxy) and the inclusion of other variables, the response function that climate scientists estimate is simply given as:

$$R_t = \beta T_t + \varepsilon_t. \quad (3.2)$$

Next, it is necessary to test whether the calibration is correct – whether, for a period outside the calibration period where we also have observations on temperature, tree ring widths can predict observed temperatures. This is the *verification period*. Given that observed temperature data are available only since 1850, the calibration period might constitute the years 1930–2000, while the verification period runs from 1850 to 1929. To determine temperatures from tree ring data requires the *transfer function*:

$$T_t = \gamma_0 R_t + \gamma_1 R_{t-1} + \gamma_2 R_{t-2} + \xi_t, \quad (3.3)$$

where ξ_t refers to the error structure of the model and the γs are parameters that might need to be estimated.

The analyst can estimate either (3.2) or (3.3). If she estimates (3.2), it is then necessary to invert the equation to get

$$T_t = (1/\beta) R_t + (1/\beta) \varepsilon_t, \quad (3.4)$$

[19] The methods discussed here are described in more detail by Montford (2010, pp. 41–48) and Auffhammer et al. (2010).

which can then be used to predict temperatures given data on tree ring widths. This is the procedure that econometricians appear to prefer (Aufhammer et al. 2010). The other method is to estimate (3.3) directly.

Both approaches have their drawbacks. The first approach (estimate the response function and then invert it) leads to upward bias in the standard errors of the reconstruction – that is, the variances of the temperature reconstructions are larger than they should be. The second approach (invert the response function and then estimate the transfer function directly) leads to a downward bias in both the reconstructed temperatures (they are lower than they should be) and variances (which are too small relative to the actual variance in temperatures).

The response and the transfer functions are assumed to be linear. If they are non-linear problems arise. For example, suppose the functions are quadratic. Then, as temperatures rise tree ring width will first increase, but as they rise further tree ring width will fall. In that case, one would observe the same tree ring width for two different temperatures, one lower than the other. Likewise, one would predict two temperatures for each tree ring width.

There has also been some discussion in the literature as to which temperatures to employ – the local temperature or a global one. By relying on global temperatures, one is essentially assuming that trees "not responding to their own local temperature can nevertheless detect a signal in a wider temperature index" (Montford 2010, p. 47). This is difficult to accept, but has been assumed by some paleoclimatologists.

Finally, if the regression model (3.4) is able to predict temperatures from tree-ring widths for the verification period with reasonable (i.e., statistical) accuracy, then the tree ring data for the historical period for which we have no observed temperatures can be used to construct a 'proxy' temperature record – the *reconstruction period*. While we have discussed how this is done with tree rings, it can also be done using information from stalagmites, lake sediment boreholes, ice cores, et cetera. It is only necessary to find some measure that is a good proxy for temperature – that is strongly correlated with temperature. For example, the depth of each organic layer in a sediment might be indicative of higher growth during warm periods and less during cold ones. Likewise, the composition of dead organisms in a lake bed sediment (as opposed to the depth of an organic layer), or isotopes of various gases (or their ratios) in ice-core samples, might be highly correlated with temperatures, and thus can serve as temperature proxies. However, the resolution in these cases will not be annual.

Sometimes several proxies can be used to develop a single historical temperature series that goes back as far as the proxy. For example, if a dendrologist has measured and dated tree ring widths for four trees at some locale, all four might be used directly to estimate the transfer function (3.3), with the dependent variable temperature and the four annual tree ring width measures constituting the regressors (explanatory variables) as follows:

$$T_t = \gamma_1 R_{1,t} + \gamma_2 R_{2,t} + \gamma_3 R_{3,t} + \gamma_4 R_{4,t} + e_t, \quad (3.5)$$

where the subscripts indicate the four tree rings used in the temperature reconstruction. Regardless, once a temperature series has been constructed from proxy variables, it is possible to combine it with other temperature series to develop a broader, perhaps global, temperature reconstruction. This task has been quite controversial.

3.3.1.2 Aggregating Temperature Proxies

Suppose one has 50 or more temperature series developed from various proxy data, such as tree ring widths, ice cores, et cetera. In addition, a researcher might include temperatures from the central England data series (Fig. 2.3), or the Swedish temperature data construction by Leijonhufvud et al. (2010). In essence, one might have numerous different series that represent various temperature reconstructions. The reconstructions are from different geographical locations and are likely of different length and resolution. If one were to plot the many reconstructed temperatures over time, or rather the deviation of temperature from some average temperature (either determined as the series temperature average or chosen exogenously, and it does not matter), which is known as the temperature anomalies, one would get what has been referred to as 'spaghetti graphs' – wiggly lines, some of which may take a definitive upward trend in the twentieth century and others not.

To make sense of all the spaghetti lines, and get something useful that might be an indicator of a global trend in temperatures, it is necessary to somehow combine the information from all of the different series. Of course, as we saw in Sect. 2.1 of the previous chapter, the easiest way to summarize the data is simply to average it (and we illustrate that below). However, averaging might obscure interesting and important things. For example, a large portion of the data series might indicate a sharp upward trend in temperatures in the twentieth century, while remaining series indicate a gentle decline in temperature. If you look only at the average, the sharp uptick might be obscured or missed altogether. Principal component analysis is a well known, long standing statistical technique that teases out the most important trends by looking at patterns in all the data from the various temperature series.

Principal component analysis combines data series so that there are just as many principal components (linear combinations of the data series) as there are original data sets. However, the principal components (PCs) are organized so that they explain a decreasing amount of the overall variation in the overall data. Each principal component is a linear combination of the spaghetti data sets, with the weights assigned to some of the data sets in a PC much greater than those of other data, with some data sets assigned such a low weight they are essentially ignored in that particular PC. Thus, a PC constitutes a weighted average of the various data series, with the first PC (PC1) constructed so as to explain the greatest underlying variability in the data. PC2 accounts for the greatest amount of the remaining underlying variability, and so on. The first PC might account for 80% or more of the variation between the various reconstructed temperatures, and the first three to five PCs might account for 95% or more of the observed differences between temperatures. In this fashion, a 100 or more data series might be reduced to only a few.

With principal component analysis, it is important that the time step is the same for each of the spaghetti data series and that they each have the same number of observations. This is unlike the situation encountered in Sect. 2.1, where some observations were missing; in that case, it was possible to construct raw averages by simply dividing the sum of observations for a period by the number of data points for that period. For paleoclimatic spaghetti data series, the time step is generally annual and the period to be covered varies from several hundred to perhaps 2,000 years. Missing observations are typical, especially at the beginning and end of a data series because the lengths of the spaghetti data series vary significantly.

In the paleoclimatology literature, there are two sources of controversy that have not been adequately addressed because each involves value judgments. First, filling in gaps where information is missing can be done by linear extrapolation, regression analysis using information from the other temperature reconstructions, or some combination of these approaches. This poses several problems: How can linear interpolation adequately address large gaps given that climate is inherently variable and nonlinear over short and long periods of time? If interpolation is not the sole method employed, does one use all the available temperature series or a subset that is based on proxy datasets geographically close to the one with observational gaps? How is the choice made?

Second, how does one fill in missing observations at the beginning or end of a temperature series? One method that has been employed, probably for convenience given the lack of scientific guidance on the issue, is to use the first temperature record to 'in fill' all of the missing years prior to that first observation, and to do the same at the end of the series using the final observation. Where this has been done, it has been a source of controversy, but no less so than some alternatives. Clearly, given that scientists believe temperatures to be increasing, the use of the average value to fill in missing temperatures at either end of the series will be avoided. However, replacing missing observations at the beginning of the series with the temperature of the first observation or temperatures from another series (or some combination of series) is fraught with the same objections as those raised concerning other method(s) used to fill gaps. Nonetheless, splicing temperatures from the instrumental record at the end of a proxy series, which has been done in some cases (see below), is considered to be invalid for obvious reasons.

There is no simple way out of the dilemma. Yet, the choices that are made affect the conclusions one reaches about temperatures during the Medieval Warm Period and Little Ice Age relative to those in the late twentieth Century. This is reflected in the so-called 'hockey stick' controversy.

3.3.2 *The Hockey Stick*

Climate scientists have combined data from various proxies to derive a temperature graph that goes back more than 1,000 years. The graph shows temperatures to be flat for some 1,000 or more years, and then rising rapidly during the past 1,000 years.

The graph takes a hockey stick shape with the long flat part of the graph analogous to the stick's shaft and the sharp uptick in the twentieth century analogous to the blade. The higher temperatures associated with the Medieval Warm Period and the lower temperatures of the Little Ice Age have essentially been eliminated by relegating them to regional phenomenon. This has resulted in quite a bit of controversy, as discussed in this section and hinted at above. Although the controversy has been ably and helpfully discussed by Montford (2010), we provide an additional overview here.

MBH98 managed to reconstruct temperatures for the period 1400 to the present, while MBH99 were able to extend the reconstruction back to 1000 AD. Other paleoclimatic reconstructions of temperature go back two millennia, while, in some cases, both the CO_2 content of the atmosphere and temperature reconstructions based on sediment and ice-core data go back several hundred thousand years. It should already be evident that attempts to reconstruct temperatures going back that far are fraught with uncertainty, and even CO_2 measurements from ice cores are beset by problems related to, among other things, the age of the samples. Temperature reconstructions that go back 1,500–2,000 years are needed, however, if we wish clearly to identify the Medieval Warm Period from something other than historical writings.

Data from 71 series that go back for two millennia have been 'collated' by Fredrik Charpentier Ljungqvist (2009), a history student at Stockholm University in Sweden. He collected all of the paleoclimatic data series he could find in the literature that provided temperature information over the past two millennia – a total of 71 separate records. Ljungqvist could not obtain seven records that appear in the literature, discovered that one was an index as opposed to a proper temperature record (Record 19 from Galicia in Spain), and found 12 records for the Northern Hemisphere that were not tested to determine if there was a statistically significant temperature signal – rather, statistical significance was assumed because these series were considered to come from a trusted source. To be included, a series had to provide data points no more than a century apart. "All records with a sample resolution less than annual [had] been linearly interpolated to annual resolution" (Ljungqvist 2009). Subsequently, Ljungqvist's data records were posted with the World Data Center for Paleoclimatology in Boulder, Colorado and the U.S. National Oceanic and Atmospheric Administration's Paleoclimatology Program.[20]

It is difficult to determine how paleoclimatic data should be analyzed. As noted above, principal component analysis can be used to find combinations of various data series that account for the majority of the variability in the underlying data. Also as discussed above, this requires that one has observations for each year in

[20] NOAA's World Data Center for Paleoclimatology also makes available hundreds of different ice-core, lake-bed sediment, coral reef and other paleoclimatic records at their website http://www.ncdc.noaa.gov/paleo/paleo.html, although it is sometimes difficult to determine what each record actually contains/means. Much of the data is in 'raw' form so it is still necessary to develop temperature or other proxies from it. The data are available at (viewed May 25, 2011): ftp://ftp.ncdc.noaa.gov/pub/data/paleo/contributions_by_author/ljungqvist2009/ljungqvist2009recons.txt.

each data series. Ljungqvist used a straight-line extrapolation between the years for which observations are available to fill in missing data, but he did not attempt to fill in missing observations at the beginning or end of a reconstructed temperature series. We use Ljungqvist's data to discuss some of the problems associated with temperature proxies. Indeed, one must take care in working with paleoclimatic data and recognize that it comes with many qualifications. It is no wonder that some climate scientists, such as Keith Briffa (quoted above), are concerned about the types of conclusions one can draw.

To shed some light on the hockey stick controversy, we begin by examining the 71 data series in more detail.[21] Three series are not public, 18 provide only the historical variation in temperatures assuming a standard normal distribution (i.e., z-scores are provided), and 50 series provide actual temperature data. Suppose we begin with the temperature data and take the average of the available observations for each year. The average temperature in any given year depends on the number of observations available for that year, and is influenced by the location – whether a data series comes from Antarctica, a temperate region or a tropical one. Thus, if an observation from Antarctica falls out for any given year while one in a tropical region remains or enters for that year, the average will be higher than warranted, and contrariwise should a tropical data point be missing while one from the Antarctic remains or enters.

Given this proviso, a plot of average temperatures for the past 2,000 years was constructed (but not provided here). In this plot, temperatures are low in the early years, but they are also unusually low and volatile at the end of the period. The reason for this relates to missing data – a situation that is most acute on either end of the series. Indeed, there are between 41 and 45 observations for the first several decades in the series, the maximum number of 50 observations occurs for the period 133 AD to 1811, and then the available observations begins to taper off. By about 1980 there are less than 20 series for which there are data, falling to only 5 by 2000! The average temperature anomaly fluctuates so much after about 1950 that no definitive trend is observable. The paleoclimatic reconstructions using these series indicate that average global temperatures after the LIA are still well below those experienced during the Medieval Warm Period, and that recent temperatures may simply have been rising as the earth came out of the Little Ice Age.[22]

What happens if we include all of the 68 available series (out of 71)? The first step in doing so requires that we construct z-scores (as described earlier) for each of the 50 series for which actual temperature estimates are provided.[23] A plot of the average z-scores is found in Fig. 3.5a for the period 1 BC to 2000 AD. Notice that

[21] See http://www.ncdc.noaa.gov/paleo/paleo.html (viewed April 17, 2010).

[22] This accords with remarks by Phil Jones in a February 13, 2010 BBC interview, in which he indicated that the MWP was warmer than anything experienced recently.

[23] This is similar to the temperature anomalies that we encountered in Chap. 2. There, for example, the CRUT3 temperature series constitute an anomaly about the 1961–1990 global average temperature. Here the average temperature of a proxy series is simply the average over all observations for that series.

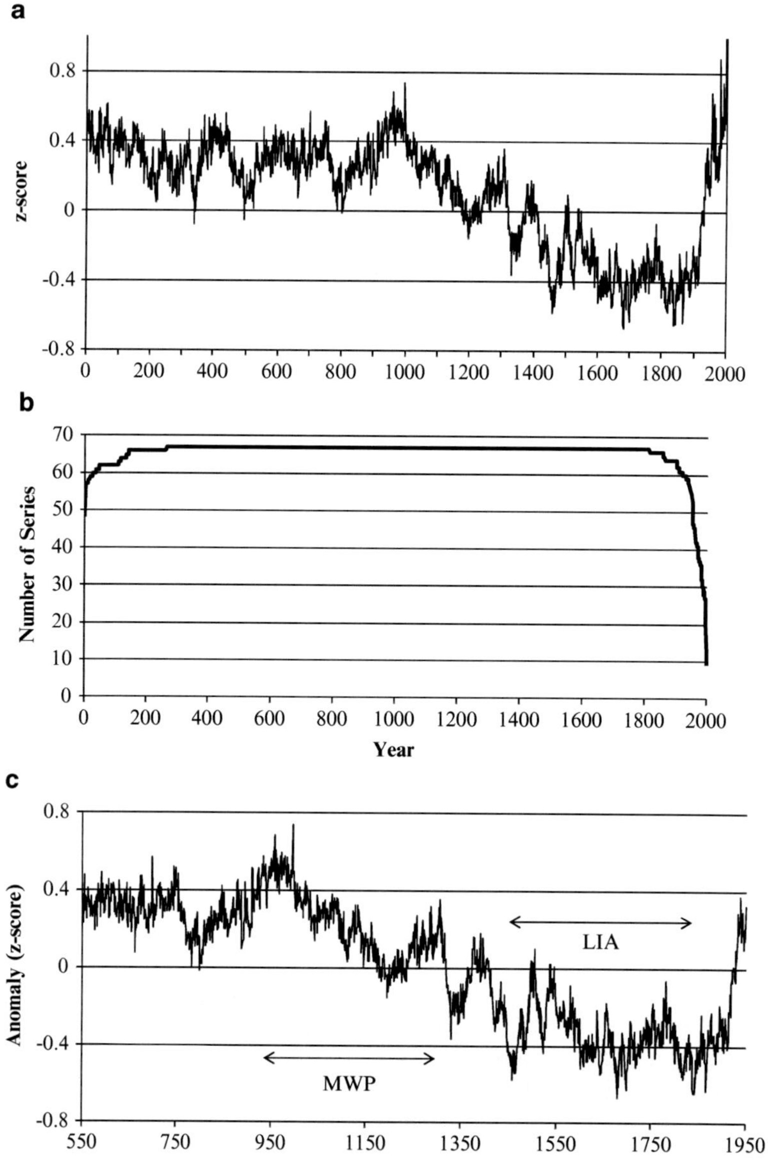

Fig. 3.5 (**a**) Average global temperature anomalies over the past two millennia. (**b**) Number of data series used to calculate average global temperature anomalies over the past two millennia, (**c**) Average global temperature anomalies for shorter period 550–1950

temperatures appear to rise rapidly at the tail end of the period, particularly towards the end of the twentieth century when the anomaly is greater than it was during the MWP. This more recent warming trend, which is sometimes referred to as the

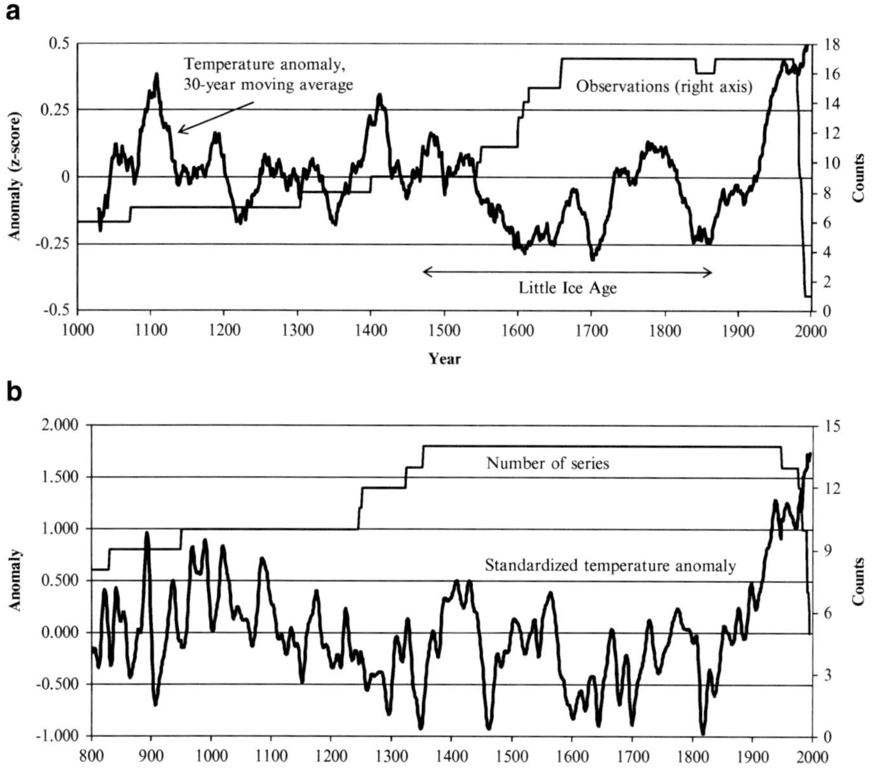

Fig. 3.6 (**a**) Average global temperature anomalies over the past millennium, based on paleoclimatic (Data from Jones et al. 1998). (**b**) Average global paleoclimatic temperature anomalies, 800–2000

Current Warm Period (CWP), did not appear when 18 of the available series developed by Ljungqvist were excluded. Yet, the most recent observations from paleoclimatic records fall off dramatically after 1950, as indicated in Fig. 3.6b.Thus the temperature rise at the tail end of the period may be an artifact of the lost observations; the number of series included in the construction of the average *z*-scores falls rapidly in the latter part of the record (Fig. 3.6b).

To get a better feel for what is happening, we re-specify the period on the horizontal axis to exclude the two extreme ends of the data series, providing the plot in Fig. 3.5c. That is, we only plot global average temperatures from paleoclimatic data for the period 550–1950, after which the number of available data series for constructing temperature averages drops off rapidly. (The period after 1950 is revisited below.) Note that the Medieval Warm Period and the Little Ice Age can now be readily identified in the figure. It is also clear from both Figs. 3.5a, c that, after the LIA, there was a rapid increase in average global temperatures determined from the paleoclimatic proxies.

The 71 data series considered above are clearly not the only ones available, even in the public domain. There are literally hundreds of different paleoclimatic data series on tree rings, ice cores, caves and so on, and these are available from the World Data Center for Paleoclimatology in Boulder, Colorado.[24] Two examples are provided in Fig. 3.6. In Fig. 3.6a, we consider 17 proxy series from Jones et al. (1998).[25] Again we constructed *z*-scores and averaged these for each year, but we had to use a 30-year moving average trend line to make sense of the data. Only upon doing so could we identify the Little Ice Age, while the Medieval Warm Period is more difficult to discern. Note that the number of series (observations) rises at the start of the LIA and then falls precipitously beginning around 1975, although 17 series is small to begin with.

In Fig. 3.6b, we employ only 14 temperature-related proxy series due to Tom Osborn and Keith Briffa (2006). Given that these were already in a standardized form, we simply averaged them and did not need to employ a moving average trend to get a better sense of the data. The number of available series or observations declines beginning around 1960, which again corresponds with an increase in average global temperatures derived from proxies. Further, the total number of observations is small.

The reduction in the number of available proxy records after about 1960 (as indicated in Figs. 3.5 and 3.6) is itself an enigma. Much of the data is based on tree rings and other proxies for which information should be more readily available in recent years than in earlier ones. Why have these records not been updated? Failure to do so has been interpreted by some to constitute an effort to hide information, namely, that the proxy data show temperatures to be falling in recent years, contrary to the instrumental record. This is discussed further by Montford (2010).

Sorting out what paleoclimatic proxy data tell us about historical temperatures is as much an art as it is science, which, unfortunately, leaves room for researcher bias regarding how information is finally reported. For example, NOAA scientists provide a graph based on 837 individual borehole records. This graph indicates temperatures rising at an increasing rate from 1500 to 2000. There is no evidence of a LIA in this reconstruction as temperatures have marched steadily upwards. Further, the borehole temperature data track very closely the instrumental record of Jones et al. (2010) that begins in 1850.[26] Yet, by eliminating the Little Ice Age, NOAA scientists find that average global temperatures have increased by only 1 °C since 1500. The graph based on NOAA borehole records strongly suggests that future temperatures will continue to rise in the future, which is the point that NOAA wishes to make. Unfortunately, it also suggests bias on the part of the graph builders.

Climate scientists tend not to present data in the simplistic form indicated above. One reason, of course, is that it is difficult to reconcile data collected from lake-bed

[24] See http://www.ncdc.noaa.gov/paleo/paleo.html

[25] Data available from http:/www.climateaudit.info/data/jser.txt (viewed April 20, 2011).

[26] See http://www.ncdc.noaa.gov/paleo/borehole/core.html (viewed April 26, 2010).

sediments in the tropics, for example, with ice-core samples from Antarctica; as explained above, dropping or adding an observation from one of these data series has a big upward or downward impact on the average. Thus, climate scientists will be selective in their use of data series, employ a variety of moving averages, rely on principal component analysis, and employ a variety of other statistical methods when summarizing and presenting proxy temperature information.

Despite the evidence in Figs. 3.5 and 3.6, and the inherent problems associated with temperature reconstructions from proxy information, some climate scientists persist in arguing that the record indicates that the Current Warm Period is somehow unusual from a historical perspective. Indeed, climate scientists persistently hold to the notion that the Earth's climate had previously been in a stable equilibrium, but that human emissions of greenhouse gases subsequently disturbed this equilibrium. This view requires that temperatures remain flat for a millennium or more before rising rapidly over the past century. As noted earlier, this depiction of events is referred to as the hockey stick. Given the extent of the so-called hockey stick wars, let us consider the issue in somewhat greater detail as it involves both the paleoclimatic record and the instrumental record.

3.3.3 The Climate Hockey Wars

A version of the MBH98-MBH99 hockey stick graph is provided in Fig. 3.7. This version came from Wikipedia and appeared in a United Nations' Environment Program report (McMullen and Jabbour 2009, p. 5) prepared in advance of the December 2009 Copenhagen meetings on climate change – COP 15 of the UN Framework Convention on Climate Change (UNFCCC). In the figure, the 'more stable' line indicates historical levels of atmospheric CO_2, rising from 280 ppm in year 1000 to about 380 ppm in 2000 (see Fig. 3.1 above); the other ('squiggly') line indicates average global temperatures, which are approximately flat from 1000 to 1900 and then rise sharply thereafter by about 0.8 °C.[27] Notice that the Medieval Warm Period (MWP) has disappeared, and even the Little Ice Age (LIA) of 1300–1850 is impossible to discern. One might immediately ask: If the relationship between CO_2 and temperature is as indicated in Fig. 3.7, what is the statistically-significant correlation between atmospheric CO_2 concentration and average global temperature? But no such correlation appears in the literature. Further, the scale of the graph has been chosen to dramatize the impact on temperatures of economic development and rising populations during the 1900s, and to highlight that current warming is unprecedented by historical standards.

[27] The graph of temperatures is nearly identical to NOAA's temperature graph based on borehole data (see previous note).

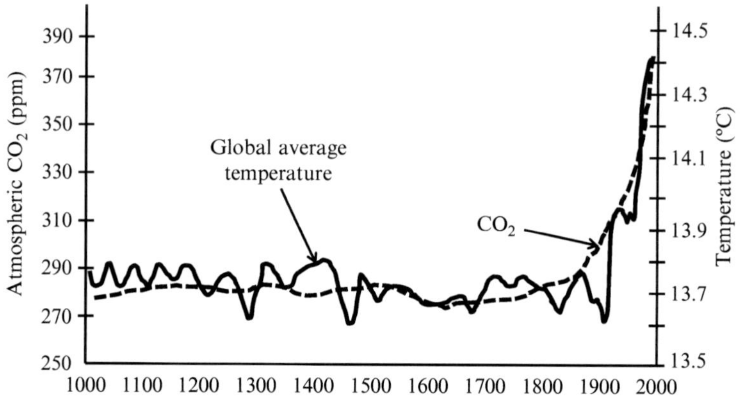

Fig. 3.7 The infamous hockey stick

Compare Fig. 3.7 with Figs. 3.5a, c, and 3.6a. Clearly, the data available from the World Data Center for Paleoclimatology in Boulder, Colorado, do not lead to a graph anywhere close to the one in Fig. 3.7. Therefore, it is not surprising that the hockey stick view of the world has been controversial and proven wrong. And, as a result of pressure by bloggers, the figure in the UN report (Fig. 3.7 here) was quietly replaced by another figure (McMullen and Jabbour 2009, p. 5).[28]

After much haranguing that included requests to the journal *Nature* to obtain the MBH98 data, McIntyre and McKitrick (2003, 2005a, b, 2009), hereafter MM, were able to investigate the claims of MBH98 and MBH99. They found a methodological flaw in the MBH analysis. In particular, MM found they could get the hockey stick result with any arbitrary data as long as they used the MBH procedure, even randomly generated data. The corrected MBH98 construct is compared to the original MBH98 temperature series in Fig. 3.8. Notice that MM's corrected series indicates where the Medieval Warm Period turns into the Little Ice Age (around 1450), while no such distinction appears in the MBH proxy, because the MWP has somehow been made to vanish.

MM discovered two problems with the MBH reconstructions. First, principal component analysis requires that, to make series compatible, it is necessary to standardize the data in each temperature series by constructing *z*-scores (subtracting from each data point the mean of the series in which it is found and dividing by the standard deviation of the series). With reconstructed data, however, it was necessary only to subtract means as the paleoclimatic temperature reconstructions had already

[28] See http://wattsupwiththat.com/2009/10/05/united-nations-pulls-hockey-stick-from-climate-report/ (viewed October 12, 2009). The book's official website http://www.unep.org/compendium2009/) was "under revision" as of October 12, 2009, but available again in February 2010. Interestingly, the graph that replaced the original figure (Fig. 3.7 in the text) begins in 1880 and goes to 2005, rather than the period 1000–2000. However, for the period after 1998, it continues to show temperatures increasing contrary to official data, as shown in Fig. 2.4 of the previous chapter.

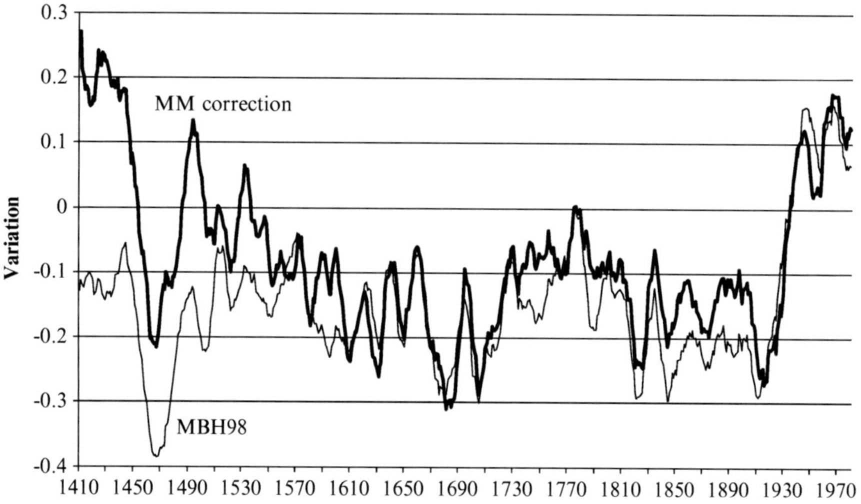

Fig. 3.8 Comparison of the McIntyre–McKitrick historical temperature reconstruction (*thick line*) with the original MBH hockey stick (*thin line*), 10-year moving average

been standardized (Montford 2010, pp. 194–195); this is referred to as 'centering the data.' Rather than center the data using the means of each of the series, MBH used the means of the calibration period. This 'short-centering' procedure was unusual and tends to bias outcomes towards the more recent calibration period.

Second, along with short centering, the algorithm used by MBH leads to a temperature reconstruction that gives precedence to any series that indicate a strong upward or downward trend during the calibration period (or twentieth century). That is, if there was one temperature series that gave a hockey stick shape, this one series would dominate as long as no other series had a strong twentieth century downward trend. McIntyre and McKitrick demonstrated this by combining a single 'hockey-stick' series with numerous randomly-constructed data series that exhibited no trend (i.e., were white noise series), and, using MBH's algorithm, obtained the hockey stick shape associated with the single 'hockey-stick' series. Regardless of anything else going on, as long as one series with a strong twentieth century uptick in temperatures was included, one obtained the hockey stick shape using short centering and the MBH algorithm. Even if a number of series included a MWP and LIA, these disappeared because of short centering.

The one proxy series that displays the uptick associated with the hockey stick is a tree ring series from bristlecone pines in the western United States collected by Graybill and Idso (1993) for the purpose of demonstrating the fertilization effect of rising atmospheric CO_2 during the 1900s on tree growth. The authors specifically stated that the twentieth century growth in these trees was not accounted for by local or regional temperatures and was hypothesized to be the result of CO_2 fertilization. Thus, it was surprising that this series was used in temperature reconstructions. Not only that, but all reconstructions leading to a hockey stick result included the

Graybill-Idso bristlecone pine series, although in some cases the bristlecone pine series was hidden.

It is important to note that a variety of instrumental temperature records, such as the Central England series, and proxies constructed from tree-rings, lake-bed sediments and so on can somehow be combined to derive a historical reconstruction of global (or regional) temperatures. For example, the 71 series from Ljungqvist were used to provide some indication of past global temperatures (see Fig. 3.5). Indeed, there are now hundreds of 'spaghetti' graphs, one for each temperature proxy. And each proxy is derived from one or a few tree ring series, lake bed sediments, ice core samples, and so on. The simplest way to combine series, say for a region or supra-region (even global level), is to average them, as we did for the Ljungqvist series. However, this is a statistically crude method and, as noted earlier, a preferred statistical method is principal component analysis. A principal component analysis of the Ljungqvist series, for example, finds that no more than 18% of the total variation in temperatures can be explained by a single PC – by the first PC.

In some cases principal components can be constructed from a whole bunch of different series for a particular region, with one or more of these PCs subsequently used in a global temperature reconstruction. In many of the paleoclimatic reconstructions of global temperatures, PCs were constructed from the International Tree Ring Database for North America. In this database, the bristlecone pine data are included in the fourth principal component (PC4), which explains only 8% of the total variation in the tree ring data. As Montford (2010) points out:

> What this means is that their hockey stick shape is a rather unimportant pattern in the database, as would be expected since bristlecones are a couple of closely related species from a small area of the western USA. However, because they correlate well to temperature in the twentieth century, they dominate the calibration results and hence the reconstruction too (p. 327).

Needless to say, the hockey stick did not disappear without a fight. As a result of the controversy, two independent review panels were struck – one by the National Academy of Sciences and the other at the request of Congress. Both supported the MM analysis. In addition, at the request of Representatives Joe Barton and Ed Whitfield, the U.S. House of Representatives commissioned an independent evaluation by Edward Wegman, Chair of the National Academy of Sciences' Committee on Applied and Theoretical Statistics (Wegman et al. 2006) – known as the Wegman Report. The Wegman Report reviewed the data and the statistical methods used by Mann (1998), MBH98 and MBH99, and McIntyre and McKitrick (2003, 2005a, b), concluding in favor of McIntyre and McKitrick. With regard to the statistical analysis, Wegman and his colleagues found that there was a deficiency in the way proxy and instrumental temperature data were analyzed: "A serious effort to model even the present instrumented temperature record with sophisticated process models does not appear to have taken place" (Wegman et al. 2006, p. 15).

In addition to the statistical evidence, the Wegman Report employed network analysis to examine relations among researchers. Wegman found that there are too few independent researchers looking into the historical temperature record, so much so that objectivity in the review process could not be guaranteed. Of course, the number one critique leveled at the Wegman Report was that it is not peer-reviewed![29]

The Wegman Report made four recommendations.

1. "Especially when massive amounts of public monies and human lives are at stake, academic work should have a more intense level of scrutiny and review... [A]uthors of policy-related documents, like the IPCC report, ... should not be the same people as those that constructed the academic papers.
2. ... federally funded research agencies should develop a more comprehensive and concise policy on disclosure... Federally funded work including code should be made available to other researchers.
3. With clinical trials for drugs and devices to be approved for human use by the FDA, review and consultation with statisticians is expected... evaluation by statisticians should be standard practice ... [and] mandatory.
4. Emphasis should be placed on the Federal funding of research related to fundamental understanding of the mechanisms of climate change".

These recommendations anticipated the climategate revelations by some 3 years, particularly as these relate to freedom of information requests.

3.3.4 The Hockey Stick Strikes Back

A new hockey stick result was published in the IPCC's Fourth Assessment Report (IPCC WGI 2007) based on tree ring data from the Yamal Peninsula in Siberia by Keith Briffa and colleagues (Briffa et al. 2008; also Briffa et al. 1996; Briffa et al. 2001; Schweingruber and Briffa 1996). This result finds that the coldest year in the previous 1,200 years occurred during the MWP, but the chronology they use inexplicably adds core counts from the Polar Urals (1995) in the absence of Yamal core counts. The data do not indicate an increase in temperatures for the twentieth century until after 1990, when the available tree ring data collapse from samples with 30+ trees to 10 trees (1990) and then 5 trees (1995).

Briffa's colleague F.H. Schweingruber produced a larger data set from Polar Urals that showed the MWP to be much warmer than the late twentieth century, but this paper and the data appear to have been suppressed. Stephen McIntyre repeatedly requested access to Briffa and Schweingruber's Yamal data, but could not get it from *Nature* or *Science*, until the authors published a paper using the data in the *Philosophical Transactions of the Royal Society* in 2008 (Briffa et al. 2008; see also Briffa et al. 1996, 2001). The results are provided in Fig. 3.9. The original temperature reconstruction employs Briffa's data based on 12 cores for Yamal, and it indicates a sharp uptick in temperatures in the last 100 years. The correction by McIntyre using Schweingruber's data from a series called Khadyta River, which is

[29] It turns out, however, that the IPCC itself relied on non-peer reviewed material for a number of its assertions (see Chap. 5).

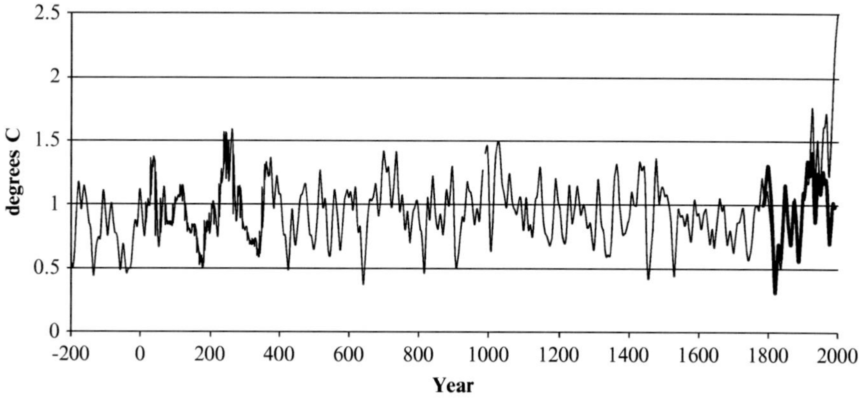

Fig. 3.9 The hockey stick strikes back, or does it?

near the Yamal site, does not indicate a similar uptick, but rather a downturn early in the past century. The corrected results dramatically change the conclusions – the Yamal Peninsula data no longer provide any evidence of warming that falls outside the historical range. The apparent 'hockey stick', indicated by the thin black line that rises sharply at the end of the time horizon, disappears to be replaced by the thick black line (Montford 2010, pp. 394–401).

Finally, Mann et al. (2008) made an attempt to reconstruct the hockey stick without tree ring data. Earlier, Loehle (2007) had demonstrated that there was no 'hockey stick' in temperature reconstructions that excluded tree rings. Mann and his colleagues relied on data from four lake bed sediments in Finland collected by Mia Tiljander as part of her PhD dissertation research. The Tiljander proxies indicated an uptick in twentieth century temperatures, but this was attributed to a disturbance caused by ditch digging. Nonetheless, Mann et al. (2008) argued that they had demonstrated that this did not matter. They did this by first showing that they could get a hockey stick result without employing the Tiljander data. However, this version of the hockey stick included the bristlecone pine data. They then demonstrated that the hockey stick result was not due to bristlecone pines by removing them from the reconstruction; however, when they did this, they again put in the Tiljander proxies (Montford 2010, pp. 362–373)!

What is most disturbing about the hockey stick debate is the difficulty that independent researchers, such as MM, have had accessing data that are the basis of results published in peer-reviewed journals.[30] After all, verification is a key element of any empirical research and journals generally have policies regarding data and the ability of others to verify results. If anything, on the face of it, the hockey stick

[30] Many papers, correspondence and other documents relating to the hockey stick debate between MM and MBH can be found at http://www.climateaudit.org/?page_id=354 (viewed April 12, 2011). Also, the climategate controversy may have resulted in greater openness in the sharing of data and computer code.

debate might be considered a black eye for science. However, difficulty obtaining data that should be made available to other researchers has taken second stage to the more pejorative 'climategate' revelations.

3.3.5 Hiding the Evidence?

One of the problems that climate scientists working with paleoclimatic data encountered was that their proxy records did not always coincide with the instrumental records. Some proxy records indicate that temperatures should have been declining rather than rising during recent decades, as indicated by instrumental obseervations. Since instrumental data are available for some 130 years, and that instrumental data are used to calibrate the temperature-proxy relationship, it is surprising to find that temperatures based on proxy data and the instrumental temperature data diverge for upwards of 40 years. Clearly, it is necessary to investigate this divergence further using the best available statistical methods, which might lead to a re-evaluation of the paleoclimatic record as the proxy response function (3.1) and the temperature transfer function (3.3) need to be revaluated. However, climate scientists dismissed the divergence by attributing it to higher environmental pollution after 1960, although evidence of this is lacking, and other factors, and chose to 'hide the decline' as one climategate email put it.[31]

The 'hide-the-decline' controversy refers to a graph based on tree-ring data from Keith Briffa that appears in the 2001 IPCC report.[32] A facsimile of the graph in question is shown in Fig. 3.10. The original tree-ring proxy record from Briffa et al. (1998) in panel (a) of the figure indicates a sharp downturn in temperatures after 1960. A similar downturn is found in other records as well, two of which are shown in Fig. 3.10. The reconstruction in panel (b) shows something quite different: the three proxy-based temperature records in panel (a) have been truncated in 1960 and instrumental data (in this case a moving average of the HadCRUT3 temperature product) substituted for the remaining years. In this way, the IPCC was able to 'hide the decline.'[33]

---

[31] See http://climateaudit.org/2009/12/10/ipcc-and-the-trick/ (viewed April 24, 2010) and http://climateaudit.org/2011/03/17/hide-the-decline-sciencemag/ (viewed April 12, 2011).

[32] The graph was digitized by Stephen McIntyre in 2010 at the internet site indicated in the preceding note. Since then, data are more readily accessible from the internet, but one must still search to find the appropriate data and instructions regarding what the data mean. The data provided below are from Briffa et al. (1998), Jones et al. (1998) and Briffa (2000), and can be accessed via McIntyre's climateaudit.org website.

[33] In two videos on YouTube, Berkeley physics professor, Richard Muller, provides an excellent overview of the issue, as well as a scathing attack on the climate scientists responsible (see: http://www.youtube.com/watch?v=8BQpciw8suk and http://www.youtube.com/watch?v=U5m6KzDnv7k (viewed April 9, 2011)). It should be noted that Fig. 3.10 is not exactly the same as the figure in the IPCC report as it is only meant to demonstrate how the 'trick' (as it was described in the climategate emails) was implemented.

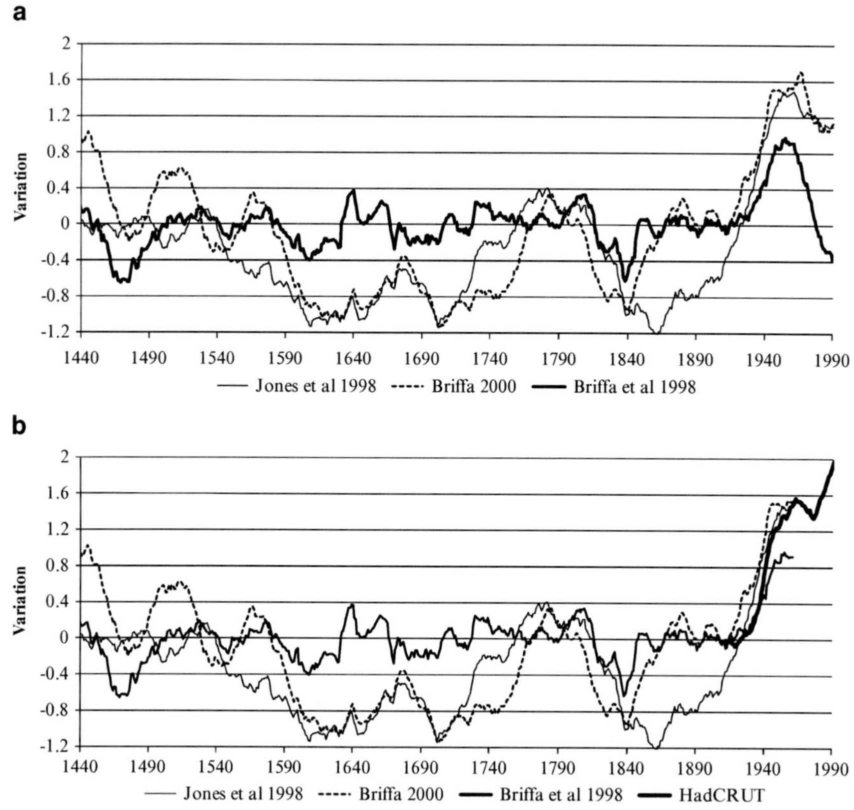

Fig. 3.10 (**a**) Tree-ring proxy records indicating a sharp downturn in temperatures at the tail-end of the record, 30-year moving average. (**b**) Hide the decline: tree-ring proxy records truncated in 1960 and replaced with temperature data, 30-year moving average

In the original IPCC figure (similar to Fig. 3.10b), the caption indicates that this reconstruction uses "… a 40-year Hamming filter with end-point padding with the mean of the closing 20 years." Along with the climategate emails, this suggests that climate scientists have spliced one dataset onto another to support their preconceived view that the latest decades are the warmest on record. As Ross McKitrick observed in testimony before a commission trying to sort out the climategate information: "The apparent agreement between the proxy records and the temperature records was achieved by the undisclosed step of replacing the ending two to four decades of the proxy records with the CRU temperature series and heavily smoothing over the splice" (McKitrick 2010, paragraph 16, p. 9). He further points out that:

> As an academic matter, scientists combine different types of data all the time for the purpose of extracting information and constructing statistical models. As long as the methods are clearly explained and the reader is given the information necessary to evaluate the quality of

> the calibration/fitting process, there is nothing wrong with this, and indeed it is often the path to important discoveries and progress. But in the case of the preparation of the WMO and IPCC diagrams, the problem is that readers were not told about the way different data sets were being trimmed and/or combined, hence materially adverse information was withheld from readers, thus exaggerating the quality of the statistical model (para 46, pp. 25–26).

What is disturbing is the way in which the authors of the IPCC Working Group I report (IPCC WGI 2007) deal with the criticisms levied at the various reconstructions of historical temperatures using paleoclimatic proxy data. The IPCC continues to adhere to the 'hockey stick' story, although it features much less prominently and more subtly than in the previous (IPCC WGI 2001) report.[34] What has been most frustrating in all of this is the reluctance of the climate scientists involved to make their data and methods available to other researchers, even if these researchers might have had a different view on the causes of global warming, and their failure to collaborate with researchers who have the necessary statistical expertise.[35] In the meantime, the hockey stick story continues to dominate the pages of public documents.

Nonetheless, change is coming. In a recent paper, econometricians McShane and Wyner (2011) use time series anlaysis and the proxy data to predict temperatures found in the instrumental record. They find that the proxies climate scientists use are no better at predicting future temperatures than "random series generated independently of temperature." Indeed, statistical models based on proxy data are unable to forecast or backcast high temperatures or rapid increases in temperature, even for in-sample predictions (i.e., where the prediction and estimation periods coincide). Upon reconstructing the Northern Hemisphere land temperatures using the climate scientists' proxies (i.e., the contentious data series), McShane and Wyner (2011) find a similar reconstruct to that of the climate scientists, but with a higher variance. They conclude that the recent high temperatures are not statistically different from those of earlier years.

[34] This is evident from the figures on pages 467–468, 475, 477 and 479 of the IPCC WGI (2007). Each of the figures still has temperatures rising rapidly during the 1900s and into the twenty-first century. The McIntyre-McKitrick critique of the hockey stick is summarily dismissed (IPCC WGI 2007, p. 466) with a reference to a paper by Wahl and Ammann (2007) that had not yet appeared and an incorrect reference to a paper by Wahl et al. in *Science* (2006) to which MM were not permitted to respond. The IPCC authors ignored the Wegman report and other research supporting MM. Indeed, one of the IPCC's review editors (gatekeepers) believed the methods used to derive the hockey stick result were biased, giving statistically insignificant results; yet, he signed off on the paleoclimatic chapter, thereby agreeing that the hockey stick constituted a 'reasonable assessment' of the evidence (Montford 2010, pp. 446–447).

[35] A colleague suggested that the CRU was simply so overwhelmed with requests to access the data under Freedom of Information (FOI) that they could not possibly respond to all such requests. This argument is specious because data, computer code, etc. could easily have been made available on the internet (available data are currently dispersed across various sites); further, climategate emails indicate that requests came before FOI became an issue and that the number of requests was not onerous. Indeed, climategate emails strongly suggest that there was a deliberate attempt to prevent 'outsiders' from accessing the data.

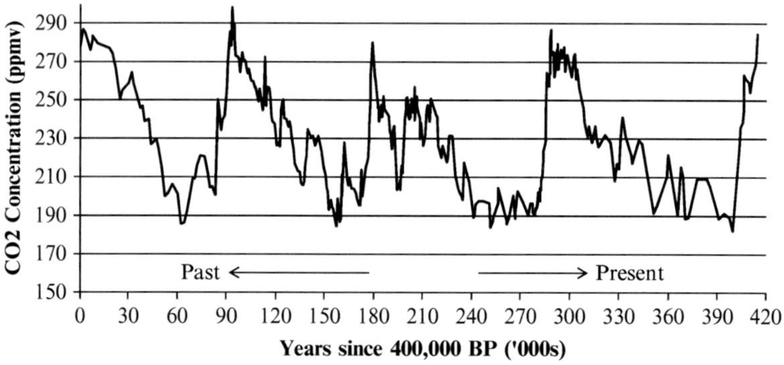

Fig. 3.11 Atmospheric concentrations of CO_2 (ppmv) over the past 400,000 years

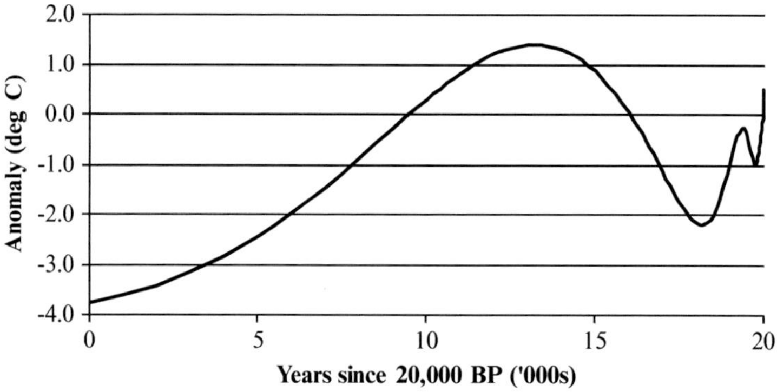

Fig. 3.12 A temperatures record of the past 20,000 years

Other economists are finding similar problems with reconstructions of historical temperatures using proxy data (e.g., Aufhammer et al. 2010). Their work confirms the criticisms levied at the hockey stick by McInyre and McKitrick, Pielke et al. (2007), the Wegman Report, and others.

3.3.6 *Temperatures and CO_2 Before Time*

Scientists are also interested in determining how current temperatures and atmospheric CO_2 concentration compare with those of the more distant past. In Fig. 3.11, we provide a reconstruction of the atmospheric concentration of CO_2 based on Vostok ice core data.[36] The CO_2 concentration is measured in parts per

[36] Source: http://cdiac.ornl.gov/trends/co2/contents.htm (viewed April 17, 2010).

million by volume (ppmv). In the chart, time begins some 400,000 years before the present (BP), which is denoted 0. Comparing this graph with Fig. 3.1, we find that current concentrations of atmospheric CO_2 are high, but certainly not outside historic experience.

Next we consider, in Fig. 3.12, a temperature reconstruction that goes back 20,000 years. Again, the origin represents 20,000 BP while 20 represents the current period. Temperature is measured in terms of the anomaly from the 1961–1990 global average. The data used in the figure are from Huang et al. (2008) and include instrumental data in the latter part of the record, which accounts for the upward trend in the last century of the record. The MWP and LIA are clearly identified by the last two turning points in the graph. Again, the story is the same: current temperatures are not outside historical experience. What one should worry about is the cold period from about 20,000 to 9,000 BP. It was only after that ice-age period ended that human civilization flourished. Clearly, humans and other animals thrive in warmer temperatures.

3.4 Discussion

The conclusion of the latest IPCC report is simple: "It is very unlikely that the twentieth-century warming can be explained by natural causes. The late twentieth century has been unusually warm. Palaeoclimatic reconstructions show that the second half of the twentieth century was likely the warmest 50-year period in the Northern Hemisphere in the last 1,300 years" (IPCC WGI 2007, p. 702). There are two problems: First, as pointed out in this chapter, there is no statistical evidence to suggest that the last 50 years of the twentieth century were statistically the warmest of the past 1,300. As shown in this chapter, many peer-reviewed publications and commissioned reports call into question the temperature reconstructions that form the basis for the IPCC's conclusion, arguing that the IPCC ignores the well-documented Medieval Warm Period which may well have seen higher average global temperatures than seen in the Current Warm Period. Indeed, a recent study by Blakeley McShane and Abraham Wyner (2011) found that recent temperatures were not statistically different from past temperatures, even when the IPCC's paleoclimatic reconstructions were used as the basis for comparison. Second, if anthropogenic emissions of greenhouse gases resulted in the warm years of the latter part of the twentieth century, what anthropogenic sources resulted in the 'cooling' since 1998 whilst atmospheric concentrations of CO_2 have continued to rise?

Paleoclimatic reconstructions of past climate are not necessary to make a scientific case for global warming. Rather, reconstructions such as the hockey stick are important only from a political standpoint, because, if it is possible to demonstrate that current temperatures are higher than those experienced in the past, it will be easier to convince politicians to fund research and implement policies to address climate change. However, the opposite may now have occurred. By hitching its wagon to the hockey stick, the IPCC may have harmed its credibility. As Montford (2010) points out: "What the Hockey Stick affair suggests is that the case for global warming, far from being settled, is actually weak and unconvincing" (p. 390). The implication for policymakers is that fears of global warming are likely overblown. This chapter and the previous one have shown how difficult it is to aggregate temperature data from various sources, determine an average global temperature, and construct historical global temperature averages from proxy data. Although instrumental and paleoclimatic temperature evidence are an important component in helping us understand climate change, they are only one part of the science.

On October 14, 1997, there was an El Niño Community Preparedness Summit in Santa Monica, California, to discuss the super El Niño event of that year. This El Niño led to the highest temperature on record in 1998 and relatively high but falling temperatures for several years thereafter (see Figs. 2.6 and 2.7 in Chap. 2). The invited speaker, Al Gore, predicted that, because of human emissions of CO_2, there would no longer be La Niña events and that, according to his fellow scientists, El Niño events "would become permanent."[37]

This incidence illustrates what is truly sad about the state of climate science, namely, that truth has been sacrificed for political expediency. No climate scientists have denounced Al Gore and his fictional depiction of the future; none questioned his membership among the scientific elite, although they are quick to question the credibility of scientists working outside the climate community (e.g., see Anderegg et al. 2010). None have cried foul when reports supporting peer-reviewed literature finding the hockey stick to be incorrect were castigated by environmentalists, and none have denounced the UNEP's climate science report (McMullen and Jabbour 2009) for attributing every weather event (cyclones, drought, torrential rains, etc.) to anthropogenic global warming (a topic discussed further in Chap. 7). And a major reason why many scientists have not spoken out is due to the trust placed in the analysis of paleoclimatic data.

There are many variants of the historical temperature graphs that can be built. Because of the ad hoc way in which data series are combined and graphs subsequently constructed (e.g., normality assumptions), none is truly representative of what the global climate was really like over the past several millennia. Is it even appropriate to use averages of various series, or should one let each series speak for itself? Should one employ principal component analysis? Are principal components useful if any one accounts for no more than 18–20% of the total variation in the data? That is, does it make sense to replace 50 data sets, for example, with 20 principal components? Does this shed further light on past climate?

As pointed out by Marc Sheppard,[38] and evident from Fig. 3.5c, the paleoclimatic proxies from tree rings tend to diverge from the instrumental (observational) record

---

[37] Reported in the *San Francisco Chronicle*, October 15, 1997 at (as viewed October 19, 2009): http://icecap.us/index.php/go/joes-blog/metsul_special_report_to_icecap_al_gores_inconvenient_mistake/

[38] http://www.americanthinker.com/2009/11/crus_source_code_climategate_r.html (viewed February 18, 2010).

after about 1960, depending on the particular proxy construct. Assuming the instrumental record is reliable from 1880 to 2000, this implies that proxy temperature data and instrumental data give opposing results for 40 out of 120 years, or one-third of the time. There is no explanation of why this is the case. Some climate scientists have simply argued that the more recent tree-ring data are unreliable for some reason, often attributed to environmental pollution. As a result, the proxy data are dropped and the instrumental data put in their place, or the proxy data are 'corrected' to accord with the observed record. Although one cannot fault climate scientists for doing this as a stop-gap measure, it is necessary to acknowledge ignorance about the climate record. Until an explanation for the difference between the proxy data and the instrumental record is found, one cannot argue that the historical record provided by the proxy reconstruction is reliable.

Clearly, the analysis of paleoclimatic data leaves much open to interpretation, which results in a great deal of uncertainty about the human role in global warming and what, if any, action governments should take to affect human activities to reduce greenhouse gas emissions. This wicked uncertainty needs to be taken into account in determining the costs and benefits of mitigating CO_2 emissions, and if mitigation is even an optimal policy. Before turning exclusively to economic issues, however, it is necessary to consider some further issues related to climate science as these affect economic analysis and the conclusions one can get from economic science.

References

Anderegg, W. R. L., Prall, J. W., Harold, J., & Schneider, S. H. (2010). Expert credibility in climate change. *PNAS, 107*(27), 12107–12109.

Aufhammer, M., Wright, B., & Yoo, S. J. (2010). *Specification and estimation of the transfer function in Paleoclimatic reconstructions*. Paper presented at the Fourth World Congress of Environmental and Resource Economists, Montreal, QC.

Bray, D., & von Storch, H. (2007). The perspectives of climate scientists on global climate change. *GKSS 2007-11*. Retrieved February 15, 2010, from http://dvsun3.gkss.de/BERICHTE/GKSS_Berichte_2007/GKSS_2007_11.pdf

Briffa, K. R. (2000). Annual climate variability in the Holocene: Interpreting the message of ancient trees. *Quaternary Science Reviews, 19*(1–5), 87–105.

Briffa, K. R., Jones, P. D., Schweingruber, F. H., Karlén, W., & Shiyatov, S. G. (1996). Tree-ring variables as proxy-climate indicators: Problems with low-frequency signals. In R. D. Jones, R. S. Bradley, & J. Jouzel (Eds.), *Climatic variations and forcing mechanisms of the last 2,000 years* (pp. 9–41). Berlin: Springer.

Briffa, K. R., Jones, P. D., Schweingruber, F. H., & Osborn, T. J. (1998). Influence of volcanic eruptions on northern hemisphere summer temperature over the past 600 years. *Nature, 393*(6684), 450–455.

Briffa, K. R., Osborn, T. J., Schweingruber, F. H., Harris, I. C., Jones, P. D., Shiyatov, S. G., & Vaganov, E. A. (2001). Low-frequency temperature variations from a northern tree-ring density network. *Journal of Geophysical Research-Atmospheres, 106*(D3), 2929–2941.

Briffa, K. R., Shishov, V. V., Melvin, T. M., Vaganov, E. A., Grudd, H., Hantemirov, R. M., Eronen, M., & Naurzbaev, M. M. (2008). Trends in recent temperature and radial tree growth spanning 2000 years across Northwest Eurasia. *Philosophical Transactions of the Royal Society B: Biological Sciences, 363*(1501), 2271–2284.

Diamond, J. (2005). *Collapse*. New York: Viking.

Fagan, B. M. (2000). *The little ice age. How climate made history 1300–1850*. New York: Basic Books.

Fagan, B. M. (2008). *The great warming. Climate change and the rise and fall of civilizations*. New York: Bloomsbury Press.

Ge, Q.-S., Zheng, J. Y., Hao, Z.-X., Shao, X.-M., Wang, W.-C., & Luterbacher, J. (2010). Temperature variation through 2000 years in China: An uncertainty analysis of reconstruction and regional difference. *Geophysical Research Letters, 37*, L03703. doi:10.1029/2009GL041281.

Gerondeau, C. (2010). *Climate: The great delusion: A study of the climatic, economic and political unrealities*. London: Stacey International.

Goddard, S. (2011). The hottest year on record? *SPPI reprint series* (20 pp). Retrieved February 15, 2011, from http://scienceandpublicpolicy.org/images/stories/papers/reprint/the_hottest_year_ever.pdf

Graybill, D. A., & Idso, S. B. (1993). Detecting the aerial fertilization effect atmospheric CO_2 enrichment in tree-ring chronologies. *Global Biochemical Cycles, 7*(1), 81–95.

Hansen, J. E., Ruedy, R., Sato, M., & Lo, K. (2010). Global surface temperature change. *Reviews of Geophysics, 48*, RG4004. doi:10.1029/2010RG000345.

Huang, S. P., Pollack, H. N., & Shen, P.-Y. (2008). A late quaternary climate reconstruction based on borehole heat flux data, borehole temperature data and the instrumental record. *Geophysical Research Letters, 35*, L13703. doi:10.1029/2008GL034187.

Idso, C., & Singer, S. F. (2009). *Climate change reconsidered: 2009 report of the Nongovernmental International Panel on Climate Change (NIPCC)*. Chicago: The Heartland Institute.

IPCC. (1990). *First assessment report of the Intergovernmental Panel on Climate Change*. Cambridge: Cambridge University Press.

IPCC WGI. (2001). *Climate change 2001. The scientific basis. Working group I contribution to the third assessment report of the Intergovernmental Panel on Climate Change*. Cambridge: Cambridge University Press.

IPCC WGI. (2007). *Climate change 2007. The physical science basis. Working group I contribution to the fourth assessment report of the Intergovernmental Panel on Climate Change*. Cambridge: Cambridge University Press.

Jones, P. D., Briffa, K. R., Barnett, T. P., & Tett, S. F. B. (1998). High-resolution palaeoclimatic records for the last millennium: Interpretation, integration and comparison with general circulation model control-run temperatures. *The Holocene, 8*(4), 455–471.

Jones, P. D., Parker, D. E., Osborn, T. J., & Briffa, K. R. (2010). Global and hemispheric temperature anomalies - Land and marine instrumental records. *Trends: A compendium of data on global change*. From http://cdiac.ornl.gov/trends/temp/jonescru/jones.html

Khim, B.-K., Yoon, H. I., Kang, C. Y., & Bahk, J. J. (2002). Unstable climate oscillations during the late Holocene in the Eastern Bransfield Basin, Antarctic Peninsula. *Quaternary Research, 58*(3), 234–245.

Ladurie, E. L. (1971). *Times of feast, times of famine. A history of climate since the year 1000* (B. Bray, Trans.). London: George Allen & Unwin.

Leijonhufvud, L., Wilson, R., Moberg, A., Söderberg, J., Retsö, D., & Söderlind, U. (2010). Five centuries of Stockholm winter/spring temperatures reconstructed from documentary evidence and instrumental observations. *Climatic Change, 101*(1–2), 109–141.

Levitt, S. D., & Dubner, S. J. (2009). *Super freakonomics: Global cooling, patriotic prostitutes, and why suicide bombers should buy life insurance*. New York: Harper Collins.

Ljungqvist, F. C. (2009). Temperature proxy records covering the last two millennia: A tabular and visual overview. *Geografiska Annaler, Series A, Physical Geography, 91*(1), 11–29.

Loehle, C. (2007). A 2000-year global temperature reconstruction based on non-tree-ring proxies. *Energy and Environment, 18*(7), 1049–1058.

Lomborg, B. (2007). *Cool it. The skeptical environmentalist's guide to global warming*. New York: Knopf.

Mann, M. E. (1998). *A study of ocean-atmosphere interaction and low-frequency variability of the climate system*. Unpublished PhD dissertation, Faculty of the Graduate School, Yale University.

Mann, M. E., Bradley, R. E., & Hughes, M. K. (1998). Global-scale temperature patterns and climate forcing over the past six centuries. *Nature, 392*(6678), 779–787.

Mann, M. E., Bradley, R. E., & Hughes, M. K. (1999). Northern hemisphere temperatures during the past millennium: Inferences, uncertainties and limitations. *Geophysical Research Letters, 26*(6), 759–762.

Mann, M. E., Zhang, Z., Hughes, M. K., Bradley, R. E., Miller, S. K., Rutherford, S., & Ni, F. (2008). Proxy-based reconstructions of hemispheric and global surface temperature variations over the past two millennia. *PNAS, 105*(36), 13252.

McIntyre, S., & McKitrick, R. (2003). Corrections to the Mann et al. (1998) Proxy data base and Northern hemisphere average temperature series. *Energy and Environment, 14*(6), 751–771.

McIntyre, S., & McKitrick, R. (2005a). Hockey sticks, principal components, and spurious significance. *Geophysical Research Letters, 32*, L03710. doi:10.1029/2004GL021750.

McIntyre, S., & McKitrick, R. (2005b). The M&M critique of MBH98 Northern hemisphere climate index: Update and implications. *Energy and Environment, 16*(1), 69–100.

McIntyre, S., & McKitrick, R. (2009). Proxy inconsistency and other problems in millennial paleoclimate reconstructions. *PNAS, 106*(6), E10. doi:10.1073/pnas.0812509106.

McKitrick, R. (2010, February 26). *Evidence submitted to the independent climate change email review (ICCER), Sir M. Russell, Chairman* (80 pp). Retrieved April 15, 2010, from http://sites.google.com/site/rossmckitrick/McKitrick-ICCER-Evidence.pdf

McMullen, C. P., & Jabbour, J. (Eds.). (2009). *Climate change 2009. Science compendium.* Nairobi: United Nations Environment Programme.

McShane, B. B., & Wyner, A. J. (2011). A statistical analysis of multiple temperature proxies: Are reconstructions of surface temperatures over the last 1000 years reliable? *Annals of Applied Statistics, 5*(1), 5–44.

Montford, A. W. (2010). *The hockey stick illusion: Climate and the corruption of science.* London: Stacey International.

Nelson, R. H. (2010). *The new holy wars: Economic religion versus environmental religion in contemporary America.* University Park: Pennsylvania State University Press.

Nova, I. J. (2010). *ClimateGate: Thirty years in the making.* Retrieved February 9, 2010, from http://joannenova.com.au/global-warming/climategate

Oreskes, N. (2004, December 3). The scientific consensus on climate change. *Science, 306*(5702), 1686.

Osborn, T. J., & Briffa, K. R. (2006). *Spatial extent of warm and cold conditions over the northern hemisphere since 800 AD. IGBP PAGES/World Data Center for Paleoclimatology Data Contribution Series #2006-009.* Boulder: NOAA/NCDC Paleoclimatology Program.

Pielke, R. A., Sr., Davey, C. A., Niyogi, D., Fall, S., Steinweg-Woods, J., Hubbard, K., Lin, X., Cai, M., Lim, Y.-K., Li, H., Nielsen-Gammon, J., Gallo, K., Hale, R., Mahmood, R., Foster, S., McNider, R. T., & Blanken, P. (2007). Unresolved issues with the assessment of multidecadal global land surface temperature trends. *Journal of Geophysical Research, 112*, D24S08. doi:10.1029/2006JD008229.

Plimer, I. (2009). *Heaven & earth. Global warming: The missing science.* Ballan: Connor Court Publishing.

Ran, L., Jiang, H., Knudsen, K. L., & Eiriksson, J. (2011). Diatom-based reconstruction of Palaeoceanographic changes on the North Icelandic Shelf during the last millennium. *Palaeogeography, Palaeoclimatology, Palaeoecology, 302*(1–2), 109–119.

Schulte, K.-M. (2008). Scientific consensus on climate change? *Energy & Environment, 19*(2), 281–286.

Schweingruber, F. H., & Briffa, K. R. (1996). Tree-ring density networks for climatic reconstruction. In P. D. Jones & R. S. Bradley (Eds.), *Climatic variations and forcing mechanisms* (NATO Series, Vol. I 41). Berlin: Springer.

Sussman, B. (2010). *Climategate. A veteran meteorologist exposes the global warming scam.* New York: WND Books.

U.S. Senate Committee on Environment and Public Works. (2010). *'Consensus' exposed: The CRU controversy* (84 pp). Retrieved April 15, 2010, from www.epw.senate.gov/inhofe

Vinther, B. M., Andersen, K. K., Jones, P. D., Briffa, K. R., & Cappelen, J. (2006). Extending Greenland temperature records into the late eighteenth century. *Journal of Geophysical Research, 111*, D11105. doi:10.1029/2005JD006810.

Vinther, B. M., Jones, P. D., Briffa, K. R., Clausen, H. B., Andersen, K. K., Dahl-Jensen, D., & Johnsen, S. J. (2010). Climatic signals in multiple highly resolved stable isotope records from Greenland. *Quaternary Science Reviews, 29*(3–4), 522–538.

Wahl, E. R., & Ammann, C. M. (2007). Robustness of the Mann, Bradley, Hughes reconstruction of Northern Hemisphere surface temperatures: Examination of criticisms based on the nature & processing of proxy climate evidence. *Climatic Change, 85*, 33–69.

Wahl, E. R., Ritson, D. M., & Ammann, C. M. (2006). Comment on "Reconstructing past climate from noisy data". *Science, 312*(5773), 529. doi:10.1123/science.1120866.

Wanliss, J. (2010). *Green dragon: Dominion not death.* Burke: Cornwall Alliance.

Wegman, E. J., Scott, D. W., & Said, Y. H. (2006). *Ad hoc committee report on the 'Hockey Stick' global climate reconstruction.* Retrieved October 16, 2009, from http:// republicans.energy-commerce.house.gov/108/home/07142006_Wegman_Report.pdf

Chapter 4
Emission Scenarios and Climate Modeling

Science is the belief in the ignorance of experts

– Richard Feynman, Nobel Physicist

Unfortunately, scientific research can be suitably slanted to support just about anything.

– William Dembski, mathematician and proponent of Intelligent Design, in The End of Christianity (Nashville, TN: BH Publishing, 2009, p. 161)

In previous chapters, we examined the historical climate record, both the instrumental record and that based on paleoclimatic reconstructions. Statistical evidence indicates that socioeconomic factors, such as population growth and increased economic activity, explained much of the recent rise in global surface temperatures. Since satellite-based temperatures were not similarly impacted by such non-climatic socioeconomic factors, it is clear that readings from surface weather stations are somehow contaminated and not entirely reliable as an indicator of global warming. It also turns out that there is no definitive proof that the Current Warm Period, especially the period 1977–1998, was the warmest humans have ever experienced; evidence to suggest that the Medieval Warm Period did not take place or that global temperatures during this period were below those of the CWP is weak at best. A great deal of uncertainty still needs to resolved before scientists can truly compare the current climate with that of the past, and it may never be possible to resolve enough uncertainty to declare that the current warming is indeed unprecedented by historical standards. In this chapter, we turn our focus from the past to the future. We examine projections of future climate by investigating how climate models are constructed and used to forecast a future climate scenario.

Climate scientists rely on computer models that link various components (often separate models) comprising the atmosphere, oceans and sometimes terrestrial systems, and sometimes even socioeconomic relations. These climate models are then used to predict future climate scenarios. Although the relationships in the climate

G.C. van Kooten, *Climate Change, Climate Science and Economics: Prospects for an Alternative Energy Future*, DOI 10.1007/978-94-007-4988-7_4,

models are based on known physical relationships, there are many parameters in these models that are set at the discretion of the modeler. Parameters are often chosen so that the model best replicates a known climate scenario or outcome – the selected parameters are those that appear to work best. This might explain why the hockey stick has been defended with such ferocity: The Medieval Warm Period poses a challenge to climate modelers. If human emissions of CO_2 are the cause of current warming, what factors explain the warming of the MWP?

The driving force behind climate models are human emissions of CO_2 and other greenhouse gas; greenhouse gas emissions constitute what is referred to as a 'forcing.' To get a handle on the extent of emissions in the future – to determine the forcing to use in the development of climate scenarios – the IPCC published a Special Report on Emission Scenarios in 2000 (IPCC 2000). The climate scenarios developed for the Fourth Assessment Report (IPCC WGI 2007) are based on these emission scenarios. (The 2000 scenarios updated the emission scenarios used in earlier IPCC reports.) Climate models include what are known as 'feedback effects' that amplify or reduce the impact of the initial emissions forcings on temperatures. If a feedback amplifies the impact of the initial forcing, it is taken to be positive (enhancing warming), while it is negative if it works opposite to the initial forcing and tends to lower temperatures. A major source of disagreement among climate scientists concerns the direction and extent of feedbacks (see Chap. 5), and thus the parameters used in models to represent them.

In this chapter, we begin by examining the emission scenarios that drive climate models. For the most part, emission scenarios assume large increases in global incomes and a dramatic closing of the gap in per capita incomes between rich and poor – something known as convergence. Even the worst case scenario has the poorest people on earth increasing their real per capita incomes by an incredible 16-fold. This raises serious questions concerning the ability of developing nations to adapt to future climate change; poor countries are assumed to be the hardest hit by global warming, but the scenarios that are used to predict such adverse consequences also assume that they will not be as poor as supposed. Rather, they are likely to be rich enough to adapt to the warmer climate regime. Thus, if action to avert climate change harms the Earth's poor, it may be better not to take action since, if the worst case scenarios develop, they are assumed to have the resources to cope with it.

Economists have also entered the climate modelling fray in ways other than those related to the development of emission scenarios. Based on their experience in economic modelling and forecasting such things as economic growth, inflation, unemployment, financial markets and so on, economists are increasingly wary of the validity of projections from climate models. As discussed in the latter part of the chapter, climate modelers often fail to follow proper criteria for insuring that their forecasts are scientifically sound.

4.1 Emission Scenarios

An important input into climate models' projections of future global temperatures is an emission scenario that indicates by how much greenhouse emissions will grow between now and 2100. Emission scenarios were developed in the *IPCC Special*

Report on Emission Scenarios (IPCC 2000); in all, 40 emission scenarios were developed along four storylines by six groups of modelers. In this section, we briefly describe the structure of the models used to generate the emission scenarios and their underlying assumptions.

4.1.1 Models

The development of any future emissions scenario relies on assumptions concerning future demographics, growth in GDP (or per capita incomes) in developed and developing countries (and thus rates of convergence in per capita GDP between rich and poor), the available resource base and lifestyles. There are also assumptions concerning the rate of technological change. To develop emission scenarios, it is necessary to feed the various assumptions through some sort of techno-socio-economic model that considers production, consumption and trade in goods and services, including energy, while addressing land use and other environmental impacts, including, importantly, emissions of CO_2 and other greenhouse gases. Because the structure of models varies greatly, a number of different models are employed to give some notion of the range of potential emissions that might be expected when the same assumptions concerning the emission drivers are employed.

The models that were used to obtain the emission scenarios are discussed in Appendix IV of the *Special Report on Emission Scenarios* (SRES) (IPCC 2000), where greater detail and model references are found. The six models are the following:

- The Asian Pacific Integrated Model (AIM) from the National Institute of Environmental Studies in Japan;
- The Atmospheric Stabilization Framework Model (ASF) by a U.S. Consulting firm;
- The Integrated Model to Assess the Greenhouse Effect (IMAGE) by the Institute for the Environment (RIVM) in the Netherlands and the Dutch government's Central Planning Bureau (CPB);
- The Multiregional Approach for Resource and Industry Allocation (MARIA) from the Science University of Tokyo, Japan;
- The Model for Energy Supply Strategy Alternatives and their General Environmental Impact (MESSAGE) developed by the International Institute of Applied Systems Analysis (IIASA) in Austria; and
- The Mini Climate Assessment Model (MiniCAM) from the Pacific Northwest National Laboratory (PNNL) in the United States.

The above are variants of integrated assessment models, which generally include a number of components. Not all models optimize some objective function, such as the sum of consumers' plus producers' surpluses (see Chap. 6), or take into account all of the interrelationships within an economy. Not all models include all of the same components, and there exist significant differences in how relationships within

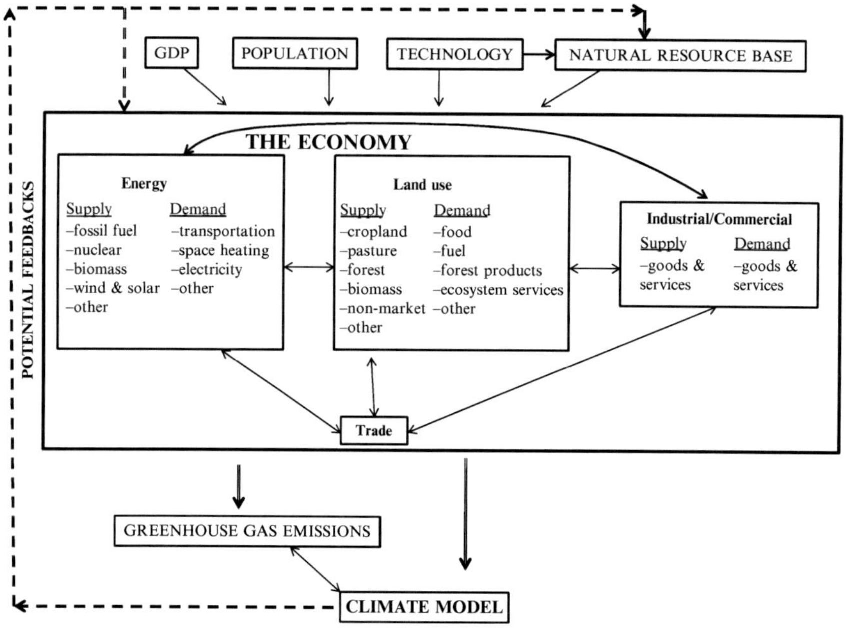

Fig. 4.1 Schematic of integrated assessment models used to generate emission scenarios

each of the components are constructed. In some models, production and consumption of goods and services are determined exogenously as fixed proportions of population or income, while in fewer instances production and consumption are endogenously determined by the interactions between supply and demand. In some cases, a model employs outputs from one or more external models as inputs; external models are sometimes integrated into the overarching model to some extent, if only to handle data interchange. One of the models used in MiniCam, for example, is Manne et al.'s (1995) Model for Evaluating Regional and Global Effects (MERGE) of greenhouse gas reduction policies, which is combined with a climate model. MERGE was originally designed as a stand-alone model (see also Manne and Richels 1992).

A schematic of the general structure of a model that is used to project future emissions is provided in Fig. 4.1. As indicated in the figure, the model components generally consist of the following:

- There is a set of initial assumptions relating to GDP, population, technology and the natural resource base. The models are static because there is no optimization over time – each time period stands on its own although results from one period may be carried into the next, possibly as a feedback (as indicated by the dashed line). For each period, however, GDP, population and technology are determined from outside the model.
- Each model has an energy sector as its most important component. The energy sector consists of demand and supply that are functions of GDP, population,

technology and available resources, with prices determined endogenously; but this is not true in all models, because required energy use may simply be exogenously driven by population, GDP and technology. Renewable, nuclear and fossil fuel sources of energy are treated separately, which facilitates the inclusion of technological change parameters; electricity is sometimes treated separately from transportation and space heating. There may or may not be links to the rest of the economy; and there may or may not be a link to a land use sector (see next point). Basic energy cost coefficients are specified in some models (MARIA), while shadow prices are calculated in others based on extraction costs (MESSAGE).

- Models also include a land use sector, partly to account for emissions from land use and land use change in forestry and agriculture, but also to take into account energy production on land. A land allocation model is required because production of food, wood products and energy, as well as ecosystem services and environmental amenities (e.g., wildlife habitat, recreation), compete for a limited land base.
- The rest of the economy needs to be represented because it deals with lifestyles that impact the demand for varying forms of energy and land uses. Although this sector is likely the largest and most complicated component of any economy, it is generally modeled with the least detail and, in some models, is simply represented by a few exogenously-given parameterizations.
- Trade is also permitted if the models are multi-regional in nature, which is usually the case.
- One component of the model needs to track greenhouse gas emissions of all kinds and not just carbon dioxide.
- A climate model may also be included to take into account feedbacks among greenhouse gases in determining emissions. A fully dynamic, integrated assessment model would require a climate model to take into account the impact of human activities and greenhouse gas emissions on future temperatures and, thus, economic damages (which are discussed in Chap. 7). However, none of the above models offers such a level of dynamic integration; rather, it is necessary to wait for the next generation of such models.

In summary, each of the six models used to generate emission scenarios has more or less detail on each of the components that one might like to include in a dynamic, integrated assessment model. It is beyond the scope of this discussion to examine models separately. Rather, the *Special Report* (IPCC 2000) should be consulted. It is worthwhile noting that, in the 'terms of reference' given to modelers by the IPCC, specific policies for reducing greenhouse gas emissions were not to be included in the modeling results. Thus, for example, any efforts to reduce emissions related to the Kyoto Protocol were not to be taken into account. In developing their scenarios, the modelers were given specific instructions regarding population growth, energy efficiency gains over time, convergence in per capita GDP between rich and poor countries, and so on. These assumptions have a tremendous impact on actual greenhouse gas emissions over the next century.

4.1.2 Emission Scenarios

We now consider the underlying assumptions of the various scenarios that were developed. As noted earlier, there are four storylines and 40 total emission scenarios. Rather than examine each of these, we briefly discuss the storylines and then provide the underlying assumptions for five commonly used scenarios.

There are four storylines that the IPCC uses to imagine the changes that will occur in the future – an attempt to project what the world will look like in 100 years. The storylines are referred to as A1, A2, B1 and B2, but within each there are alternative scenarios that provide more detail.[1] It is important to note that the storylines may not be at all realistic, but that the assumptions underlying them are the result of a political process that involves the United Nations and member countries. For example, the stated goals of rich countries are to maintain their own growth, while ensuring economic development in poor countries so that they will eventually 'catch up' with the richer nations – known as convergence. Developed countries cannot neglect their own citizens, so there is an emphasis on continued economic growth to prevent recessions, such as that of 2008–2009, which result in high unemployment and social unrest. At the same time, the millennium development goals of the United Nations and World Bank commit rich nations to aid in reducing poverty in developing countries. Needless to say, these political objectives are reflected in the emission storylines.

The A1 storyline assumes very rapid economic growth, a global population that peaks halfway through the 100-year forecast period, and rapid introduction of new and more efficient technologies. In addition, it is assumed that increased cultural and social ties, and ability to improve governance in developing countries, will lead to a marked reduction in regional income disparities that will, in turn, lead to a declining global population. Two main bylines in this storyline differ according to their assumptions regarding the underlying technological change as it relates to the energy sector. In the A1F1 scenario, the world community continues to rely on fossil fuels, but technology focuses on improving the efficiency in their use. In the A1T scenario, technological change results in the displacement of fossil fuels with non-fossil fuel sources of energy. There is also a balanced scenario (A1B) that provides an intermediate between the A1F1 and A1T scenarios, but it is not considered further here.

In the A2 storyline, the world continues to be a highly heterogeneous place. Convergence in per capita incomes and fertility rates is slow, the global population continues to grow throughout the twenty-first century, and technological change and its adoption are not homogeneous across regions.

The B1 storyline is much like A1 in that regional convergence in incomes and fertility rates, and technological improvements and their adoption, are rapid. World population peaks mid century and declines thereafter. In this storyline, economies shift from a material basis toward economies restructured to emphasize services and information. There are major developments in the introduction of clean and resource-efficient technologies. This storyline assumes global cooperation in solving economic, social and environmental problems, and such cooperation is taken to be effective. However, as required by the terms of reference, no additional climate initiatives are

[1] The discussion of the storylines is based on the Summary for Policymakers (IPCC 2000).

Table 4.1 Assumptions regarding GDP and population for selected IPCC emission scenarios

Year	A1F1	A1T	A2	B1	B2
Global population (×10⁹)					
1990	5.3	5.3	5.3	5.3	5.3
2020	7.6	7.6	8.2	7.6	7.6
2050	8.7	8.7	11.3	8.7	9.3
2100	7.1	7.0	15.1	7.0	10.4
Global GDP (10^{12} 1990 US dollars)					
1990	21	21	21	21	21
2020	53	57	41	53	51
2050	164	187	82	136	110
2100	525	550	243	328	235
Ratio of rich to poor per capita incomes					
1990	16.1	16.1	16.1	16.1	16.1
2020	7.5	6.2	9.4	8.4	7.7
2050	2.8	2.8	6.6	3.6	4.0
2100	1.5	1.6	4.2	1.8	3.0
Average global per capita income ($US1990)					
1990	3,962	3,962	3,962	3,962	3,962
2020	6,974	7,500	5,000	6,974	6,711
2050	18,851	21,494	7,257	15,632	11,828
2100	73,944	78,571	16,093	46,857	22,596
Average per capita income of poorest countries ($US1990)					
1990	246	246	246	246	246
2020	930	1,210	532	830	871
2050	6,732	7,677	1,099	4,342	2,957
2100	49,296	49,107	3,832	26,032	7,532

Source: Based on IPCC 2000: Special Report on Emissions Scenarios. Prepared by Working Group III of the Intergovernmental Panel on Climate Change, Table SPM-1a. Cambridge University Press

implemented to help reduce greenhouse gas emissions. Overall, this storyline is much more optimistic than the A1 storyline.

In the B2 storyline, solutions to economic, social and environmental problems are not addressed at the global level, but, rather, at the local or regional level. Global population continues to rise throughout, but at a rate below that of the A2 storyline. This results in somewhat better economic development than in the A2 case, but technological change is somewhat less rapid but more diverse than in the B1 and A1 storylines. In summary, while the B2 story is oriented towards enhanced environmental protection and social equality, the impetus for this occurs at the regional and not global scale.

To provide some indication of the emission scenarios upon which the IPCC bases it projections of future climate change, we provide an abbreviated overview of the A1F1, A1T, A2, B1 and B2 scenarios in Tables 4.1, 4.2, and 4.3. The situation that existed in 1990 is taken as the base year as this is the base year for emissions reduction under the Kyoto Protocol. In Table 4.1, underlying assumptions are provided regarding population growth, change in GDP, and, importantly, the rate of convergence in per capita incomes between rich and poor countries.

Table 4.2 Assumed rates of technical change in energy se for selected IPCC emission scenarios

Year	A1F1	A1T	A2	B1	B2
Final energy intensity (10^6 J per US$)					
1990	16.7	16.7	16.7	16.7	16.7
2020	9.4	8.7	12.1	8.8	8.5
2050	6.3	4.8	9.5	4.5	6.0
2100	3.0	2.3	5.9	1.4	4.0
Primary energy use (10^{18} J per year)					
1990	351	351	351	351	351
2020	669	649	595	606	566
2050	1,431	1,213	971	813	869
2100	2,073	2,021	1,717	514	1,357
Share of coal in primary energy (%)					
1990	24	24	24	24	24
2020	29	23	22	22	17
2050	33	10	30	21	10
2100	29	1	53	8	22
Share of zero carbon in primary energy (%)					
1990	18	18	18	18	18
2020	15	21	8	21	18
2050	19	43	18	30	30
2100	31	85	28	52	49

Source: Based on IPCC 2000: Special Report on Emissions Scenarios. Prepared by Working Group III of the Intergovernmental Panel on Climate Change, Table SPM-2a. Cambridge University Press

What is most striking about the underlying assumptions is the tremendous growth in per capita incomes of the poorest countries. Even in the worst case scenario in terms of the income prospects of the least advantaged (scenario A2), the average real (inflation-adjusted) per capita income of the poorest countries is projected to be some 15½ times greater by 2100 than it is now. Global *per capita incomes* are expected to rise by about 400 % in scenario A2, which translates into a low annual growth in per capita income of 1.1 %. While *total income* is expected to grow at an annual rate of 2.4 % (or by 1,100 %), per capita incomes grow much slower due primarily to a high rate of population growth. In the most optimistic scenarios (A1F1 and A1T), the per capita incomes of the poorest peoples are expected to increase by some 5.4 % annually throughout this century, or nearly 200-fold. Total global incomes are expected to rise by a factor of 25 or more, or more than 3.2 % annually. Per capita incomes are likewise expected to increase because fertility rates fall dramatically followed by a decline in global population.

Clearly, the assumptions underlying the emission scenarios and thereby the projections from climate models suggest that, at the time the worst impacts of a warmer climate are expected, there will be very few poor people. This effectively nullifies concerns about the negative impact of climate change on the poor; indeed, as evident from Table 4.1, it would appear that there will be no countries with insufficient resources to

Table 4.3 Expected emissions of carbon dioxide for selected IPCC emission scenarios[a]

Year	A1F1	A1T	A2	B1	B2
CO_2 from fossil fuels (tons × 10^9)[b]					
1990	22.0	22.0	22.0	22.0	22.0
2020	41.1	36.7	40.3	36.7	33.0
2050	84.7	45.1	60.5	42.9	41.1
2100	111.1	48.0	106.0	19.1	50.6
Cumulative[b]	7,803	3,806	6,501	3,626	4,253
CO_2 from land use (tons × 10^9)[b]					
1990	4.0	4.0	4.0	4.0	4.0
2020	5.5	1.1	4.4	2.2	0.0
2050	2.9	0.0	3.3	−1.5	−0.7
2100	−7.7	0.0	0.7	−3.7	−1.8
Cumulative[b]	224	227	326	−22	15
Cumulative carbon dioxide TOTAL (tons × 10^9)[b, c]					
1990–2100	2,189	1,068	1,862	983	1,164

Source: Based on IPCC 2000: Special Report on Emissions Scenarios. Prepared by Working Group III of the Intergovernmental Panel on Climate Change, Table SPM-3a. Cambridge University Press

[a] For other greenhouse gases, see (IPCC WGI 2001, pp. 17–18)

[b] Cumulative undiscounted CO_2 emissions for 1990–2100

[c] 1 gigaton = Gt = tons × 10^9

adapt to climate change. Further, as pointed out in Chap. 7, the expected increase in incomes will lead to great improvements in health and living conditions, so much so that, upon comparing the overall wellbeing of citizens with and without climate change, they are much better off avoiding costly action to mitigate climate change.

Assumptions about population and income growth for the B1 and B2 scenarios fall between those of the A1 and the A2 scenarios. Overall, however, it is clear that there is significant growth in global GDP and per capita incomes, but not unlike that experienced in the past. What is unusual in these scenarios is the rate of convergence between rich and poor. Quite rapid convergence is assumed, which has not been the experience of the past. Such a convergence can only come about by changes in governance structures in developing countries, and not through transfer of income between rich and poor, which has not worked in the past (see, e.g., Moyo 2009).

The scenarios vary greatly regarding energy use. Assumptions regarding energy use and expected changes in energy technologies are summarized in Table 4.2. These indicate that energy use will increase between 1990 and 2100 by 590% in the A1F1 scenario but by only 150% in the B1 scenario, although in the latter case it will first increase by more than that amount before declining. The share of coal in primary energy production is assumed to decline to 1% from 24% for the A2 case, but rise to 53% from 24% in the B1 case. The carbon component of energy production is thus expected to fall from 82% in 1990 to 15% by 2100 in A2 but only to 72% in B1.

Finally, the background economic assessment models for generating emissions also tell us something about the amount of carbon dioxide and other greenhouse gases that can be expected to enter the atmosphere. In Table 4.3, we provide only the emissions for CO_2 (and not CO_{2e}). These are derived from the models on the basis of economic activity, assumptions concerning technological change, and the economic drivers of land use and land-use change. In 1990, total annual carbon dioxide emissions from fossil fuels amounted to 22.0 Gt, with another 4 Gt coming from land use changes. By 2100, annual emissions from fossil fuel are expected to fall to 19.1 Gt in the B1 scenario, but rise to 111.1 Gt CO_2 in the A1F1 scenario – a fivefold increase. They are expected to more than double for the A1T and B2 scenarios and almost increase fivefold for the A2 case. With the exception of the A2 scenario, emissions of CO_2 from land use and land-use change are expected to level off or decline by 2100, although rates of decline vary across scenarios with the greatest reduction in emissions coming in the B2 scenario. Of course, these results are predicated on socioeconomic factors and technological changes that are inherently unpredictable, even with the use of the best possible models. That is the Achilles heel of modeling: the inherent and large unpredictability of the economic and social changes that might occur.

4.2 The Theory of Climate Modeling: Energy Balance

Climate models are essentially energy balance models. They calculate the amount of energy reaching the earth from the sun, absorbed by the oceans, the terrestrial surface and ecosystems, and the atmosphere, and transmitted back to space. The difference between what arrives and what leaves results in a warming that is sometimes erroneously (as argued later in this section) referred to as the 'greenhouse effect.' In this section, we present the basic notion of energy balance that underlies all climate models, followed by a discussion of feedback effects and the controversy this evokes. Then, in Sect. 4.3, we provide an in-depth discussion of the types of climate models and how they are constructed. Insights regarding the interpretation of the outcomes of climate models are provided in Sect. 4.4, while alternative explanations of climate change are left to Chap. 5.

4.2.1 Radiative Transfer

We can illustrate the concept of energy balance with a pot of water that is being heated on a stove.[2] The heat from burning natural gas is absorbed by the water (and pot) and constitutes a gain in energy, but, at the same time, the water loses heat to the surrounding air. Heat is lost because (i) as the water gets hotter, evaporation occurs at a faster rate and the energy required to change the liquid water into a gas

[2] This illustration comes from Spencer (2010).

Table 4.4 Disposition summary of the Sun's energy reaching earth

Item	Approximate proportion of incoming sunlight (%)
Absorbed by the earth and oceans	50
Scattered and reflected by clouds and aerosols	20
Absorbed by the atmosphere and clouds	20
Scattered from the atmosphere by air molecules	6
Reflected by the Earth's surfaces (albedo)	4
Total	**100**

Source: Author's calculation based on various sources

is lost to the air with the vapor – energy is always gained or lost when a phase change occurs. (ii) Convection of the surrounding air takes away heat, and (iii) the pot and water lose energy through infrared radiation. Indeed, objects almost always give off infrared radiation regardless of the temperature, although the amount of radiation given off rises as the object becomes warmer. As long as an object gains more heat than it loses, its temperature (stored energy) will rise; if it loses more heat than is gained, its temperature falls. This is what is meant by energy balance.

Now consider the Earth and its temperature. Although climate models consider the convection of air and the various phases of water (ice, liquid and vapor) as these affect cloud formation and the reflectivity of the globe's surface, the main determinant of the Earth's temperature is the energy balance – the energy absorbed by the planet and that radiated into space. If the energy received by the planet exceeds that leaving the planet, the temperature will inevitably rise. With the exception of the energy related to tides and an insignificant (from a global warming standpoint) amount of energy coming from deep within the Earth's crust (viz., geysers, volcanoes and natural radioactive decay), the energy received by Earth comes from the sun, while that emitted is the result of reflection or infrared radiation.

The sun's ultraviolet light has a wavelength of 0–380 nm (nm = 10^{-9} m), visible light from the sun has a wavelength of 380–760 nm, and the sun's infrared light has a wavelength of more than 760 nm; these account for 10.0, 44.8 and 45.2%, respectively, of the total radiation from the sun. It is mainly visible light that reaches the Earth's surface, however, with much of the light in the ultraviolent and infrared spectrum reflected by the Earth's atmosphere; indeed, about 30% of the light gets reflected or scattered back to space (see Table 4.4).

Averaged over the entire surface of the Earth, incoming energy represents about 342 W per square meter (W/m^2) of incident solar radiation, with some 168–175 W/m^2 actually reaching the surface. Outgoing longwave radiation emitted by the Earth and its atmosphere represents an energy loss to space of some 235 W/m^2 (or 68.7% of incoming solar energy). These numbers are approximations as they depend on a large number of factors, including the location of the Earth relative to the sun (i.e., the distance they are apart), solar activity, the Earth's reflectivity (or albedo), et cetera. The difference between the incoming (mainly shortwave) radiation and the outgoing (longwave) radiation is required to keep the Earth's temperature warm enough for life. In the greenhouse story, the sunlight reaching the Earth heats the ground, which then emits infrared light of a long wavelength, and this infrared radiation gets trapped by the greenhouse gases thereby causing the globe to warm.

Consider the physics involved in determining the radiative balance of a blackbody, which refers to any object (including a gaseous one) that does not reflect incoming radiation (e.g., coal), and then the radiative balance of the Earth. We begin with the Planck function which is an experimentally derived relation that provides the *intensity* of radiation given off by any blackbody (Wallace and Hobbs 2006, p. 117):

$$\mathrm{I}_{\mathrm{B}\omega}(\mathrm{T}) = \frac{b_1 \lambda^{-5}}{\pi (e^{b_2/\lambda \mathrm{T}} - 1)}, \tag{4.1}$$

where λ is wavelength measured in micrometers ($\mu\mathrm{m} = 10^{-6}$ m), T is absolute temperature in degrees Kelvin (K), $b_1 = 3.74 \times 10^{-16}$ W·m², and $b_2 = 0.0145$ m · K. By integrating the Planck function $\pi \mathrm{I}_{\mathrm{B}\omega}(\mathrm{T})$ over all wavelengths, we obtain the Stefan-Botzmann law that gives the flux density or irradiance from any blackbody (Pierrehumbert 2011):

$$F(\mathrm{T}) = \sigma \mathrm{T}^4 \mathrm{W/m}^2, \tag{4.2}$$

where

$$\sigma = \frac{2\pi^5 k^4}{15 c^2 h^3} = 5.67 \times 10^{-8}\,\mathrm{W} \cdot \mathrm{m}^{-2} \cdot \mathrm{K}^{-4} \tag{4.3}$$

is the Stefan-Boltzmann constant. In Eq. (4.3), *k* is the Boltzmann constant, *h* is the Planck constant and *c* is the speed of light.[3]

Now apply the notion of radiative balance to the Earth which, unlike a pure blackbody, is characterized by clouds that reflect light and an atmosphere that reflects and scatters light, as well as a so-called 'natural greenhouse effect.' Calculating the radiative balance is not straightforward in this case, because there are different ways to calculate the amount of solar energy that gets absorbed by the Earth's surface (e.g., see Gerlich and Tscheuschner 2009).

The surface area of a sphere is equal to $4\pi r^2$, where *r* is the radius of the sphere. However, only a proportion of the total area of the Earth's surface is exposed to the sun at any given time; indeed, about one-quarter of the Earth is directly exposed to the sun at any instant. Therefore, the amount of solar power that reaches the earth at any time is given by $S_0\, \pi\, r_e^2$, where S_0 is referred to as the solar energy flux and r_e

[3] The values of the constants are as follows: c=299,792,458 m/s; h=6.626068×10⁻³⁴ m²·kg/s; k=1.3806504×10⁻²³ J/K, where J refers to Joules (1 J=1 W·s) and K to Kelvin (−273.15 K=0 °C). The Boltzmann constant equals the gas constant (= 8.314472 J/K/mol, where mol refers to moles) divided by the Avogadro constant (= 6.626068×10²³ mol⁻¹), and links the macroscopic and microscopic worlds. Unlike *k* and *h*, however, σ is not a universal constant of physics as it depends on the geometry of the situation (see Gerlich and Tscheuschner 2009, p.21).

refers to the radius of the earth (6,378 km). (The overall surface of the Earth is approximately 5.114×10^{14} m^2.) The solar energy flux varies with the distance from the sun according to:

$$S_0 = T^4_{sun} \frac{r^2_{sun}}{r^2_{planet\ distance\ from\ sun}}. \tag{4.4}$$

Given that the average surface (but not core) temperature of the sun is 5,780°K, the radius of the sun is 695,500 km and the average distance from the earth to the sun is 149.6 million km, it is possible to determine $S_0 = 1{,}367$ W per m^2. This is the solar constant for the Earth, although it will vary throughout the year as the distance between the sun and the Earth varies. It also varies with the solar cycle – a fluctuation of –0.10 to +0.12 % (IPCC WGI 2007, p. 189).[4] Upon dividing by four to determine the incident solar radiation over the entire globe, we get the average radiation of about 342 W/m^2, as noted above.

If the Earth was a perfect blackbody, it would absorb energy from the sun according to $S_0\,\pi\,r_e^2$, and radiate energy according to the Stefan-Botzmann law (4.2), but then from the entire sphere and not just that part exposed to the sun. Equating these two relations gives:

$$S_0\pi r_e^2 = \sigma \mathrm{T}^4 4\pi r_e^2 \rightarrow S_0 = 4\sigma \mathrm{T}^4. \tag{4.5}$$

The Earth is not a blackbody, however, but has some reflective capacity – it is a 'greybody' that reflects incoming sunlight (see Table 4.4). This reflective capacity is referred to as albedo, which we denote *A*. Therefore, the fraction of solar energy that is absorbed is (1–*A*), and Eq. (4.5) therefore needs to be adjusted as follows:

$$S_0\,\pi r_e^2(1-A) = \sigma \mathrm{T}^4 4\pi r_e^2 \rightarrow (1-A)S_0 = 4\sigma \mathrm{T}^4. \tag{4.6}$$

Solving for T^4 yields:

$$\mathrm{T}^4 = \frac{S_0}{4\sigma}(1-A) \rightarrow \mathrm{T} = \sqrt[4]{\frac{S_0}{4\sigma}(1-A)}. \tag{4.7}$$

The albedo of the Earth is assumed to be $A = 0.3$ (see Table 4.4), so the effective temperature of the Earth would be T = 254.86 K = −18.3 °C. However, since the average temperature of the Earth is about 15 °C, it is argued that the atmosphere acts as a greenhouse that results in approximately 33 °C warming. If $A = 0.4$, then the effective temperature of the Earth would be –27.9 °C rather than –18.3 °C; and, if $A = 0.5$, it would be –38.8 °C. These are significant differences and indicate that an increase in cloud cover that increases albedo could have a significant temperature lowering effect, and vice versa if cloud (or snow/ice) cover declines.

---

[4] Other than the 11-year solar cycle, the IPCC does not identify any other event or cause that might contribute to solar flux. This is examined further in Chap. 5.

Given that the atmosphere absorbs infrared radiation that is emitted from the Earth, how much does get absorbed and what does this do to temperature? One way to find out is simply to use Eq. (4.7) with a temperature T=288 K, but include an atmosphere heat absorption parameter θ as follows:

$$\theta T^4 = \frac{S_0}{4\sigma}(1-A). \tag{4.8}$$

Using the current average global temperature, T=287.5 K, $A=0.3$, $S_0=1{,}367$ W/m^2 and $\sigma=56.7\times10^{-9}$ W/m^2/K^4, we calculate θ=0.61755, which is the infrared transmissivity of the atmosphere.[5] This does not address the impact of carbon dioxide on global warming, however.

To determine the impact of human emissions of CO_2 on climate, we proceed by first figuring out the feedback effects from any 'forcing,' in this case the forcing due to incoming solar flux. Then we want to examine the forcing from human emissions of CO_2 and consider the related feedbacks as it is the feedbacks that are likely the most important aspect of climate change.

Before examining feedbacks, we also want to point out that there is a non-radiative forcing associated with the interchange of heat between the oceans and the atmosphere – the exchange of energy between the oceans and atmosphere. Such non-radiative forcing is related to variations in the rate of evaporation of water from the ocean, a process that, in turn, is impacted by differences in the temperature between the ocean and atmosphere, changes in average wind speeds along the surface of the ocean, and by changes in the humidity of the air flowing across the surface waters. The non-radiative forcing is observed in the Pacific Decadal Oscillation and El Niño events, for example. In such events large amounts of energy may be transferred to the atmosphere from the ocean. These cause the temperature of the atmosphere to rise significantly, while ocean temperature declines by only a very small amount because the oceans are such an enormous sink. Non-radiative forcing confounds the carbon dioxide fingerprint on climate in ways that are difficult to measure or predict, and thus are impossible to include in climate models in any meaningful way.

4.2.2 Feedbacks

Suppose there is an increase in the net downward solar flux density or irradiance equal to d*S*. (The use of d*S* is typical in calculus to denote a very small, or infinitesimal, change in the solar flux, as opposed to Δ*S*, where Δ is a mathematical operator that refers to difference so that Δ*S* denotes a 'large' change in the solar

---

[5] More correctly, $\theta=\varepsilon\tau_a$, where ε is the infrared transmissivity of the atmosphere (=1 for a blackbody and 0.67 for water, e.g.) and τ_a is the optical depth [McGuffie and Henderson-Sellers (2005, pp.84–85) are unclear, but see Wallace and Hobbs (2006, p.130)].

flux.) Surface air temperature T_S would rise over time in response to this imbalance, until the increase in outgoing radiation again balances the incoming radiation. The temperature response is denoted dT_S.

We denote the total feedback factor by λ, as is done in the literature. The sensitivity of the Earth's surface temperature to radiative forcing *F* is $\lambda = dT_S/dF$, which is referred to as the *climate sensitivity* (IPCC WGI 2007, p. 133; Wallace and Hobbs 2006, p. 444). If there are auxillary variables, such as increased water vapor, snow and ice, and/or clouds that affect the climate, then the total derivative (dT_S/dF) would need to be expanded to take into account the other factors affecting the climate sensitivity. The expanded total derivative is written as:

$$\frac{dT_s}{dF} = \frac{\partial T_S}{\partial F} + \frac{\partial T_S}{\partial y_1}\frac{dy_1}{dF} + \frac{\partial T_S}{\partial y_2}\frac{dy_2}{dF} + \cdots + \frac{\partial T_S}{\partial y_n}\frac{dy_n}{dF} = \lambda, \tag{4.9}$$

where there are *n* auxillary parameters that could affect the climate sensitivity λ.[6] In the absence of any feedbacks, including Earth's albedo and feedbacks from the atmosphere,

$$\lambda_0 = \frac{\partial T_s}{\partial F} \approx \frac{\partial T_e}{\partial F}, \tag{4.10}$$

where T_e is the Earth's blackbody temperature. Both λ and λ_0 have dimension $K \cdot W^{-1} \cdot m^2$.

The changes in the auxillary variables in (4.9) are a consequence of their dependence on the change in radiative forcing, *F*, or the change in temperature that *F* causes. That is, the cryosphere albedo (reflection from ice and snow) and the amount of water vapor in the atmosphere, and cloud formation, are each functions of the radiative forcing – they are each a function of the additional energy entering into the atmosphere. Thus, for any y_i, $\frac{dy_i}{dF} = \frac{dy_i}{dT_S}\frac{dT_S}{dF}$, which, upon substituting into (4.9), yields:

$$\lambda = \frac{dT_S}{dF} = \frac{\partial T_S}{\partial F} + \frac{dT_S}{\partial F}\sum_i^n f_i, \tag{4.11}$$

where $f_i = \frac{\partial T_S}{\partial y_i}\frac{dy_i}{dT_S}$ are feedback factors or multipliers that have no dimension (Wallace and Hobbs 2006, pp. 444–445). The sign or direction of a feedback effect depends on the signs of the two derivatives. If both are positive or both are negative, then an initial change in the temperature caused by the radiative forcing *F* results in an overall positive feedback – a reinforcement of the temperature increase. If one derivative is negative while the other is positive, the effect of temperature changes working through y_i causes a negative feedback – a lowering of the initial rise in temperature.

[6] The choice of λ (climate sensitivity) here follows standard practice but should not be confused with the λ (wavelength) used in equation (4.1), which also follows standard practice.

The feedback factors are additive in Eq. (4.9) and (4.11), so the total feedback $f = \sum_{i=1}^{n} f_i$. Then, upon solving (4.11), we get

$$\frac{dT_S}{dF} = \frac{\frac{\partial T_S}{\partial F}}{1-f} = \frac{\lambda_0}{1-f} = \lambda. \tag{4.12}$$

The proportional gain due to the presence of climate feedbacks is then $f_B = \lambda/\lambda_0 = 1/(1-f)$, with $f < 1$.[7] Values of $f \geq 1$ are unrealistic, because these would result in infinite feedback gain and a spiraling ever upward temperature, which is clearly not about to happen.

Suppose we are interested in a doubling of atmospheric carbon dioxide. In that case, we are no longer dealing with incremental changes. Consider the situation where there is an initial change in temperature ΔT, regardless of its cause (although we can presume it to be the result of CO_2 forcing). Then the final change in temperature is determined by the feedbacks in the system as follows:

$$\Delta T_{Final} = \Delta T + \Delta T_{feedbacks}. \tag{4.13}$$

Using the preceding results, we can rewrite Eq. (4.13) as:

$$\Delta T_{Final} = f_B \Delta T, \tag{4.14}$$

where f_B is defined as above.

Feedbacks are crucial to the determination of the final rise in average global temperature. Suppose that there has been no change in global average temperatures for a very long time (e.g., as evidenced by the hockey stick story), but that there is now a radiative forcing F introduced into the system because of human emissions of CO_2. Over time, the system will come back to a new (higher) temperature equilibrium given by the following equation (McGuffie and Henderson-Sellers 2005, p. 39; Spencer 2010, p. 168)[8]:

$$\frac{d\Delta T}{dt} = \frac{1}{C_p}(F - \lambda \Delta T), \tag{4.15}$$

where ΔT is the temperature departure from equilibrium (°C or K), t represents time, F is the externally prescribed net radiative forcing (W/m^2), λ is the total feedback parameter (W/m^2/K), and C_p is the total heat capacity of the system and is dominated by the depth of the ocean.

The heat capacity of a system is simply given by its mass multiplied by its specific heat. For Earth, the heat capacity is determined by the oceans, which cover about

[7] Factor f_B cannot be added or multiplied, so it has no real mathematically-useful function.

[8] Notice that equation (4.15) is identical to (1.4) in McGuffie and Henderson-Sellers.

70% of the globe's area. Assume that energy gets absorbed in the top 70 m of the ocean (the globally averaged depth of the top or mixed layer). Then,

$$C_P = 0.7 p_w c_w d A_e = 1.05 \times 10^{23}\,\mathrm{JK^{-1}} \tag{4.16}$$

where p_w is the density of water (1,000 kg/m^3), c_w is the specific heat capacity of water (4.187 kJ kg^{-1} K^{-1}), d is depth (70 m), and A_e is the area of the Earth (511.4×10^{12} m^2) (McGuffie and Henderson-Sellers 2005, p. 84). If the depth of the ocean is taken to be 200 m, then the heat capacity of the system is $C_P = 3.00 \times 10^{23}$ J/K.

The radiative efficiency of CO_2 is 0.0155 W/m^2 per one part per million by volume increase in atmospheric concentration (Wallace and Hobbs 2006, p. 454). Thus, an increase in CO_2 concentration of 280 ppmv, which constitutes a doubling of carbon dioxide in the atmosphere over pre-industrial concentrations, should result in a radiative forcing of 4.34 W/m^2, but climate models find a slightly smaller forcing. The forcing attributable to a doubling of CO_2 quoted in the IPCC's Third Assessment Report (IPCC WGI 2001, p. 358) is 3.71 W/m^2, while that in the IPCC's Fourth Assessment Report (IPCC WGI 2007, p. 758) is somewhat lower at 3.67 W/m^2; the latter consists of the forcing due to longwave radiation (3.80 W/m^2) plus the negative forcing from shortwave radiation (−0.13 W/m^2). A more recent value estimated for CO_2 doubling using the Hadley Centre Coupled Model, version 3 – or HADCM3 climate model – provides a value of 3.74 W/m^2.[9] Clearly the forcing that is caused by human emissions of carbon dioxide is extremely small compared to the 235 W/m^2 of outgoing radiation from Earth back to space. The energy absorbed by carbon dioxide if there was double the amount of CO_2 in the atmosphere than in 1750 would amount to 1–2% of the estimated outgoing radiation, an amount well within the measurement error of the outgoing flux.

To provide some further notion of the radiative forcing of CO_2 and other factors, consider the estimates of radiative fluxes provided in Table 4.5 for the period 1750 (pre-industrial) to 2005. Notice that, over this period, the radiative forcing from long-lived greenhouse gases (LLGHGs), including CO_2, is 2.63 W/m^2, but that it is somewhat less for all anthropogenic factors (1.6 W/m^2). This is because there are several negative forcers such as aerosols in the atmosphere that reflect sunlight (while soot from burning results in black carbon that falls on snow, reducing its albedo and contributing to warming). The eruption of Mount Pinatubo in 1991 led to a forcing of about −0.4 W/m^2, temporarily cooling the Earth by approximately 0.5 °C (McGuffie and Henderson-Sellers 2005, p. 32).

Now, in Eq. (4.15), we can set the left-hand-side to zero by assuming no further change in temperature over time. Then we have $F = \lambda\, \Delta T$, or $\lambda = F/\Delta T$. From the HadCRUT3 temperature series, we find that $\Delta T = 0.9$ °C since 1850, while the overall anthropogenic forcing has been 1.6 W/m^2 since 1750; therefore, $\lambda = 1.78$ W m^{-2} K^{-1}, which is a negative feedback as it dampens the increase in temperature from anthropogenic forcing. Spencer (2010) employs Eq. (4.15) as a climate model, along with

[9] See http://unfccc.int/resource/brazil/climate.html (viewed June 17, 2010).

empirical data from satellites and assumptions of ocean absorption of energy (also based on empirical data), to demonstrate how various feedbacks impact temperatures over time. We examine this equation in more detail at the end of Sect. 4.3.

Finally, it is interesting to determine just how fast the Earth's temperature would drop should the energy available from the sun be unavailable. The following equation can be used to determine this rate of change (Wallace and Hobbs 2006, p. 120):

$$\frac{dT}{dt} = \frac{1}{C_P} 4\pi r^2 \sigma T_e^4 = \frac{A_e \sigma T_e^4}{C_P}. \tag{4.17}$$

Recall that J = W·s, dT/dt = 1.9×10^{-6} K/s. Then, without the sun, the Earth's temperature would decline at a rate of approximately 5 °C per month.

Our interest, however, is in knowing the extent to which radiative forcing from human emissions of CO_2 will result in global warming. In particular, we want to know if there is any scientific uncertainty regarding the link between human emissions of greenhouse gases and global warming, because uncertainty regarding the science has important implications for economic policy.

4.2.3 The Role of Carbon Dioxide

We begin by considering the role of carbon dioxide and other greenhouse gases as a driver (forcer) of climate change. It is clearly politically incorrect to question the belief that higher levels of atmospheric CO_2 will result in potentially catastrophic global warming, but not doing so would be contrary to the spirit of scientific inquiry. Hence, we begin by examining some recent studies that question the dominant role of greenhouse gases in forcing climate change. Clearly, as evident from Table 4.5, CO_2 is a minor contributor to climate change[10]; the argument is that the rise in CO_2 causes an initial warming that triggers greater evaporation of water from the oceans, thereby increasing the height of moist air in the atmosphere. The climate models predict that, as the planet warms, the layer of moist air expands upwards into the cool dry air above, and that this will occur firstly over the tropics to a height of about 10 km. This is where the initial rise in temperature will be observed. Hence, it is this increase in the atmosphere of the potent greenhouse gas water vapor that results in significant warming – the initial CO_2-induced warming is insignificant by comparison.

As discussed in the latter part of Sect. 4.3 below, this positive feedback from the initial CO_2 forcing is important in climate models. However, while increased

[10] Pierrehumbert (2011) argues to the contrary, attributing one-third of the greenhouse effect to the small amount of carbon dioxide in the atmosphere. But he ignores entirely cloud albedo, indicating that a reduction in atmospheric CO_2 would "ultimately spiral Earth into a globally glaciated snowball effect" as clouds disappeared, while rising CO_2 would do the opposite.

amounts of water vapor contribute to warming, some of the water vapor may form clouds. Seen from below, clouds reflect infrared radiation back to Earth (a warming effect), but seen from above they reflect infrared radiation into space (an albedo feedback that reduces temperatures). It is an empirical question as to which of these processes is of greater importance. As pointed out by the IPCC WGI (2007, p. 637), "... the sign of the climate change radiative feedback associated with the combined effects of dynamical and temperature changes on extratropical clouds is still unknown."

Does CO_2 lead to a greenhouse effect? This question might seem startling, but several recent papers by physicists and engineers question whether CO_2 even has a significant role in forcing climate change (Essenhigh 2006; Gerlich and Tscheuschner 2009; Miskolczi 2010). As noted earlier and again in the next section, climate models are essentially complicated energy balance models, and the mechanism of global warming is termed the 'greenhouse effect.' Mainly visible light from the sun reaches and heats the Earth's surface, which then emits infrared light of a long wavelength. It is this infrared radiation that gets trapped by the greenhouse gases causing the globe to warm. At least this is the story. To what extent is it true?

Before answering this question, it is worth pointing out that:

> [I]t is *impossible* to find a book on non-equilibrium thermodynamics or radiation transfer where this effect is derived from first principles. ... [The] 'atmospheric greenhouse effect' does not appear in any fundamental work of thermodynamics, in any fundamental work of physical kinetics, [or] in any fundamental work of radiation theory (Gerlich and Tscheuschner 2009, pp. 37, 44; emphasis in the original).

In a real greenhouse (or glass house), the extent to which sunlight of various wavelengths is permitted through the glass determines the extent and speed at which the ground of the greenhouse warms. The ground acts as a black body that emits radiation of wavelengths above 3,000 nm, which are trapped by the glass. The net radiation from a plane surface of the ground is given by Eq. (4.2) – the Stefan-Botzmann law for radiation from a black body.

But there is something else at play. The radiation from the ground cannot possibly explain the warming of the greenhouse. Rather, the warming occurs due to the suppression of convection. That is, hindered heat transmission due to the radiation trapping effect of the glass is less important than the role of convection. A greenhouse warms because there is no turbulent heat loss or cooling effect, not because of trapped infrared light. This has been demonstrated experimentally as early as 1909 (Gerlich and Tscheuschner 2009; Wood 1909).

Upon examining radiative balance equations and applying them to the atmospheres of Venus, Mars and the Earth, Miskolczi (2007) likewise concludes that an unconstrained greenhouse effect is not possible because it contradicts the energy balance equations. He also finds a significantly reduced warming sensitivity to optical depth perturbations and that, on a global scale, "there cannot be any direct water vapor feedback mechanism, working against the total energy balance requirement of the system."

What is going on? We saw earlier that there was a so-called 'natural greenhouse effect' that, for given parameterizations of things like albedo, had to result in approximately 33 °C warming. This conclusion was reached because, from Eq. (4.8), where $T = \sqrt[4]{\frac{S_0}{4\sigma}(1-A)}$, and an albedo assumed to be $A=0.3$, the effective temperature of the Earth would be T=254.86 K=−18.3 °C. Given the observed average temperature of the Earth is about 15 °C, the natural greenhouse effect must be 33 °C. This result is not as straightforward as it first appears, however, because there are different ways to calculate the amount of solar energy that gets absorbed by the Earth's surface and how to calculate the average temperature of the Earth.

Gerlich and Tscheuschner (2009) argue that the derivation from first principles of expression (4.8) – the above formula for temperature – is wrong because the integration for deriving the average temperature is not correct. Let's see why this might be the case.

To determine the average temperature, it is necessary to integrate temperatures across the Earth's surface. Think of it this way: Given that the Earth is a sphere, the energy reaching the Earth's surface from a distant source like the sun will be uneven. This needs to be taken into account by integrating across all the different temperatures. In the derivation of Eq. (4.8), the integration is done over T^4 values rather than the T values. However, integrating over temperatures is more appropriate than integrating over the quadratic of temperatures, and doing so leads to a different result. Upon integrating over T rather than T^4, the correct temperature is determined as follows:

$$T' = \frac{2}{5}\sqrt[4]{\frac{S_0}{\sigma}(1-A)}. \tag{4.18}$$

Clearly,

$$T' = \frac{2\sqrt{2}}{5}T \rightarrow T' < T. \tag{4.19}$$

In this case, the average temperature of the Earth would be T′=144.17 K= −129.0 °C if $A=0.3$ and T′=−140.6 °C if $A=0.5$. Clearly the natural greenhouse effect cannot be on the order of 144 °C to 155 °C! If Gerlich and Tscheuschner (2009) are correct, then there could be something fundamentally wrong in the way the physics is applied in climate models.

Essenhigh (2006) provides an analytic solution to the original (1905–1906) Schuster-Schwartzchild (S-S) Integral Equations of Radiative Transfer. These equations relate temperature, atmospheric pressure and density across different vertical profiles of the atmosphere – a one-dimensional climate model (reviewed in Sect. 4.3). Essenhigh's solutions to the S-S equations correspond extremely well with empirical data on temperature, pressure and density for altitudes ranging from ground level to 30 km, and for the corresponding molecular ratios of atmospheric water vapor to carbon dioxide, which ranges from 25:1 at ground level to one-to-one at 10 km and higher. The median ratio for the atmosphere is 4:1.

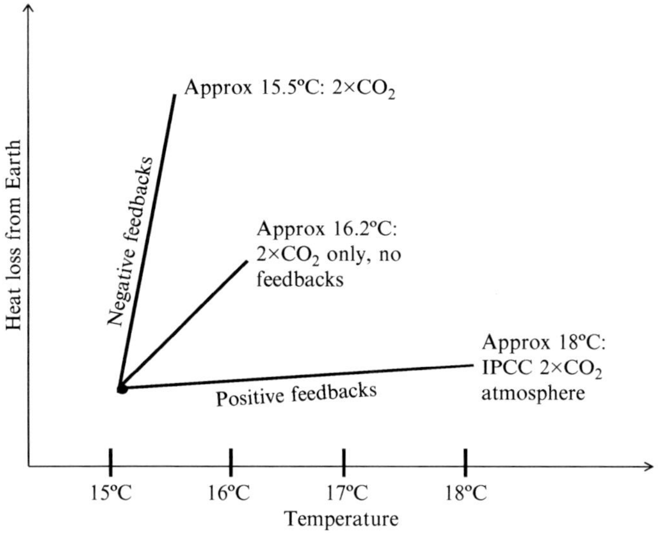

Fig. 4.2 Importance of feedbacks in determining temperatures under 2×CO2 atm, with current temperature of approximately 15 °C

Climate models solve the S-S equations numerically, which is of limited and arguable value, particularly if prediction is outside the range of observations or "agreement between the prediction and experiment is obtained by curve fitting, using parameter-value selection and adjustment which does not correspond to true model validation" (Essenhigh 2006, p. 1058). As discussed in Sect. 4.3, this is a particular problem in climate modeling.

The solutions to the S-S equations of radiative transfer do not support the concept of a CO_2 forcing that is then amplified by an increase in water vapor. Rather, they support an alternative conclusion: "it is the rising temperature that is driving up the carbon dioxide [in the atmosphere] by reducing the temperature-dependent saturation level in the sea and thus increasing the natural emissions" (Essenhigh 2006, p. 1067).

Finally, there is the question of CO_2 residency in the atmosphere, which is often assumed to be on the order of 100 years (or more). Given that there are some 750 Gt of carbon dioxide in the atmosphere and the annual flux from natural sources, primarily the oceans and terrestrial ecosystems, is about 150 Gt, this suggests that the average lifetime of a CO_2 molecule is about 5 years. Further, fossil-fuel emissions of CO_2 amount to some 5–6 Gt per year, or about 4 % of the CO_2 flux from natural sources. The human contribution to atmospheric carbon dioxide is marginal at best and is certainly swamped by the contribution to warming from water vapor.

4.2.4 Carbon Dioxide, Feedbacks and Cloud Formation

The effect and importance of feedbacks can be illustrated with the aid of Fig. 4.2. As carbon dioxide increases in the atmosphere, there is a small radiative forcing

Table 4.5 Global mean radiative forcings, 1750–2005, and efficacy parameters

Source	RF (W/m^2)	Range (W/m^2)	Efficacy
Combined anthropogenic forcing	+1.6	[+0.6, +2.4]	
LLGHGs, of which	+2.63	[+2.37, +2.89]	
– CO_2	*+1.66*	*[+1.49, +1.83]*	*1.0*
– CH_4	*+0.48*	*[+0.43, +0.53]*	*4.0–1.2*
– N_2O	*+0.16*	*[+0.04, +0.18]*	
– *Halocarbons*	*+0.34*	*[+0.31, +0.37]*	
Stratosphere ozone	–0.05	[–0.15, +0.10]	0.5–2.0
Troposphere ozone	+0.35	[+0.34, +0.38]	0.5–2.0
Stratosphere water vapor	+0.07	[+0.02, +0.12]	~1.0
Total direct aerosols	–0.5	[–0.9, –0.1]	0.7–1.1
Cloud albedo effect	–0.7	[–1.1, +0.4]	1.0–2.0
Surface albedo (land use)	–0.2	[–0.4, 0]	
Surface albedo (black carbon aerosol on snow)	+0.1	[0, +0.2]	1.3 [a]
Persistent linear contrails	+0.01	[–0.007, +0.02]	~0.6
Solar irradiance	+0.12	[–0.06, +0.18]	0.7–1.0

Source: Adapted from Climate Change (2007, pp. 203–204) : The Physical Science Basis. Working Group I Contribution to the Fourth Assessment Report of the Intergovernmental Panel on Climate Change, Figure 2.20. Cambridge University Press

[a]Indicates no consensus

(about 1.66 W/m^2 from Table 4.5) that causes a rise in temperature that, in turn, triggers feedbacks that might enhance the initial warming or reduce it. (The IPCC's estimates of the forcings from various sources are provided in Table 4.5, but these are based on results from climate models.) The impact of feedbacks can be illustrated with the aid of Fig. 4.2. Without any feedbacks, the globe might warm to 16.2 °C as a result of a $2 \times CO_2$ atm, but with positive feedbacks there is less heat loss from the Earth (the radiative forcings are positive) and temperature might rise by 3 °C rather than only 1.2 °C. However, if there are negative feedbacks in the system, the increase in temperature might be much less as these feedbacks result in the loss of heat to space. Then the temperature might only rise to 15.5 °C, say, rather than 16.2 °C.

Climate scientists consider two feedbacks. The first concerns what happens when carbon dioxide causes the initial warming. As the amount of CO_2 in the atmosphere increases, the atmosphere warms and is capable of holding more water vapor, which is itself a potent greenhouse gas. However, greater water vapor in the atmosphere causes more clouds (and ice crystals) to form, thereby reflecting more of the sun's rays into space and cooling the Earth. CO_2-induced warming also causes ice to melt, thereby reducing the Earth's albedo – more sunlight is absorbed rather than reflected back into space, thereby heating the globe. Climate scientists argue that, once all the feedbacks involving water vapor, clouds and ice-albedo are taken into account, the overall impact will be enhanced warming. Thus, climate models project warming for a doubling of CO_2 of 1.5–4.5 °C. We denote this feedback effect as f_1.

There is a second feedback that builds on the first; that is, without the first feedback, the second does not materialize. This second feedback occurs because warming results in the release of additional greenhouse gases, over and above water vapor, the reduction in ice albedo and some other auxiliary variables (some of which are included in Table 4.5). Higher temperatures that result from anthropogenic emissions of CO_2 and increased levels of atmospheric water vapor (and reduced albedo) will, in turn, lead to greater emissions of CO_2 from non-anthropogenic sources (e.g., release of carbon stored in wetlands that dry up), release of methane, CH_4, as the tundra melts, and so on (Matthews and Keith 2007; Scheffer et al. 2006).[11] The second feedback is due, in other words, to the release of additional greenhouse gases that the initial CO_2-induced warming triggers. These emissions lead to further warming, which then results in more water vapor that results in even more warming, and so on. We denote this feedback gain (sum of independent feedbacks) related to other greenhouse gases (second feedback) by f_2.

As noted above, the climate sensitivity parameter in the absence of feedback effects is $\lambda_0=\partial T_S/\partial F$, where T_S denotes surface temperature and λ_0 has dimension $K\,W^{-1}\,m^2$. This holds true for any forcing factor. Given our interest in carbon dioxide, it might be useful to convert the climate sensitivity parameter to dimension K per ppm; in that case, the climate sensitivity parameter would provide the instantaneous change in temperature associated with the forcing from a very small increase in CO_2, say one part per million by volume increase in atmospheric concentration. Since the radiative efficiency of carbon dioxide has dimension W/m^2 per ppm(CO_2), we could multiply λ_0 by the radiative efficiency to get a climate sensitivity parameter, which we denote s_0, measured in °C per ppm(CO_2). Using an earlier value of radiative efficiency, for example, $s_0=0.0155$ °C/ppm(CO_2); but this value leads to a temperature rise of more than 3.5 °C for a doubling of CO_2, without feedbacks, which is considered too high once feedbacks are introduced. Indeed, climate scientists project a climate sensitivity of 1–1.2 °C for a doubling of atmospheric CO_2 and an increase of 1.5–4.5 °C (or more if there are second-order feedbacks).

Martin Weitzman (2009) simply assumes $s_{2\times CO2}=1.2$ °C and then employs the following approximation to convert atmospheric CO_2 into temperature increases:

$$\Delta T = \frac{s_{2\times CO2}}{\ln(2)} \Delta \ln(CO_2), \tag{4.20}$$

where ΔT again refers to a discrete change in temperature and ln is the natural logarithm. Notice that, for a doubling of atmospheric CO_2 from 250 to 500 ppm, say, $\Delta\ln(CO_2)=\ln(500\text{ ppm})-\ln(250\text{ ppm})=\ln(500\text{ ppm}/250\text{ ppm})=\ln(2)$. The level of CO_2 in the atmosphere has increased from about 275 ppm in 1750 to 390 ppm today (Fig. 3.1); then, using (4.20), ΔT associated with the increase in carbon dioxide and ignoring feedbacks equals 0.6 °C.

[11] It should be noted that this is extremely speculative; indeed, Beilman et al. (2009) find that thawing of the permafrost might actually promote an increase in peat carbon sequestration.

With feedbacks, formula (4.20) needs to be adjusted as follows:

$$\Delta T = \frac{\frac{s_{2\times CO2}}{1-f_1}}{\ln(2)} \Delta \ln(CO_2) \times \frac{1}{1-f_2} = \frac{\Delta \ln(CO_2)}{\ln(2)} \times \frac{s_{2\times CO2}}{(1-f_1)(1-f_2)}, \tag{4.21}$$

where f_1 is the feedback involving water vapor, clouds and albedo, and f_2 the secondary feedback from the release of additional greenhouse gases (e.g., due to melting permafrost). The value of f_1 is determined from climate models – it is not based on empirical evidence but on the hypotheses built into the climate models, just like the forcings in Table 4.5. The climate models predict a warming of 1.5–4.5 °C rather than the 1.2 °C warming associated a CO_2 forcing in the absence of (positive) feedbacks. This is the extent of the feedback, and, using the relation $f_1 = (\Delta T - s_{2\times CO2})/\Delta T$, one finds that $f_1 = 0.20$ to $f_1 = 0.73$, depending on whether the projected change in temperature is 1.5 or 4.5 °C, respectively. The value of f_2 has to be derived from other sources. For example, based on ice core data, scientists have found the second feedback parameter to vary depending on whether the impact is local or hemispheric; thus, Torn and Harte (2006) report the following derived values: $f_{2,\text{local}} \approx 0.042$ and $f_{2,\text{hemispheric}} \approx 0.067$.

Clearly, there is much disagreement concerning these parameters. For example, the temperature change from a doubling of atmospheric CO_2 (i.e., the sensitivity parameter) is likely to be no more than 1.0 °C rather than 1.2 °C, partly because of the higher value underestimates the urban heat island effect in the temperature record (Idso and Singer 2009, pp. 95–106; see Chap. 2).

There is greater disagreement about the role of clouds. As pointed out in one text devoted to the subject, "a complete in-context understanding of cloud micro-physics including dynamic, electrical and chemical effects must await some sort of grand synthesis, an elusive and distant goal even from the point of view of presently available models" (Pruppacher and Klett 1997, p. 9). The IPCC WGI (2007, pp. 200–206) downplays the direct role of clouds, arguing that aerosols result in a greater cloud albedo, which leads to slight cooling; however, the overall effect of clouds is to trap heat causing greater warming, with lack of clouds resulting in cooling. Spencer (2008, 2010), Gray and Schwartz (2010), Lindzen and Choi (2009, 2011), Spencer and Braswell (2008, 2010), Lindzen et al. (2001), and others (see Chap. 5) provide evidence that the climate system has a negative feedback effect on carbon dioxide-caused warming.[12] This results, in part, because, as the atmosphere contains more water vapor, more clouds are formed and these reflect sunlight back into space (the albedo effect). Overall, the albedo effect of clouds more than offsets the greenhouse effect of the water in clouds. The overall effect, according to this view, is that a doubling of atmospheric CO_2 will produce a warming of no more than 0.5 °C, well

[12] Dessler (2010) refutes Spencer and Braswell's (2010) notion that clouds provide a strong negative temperature feedback, but Spencer and Braswell (2011) demonstrate the correctness of their position, pointing out that the Earth loses more energy than indicated in climate models.

within natural variation (see Fig. 4.2). This view is contrary to that of the IPCC (e.g., Dessler et al. 2008).[13]

The forgoing analyses do not dispute the fact that, without the sun and an atmosphere, the Earth would be a much, much colder place to live. It also does not refute the fact that carbon dioxide absorbs some of the outgoing infrared radiation, thereby warming the atmosphere. The mathematical analyses provided in the foregoing section only question the underlying physics that climate scientists use to determine radiative flux, and the relations subsequently employed in climate models.

4.3 Climate Models

The climate models that scientists use are generally so complex that they take many days and sometimes many months to solve on a 'super computer.' This is mainly because scientists use climate models to project conditions using a very short time step and a fine grid that includes several atmospheric and ocean layers. However, as discussed in the first section, there are much simpler models that can be are used to investigate various aspects of climate change. Indeed, as shown in the second subsection, an energy balance model based on Eq. (4.15) can be used to predict future average global temperatures and investigate various aspects related to such predictions.

4.3.1 Climate Models: A Brief Description

There are two approaches to climate modeling, known as 'bottom up' and 'top down.' Bottom-up models are much more detailed, and include the physical and chemical equations that affect ocean and atmospheric circulation and other relationships influencing the climate system. Bottom-up models tend to be multi-layered, with information passed from lower levels to higher ones. Any uncertainty at one level is thus passed onto another level and, due to the model structure, subsequently gets amplified. This makes the final model outcomes – the macro scale outcomes – less reliable than outcomes at lower levels – the micro scale outcomes. The extent to which this is a problem is difficult to judge and will vary from one model to another and from one simulation to another.

Many bottom-up climate models that are used to predict temperatures (and precipitation) 20, 50 or 100 years into the future started out as computer models for

[13]A discussion can be found at http://www.drroyspencer.com/2009/02/what-about-the-clouds-andy/ (posted February 21, 2009, viewed September 2, 2010). There Roy Spencer points out that there are two components to the energy radiative balance: (1) absorbed/reflected solar, shortwave (SW) radiation; and (2) emitted infrared, long-wave (LW) radiation. He argues that Dessler et al. (2008) and the IPCC models only take into account the LW radiation (in which case he gets identical results). However, they ignore SW radiation, which leads to the negative as opposed to positive feedback from water vapor.

forecasting weather, or at least have much in common with short-term weather forecasting models. In many cases, more layers have been added. This provides an explanation for the inherent problems of reliability: "The computer models that are used for generating projections of future climate are in many respects similar to the models used for weather prediction on a daily-to-biweekly time-scale. Yet, the reliability of long-term climate change projections is much harder to estimate than that of weather forecasts" (Räisänen 2007).

Top-down climate models employ an entirely different methodological approach. They do not specify detailed physical relationships, because the objective is not to simulate micro-level relations, but, rather, to simulate the overall behavior of the climate system.

Climate models can be classified into four types depending on their complexity and the way they address five components of the climate system: radiation, surface processes, chemistry, dynamics and resolution (McGuffie and Henderson-Sellers 2005, pp. 49–63). (i) Radiation refers to the way models handle input and absorption of solar radiation by the atmosphere and oceans, and subsequent emission of infrared radiation to space. (ii) Surface processes refer to the factors affecting albedo and the exchanges of energy and moisture between the Earth's surface/oceans and the atmosphere. (iii) The chemical composition of the atmosphere and how it is affected by, for example, carbon exchanges between the ocean and atmosphere and Earth and atmosphere are the chemistry component of models. (iv) The dynamic component of climate models deals with the movement of energy around the globe by winds and ocean currents, both horizontally and vertically. (v) Finally, the resolution of models is related to four dimensions – the three dimensions of space plus time.

Before discussing resolution, it is well to point out and briefly describe the four types of models that climate scientists employ.

1. Energy balance models (EBMs) constitute the simplest genre of climate models. Yet, they serve an important purpose because they enable climate scientists to investigate particular aspects of the climate, such as the effect on temperature due solely to CO_2 in the absence of feedbacks. That is, they facilitate the development of key parameters that are also used in more complex models. Zero-dimensional EBMs consider the earth to be a single object in space; the energy from the sun, radiation from earth and the role of the atmosphere (including clouds and water vapor) are modeled in detail. A one-dimensional EBM takes into account energy exchanges across latitudes (e.g., by including an eddy-diffusion process), thereby permitting scientists to study the effect of changes in albedo caused by snow and ice, for example.
2. One-dimensional, radiative-convective (RC) climate models include altitude, but do not permit horizontal exchanges of energy. In essence, they treat the entire surface of the earth as a single grid cell with a column representing the atmosphere and the land/ocean surface with its various characteristics (water or depth of water, types of vegetation, etc.) considered as one. They "operate under the constraints that at the top of the atmosphere there must be a balance of shortwave and longwave fluxes, and that surface energy gained by radiation equals that lost by convection" (McGuffie and Henderson-Sellers 2005, p. 53).

3. Dimensionally-constrained climate models add a horizontal dimension to the vertical column – combining the latitudinal dimension of the one-dimension energy balance models with the vertical one of the radiative-convective models. Energy transport across latitudes is based on statistical summaries and eddy diffusion processes such as those in the EBMs. The resulting two-dimension, statistical dynamic models are a precursor to a class of models referred to as Earth modeling with intermediate complexity (EMIC). EMIC models lack the complexity of General Circulation Models (GCMs), considered under point 4, because they incorporate only two rather than three physical dimensions, although they do include a time element. This is why their complexity is 'intermediate.' However, they sacrifice greater complexity so that they can incorporate human systems and their impacts.
4. General circulation models constitute the most complex class of climate models. These are three dimensional in the true sense. As noted in Chap. 2, the globe's surface can be divided into grids that, in GCMs, have a resolution as small as 2° latitude by 2° longitude, although coarser grids of 5° latitude by 5° longitude (and 5° latitude × 2° longitude) are also employed, depending on purpose, modeling design, and computing and data constraints. Thus there are over 32,000 grids in some models. For each grid, there is a vertical column similar to that used in radiative-convective models. Depending on the model, however, the atmosphere in each column consists of between 6 and 50 layers (20 are usual), while the water column (for ocean grids) consists of fewer layers as does the terrestrial component of the column (which may have no separate layers). GCMs take into account energy flows between layers in a column and, horizontally, between neighboring columns (e.g., by modeling regional-scale winds), and a time step of 20–30 min (approximately 17,500–26,000 periods per year) is employed.[14]

While we employ the term General Circulation Model, modern climate models are more aptly described as complex, integrated ocean–atmosphere-terrestrial energy circulation models. Not all climate models include a highly-developed terrestrial component, and some models consist only of an atmosphere; in these cases, the ocean and/or terrestrial components are taken into account through some sort of exogenous parameterization that might be derived from a relevant stand-alone model.

Model resolution is perhaps the most difficult aspect to address in climate models. First it is necessary to determine the size of the grid. While numerical weather forecasting models are predictive tools, GCMs can only predict probable conditions. Hence, the former employ a finer grid than the latter. But the use of a coarser grid creates problems. Flows of a liquid or gas that are contrary to the main current, which are known as eddies, are important in the atmosphere and in the ocean because they transfer energy in both a horizontal and vertical direction. In the atmosphere, eddies are of a much larger scale than in the ocean. Eddies that are found in the atmosphere

[14] This implies that a fine-grid (2° latitude × 2° longitude) model, with 20 vertical layers, a 20-min time step, and projecting climate 50 years into the future must keep track of 851,472,000,000 different values of one variable alone!

can be 'resolved' in a 2°×2° latitude-longitude grid, but a much finer grid of 1/6°×1/6° resolution can take into account ocean eddies – wind eddies occur on a scale of 1,000 km, while ocean eddies are around 10–50 km (McGuffie and Henderson-Sellers 2005, p. 58). Given that the grids used in atmospheric models are insufficiently fine from the perspective of ocean modeling, a coupled-atmosphere–ocean climate model must ignore ocean eddies or model these at a sub-grid level (in which case they are essentially exogenous to the larger climate model).

A similar but perhaps more serious problem of climate model resolution relates to differing scales of motion and interaction that vary from the molecular to the planetary, and time scales that range from nanoseconds to geological epochs. We have already noted that GCMs use a time step of under 1 h, but this is much too fast for processes such as the growth of vegetation and consequent uptake of CO_2. Suppose we wish to investigate the climatic response to a major volcanic eruption using a climate model. The climate model needs to be perturbed from its current equilibrium by increasing the amount of sulfur dioxide and particulates entering the atmosphere. Once the volcanic eruption ends, it takes atmospheric variables some 10 days to regain equilibrium, but it could take the ocean layers months to many years to regain equilibrium, depending on the extent to which the volcanic eruption actually impacted the ocean to begin with. As we have seen, this impact is determined not through the interface between the atmosphere and ocean surface layers, but through an external model. Vegetation may also be impacted, but the time scale might be on the order of one or more years.[15] Further, volcanic ash could darken ice, thereby reducing albedo for many years to come. Climate models need to address slow- and fast-responding processes and variables, but this involves making judgments about their importance in the climate system.

It is also important to note that there is no straightforward way to model the interaction between land or ocean surface and the near-surface layer of the atmosphere. To capture the physical energy transfers across these surface interfaces involves second-order differential equations that are impossible to solve numerically (Gerlich and Tscheuschner 2009; McGuffie and Henderson-Sellers 2005, p. 59). Therefore, it is necessary to parameterize the interaction across the atmosphere–ocean and atmosphere-land surface boundaries.

It is not surprising, therefore, that, in addition to the four general categories of model described above, climate models take on a large number of forms, consisting of a variety of inter-linked components, external sub-grid and internal parameterizations, and so on. There now exist at least 15 large climate models, but there are likely quite a few more depending on what one considers a climate model. Some models are simple energy balance models that rely on knowledge of the energy exchanges between the earth and space, while others are massive coupled ocean–atmosphere models that may even have links to terrestrial ecosystems and the

[15] For example, the Mount St Helen's eruption of May 18, 1980, led to enhanced 1980–1981 crop yields in the Palouse region of eastern Washington. Volcanic ash absorbed and held moisture, which aided crop growth in this moisture constrained region.

economy. It is clear that climate modeling is as much an art as a science (McGuffie and Henderson-Sellers 2005, p. 73).

In the end, most climate models are redundant when it comes to the simple question that everyone seeks to answer. If anthropogenic emissions of greenhouse gases cause climate change, what will the average global temperature do in the future? Will the Earth's temperature rise and to what extent? Are there non-anthropogenic and external factors that affect the Earth's temperature and to what extent? All of the basic questions can be answered using a 'simple' energy balance model – an accounting of the energy arriving and leaving the Earth, and its effect on temperature. Large GCMs look at energy exchanges within the various atmospheric columns, from one column to its neighbors (albeit at various vertical levels), and exchanges with the land and ocean surfaces, with each of the terrestrial and ocean systems modeled separately and often with great complexity. Although this adds tremendous complexity in the way of mathematics and computing needs, the essential element remains the radiation flux coming from the sun, the energy retained by the Earth, and the energy flux from the Earth to outer space, and the factors that affect these fluxes.

4.3.2 A Simple Energy Balance Model

A simple zero-dimensional, energy-balance model is just as capable of predicting future trends in global average temperatures as a more-complicated GCM. In this sub-section, we employ such a model to examine the impact on predicted temperatures of assumptions regarding several parameters that are found in one way or another in all climate models. The purpose here is illustrative only: we wish to demonstrate how sensitive model outcomes are to the explicit (or implicit) assumptions that are employed and, in particular, the importance of randomness on the path of future temperatures. We show that randomness associated with non-CO_2 forcing can have a large influence on the temperature, even to the point of obscuring anthropogenic warming entirely and the associated need for expensive policies to lower greenhouse gas emissions.

The assumptions or model parameters that can be adjusted are, firstly, the assumed change in the net radiative forcing. A radiative forcing is generally defined to mean those factors that affect the balance between incoming solar radiation and outgoing infrared radiation within the Earth's atmosphere, with positive forcing warming the Earth's surface and negative forcing cooling it. Thus, it includes changes in the concentrations of various greenhouse gases that affect the energy balance of the surface-atmosphere system.

Second, in addition to radiative forcing attributable to changes in the solar flux, there are non-radiative forcings, such as those that are related to the exchange of heat between the ocean and atmosphere resulting from climatic events such as an El Niño. The depth of the ocean layer impacts heat storage and thus the transfer of heat to the atmosphere. The relationship between the depth of the ocean layer and the heat capacity of the system is provided in Eq. (4.16): $C_P = 0.7\, p_w\, c_w\, d\, A_e$. If the heat capacity of the system is known, then it is possible to determine the ocean depth involved in heat storage and transfers from the ocean to the atmosphere. Further, non-radiative forcings, such as an El Niño or the Pacific Decadal Oscillation, should affect cloud cover (negative or positive feedback), because a heat exchange between the ocean and the atmosphere involves the evaporation or condensation of water.

Third, the total feedback parameter determines whether, when temperatures rise as a result of some forcing, the model reduces the warming (negative feedback) or enhances it (positive feedback). Cloud cover is related to this. For example, a forcing due to increased emissions of carbon dioxide warms the atmosphere causing an increase in water vapor (positive feedback), but does this also lead to more cloud cover and to what extent (negative feedback)?

Finally, the initial temperature departure from the norm (the temperature anomaly) will affect the sequence of temperature projections from a climate model because, without additional forcing, temperatures should trend toward normal. Likewise, the initial level of CO_2 forcing in the model is based on the current concentration of CO_2 in the atmosphere, with assumptions regarding this level of forcing impacting the climate model results. This initial forcing is generally assumed to take a value between 0.2 and 0.6 W/m^2.

The climate model we use is given by the energy balance Eq. (4.15). We can rewrite Eq. (4.15) in discrete form as follows:

$$\Delta T_{t+1} - \Delta T_t = \frac{1}{C_P}(F - \lambda \Delta T_t)\Delta t, \tag{4.22}$$

where Δt is the size of the time step used in the model. Before proceeding, it is necessary to check the appropriateness of the physical dimensions in Eq. (4.22):

$$\text{K} = \text{J}^{-1}.\ \text{K}.\ [\text{W}.\ \text{m}^{-2}\text{-W}.\ \text{m}^{-2}.\ \text{K}^{-1}.\ \text{K}].\text{s}. \tag{4.22$'$}$$

The unit of measure on the left-hand side of (4.22′) is K or °C. For the right-hand side, recall that J = W · s; then the right-hand side of the equation is K/m^2, or °C/m^2. Therefore, it is necessary to multiply the right-hand side of Eq. (4.22) [or (4.22′)] by the surface area of the Earth, measured in m^2.

For the current application, we employ a time step of 1 month or approximately 2,626,560 s (assuming a bit more than 30 days per month). We also assume that there are several forcings, so that $F = F_{CO2} + F_R + F_O$, where subscripts *CO2*, *R* and *O* refer to human emissions of carbon dioxide as a forcing, and solar radiative and other forcings, respectively. Rewriting (4.22) gives:

$$\Delta T_{t+1} - \Delta T_t = \frac{0.89616 \cdot m^3 \cdot K \cdot J^{-1} \cdot s}{d \cdot m} \times [F \cdot W \cdot m^{-2} - \lambda \cdot W \cdot m^{-2} \cdot K^{-1} \times \Delta T_t \cdot K], \tag{4.23}$$

where d refers to the ocean depth, $F = F_{CO2} + F_R + F_O$, and λ is the total feedback, which consists of various components such as ice-albedo and cloud feedbacks. The physical units are provided in the above equations and need to be multiplied and divided separately to ensure that the same units occur on both sides of the equation.

It is easy to verify that only the temperature measure K (or °C) remains on both sides of the equation.

Now consider how (4.23) evolves over time. Suppose we have a starting temperature anomaly given by ΔT_0. Then, given a forcing F, feedback λ and ocean depth d, the temperature anomaly evolves on a monthly time step as follows:

Period 1: $\Delta T_1 = \Delta T_0 + 0.89616 \times (1/d) \times [F - \lambda \Delta T_0]$

Period 2: $\Delta T_2 = \Delta T_1 + 0.89616 \times (1/d) \times [F - \lambda \Delta T_1]$

…

Period M: $\Delta T_M = \Delta T_{M-1} + 0.89616 \times (1/d) \times [F - \lambda \Delta T_{M-1}]$

The change (increase or decrease) in temperature after M months is therefore given by ΔT_M. It is clear from the above iterations that there are four parameters that drive this simple climate model – the starting value of the anomaly (which might be zero), ocean depth, the externally prescribed change in the net radiative plus non-radiative flux (F), which includes an assumption about the initial or current CO_2 forcing (to which the projected increase in CO_2 forcing is added), and the total feedback. The total feedback might be determined as follows: $\lambda = \lambda_{BB} + \lambda_{water\ vapor} + \lambda_{ice\text{-}albedo} + \lambda_{clouds}$. (In the model, a negative value indicates a positive feedback and a positive value a negative feedback.)

We employ the model in Eq. (4.23) to see how global average temperatures might evolve over time under different assumptions about the model parameters.[16] Our interest is simply to illustrate how sensitive modeled global temperature outcomes are to model parameters. We assume a 100-year time with a monthly time step. We associate this with a doubling of atmospheric CO_2, but this is only approximate and is not to be confused with 'doubling of CO_2 relative to pre-industrial levels.' In particular, we assume an initial CO_2 forcing of 0.6 W/m^2, which, on the basis of data provided in Sect. 4.2, assumes a low concentration of CO_2.

We also assume that the CO_2 forcing increases by 0.2 W/m^2 per decade. Further, an ocean depth of 500 m and fixed black body radiation are employed throughout; also assumed are a solar and an ocean forcing of 2 W/m^2 each, with both randomly determined in each month by multiplying the forcing by a value drawn from a uniform normal distribution. Thus, the global average temperature is modeled to trend slightly upward over time in the absence of any human forcing, simply to mimic an overall post-ice age warming. To examine the potential impact of non-CO_2 climate factors on temperatures, the ocean forcing is modeled as a mean-reverting stochastic process in two scenarios.[17] The following five scenarios are considered:

1. Doubling of CO_2 without any feedbacks, except black body radiation;
2. Doubling of CO_2 with feedbacks, but the feedback from clouds is neutral (zero);

---

[16] Although our approach is similar to that of Spencer (2010), we employ somewhat different assumptions and a more complex method for addressing randomness of solar and ocean radiative fluxes.

[17] An excellent discussion of a mean-reverting stochastic process is provided in Dixit and Pindyck (1994, pp.60–79).

3. Doubling of CO_2 with feedbacks, with clouds having a negative feedback effect;
4. No CO_2 forcing, but heat exchange between the atmosphere and oceans results in a forcing that follows a random mean-reverting stochastic process with mean 2 W/m^2;
5. Same as 4, except there is a forcing from doubling of CO_2 with feedbacks of which the feedback from clouds is neutral.

The model results are found in Figs. 4.3 and 4.4. In each figure, the evolution of global average temperature is provided; in Fig. 4.3, the net radiative flux is provided along with temperatures; in Fig. 4.4 the ocean flux is provided rather than the net radiative flux. In each figure, a 2-year moving average is used for temperature and radiative flux. A doubling of atmospheric CO_2 is projected by our model to increase temperature by 1.13 °C when there is no positive feedback from an increase in water vapor and from reduced ice-albedo (as warming reduces the Earth's ice surface area). This increase in temperature is approximately midway between the lower and upper estimates of the total expected increase in temperature attributed solely to CO_2. When feedbacks are taken into account, the projected increase in temperature is 4.45 °C, which is reduced in the model to only 0.58 °C if there is a negative feedback from cloud formation. These three projections of temperature changes are very close to those in Fig. 4.2.

When the ocean–atmosphere forcing is modeled as a stochastic process and no CO_2 forcing is assumed, the average global temperature first rises but then falls (Fig. 4.4a), declining by 0.39 °C by the end of the 100-year time horizon. This provides some indication of the powerful effect that events such as El Niño and the PDO could have on the globe's climate, and that such events can have long-lasting impact.[18]

Even when the ocean–atmosphere forcing is imposed on a CO_2 forcing (with no negative feedback from clouds but a positive one from water vapor and ice albedo), it has a significant impact (Fig. 4.4b), although it is not sufficient to prevent CO_2-induced global warming. In this case, the global average temperature rises to 3.08 °C after 100 years, which is nearly 1.4 °C below what the temperature increase would be if oceans had no significant impact on global climate.

It is important to recognize the influence of randomness in the climate system. Upon comparing the evolution of the forcing in the upper part of Fig. 4.4 with that in the lower diagram, one recognizes that the rise in temperatures with a CO_2 forcing would have been lower if the ocean forcing had followed the stochastic process in the upper part of the diagram as opposed to that in the lower part.

The evolution of the climate system often looks random, but that is because it is inherently nonlinear or chaotic. As discovered by Edward Lorenz, even modeling of the climate is inherently random even though the underlying nonlinear equations give the appearance of predictability. What the mathematical models demonstrated

[18] As discussed in Chap. 5, some solar physicists argue that changes in the sun's activities and its magnetic field can impact the Earth's climate. These cycles operate much like a mean-reverting stochastic process and could be modeled that way, but the rate of reversion would be longer as would the time horizon required to investigate some of the sun's cycles.

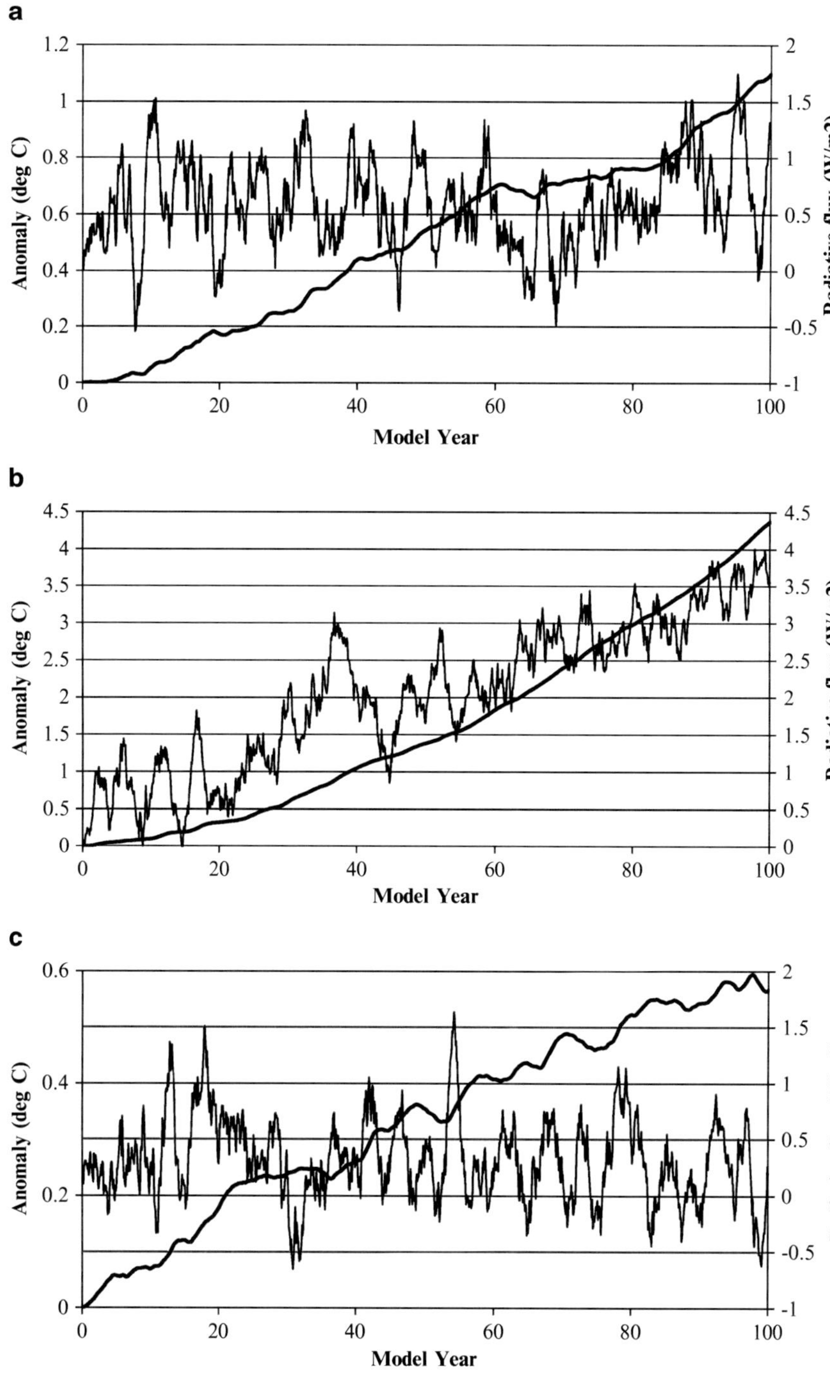

Fig. 4.3 Climate model outcomes, temperature (*dark line, left scale*) and net radiative flux (*thin line, right scale*), 2-year moving average (**a**) CO_2 forcing without any feedbacks (**b**) CO_2 forcing with positive feedbacks (clouds neutral) (**c**) CO_2 forcing with negative cloud feedback

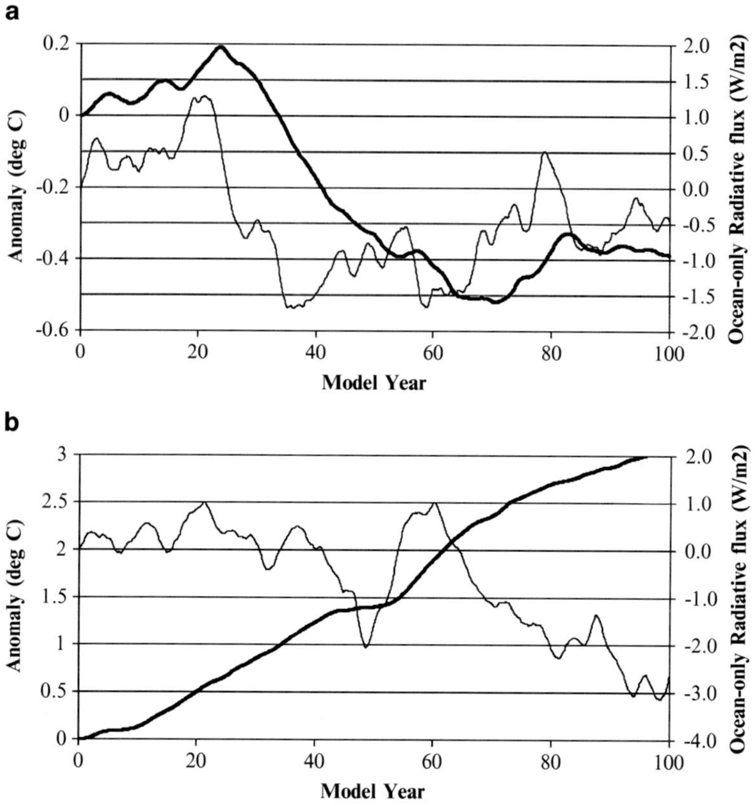

Fig. 4.4 Mean-reverting Stochastic Process as a Climate Forcing, Temperature (*dark line, left scale*) and Ocean-only Radiative Flux (*thin line, right scale*), Two-year Moving Average (**a**) With no CO_2 forcing (**b**) With CO_2 forcing accompanied by positive feedbacks

was that even the flapping of a butterfly's wings in one place could lead to major weather disturbance elsewhere – a phenomenon known as the 'butterfly effect' (Foden 2009; Gleick 1987, pp. 11–31). In modeling the climate system, therefore, it makes sense to account for chaos or, at least, treat the evolution of the climate system as random (which is not the case with complex climate models). However, given the sensitivity of climate outcomes to the model parameters, imposing randomness upon the system would simply exasperate attempts to interpret what is going on.

The real-world climate system is impacted by what can only be considered black swan events (Taleb 2010) that are totally unpredictable, just as Lorenz found with his nonlinear equations. Determining appropriate policy responses to climate change under these conditions is therefore an inherently difficult task.

4.4 Are Climate Models Reliable?

Meteorologists are among the most vocal in expressing their doubts about the ability of sophisticated ocean–atmosphere coupled climate models to predict future temperatures (e.g. Miskolczi 2007; Sussman 2010). These weather specialists are uncomfortable predicting weather beyond one or two days in advance, let alone a week or two in advance. Nonetheless, using similar types of models and ones that include terrestrial and even socioeconomic components, climate modelers are willing to predict temperatures 100 or more years into the future. Of course, the objective of the predictions is quite different. The meteorologist is asked to predict *local* climate conditions on a given day or during a given week, while the climate modelers are projecting trends in global average temperatures. The two are not really comparable, while the complexity, scale, resolution, et cetera have recently diverged. Nonetheless, as climate scientists are increasingly asked to 'scale down' global projections of future temperatures and precipitation to ever smaller regions, the line between weather forecasting and global modeling gets increasingly blurred.

4.4.1 The Scientificity of Climate Models

It is not only meteorologists who have concerns about climate models. Physicists have been among the most vociferous critics.[19] Climate scientists claim that their models are correct because they are based on the known principles of physics and chemistry – the equations in climate models are laws governing everything in the universe. There is no doubt that this is generally true for many of the equations in the models. What is neglected, however, is the fact that climate models also have a host of parameters that do not come from physics or chemistry, but come from the

[19] Examples include Norway's Nobel physicist Ivar Giaever and a group of 54 noted physicists led by Princeton's William Happer (see 'The Climate Change Climate Change' by K.A. Strassel, Wall Street Journal, June 26, 2009 at http://online.wsj.com/article/SB124597505076157449.html, viewed August 26, 2010). Happer provided testimony before the U.S. Senate Environment and Public Works Committee (chaired by Barbara Boxer) on February 25, 2009. Highly regarded physicist Harold Lewis withdrew his membership in the American Physical Society (APS) because of their 2007 pro-anthropogenic global warming stance; Giaever resigned in 2011 for the same reason. Both recommend that APS withdraw the statement as it is blatantly false and a black mark on the association (see http://wattsupwiththat.com/2010/11/06/another-letter-from-hal-lewis-to-the-american-physical-society/#more-27526 viewed November 14, 2010). Other physicists include Freeman Dyson, James Wanliss (2010), a group led by Danish physicist Henrik Svensmark (see Chap. 5), Russian solar physicist Vladimir Bashkirtsev (and almost all other solar physicists), German physicists Gerlich and Tscheuschner (2009), the Italian Nicola Scafetta (2010), Dutch physicist Cornelis de Jager, and the Hungarian Miskolczi (2007). Other scientists have also been critical of climate models, including geologists such as Ian Plimer (2009).

life sciences and geology. For example, many processes are modeled at the sub-grid level, while interactions between the atmosphere and ocean, and atmosphere and land surface, are based on parameterizations that come from outside the climate models. Parameters from geology, biology and the life sciences are not verified by meticulous scientific experiments, such as those that led to measures of the speed of light, the Planck constant or many other known scientific constants. Rather, the equations and the parameters used in the life sciences, geology and other fields, and employed in climate models, are ones that 'work' – they are the ones that appear to replicate observations best. For example, the models used to predict shoreline erosion in the face of rising sea levels are ad hoc, at best, and certainly not accurate by any means (Pilkey and Pilkey-Jarvis 2007).

Clearly, then, despite what one might understand to the contrary, climate models are not driven only by the laws of physics and chemistry. While physics and chemistry play a very important role in the development of climate models, climate modelers make many choices that have nothing to do with physics, but rather with the modelers' preferences.

For example, when it comes to the radiative transfer equations, two possible choices are available (Gerlich and Tscheuschner 2009). First, the modeler can adopt the assumption of a *Local Thermodynamic Equilibrium*. This approach is justified by the Kirchoff-Planck function (written in terms of the wave frequency of light as it moves through a medium) that underlies the derivation of the Stefan-Boltzmann law, as well as a generalized Kirchoff law. This is the glass house effect. The second approach assumes a *scattering atmosphere* that allows for turbulence and movement of air that serves to cool the atmosphere, much like the air outside a greenhouse is cooled. Climate models used in the IPCC reports employ the first approach, but, given that there are two approaches for determining the radiative transfer, the choice is somewhat arbitrary.

Climate models employ a large number of other assumptions, including assumptions related to emissions (as discussed in Sect. 4.1). In this regard, modelers must make decisions regarding the forcing agents to include and how forcings change over time. As noted by the IPCC WGI (2007, p. 208), climate models

> compute the climate response based on the knowledge of the forcing agents and their temporal evolution. While most current [models] incorporate the trace gas RFs [radiative forcings], aerosol direct effects, solar and volcanoes, a few have in addition incorporated land use change and cloud albedo effect. While LLGHGs [long-lived greenhouse gases] have increased rapidly over the past 20 years and contribute the most to the present RF, … the combined positive RF of the greenhouse gases exceeds the contributions due to all other anthropogenic agents throughout the latter half of the 20th century.

That is, assumptions need to be made about which forcing agents to include, although the overriding assumption is that greenhouse gases are the predominant driver of climate change.

The fact that climate models are not based soley in reality has been expressed clearly in an oft-cited quote by the renowned physicist Freeman Dyson (2010)[20]:

[20] A wonderful novel by Giles Foden (2009) gives some notion of the problems forecasting future climate because of difficulties in measuring and predicting turbulence in the real world.

> The [climate] models solve the equations of fluid dynamics, and they do a very good job of describing the fluid motions of the atmosphere and the oceans. They do a very poor job of describing the clouds, the dust, the chemistry and the biology of fields and farms and forests. They do not begin to describe the real world that we live in. The real world is muddy and messy and full of things that we do not yet understand.

Because they presume to rely on known laws of physics and chemistry, climate models are deterministic; the only uncertainty that is generally admitted relates to the inputs into the models. Even so, climate models are not paradigms of accuracy. The claim that they are based on science is correct, but not the claim that the relations therein are immutable. Climate models cannot be run forwards and backwards to give us even a remotely accurate representation of the average climate of the globe, let alone that of various regions. One reason is that not all processes that affect climate can be modeled. Thus, climate models are useful guides in helping us understand climate, but they are not and cannot predict future climates. Large-scale computer models of climate are simply not up to the task of predicting quantitative climates, and they likely never will.

4.4.2 Targeting the Answer

Experienced model builders also question the climate models, not because the equations might be right or wrong, but on the basis of the solution algorithms. This is particularly true of the more complex GCMs. Anyone working with models that require simultaneous solution of hundreds of thousands of equations knows that the answers are suspect. Indeed, without knowing the answer a priori, it is impossible to know whether the solution the computer model arrives at is correct – that is, one needs to know what the expected solution should look like, even approximately, before one can ascertain its correctness and usefulness. The problem is greatly magnified when nonlinear equations are involved. Nonlinear equations will usually need to be linearized about the point(s) where the solution is expected to lie. Even for a small system of five nonlinear equations, say, that are to be solved simultaneously for the five unknowns, a powerful nonlinear solution algorithm can get stuck, unable to find a numerical solution to the system of equations. Algorithms for solving nonlinear models frequently cannot be found, or the algorithm gravitates to a local albeit stable result (known as an 'attractor'), but the solution at that point may not be 'correct,' because the solution represents an unreasonable climate outcome (e.g., an ice age). To move away from such an attractor often requires fiddling with one or more of the model parameters.

Finding a numerical solution is orders of magnitude more difficult when there are discontinuities in one or more of the equations. There are very few computer algorithms that can solve for variables that are binary or discrete (i.e., not continuous); even those 'solvers' that can find solutions in these circumstances are limited in their capacity to find numerical solutions if some of the functions describing the variables are also nonlinear.

In essence, experienced model builders realize that finding numerical solutions to models often requires fine-tuning of the models, whether somehow linearizing equations or permitting discrete variables to take on non-discrete values. They also know that the solution can change quite dramatically when certain model parameters are adjusted. And there are always parameters that are set at the discretion of the modeler and which can also affect the results. Sometimes by adjusting a parameter only slightly (say from 0.00000240 to 0.00000243), the solution algorithm can go from not being able to obtain a numerical solution to finding one that makes sense – one that gives a reasonable climate outcome, although it might not be the correct one. Likewise, changing a parameter only slightly can take the solution from what appears to be a realistic result to nonsense.[21] Climate models can involve upwards of 1,000 parameters that can be adjusted by the modeler.

With nonlinear models, 'chaos' is difficult to avoid. Chaos is used to describe uncertainty related to the existence of multiple solutions to climate models. Multiple outcomes can result simply by providing the algorithms used to solve the models with different starting values for the variables for which the model has to solve. What does this mean? Why does the modeler have to supply starting values?

Consider a very simple example. Suppose we want a computer to solve the following following equation for $x > 0$: $x^2 - 9 = 0$. The answer is clearly $x = 3$. For a computer to solve this nonlinear problem requires that the modeler provide a starting value for x. Most algorithms have default starting values and, in this example, the use of the default value would suffice – the problem is simple enough. If x is not constrained to be positive, and a simple solution algorithm is employed, the solution may be $x = -3$ for one starting value and $x = 3$ for another. For much more complicated models that involve multiple, higher-order nonlinear equations, outcomes are very sensitive to the starting values.

Climate modelers consider a system to be a 'transitive system' if two initial (starting) states of the system evolve to the same 'final' state after a period of time, even though this is no guarantee than some other initial values will result in the same final state. However, if different starting values lead to different plausible outcomes, the system is considered to be an intransitive one. What happens if the same initial state can evolve into two or more 'final' states? This is the quandary raised by complexity: dynamic systems can evolve into multiple final states. The problem is described by (McGuffie and Henderson-Sellers 2005, p. 70) as follows:

> Difficulty arises when a system exhibits behavior which mimics transitivity for some time, then flips to the alternative state for another (variable) length of time and then flips back again to the initial state and so on. In such an almost intransitive system it is impossible to determine which is the normal state, since either of two states can continue for a long period

[21] An example is found in van Kooten et al. (2011). The problem involves finding the optimal level of ducks to hunt given various degrees of wetland protection. As discussed by the authors, a very slight change in the estimated parameter on a double-logarithmic function led to a difference in the optimal number of ducks that the authority might permit hunters to harvest in a season from about 1.5 million to over 30 million. A change in functional form, on the other hand, prevented any solution from being realized. Yet, the model involved no more than three nonlinear equations.

> of time, to be followed by a quite rapid and perhaps unpredictable change to the other. At present, geological and historical data are not detailed enough to determine for certain which of these system types is typical of the Earth's climate. In the case of the Earth, the alternative climate need not be so catastrophic as complete glaciation or the cessation of all deep ocean circulation. It is easy to see that, should the climate turn out to be almost intransitive, successful climate modeling will be extremely difficult.

The problem is not with the Earth's systems – that they tend to flip flop back and forth, which they may well do. Rather, the problem lies with the climate model itself. It is well known that solutions to nonlinear systems of equations can converge to a single equilibrium regardless of the starting values employed – the transitive system of the climate models. However, depending on the model parameters, they can also lead to solutions that flip flop in the manner discussed in the above quote – remaining stable for long periods and then, unexpectedly and rather rapidly, converging upon another attractor. It is also possible, indeed quite likely, that the system appears altogether chaotic, going from one state to the next in a seemingly unexplainable fashion. Yet, the mathematics of a climate model's equations lead to deterministic outcomes that, on the face of it, appear inexplicably and uncharacteristically random. McGuffie and Henderson-Sellers (2005) are probably not far from the truth in pointing out that "successful climate modeling will be extremely difficult," because the climate system is probably more chaotic (intransitive) than we realize and many climate models have chaos inherently built into them.

4.4.3 Comparing Climate Model Predictions with Actual Temperatures

In the case of climate models, the modeler will typically make several simulations with the same model, but report only the 'ensemble-mean' (EM) rather than the individual trend values. The differences between the simulations are accounted for by the use of different starting values and not necessarily different values of the model parameters. With nonlinear equations, there are multiple attractors, some of which are ruled out for any number of reasons, including that the solution does not agree with what the modeler expects. That is, the set of values that solve the equations changes if the algorithm used to find the solution starts at a different place – a different set of starting values. Only rarely will a modeler show the individual runs and trend values. For example, the Japanese Meteorological Research Institute provides results for each of five runs of its climate model, before forming the ensemble-mean. The individual trend values range from 0.042 to 0.371 °C per decade, a huge difference from a policy perspective as the lower value implies insignificant warming of only 0.4 °C over the next century. What would the results look like if the modeling exercise had considered 100 or 1,000 runs instead of only five, each with a different initial value? Clearly, the range would likely have been even greater, but the average trend might have fallen. How many runs are needed to obtain a reliable trend that can then be compared to observed trends? How reliable is any trend?

Climate models fail to address the increase in uncertainty that occurs as one takes a trend further out in time. Statistical uncertainty regarding future temperatures, say, is a function of the number of periods (whether years or decades) into the future that the climate model predicts temperatures, and predictions are conditioned by possible autocorrelation. The variance of any prediction of future temperature increases the farther out that the prediction is made, but this is ignored in climate models because such models are deterministic. Probability does not play a role, although everyone knows that predictions of future temperatures are fraught with uncertainty. While modelers talk about averages (the ensemble means), no effort is made in a statistical fashion to address the variance associated with predictions of future warming.

So, how good are the climate models at forecasting global warming? In a recent study, McKitrick et al. (2010) compare predictions from climate models with actual temperatures, using model generated data and satellite data for the mid and lower troposphere, the region in the Earth's atmosphere that is most sensitive to global warming. Using data for the period 1979 through 2009, they find that projected temperatures from climate models are two to four times higher, depending on the model and location (lower or mid troposphere), than the observed data, with the differences statistically significant at the 99 % level. Thus, climate models fail to track observed data over the 31-year period 1979–2009 – and they fail miserably.

In a subsequent study, (McKitrick and Vogelsang 2012) employ monthly temperature data for the period 1958–2010 to determine whether or not the data exhibit a definitive trend. The temperature data are for the lower- and mid-troposphere layers in the tropics (20°N to 20°S latitude), because this is where climate models predict evidence of global warming will first appear. Two observational temperature series from the Hadley Center (HadAT) for the two layers are chosen, as are the monthly temperatures predicted by 23 climate models for the same period and location. The researchers come to two very interesting conclusions. First, their statistical approach indicates that there is a level shift in the data in December 1977. If a level-shift term for this date is included in the time trend regression, no statistically significant trend in the two data series can be detected. Second, if no shift term is included, the trend in the observational data is a statistical mismatch with the average trend in temperatures from the climate models – the temperature trend in the climate models is greater than that in the observational data. If a shift term is included, the hypothesis that the trends in the observed data and the climate models are equal is rejected at an even greater level of significance.

How reasonable, then, are IPCC predictions of future global warming (IPCC WGI 2007, Chapters 8 and 10)?[22] How accurate might they be? Are they even meaningful?

[22] Green and Armstrong (2007) examine the climate models described in Chapter 8 of the IPCC's Working Group I report, and conduct a forecasting audit. They chose this chapter because, compared to Chapter 10, it provides more "useful information on the forecasting process used by the IPCC to derive forecasts of mean global temperatures" (p.1006). Despite this, Chapter 8 was "poorly written, … writing showed little concern for the target readership … [and] omitted key details on the assumptions and the forecasting process that were used" (p.1007). While the authors of Chapter 8 (IPCC WGI 2007) claimed that the forecasts of future global temperatures are well founded, the language used through the chapter was imprecise and the message conveyed lack of confidence in the projections (p.1012).

A forecasting audit was conducted by Green and Armstrong (2007) to determine whether the IPCC's authors followed standard scientific forecasting procedures as laid out, for example, by the International Institute of Forecasters (available at www.forecastingprinciples.com). These authors were "unable to find a scientific forecast to support the currently widespread belief in 'global warming'" (p. 1015). Indeed, Green and Armstrong (2007) discovered that "those forecasting long-term climate change have no apparent knowledge of evidence-based forecasting methods" (p. 1016).

Green and Armstrong (2007) found that several important forecasting principles were simply ignored. In particular, they found that no cause and effect had been established between rising atmospheric CO_2 and higher temperatures, with causality just as likely in the opposite direction so that rising temperatures result in increased atmospheric CO_2 – correlation is not the same as causation. Thus, it might not even be possible to forecast future temperatures, or that a naïve model (such as a linear trend or a random process with drift) might do as well in forecasting future temperatures. The climate modelers needed to demonstrate that the climate models performed better than a naïve model, and they needed to recognize that, as the forecast horizon increased, the uncertainty associated with the forecast rises rapidly.

Forecasting methods should be kept as simple as possible. However, climate modelers take the view that models with larger numbers of variables, more complex interactions, and highly nonlinear relations are somehow better than simpler models. Complex forecasting methods are only accurate if the current and future relations in the model are known with a great deal of certainty, data are subject to little error, and causal variables can be accurately forecast, which is certainly not the case for climate change forecasts. Research has shown that, by increasing complexity, a model can better fit the known data – the more complex a climate model, the better is its ability to track past climate. But this has little to do with forecast accuracy; indeed, by increasing model complexity to improve the fit, one actually decreases the accuracy of forecasts (see Green and Armstrong 2007, p. 1013). That is, the better a climate model is able to track past climate, the less likely it is able to provide good forecasts of future climates.

As noted earlier in this chapter, climate models assume a positive feedback warming mechanism whereby increasing CO_2 in the atmosphere results in higher temperatures, which, in turn, lead to higher levels of water vapor and precipitation that induce greater blockage of infrared energy to space – a positive feedback warming mechanism. This effect is about twice as large as the additional rainfall needed to balance the increased CO_2. However, Gray and Schwartz (2010) find that, where precipitation occurs, cloud cover is greater leading to an albedo effect that exceeds the infrared effect that prevents energy from escaping. Where there are few or no clouds, the infrared radiation escapes to space, so the net effect is nearly zero as there is no albedo from clouds. For a doubling of CO_2, GCMs predict a 2 °C rise in temperature due to a change in the water vapor feedback (total increase 3 °C), but, based on empirical data, Gray and Schwartz (2010) find a negative water vapor feedback of 0.6 °C. Thus, for a $2\times CO_2$ atm, the total increase in temperature is 0.5 °C rather than 2 °C.

It is little wonder that some climate scientists, such as Kevin Trenberth who is an IPCC coordinating lead author (IPCC WGI 2007, Chapter 3), have argued that climate models do not represent forecasts, but, rather, are stories regarding possible future climates.

> In fact there are no predictions by IPCC at all. And there never have been. The IPCC instead proffers 'what if' projections of future climate that correspond to certain emissions scenarios. There are a number of assumptions that go into these emissions scenarios. They are intended to cover a range of possible self consistent 'story lines' that then provide decision makers with information about which paths might be more desirable. But they do not consider many things like the recovery of the ozone layer, for instance, or observed trends in forcing agents. There is no estimate, even probabilistically, as to the likelihood of any emissions scenario and no best guess.[23]

If this is truly the case, then the stories represent no more than scientists' ideological views regarding the environment, human lifestyles and so on. The stories are not the result of scientific inquiry and certainly not evidence-based forecasting. This view of the future is a value judgment, not more or less valid than one that paints a rosy future or an apocalypse as a result of nuclear holocaust or some natural event (viz., super volcano, large meteorite striking earth).

4.5 Discussion

A well-known shibboleth says that you will know whether someone is a true prophet when their prophecy becomes a reality. Economists have long judged econometric and other economic models by their ability to predict the outcomes of policy with some degree of accuracy. Large econometric models that sought to predict the outcomes of economic events and government policies were found wanting in this regard. While providing useful insights to model builders and perhaps users, they remain in use partly because they are rooted in macroeconomic theory and partly because governments demand that economists predict future levels of unemployment, inflation and other macro variables – they pay staff or consultants to provide these predictions.

Quantitative macroeconometric models have become increasingly sophisticated, but many analysts have nonetheless abandoned such models in favor of large computable general equilibrium (CGE) models. General equilibrium models rely more on a microeconomic perspective of the economy, and assumptions that the actions of individual consumers and producers cause markets to clear – to move continuously towards equilibrium, as do climate models. Yet, CGE models are no better at predicting what is happening to a national or global economy than econometric models.

CGE models and even macroeconometric models are reasonably good at predicting economic outcomes in the short term – several months to perhaps two years into the future – but recent experience with the 2008 financial crisis indicates that even such short-term predictions are quite capricious (see Taleb 2010). Economic intuition as opposed to economic modeling is clearly better at projecting economic thresholds or downturns. Quantitative modeling of macroeconomic systems continues on a large scale because politicians demand it. They rely on quantitative outputs from

[23] K.E. Trenberth, 'Predictions of climate', Climate Feedback at (viewed July 21, 2010): http://blogs.nature.com/climatefeedback/2007/06/predictions_of_climate.html

models even if they are wrong simply because they can be used to justify intervention of various forms.

Economists and politicians do recognize that attempts to project economic conditions decades into the future, say to anticipate inflation induced by resource scarcity, is akin more to gazing into a crystal ball than it is science, albeit science disguised as modeling. Yet, this is exactly what is expected of integrated assessment models, which are nothing more than a derivative of the aforementioned types of quantitative economic models. The results of integrated assessment models are then used in climate models to predict global warming 50–100 or more years into the future. And climate models are themselves highly speculative and unreliable as a means of predicting what will happen a century from now.

Many governments have spent billions of dollars on climate related activities (including money for research). Many private companies, speculators and shysters have spent large amounts of money promoting, opposing or attempting to benefit from policies to make the world environmentally greener. Many governments have implemented climate change policies (most of which actually do little to mitigate emissions of greenhouse gases as we will see in Chaps. 9, 10, and 11), school curricula have been changed, and universities have taken action to implement climate programs and hire climate scientists and policy analysts. Large pension funds and financial institutions are promoting and participating in markets where dubious carbon credits are traded (see Chap. 9). The world has changed irrevocably as a result of paleoclimatic research and projections from climate models that have no greater power to predict future climate than economic models are capable of forecasting future economic conditions.

How have climate models and climate projections fared in their ability to predict climate change? For the most part, climate models have been found wanting! Reconstructions of past temperatures have been accomplished only by fiddling with the models' parameters until the 'backcasts' correspond with some degree of accuracy to realized temperatures. We have already noted that climate models could not come close to forecasting temperatures for the 30-year period beginning in 1979. Nor have models been able to reconstruct the Medieval Warm Period, perhaps as a result of the distraction caused by ongoing efforts in the climate science community to refute this period or, more likely, because modelers believe there is nothing to replicate.

Climate modelers have not done much better in projecting more recent global movements in temperature. The flattening and even the decline of global average temperatures since 1998 has not been anticipated or duplicated by the modelers; instead, there has been a focus on denying the fall in temperature. Every year for the past decade, the United Kingdom's Met Office Hadley Climate Centre, which is the UK's foremost climate change research center, has predicted that the next year would be the warmest on record.[24] Yet, the *Farmers' Almanac* seems to have done a better job predicting long-run temperatures.

---

[24] On December 10, 2009, the Met Office predicted that 2010 would be the warmest year on record (see http://www.metoffice.gov.uk/corporate/pressoffice/2009/pr20091210b.html viewed February 18, 2010). But, as we have seen, NASA attempts to refute the notion that recent temperatures are flat or declining.

Climate science and climate modeling have provided narratives of past climates and numerical predictions of the future climate. The problem is that the historical narratives and 50–100 year climate model projections have done more to instill confidence in those who make such predictions than to measure aptitude. Numerical projection from a mathematical model has replaced empirical evidence. The problem is that, just like economic models, climate models exclude the possibility of black swan explanations – they exclude the unknowable unknown that will impact the future more than anything a model can ever project. The problem is that, while climate research definitely increases knowledge (and is therefore worthwhile undertaking), such knowledge "is threatened by greater increases in confidence, which make our increase in knowledge at the same time an increase in confusion, ignorance, and conceit" (Taleb 2010, p. 138). It is likely that a taxicab driver is just as good at predicting the average temperature of the globe in 2050 and 2100 as a climate scientist using a computer model. The main difference, however, is that the former indicates how ignorant she is about her prediction, while the expert exudes unwarranted confidence that ignores the error associated with a prediction that far into the future (Taleb 2010).

Overall, computer models are not up to the task of providing accurate quantitative replication of ecosystems. "Each step in the direction of understanding ecosystems reveals more and more complexities, and in any complex system in nature we can never obtain quantitative modeling answers at the level that society needs" (Pilkey and Pilkey-Jarvis 2007, p. 21). Because of the enormous amounts of money involved and political pressure for answers that suit preconceived notions, outputs from computer models will tend to support the 'consensus' even if observational evidence and intuition suggest otherwise (Pilkey and Pilkey-Jarvis 2007). But one cannot fool reality. In the end, whether catastrophic, anthropogenic global warming is currently taking place will be shown true or false by the very systems that scientists claim to model so accurately. As will be shown in later chapters, it is too late in the game to bring climate change under control, as if that could ever be the case.

References

Beilman, D. W., MacDonald, G. M., Smith, L. C., & Reimer, P. J. (2009). Carbon Accumulation in Peatlands of West Siberia over the Last 2000 years. *Global Biogeochemical Cycles, 23*, GB1012. doi:10.1029/2007GB003112.

Dessler, A. E. (2010, December 10). A determination of the cloud feedback from climate variations over the past decade. *Science, 330*(6010), 1523–1527.

Dessler, A. E., Zhang, Z., & Yang, P. (2008). Water-vapor climate feedback inferred from climate fluctuations, 2003–2008. *Geophysical Research Letters, 35*, L20704. doi:10.1029/2008GL035333.

Dixit, A. K., & Pindyck, R. S. (1994). *Investment under uncertainty*. Princeton: Princeton University Press.

Dyson, F. J. (2010). *Many colored glass: Reflections on the place of life in the universe*. Charlottesville: University of Virginia Press.

Essenhigh, R. H. (2006). Prediction of the standard atmosphere profiles of temperature, pressure and density with height for the lower atmosphere by solution of the (S-S) integral equations of

transfer and evaluation of the potential for profile perturbation by combustion emissions. *Energy & Fuels, 20*(3), 1057–1067.
Foden, G. (2009). *Turbulence*. London: Faber and Faber.
Gerlich, G., & Tscheuschner, R. D. (2009). Falsification of the atmospheric CO2 greenhouse effects within the frame of physics. *International Journal of Modern Physics B, 23*(3), 275–364.
Gleick, J. (1987). *Chaos. Making a new science*. New York: Viking.
Gray, W. M., & Schwartz, B. (2010). The association of outgoing radiation with variations of precipitation – Implications for global warming. *American Meteorological Society, 15A.4*, 12. Retrieved from http://climaterealists.com/attachments/ftp/AMS-Final5-10.pdf
Green, K. C., & Armstrong, J. S. (2007). Global warming: Forecasts by scientists versus scientific forecasts. *Energy & Environment, 18*(7&8), 997–1021.
Idso, C., & Singer, S. F. (2009). *Climate change reconsidered: 2009 Report of the Nongovernmental International Panel on Climate Change (NIPCC)*. Chicago: The Heartland Institute.
IPCC. (2000). Emissions scenarios: A special report of the IPCC Working Group III. In N. Nakicenovic & R. Swart (Eds.), *Summary for policymakers* (p. 570). Geneva: Intergovernmental Panel on Climate Change.
IPCC WGI. (2001). *Climate change 2001. The Scientific basis. Working Group I contribution to the third assessment report of the Intergovernmental Panel on Climate Change*. Cambridge: Cambridge University Press.
IPCC WGI. (2007). *Climate change 2007. The physical science basis. Working Group I contribution to the fourth assessment report of the Intergovernmental Panel on Climate Change*. Cambridge: Cambridge University Press.
Lindzen, R. S., & Choi, Y.-S. (2009). On the determination of climate feedbacks from ERBE data. *Geophysical Research Letters, 36*, L16705. doi:10.1029/2009GL039628.
Lindzen, R. S., & Choi, Y.-S. (2011, February 12). On the observational determination of climate sensitivity and its implications. *Journal of Geophysical Research* (Submitted).
Lindzen, R. S., Chou, M.-D., & Hou, A. Y. (2001). Does the earth have an adaptive infrared iris? *Bulletin of the American Meteorological Society, 82*(3), 417–432.
Manne, A. S., & Richels, R. G. (1992). *Buying greenhouse insurance: The economic costs of CO2 emissions limits*. Cambridge, MA: MIT Press.
Manne, A. S., Mendelsohn, R., & Richels, R. G. (1995). MERGE – A Model for Evaluating Regional and Global Effects of GHG reduction policies. *Energy Policy, 23*(1), 17–34.
Matthews, H. D., & Keith, D. W. (2007). Carbon-cycle feedbacks increase the likelihood of a warmer future. *Geophysical Research Letters, 34*, L09702.
McGuffie, K., & Henderson-Sellers, A. (2005). *A climate modeling primer*. Chichester: Wiley (December 2009 reprint, 3rd ed.)
McKitrick, R., & Vogelsang, T. J. (2012). *Multivariate trend comparisons between autocorrelated climate series with possible intercept shifts* (Working Paper). Guelph: Department of Economic, University of Guelph.
McKitrick, R., McIntyre, S., & Herman, C. (2010). Panel and multivariate methods for tests of trend equivalence in climate data series. *Atmospheric Science Letters, 11*(4), 270–277. doi:10.1002/asl.290.
Miskolczi, F. M. (2007). Greenhouse effect in semi-transparent planetary atmospheres. *Quarterly Journal of the Hungarian Meteorological Service, 111*(1), 1–40.
Miskolczi, F. M. (2010). The stable stationary value of the earth's global average atmospheric Planck-weighted greenhouse-gas optical thickness. *Energy & Environment, 21*(4), 243–263.
Moyo, D. (2009). *Dead aid. Why aid is not working and why there is a better way for Africa*. Vancouver: Douglas & McIntyre.
Pierrehumbert, R. T. (2011). Infrared radiation and planetary temperature. *Physics Today, 64*(1), 33–38.
Pilkey, O., & Pilkey-Jarvis, L. (2007). *Useless arithmetic. Why environmental scientists can't predict the future*. New York: Columbia University Press.
Plimer, I. (2009). *Heaven & earth. Global warming: The missing science*. Ballan: Connor Court Publishing.

Pruppacher, H. R., & Klett, J. D. (1997). *Microphysics of clouds and precipitation. Second revised and enlarged edition with an introduction to cloud chemistry and cloud electricity*. Dordrecht: Kluwer Academic Publishers.

Räisänen, J. (2007). How reliable are climate models? *Tellus, 59A*(1), 2–29.

Scafetta, N. (2010, March 18). *Climate change and its causes: A discussion about some key issues* (Science and Public Policy Institute (SPPI) Paper). Retrieved August 24, 2010, from http://scienceandpublicpolicy.org/images/stories/papers/originals/climate_change_cause.pdf

Scheffer, M., Brovkin, V., & Cox, P. M. (2006). Positive feedback between global warming and atmospheric CO2 concentration inferred from past climate change. *Geophysical Research Letters, 33*, L10702. doi:10.1029/2005GL025044.

Spencer, R. W. (2008). *Global warming as a natural response to cloud changes associated with the Pacific Decadal Oscillation (PDO)*. Retrieved August 13, 2009, from http://www.drroyspencer.com/research-articles/global-warming-as-a-natural-response/

Spencer, R. W. (2010). *The great global warming blunder. How mother nature fooled the world's top climate scientists*. New York: Encounter Books.

Spencer, R. W., & Braswell, W. D. (2008). Potential biases in feedback diagnosis from observational data: A simple model demonstration. *Journal of Climate, 21*(21), 5624–5628. doi:10.1175/2008JCLI2253.1.

Spencer, R. W., & Braswell, W. D. (2010). On the diagnosis of radiative feedback in the presence of unknown radiative forcing. *Journal of Geophysical Research, 115*, D16109. doi:10.1029/2009JD013371.

Spencer, R. W., & Braswell, W. D. (2011). On the misdiagnosis of climate feedbacks from variations in earth's radiant energy balance. *Remote Sensing, 3*(8), 1603–1613. doi:10.3390/rs3081603.

Sussman, B. (2010). *Climategate. A veteran meteorologist exposes the global warming scam*. New York: WND Books.

Taleb, N. N. (2010). *The black swan: The impact of the highly improbable*. New York: Random House Trade Paperbacks.

Torn, M. S., & Harte, J. (2006). Missing feedbacks, asymmetric uncertainties, and the underestimation of future warming. *Geophysical Research Letters, 33*, L10703. doi:10.1029/2005GL025540.

van Kooten, G. C., Withey, P., & Wong, L. (2011). Bioeconomic modelling of wetlands and waterfowl in Western Canada: Accounting for amenity values. *Canadian Journal of Agricultural Economics, 59*(2), 167–183. doi:10.1111/j.1744-7976.2010.01216.x.

Wallace, J. M., & Hobbs, P. V. (2006). *Atmospheric science: An introductory survey* (2nd ed.). New York: Academic Press.

Wanliss, J. (2010). *Green dragon: Dominion not death*. Burke: Cornwall Alliance.

Weitzman, M. L. (2009). On modeling and interpreting the economics of catastrophic climate change. *Review of Economic and Statistics, 91*(1), 1–19.

Wood, R. W. (1909). Note on the theory of the green house. *Philosophical Magazine, 17*, 319–320.

Chapter 5
Alternative Explanations

The climate world is divided into three: the climate atheists, the climate agnostics, and the climate evangelicals.

– Jairam Ramesh, India's Minister of Environment and Forests (quoted in the Wall Street Journal, March 11, 2010)

[Climate] scientists must acknowledge that they are in a street fight, and that their relationship with the media really matters. Anything strategic that can be done on that front would be useful, be it media training for scientists or building links with credible public relations firms.

– Editorial in Nature (March 11, 2010, p. 141)

You don't need 100 famous intellectuals to disprove [my] theory. All you need is one simple fact.

– Albert Einstein (quote attributed to him by Michio Kaju, Wall Street Journal, September 26, 2011)

As we pointed out in Chap. 3 (Sect. 3.3), the IPCC considers the climate system to be in equilibrium except for volcanic, solar and anthropogenic forcing factors. Of these, volcanic activity and sunspot cycles are considered minor, leaving human emissions of carbon dioxide and other greenhouse gases as the principal driver of climate change. In the last several chapters, we examined the proposition that rising levels of carbon dioxide in the atmosphere would lead to rising temperatures. The simple correlation between monthly concentration of atmospheric CO_2 and temperature for the period 1958–2008 is relatively high, about 82%, which makes the CO_2 story of global warming pretty appealing. On the other hand, temperatures and sunspots are absolutely not correlated, with a simple correlation coefficient of 0%. The lack of correlation between sunspots and temperature is rather odd as one might have expected at least some spurious correlation given there are 612 observations over this period, particularly as there is a negative correlation of 18% between atmospheric CO_2 and sunspots; this suggests that sunspots will have some negative impact on the amount of CO_2 in the atmosphere and that, given the

G.C. van Kooten, *Climate Change, Climate Science and Economics: Prospects for an Alternative Energy Future*, DOI 10.1007/978-94-007-4988-7_5,

high correlation between atmospheric CO_2 and temperature, there should be some correlation between sunspots and temperature. Maybe Andrew Weaver (2008) is correct: The sunspot explanation of climate change is a bunch of nonsense. However, this cursory bit of evidence tells us nothing about cause and effect. About all we can know for certain is that atmospheric concentrations of CO_2 and global temperatures are not a cause of sunspots – either sunspots cause changes on Earth or they do not, but changes on Earth cannot plausibly have an effect on sunspots.

Correlation is not the same thing as causality. It may well be that the high correlation between atmosphere CO_2 and temperature is simply a function of the time period we have chosen – the period covering the instrumental record for CO_2. Over this period, both atmospheric CO_2 and temperatures generally rose. It may also be the result of temperatures that have been incorrectly homogenized (or perhaps more correctly reconstructed) to remove non-climate influences. As noted in Chap. 2, there remains a significant correlation between homogenized temperatures and non-climate influences such as levels of economic activity.

If rising temperature is indeed the result of rising CO_2 levels in the atmosphere, then one should be able to find evidence for this. A simple linear (ordinary least squares) regression of monthly temperatures on CO_2 levels for the period 1958–2008 results in a slope coefficient of 0.009. However, the standard error is 0.133 indicating that the estimated slope parameter is not statistically different from zero. Plain and simple, there is no straightforward statistical evidence that suggests rising atmospheric CO_2 causes warmer global temperatures. Consequently, there may be other explanations for climate change. Yet, the prevailing one that CO_2 is a driver of climate change through its forcing effect on cloud formation cannot be ruled out entirely.

It is important to recognize that CO_2 is a greenhouse gas that contributes to global warming. This is not in doubt, although some physicists have even questioned this aspect of climate change (see Sect. 5.1 below). What is mainly disputed is the extent to which carbon dioxide is responsible for climate change and the degree to which human activities cause the observed increase in atmospheric CO_2. After all, carbon dioxide is a minor greenhouse gas and humans contribute only a very small amount of the total CO_2 that enters the atmosphere each year, perhaps 3%. There is a huge exchange of CO_2 between the oceans and the atmosphere, and between terrestrial ecosystems and the atmosphere, with the interchange going in both directions. Indeed, approximately 120 gigatons (Gt) of carbon are exchanged between the atmosphere and terrestrial ecosystems each year, while another 90 Gt is exchanged between the atmosphere and the oceans. Fossil fuel burning releases some 6.4 Gt of carbon into the atmosphere, with (mainly tropical) deforestation releasing another 1.6 Gt. Of these amounts, an average of some 2.2 Gt of carbon gets absorbed by the oceans and 2.6 Gt by terrestrial ecosystems (primarily growing trees).[1] The difference between what enters the atmosphere and what leaves is positive (amounting to

[1] There is debate about how much carbon sequestration can be attributed to northern forests. Initially scientists used carbon-uptake models to suggest that some 90 % of the carbon went into boreal forest sinks, but on the ground measuring indicated it might be closer to 30 %. This implies that tropical forests account for more uptake than originally thought. See Burgermeister (2007).

about 3.2 Gt of carbon), and is attributed to anthropogenic emissions. Scientists conclude that humans must be responsible for the increasing level of CO_2 in the atmosphere.

There is indeed support for the notion that the origin of CO_2 in the atmosphere is due primarily to fossil fuel burning. The arguments are twofold:

- The ratio of oxygen (O_2) to nitrogen (N_2) in the atmosphere has been declining as CO_2 rises (Manning et al. 2003). Why? When fossil fuels are burned, oxygen is required, lowering the O_2/N_2 ratio and raising CO_2. Of course, this assumes that the level of N_2 in the atmosphere remains unchanged. It is possible that N_2 has not remained fixed and, indeed, should have risen as land use changes have reduced the area of nitrogen fixing plants (grasslands, alfalfa, etc.). An increase in N_2 ever so slight would have caused O_2/N_2 to fall.
- Further support for attributing the rise in atmospheric CO_2 to fossil fuel burning comes from carbon isotopes. The amount of the isotope ^{13}C in CO_2 molecules has declined relative to ^{12}C as atmospheric CO_2 has risen. The decline has been ongoing since the 1200s (Ghosh and Brand 2003). Plants discriminate against ^{13}C and thus have less ^{13}C relative to that found in atmospheric CO_2. As a result of photosynthesis, the ratio of ^{13}C to ^{12}C in the atmosphere should rise; offsetting this, however, is a decline in the ^{13}C to ^{12}C ratio resulting from fossil fuel burning, which emits more $^{12}CO_2$ because fossil fuels are based on carbon sequestered in plants many eons ago. A declining ^{13}C to ^{12}C ratio in atmospheric CO_2 is therefore an indicator that fossil fuels are the culprit.

While fossil fuels may be a contributing factor to increasing concentrations of CO_2 in the atmosphere, this is somewhat irrelevant if there is no direct link between atmospheric CO_2 and global warming.

While no one denies that atmospheric CO_2 plays a role in the Earth's climate, there are a significant number of scientists who feel there are other explanations for changes in climate. These other explanations are not denied by those who attribute global warming to rising CO_2 in the atmosphere, they simply do not believe that other factors play a significant role compared to that of CO_2. And clearly, as the crude correlations between sunspot activity and temperatures seem to indicate, neither can sunspots be an all sufficing explanation. Yet, they too may play a role, indirectly forcing other changes that do affect climate, much as emissions of CO_2 are thought to affect cloud formation. In that case, simple correlations between temperature and CO_2 or sunspots, or a simple regression of temperature on CO_2 or sunspots, cannot tell us anything about actual cause and effect since the effect may appear some time after the cause. Sorting these things out is not an easy task.

The chore of sorting out the underlying factors that affect long-term climate is influenced by belief – scientists' beliefs get in the way. If the carbon dioxide explanation dominates, this strongly suggests that humans have the ability to somehow control the Earth's climate. For some, evidence suggesting that increases in atmospheric CO_2 are the result of anthropogenic emissions of greenhouse gases is particularly appealing – such evidence provides empirical support for one's beliefs concerning human destiny (or something else). Further, there are environmentalists and others who are disturbed (perhaps rightfully so) by the wanton materialism that they see in modern society. Again, the notion that greenhouse gas emissions resulting from such gratuitous consumption are the cause of climate calamity is appealing, because it leads to the inevitable conclusion that such consumption must be scaled back, or that the population of the Earth should perhaps be reduced.

Others simply cannot conceive of the notion that humans can control the climate, as this is something only God can do. Yet others fear the loss of freedom that likely accompanies government control over greenhouse gas emissions. A greater government role is likely inevitable in the real world of climate policy as regulation and emission trading are preferred to straightforward taxes and revenue recycling (see Chaps. 8 and 9). These folks fear big government, and they are concerned that policy (government) failure may become a bigger threat to society and the climate than the potential warming that is projected by climate models.

It is inevitable that there are multiple theories concerning the factors causing climate change. Even Phil Jones of the Climate Research Unit at East Anglia University admitted in his February 13, 2010 BBC interview that the majority of scientists likely do not support the view that carbon dioxide is the principal factor – despite what Oreskes (2004) might have found, there is no scientific consensus. In this chapter, therefore, we investigate alternative explanations of climate change and consider some other issues of contention.

If you listen to the media or to some committed climate scientists, you would get the impression that alternative explanations for global warming and reasons as to why the globe may not warm as much as indicated by climate models are devised by quacks, existing only in the so-called 'blogosphere' of the internet. Where do these bloggers get their ideas? Contrary to what some might think, alternative theories of climatic change, and explanations regarding feedbacks that serve to cool rather than warm the Earth, are found in the scientific literature. These ideas appear in respectable journals, often the same journals that publish the works of climate scientists who only accept the view that warming is the result of human activities. In this chapter, we look at these alternative explanations for climate change, and theories as to why warming is highly unlikely to be catastrophic – current and projected future temperatures are well within the historical experience of humans.

In the next sections, we examine alternative theories of climate change that relate directly to the role of greenhouse gases, primarily carbon dioxide, the non-cosmological regulators of the climate system, and cosmological explanations for climate change. The citations provided in the next three sections are to genuine scientific papers, published in peer-reviewed journals by serious scientists. Because there has been much ado about lack of peer-review and the role of blogging in 'promoting' alternative explanations of climate change, we end the chapter with a discussion of climate science, climategate and the IPCC.

5.1 Explanations that Exclude the Sun and the Cosmos

Natural variations of the climate system are also important to climate change, but are ignored in climate models.[2] For example, the origins of the Pacific Decadal Oscillation are not understood. What causes El Niños similar to the one that, in mid 2010, appears to be ending? The El Niño caused the global temperatures of the first part of 2010, as measured from satellites, to be significantly warmer than normal. While some interpret the warmer temperatures on early 2010 as evidence of human-caused global warming, the temperature record strongly suggests that higher temperatures (such as those of 1998) are correlated with El Niño events (see Fig. 2.8). What causes such events is simply unknown.[3]

In recent years, El Niño events related to the Southern Oscillation (SO) have been more frequent. The El Niño/Southern Oscillation (ENSO) is the most important coupled ocean-atmosphere phenomenon to cause global climate variability on inter-annual time scales.[4] An El Niño tends to warm the entire globe (Fig. 2.8). In addition to the ENSO, other natural weather events affect the Earth's climate, including the Pacific Decadal Oscillation (PDO) and Atlantic Multi-decadal Oscillation (AMO), both of which have similar time scales as the warming and cooling periods observed during the twentieth century (see Chap. 3). Roy Spencer regressed the change in temperature on the annual average values of the PDO, AMO and SO for the period 1900 through 1960.[5] The subsequent regression model was then used to predict temperatures for the period 1961–2009 using the available values of these natural weather events. The predicted temperatures closely track the CRUT3 temperatures for the Northern Hemisphere. Spencer concludes that the regression results provide

---

[2] I want to thank Joseph L Bast for permission to use material from his 2010 paper 'Seven Theories of Climate Change' (Chicago, IL: Heartland Institute). This paper can be found on the Heartland Institute's website (www.heartland.org). In some instances in this section and the next, I follow Bast quite closely, but, at other times, the research is solely that of the current author. Additional information can also be found in Idso and Singer (2009).

[3] The Australian geologist Ian Plimer speculates that El Niño events may be related to undersea volcanoes (see http://www.sourcewatch.org/index.php?title=Ian_Plimer as viewed September 1, 2010), and has been ridiculed for this. Yet, Bill McGuire of the University College London and David Pyle of Oxford University, among others, have argued the opposite – that climate and, particularly, El Niño events could trigger earthquakes and thus volcanoes; see 'New Scientist – Climate change may trigger earthquakes and volcanoes' by Richard Fisher September 23, 2009 at http://asnycnowradio.wordpress.com/2010/03/04/climate-change-may-trigger-earthquakes-and-volcanoes/ (viewed September 2, 2010). It seems that the issue is one of cause and effect, but more evidence is required to ascertain either viewpoint. If the latter viewpoint is correct, however, this suggests that a warming earth triggers volcanoes that then cause cooling – a built-in regulator.

[4] See http://www.esrl.noaa.gov/psd/people/klaus.wolter/MEI/ (viewed November 26, 2010).

[5] "Warming in last 50 years predicted by natural climate cycles," June 6, 2010, at: http://www.drroyspencer.com/2010/06/warming-in-last-50-years-predicted-by-natural-climate-cycles/ (viewed June 18, 2010).

evidence that most of the warming that the IPCC has attributed to human activities over the last 50 years could simply be due to natural, internal variability in the climate system. If true, this would also mean that (1) the climate system is much less sensitive to the CO_2 content of the atmosphere than the IPCC claims, and (2) future warming from greenhouse gas emissions will be small.

The ENSO is one of the most important natural events to affect global climate. In the remainder of this section, we examine other natural events and factors that have an impact on climate. Some of these natural factors are important because they may offset warming attributed to carbon dioxide and other greenhouse gas emissions. Such factors are a self-regulatory means by which the Earth moderates the impacts of humans, without the need of human intervention. However, we also consider some anthropogenic forcers other than greenhouse gas emissions.

5.1.1 Bio-thermostat

One theory holds that negative feedbacks from biological and chemical processes offset positive feedbacks that might be caused by rising carbon dioxide in the atmosphere. In essence, the globe's biological and other natural processes serve to offset CO_2-induced warming so that temperatures are kept in equilibrium. Here we consider five such feedbacks, although many more have been identified, including cloud formation (discussed previously and in Sect. 5.2 below).

5.1.1.1 Carbon Sequestration

As discussed in Chap. 9, forest and other terrestrial ecosystems remove CO_2 from the atmosphere by sequestering it in biomass. Since the productivity of many plants and trees is enhanced by a rise in atmospheric CO_2, more carbon will be sequestered as more CO_2 enters the atmosphere. Further, higher temperatures also tend to increase carbon uptake by plants and trees. Together these biological processes serve to offset some of the temperature enhancing impact of rising atmospheric carbon dioxide.

How powerful is this negative feedback? The answer depends on the size, growth rate, and duration of the terrestrial sinks in which carbon is stored, and that depends on a large number of factors. Wolfgang Knorr of Bristol University in England finds evidence that sinks are growing in pace with human CO_2 emissions, "having risen from about 2 billion tons a year in 1850 to 35 billion tons a year now" (Knorr 2009). In addition, new carbon sinks appear to be discovered every few years. Climate models generally ignore these types of sinks because they lack a link to a good terrestrial model.

5.1.1.2 Carbonyl Sulfide

Carbonyl sulfide (COS) is a biologically produced sulfur gas emitted from soils. COS eventually makes its way into the stratosphere where it is transformed into sulfate aerosol particles, which reflect solar radiation back into space and thus cool the earth. The rate at which COS is emitted increases as vegetation responds to the ongoing rise in the atmosphere's CO_2 content – a negative feedback to global warming. Research indicates that the COS-induced cooling mechanism also operates at sea, because higher CO_2 and higher temperatures increase surface-water chlorophyll concentrations.

Ice core samples reveal that tropospheric COS concentration has risen approximately 30 % since the 1600s, from a mean value of 373 parts per trillion (ppt) over the period 1616–1694 to about 485 ppt today, with only about one quarter attributable to anthropogenic sources. Climate models ignore the possible effect of COS on climate, which implies they likely overstate the warming due to increased CO_2.

5.1.1.3 Diffuse Light

Plants emit gases that are converted into aerosols known as 'biosols.' Since plant productivity increases with higher levels of atmospheric CO_2, more biosols are released as well. These act as cloud condensation nuclei, helping to create new clouds that reflect incoming solar radiation back to space, thereby cooling the planet. Further, biosols diffuse solar radiation close to the ground, reducing shade under plant canopies and thereby enhancing photosynthesis that increases the amount of CO_2 plants absorb from the air – a cooling feedback effect discussed above.

How significant is this negative feedback? Niyogi et al. (2004) found that diffused light increased net CO_2 uptake by a broadleaf deciduous forest by between 30 and 50 %. Once again, these effects are not adequately included in any computer model of the Earth's climate system.

5.1.1.4 Iodocompounds

Iodinated compounds, or iodocompounds, are particles formed in sea air from iodine-containing vapors emitted by marine algae. These compounds help create clouds, which reduce the amount of solar radiation reaching the Earth's surface. They are also stimulated by rising atmospheric CO_2 levels and warmer temperatures. Indeed, emissions of iodocompounds from marine biota "can lead to an increase in global radiative forcing similar in magnitude, but opposite in sign, to the forcing induced by greenhouse gases" (O'Dowd et al. 2002). In other words, this one biological process could offset all of the warming caused by rising CO_2 levels.

5.1.1.5 Dimethyl Sulfide

The amount of biologic dimethyl sulfide (DMS) emitted by the world's oceans is closely related to sea surface temperature: the higher the sea surface temperature, the greater the sea-to-air flux of DMS (Wingenter et al. 2007). DMS is a major source of cloud condensation nuclei, which generate clouds with greater cloud albedo.

The greater the cloud albedo, the more incoming solar radiation gets blocked and reflected out to space. Sciare et al. (2000) found that a 1 °C increase in sea surface temperatures could increase the atmospheric concentration of DMS by 50 %. Greater DMS in the atmosphere results in an important negative feedback that offsets the original impetus for warming, a further feedback ignored in climate models.

There are many other kinds of aerosols created or destroyed as a result of biological and chemical processes that are impacted by atmospheric CO_2 and changes in temperature. Many are counter-cyclical to the CO_2 forcing, thus serving to offset the impact of carbon dioxide. As noted in Chap. 4, it is difficult to include complex, nonlinear biological and Earth processes in climate models, which is why they are not included. As a result, the IPCC indicates that such biological-earth feedbacks are inconsequential. This is the case despite the growing evidence to the contrary. It appears that there is a built-in bio-thermostat that prevents humans from overheating the earth; as a result, anthropogenic emissions of greenhouse gases are not likely to bring about catastrophic global warming.

5.1.2 Cloud Formation and Albedo

Clouds play an important role in climate. Some scientists postulate that changes in the formation and albedo of clouds create negative feedbacks that cancel out all or nearly all of the warming effect of higher levels of CO_2. Rather than based on computer models, this theory relies largely on observational data reported by a series of researchers. A team of NASA scientists found that changes in cloud coverage in the tropics act as a natural thermostat (a 'thermostat-like control') to keep sea surface temperatures (SSTs) between approximately 28 and 30 °C (Sud et al. 1999). Their analysis suggested that, as SSTs rise, air at the base of the clouds is charged with the moist static energy needed for clouds to reach the upper troposphere, at which point the cloud cover reduces the amount of solar radiation received at the surface of the sea. The subsequent cool, dry downdrafts promote ocean surface cooling. This phenomenon would also be expected to prevent SSTs from rising higher in response to enhanced CO_2-induced radiative forcing.

Subsequently, Richard Lindzen, a professor of meteorology at the Massachusetts Institute of Technology, and two colleagues examined upper-level cloudiness data and SST data, discovering a strong inverse relationship between upper-level cloud area and the mean SST of cloudy regions in the eastern part of the western Pacific (Lindzen et al. 2001). The area of cloud cover increased about 22 % for each 1 °C increase in SST. The sensitivity of this negative feedback was calculated by the researchers to be significant enough to "more than cancel all the positive feedbacks" in the climate models.

The Lindzen et al. (2001) results were verified by Spencer et al. (2007) who used new satellite data to find that the net "radiative effect of clouds during the evolution of the composite ISO [tropical intra-seasonal oscillations] is to cool the ocean-atmosphere system during its tropospheric warm phase, and to warm it during its cool phase." Subsequently, Lindzen and Choi (2009) found that, "for the entire tropics, the observed outgoing radiation fluxes increase with the increase in sea surface temperatures (SSTs). The observed behavior of radiation fluxes implies negative feedback processes associated with relatively low climate sensitivity. This is the opposite of the behavior of 11 atmospheric models forced by the same SSTs." Lindzen and Choi (2011) responded to critics with a new study accounting for orbital drift by NASA's Earth Radiation Budget Experiment (ERBE) satellites and other data issues. They once again found negative feedback by clouds in the tropics that implied climate models were exaggerating climate sensitivity.

The forgoing results, if correct, indicate that clouds act as a negative feedback to the warming that would otherwise be caused by human emissions of CO_2 and other greenhouse gases. Indeed, this feedback might even eliminate the net warming due to anthropogenic greenhouse gas emissions. The role of clouds is discussed further in Sect. 5.2 in conjunction with the impact that cosmic rays have on cloud formation.

5.1.3 Human Forcings Other than Greenhouse Gases

Roger Pielke, Sr., a climatologist at the University of Colorado in Boulder, argues that observed increases in temperature may only be partly the result of human emissions of CO_2 and other greenhouse gases. Other human activities that have transformed the Earth's surface and constitute an anthropogenic forcing have been largely ignored (Pielke et al. 2009). In Chap. 2, we noted that there is an urban heat island effect, with cities tending to be warmer than suburbs and suburbs warmer than rural areas. De Laat and Maurellis (2004) concluded that the global mean surface temperature trends provided by the CRU-Hadley Center, for example, are very likely smaller than indicated because of this effect. This helps account for the argument that the climate sensitivity parameter is 1 °C rather than 1.2 °C.

Anthropogenic aerosols and ozone have shorter lifetimes than greenhouse gases, and therefore their concentrations are higher in source regions and downwind. Matsui and Pielke (2006) estimate that the effect of human aerosols on temperatures at the regional scale is some 60 times that of the mix of greenhouse gases. With many surface-based temperature stations located in urban or near-urban areas, it is likely they are registering the warming effects of these aerosols and ozone, not those of carbon dioxide.

In developing countries, deforestation by burning trees is a common practice used to convert forestland to pastures and cropland. This releases large amounts of CO_2 that is not subsequently sequestered by growing trees although, in cases where trees have been planted for palm oil, carbon uptake could be significant. Further, where pasture or cropland replaces the forest, the land tends to be warmer due to lost shade created by a forest canopy. Estimates suggest that one-quarter to one-third of anthropogenic CO_2 emissions are due to deforestation.

In coastal areas, anthropogenic activities such as logging, agriculture, construction, mining, drilling, dredging and tourism can increase or (more rarely) decrease surface temperatures of nearby bodies of water (National Research Council 2005). For example, storm runoff from city streets following heavy rains can result in seawater dilution and temperature increases. Development can produce sediment that reduces stream flow and damages coral reefs by reducing the penetration of sunlight or by direct deposit on the coral, causing damage mistakenly attributed to global warming.

Airliners often leave behind condensation trails, referred to as contrails. Composed of water vapor, they precipitate the creation of low clouds that have a net warming effect. According to Travis et al. (2007), contrails may result in net warming in certain regions that rivals that of greenhouse gases. In essence, because of the large amounts of fossil fuels burned, jet travel results in a double whammy when it comes to global warming.

Several of these human forcings have local and regional effects on climate equal to or even exceeding that of anthropogenic greenhouse gas emissions. In some cases, this might leave little or no warming left to be explained by the emission of greenhouse gases. However, the IPCC WGI (2007) places too little importance on these forcings, while global and regional models tend to ignore them altogether, again because of the difficulty of modeling such forcings.

5.1.4 Ocean Currents

William Gray, professor emeritus of atmospheric science at Colorado State University and head of the Tropical Meteorology Project at the university's Department of Atmospheric Sciences, is the leading proponent of the theory that global temperature variations over the past 150 years, and particularly the past 30 years, were due to the slow-down of the ocean's Thermohaline Circulation (THC).[6] He argues that ocean water is constantly transferred from the surface mixed layer to the interior ocean through a process called ventilation. The ocean fully ventilates itself every 1,000–2,000 years through a polar region (Atlantic and Antarctic) deep ocean subsidence of cold-saline water and a compensating upwelling of warmer less saline water in the tropics. This deep ocean circulation, called the Meridional Overturning Circulation (MOC), has two parts – the primary Atlantic Thermohaline Circulation (THC) and the secondary Surrounding Antarctica Subsidence (SAS). The average strength of the Atlantic THC varies by about one to two Sverdrups (a unit of measure of volume transport equal to about 10^9 liters per second) from its long-term average of about 14 Sverdrups.

---

[6] See his 2009 papers and presentations (viewed August 24, 2010): 'Global warming and hurricanes' (http://icecap.us/docs/change/GlobalWarming&HurricanePaper.pdf) and 'Climate change: Driven by the ocean, not human activity' (http://tropical.atmos.colostate.edu/Includes/Documents/Publications/gray2009.pdf).

Paleo-proxy data and meteorological observations show there have been decadal to multi-century scale variations in the strength of the THC over the past 1,000 years. When the THC circulation is stronger than normal, the Earth-system experiences a slightly higher level of evaporation-precipitation (~2 %). When the THC is weaker than normal, as it is about half the time, global rainfall and surface evaporation are reduced about 2 %.

It requires extra energy (29 W/m^2) from the ocean surface to evaporate or turn 1 mm of liquid water into water vapor. This energy depletion during periods of high Atlantic THC conditions acts together with the enhancement of the upwelling of deep ocean cold water into the tropical ocean upper level mixed region to bring about additional upper-level ocean energy depletion and, finally, with a lag of 5–10 years, reduced ocean surface temperatures. When the THC is relatively weak (as it was during the periods 1910–1940 and 1970–1994), the Earth-system typically has less net evaporation cooling and less deep ocean upwelling of cold water. At these times, energy accumulates in the ocean's upper mixed layer and over a period of a decade or two the global ocean begins to warm.

The average THC appears to deplete energy continuously from the ocean at a rate of about 3 W/m^2. This long-period energy loss is balanced by a near-constant extra solar energy gain. When the THC is stronger than average, this upwelling of colder deeper water into the tropical mixed layer brings a general energy depletion of the upper 50–100 m of mixed tropical ocean of about 4 W/m^2. When the THC is weaker than average, the energy depletion drops to about 2 W/m^2. These ocean energy depletions and accumulations, acting over periods of 20–30 years, can lead to significant sea surface temperature differences.

Besides this deep ocean global THC circulation, there are also up-and-down-welling ocean areas that are a product of the ocean's horizontal surface wind configurations. These so-called 'Ekman' patterns can also contribute to local and global temperature change depending on where they occur. The combined THC and Ekman changes have no known association with anthropogenic greenhouse gas increases. A slowdown of the global THC circulation that occurs when Atlantic Ocean salinity declines typically brings about a few decades of reduction in Antarctic deep-water formation.

How powerful is the effect on climate of these natural changes in ocean currents compared to estimates of the effect of human-made greenhouse gases? According to Gray, pre-industrial amounts of CO_2 have been estimated at 290 ppmv. The energy gain from a doubling of CO_2 to 580 ppmv with all other processes held fixed has been calculated to be 3.7 W/m^2. Mauna Loa Observatory measurements of CO_2 were about 390 ppmv in 2010. The change in CO_2 energy forcing from pre-industrial conditions of 290 ppmv to today's value of about 390 ppmv gives an idealized outgoing long-wave radiation (OLR) blocking of energy to space of 1.3 W/m^2 (=100/290 × 3.7). This is less than the 2 W/m^2 energy flux that occurs from the ordinary alteration of the thermohaline circulation. According to Gray, changes of the Meridional Overturning Circulation since 1995 led to the cessation of global warming since the 1998–2001 period and triggered the beginning of a weak global cooling trend since 2001. Gray projects this weak cooling to continue for the next couple of decades, with Craig Loehle of the National Council for Air and Stream Improvement having found evidence to indicate that the oceans have been cooling since 2003 (Loehle 2009).

The work of a number of scientists, such as Don Easterbrook, a geologist at the University of Western Washington, suggests we are entering into a cold period (Easterbrook 2008). If so, there is significant reason to be concerned. The works of climate pioneer H.H. Lamb (1995) and Brian Fagan (2000) demonstrated that cold periods are harmful to humankind and warm periods are generally beneficial.

5.2 Cosmological Climate Change

Climate scientists and the Intergovernmental Panel on Climate Change have essentially treated the Earth as a closed system, effectively ignoring the impact that factors outside the Earth might have on global climate. As noted in the introduction to this chapter, there is no straightforward correlation between the sun's activity (as measured by sunspots) and temperature. This does not imply that there is no such relation. The relation may be more complex than indicated by a simple linear regression model (de Jager 2008). Research suggests that there may indeed be a relation between times when the sun is more or less active and temperatures on Earth (Duhau and de Jager 2010). The rotation, tilt and orbit of the Earth also impact climate, and each of these is impacted by the sun and planets. Further, it appears that cosmic rays that originate outside our solar system affect cloud formation and thus temperatures on Earth. It also appears that the sun acts to block cosmic rays from reaching Earth, and that the intensity of cosmic rays entering the atmosphere depends on how active the sun is at any time. These cosmological effects are the subject of this section, because they provide a compelling alternative explanation for observed changes in the Earth's climate.

5.2.1 Solar Variability: Evidence of a Solar Effect

Some scientists argue that solar variability accounts for most or all of the warming in the late twentieth century and that it will dominate climate in the twenty-first century regardless of anthropogenic greenhouse gas emissions. Changes in the brightness of the sun are caused by sunspots – bursts of energetic particles and radiation – that vary in frequency in cycles of roughly 11, 87 and 210 years. These cycles cause changes in the amount of electromagnetic radiation (referred to as solar wind) that reaches the Earth's atmosphere, and impacts the Earth's climate. Most proponents of the theory that solar variability drives changes in climate believe positive feedbacks occur either by a process involving the influence of the solar wind on cosmic rays, which affects cloud formation, or on the oceans' thermohaline circulation, which affects sea surface temperatures and wind patterns.

Nicola Scafetta and Bruce West, physicists at Duke University, investigated the role of the sun and climate change in a series of research papers (Scafetta and West 2007, 2008). They argue that there are two distinct aspects of the sun's dynamics: the short-term statistical fluctuations in its irradiance and the longer-term solar cycles. Climate models do not attempt to explain observed fluctuations in global average temperatures from one year to the next, treating these as noise (e.g., the result of erroneous measurements or missing data) rather than the result of fundamental physical forces. Scafetta and West (2007, 2008) demonstrate that the sun might indeed have some impact on climate, perhaps accounting for as much as 69 % of the observed rise in temperature.

Solar physicist Cornelis de Jager and his colleagues have investigated a number of different solar cycles, concluding that these do indeed affect temperatures on Earth. An intriguing aspect of this research is its conclusions. The researchers predict that the Earth is about to experience a lengthy period of cooling, similar to the Maunder Minium (ca. 1645–1715) that took place during the Little Ice Age and led to bitterly cold winters in Europe and North America. Using nonlinear dynamic theory and observations on the heliospheric drivers of the sun-climate interaction, they identify three types of 'great climate episodes' that characterized the past 500 years or so. Using historical observations on sun activity, they use their theoretical model to identify precisely a Grand Minimum (1620–1724), known as the Maunder Minimum, a period of Regular Oscillations (1724–1924), and a Grand Maximum (1924–2009), sometimes referred to as the Current Warm Period. The dates are precise because they are associated with an abrupt change in some boundary condition (Duhau and de Jager 2010).[7]

The conclusion of the solar physicists appears dramatic: "Solar activity is presently going through a transition period (2000–2013). This will be followed by a remarkably low Schwabe [11-year sunspot] cycle, which has started recently. In turn that cycle precedes a forthcoming Grand Minimum, most likely of the long type" (Duhau and de Jager 2010). If nothing else, this conclusion might require climate scientists to reassess some of their conclusions regarding the hockey stick, how solar activities affect the Earth's climate, and whether the role of anthropogenic emissions of carbon dioxide is as dominant as indicated in climate models.

[7] Boundary conditions are determined rather precisely in a phase plane diagram that plots known proxies of the toroidal and poloidal magnetic field components of the tachocline, a layer of some 30,000 km situated 200,000 km below the solar surface. A cycle within the phase plane diagram is known as a Gleissberg cycle (which is approximately 100 years in length), and it contains several other cycles (such as the 22-year Hale cycle) that have a smaller effect on Earth's climate (see de Jager 2008; Duhau and de Jager 2010; references in these citations). The Maunder Minimum is often dated 1645–1715 and the Dalton Minimum 1790–1830 (see *The Economist*, June 18, 2011, p. 87). These datings are much shorter suggesting that any future cool period will not be too lengthy.

Some climate scientists dismiss the impact of solar cycles outright (e.g., Weaver 2008), and there has been significant debate about Scafetta and West's finding.[8] A principal argument relates to total solar irradiance (TSI). The IPCC WGI (2007) and climate modelers claim that TSI has remained relatively constant since 1980 and has been small in any event. Thus, the solar impact on the Earth's climate has been small, and, as result, the observed warming that has taken place is the result solely of anthropogenic emissions of greenhouse gases. Thus, climate models estimate the impact of TSI on future climate to be no more than "a few tenths of a degree Celsius" (IPCC WGI 2007, pp. 107, 189; Table 4.5). However, there are several measures of TSI that can be employed. Climate modelers use a proxy of TSI that is based on solar-related measures, such as sunspots, ground-based spectral line width records, and ^{14}C and ^{10}Be cosmogenic isotope production, to reconstruct TSI (Scafetta 2010). Others use data from satellites. There is no theory or empirical reason to pick one proxy of TSI over another, so any claim that one is scientifically sound compared to the other is weak at best.

According to the IPCC, changes in solar irradiance over the past 250 years have resulted in a radiative forcing of +0.12 W/m^2, with a range of +0.06 to +0.30 W/m^2 (IPCC WGI 2007, p. 30). This is an order of magnitude smaller than the IPCC's estimated net anthropogenic forcing of +1.66 W/m^2 from CO_2 emissions since pre-industrial times – since 1750. However, many scientists believe the IPCC got it backwards, that proxy data from ice cores, drift ice debris and other sources reveal that the sun's influence was ten times as important as CO_2 in influencing global temperatures in the past.

Paleo-oceanographer Gerard Bond and colleagues at Columbia University's Lamont-Doherty Earth Observatory found changes in global temperatures occurred in cycles of roughly 1,500 years over the past 12,000 years, with virtually every cooling period coinciding with a solar minimum (Bond et al. 2001). Subsequently, a team of researchers from the Heidelberg Academy of Sciences, the University of Heidelberg, the Potsdam Institute for Climate Impact Research, and the Alfred Wegener Institute for Polar and Marine Research demonstrated that the known 210-year and 87-year cycles of the sun could combine to form a 1,470-year cycle (Braun et al. 2005). Craig Loehle (2004) used a pair of 3,000-year proxy climate records to demonstrate a similar connection, and Willie Soon (2005), an astrophysicist at the Harvard-Smithsonian Center for Astrophysics, has similarly documented close correlations using different temperature records and measures of solar radiation. Clearly, the effect of the sun on climate is important and scientists have now demonstrated plausible mechanisms linking variation in solar radiation to decadal changes in global temperature.

---

[8] The paper in *Physics Today* (Scafetta and West 2008) was vigorously attacked by Philip Duffy, Benjamin Santer and Tom Wigley in the January 2009 issue of the Journal. Responses to the Duffy et al. paper were provided by Scafetta and West, and Brian Tinsley, a University of Texas physicist, with a rejoinder by Duffy et al., in the November issue of *Physics Today*. The debate clearly remains unresolved for reasons noted in the text below. On the other hand, the current author could not find research countering that of de Jager and colleagues, with even *The Economist* (June 18, 2011, p. 87) commenting that a long period of cooling now appears likely.

5.2.2 *Solar Wind Modulation of Cosmic Rays*

Henrik Svensmark and Eigil Friis-Christensen, astrophysicists at the Center for Sun-Climate Research at the Danish National Space Center, proposed a theory of cosmic ray-sun interaction that fits the observed data. Their theory is straightforward and constitutes a strong challenge to the prevailing notion that CO_2 is the main driver of climate change. They argue that cosmic rays play a role in cloud formation. Although water vapor is the most abundant greenhouse gas, the effect of clouds is somewhat different. In the first instance, when there are many large clouds, temperatures might be higher than otherwise, especially night time temperatures. People living on the west coast of Canada know that during winter a clear sky could mean that temperatures will fall below freezing during the night or early morning, while clouds will keep temperatures above 0 °C, often substantially above. But clouds also reflect incoming sunlight, thereby cooling the planet.

Overall, increased cloud cover cools the globe, except for thin clouds (high cirrus clouds) that are so cold they radiate less heat into space than they block going out from the earth. The clouds that cool the most are thick clouds at middle altitudes, as these radiate more heat into space (reflect more sunlight) than they trap heat or prevent heat from escaping the Earth (Svensmark and Calder 2007, p. 67). Removing all the clouds would raise the planet's average temperature by 10 °C, while increasing low clouds by a few percent would chill it considerably.

As noted, cosmic rays aid cloud formation, and Svenmark and Friis-Christensen (1997) explained how this could happen. Cosmic rays entering the Earth's atmosphere release electrons that encourage molecules of water to clump together. These micro-specks of water molecules constitute the building blocks of cloud condensation nuclei. During periods of greater solar magnetic activity, the stronger solar wind blocks some of the cosmic rays from penetrating the lower atmosphere, resulting in fewer cloud condensation nuclei being produced. The result is the creation of fewer and less reflective low-level clouds, leading to increasing near-surface air temperatures and thence global warming.

The theory was first tested using crude devices in the late 1990s. Subsequently, scientists at Aarhus University and the Danish National Space Institute directly demonstrated that cosmic radiation can create small floating particles – so-called aerosols – in the atmosphere. These aerosols are a catalyst in cloud formation that cools the planet. The experiments conducted by Enghoff et al. (2011) substantiate the connection between the sun's magnetic activity and the Earth's climate. That is, solar radiation is not the only manner by which the sun impacts the Earth's temperature. A further test of this proposition was provided by researchers associated with the CLOUD (Cosmics Leaving Outdoor Droplets) project at the European Organization for Nuclear Research's (CERN) Proton Synchrotron located at Europe's particle physics laboratory near Geneva.[9] Various obstacles prevented

[9] An explanation of the CLOUD experiment is provided at the CERN website (viewed March 11, 2010): http://public.web.cern.ch/public/en/Research/CLOUD-en.html

CERN from testing the influence of cosmic rays on cloud formation, so it was not until August 25, 2011 that results from the first CLOUD experiments were reported in *Nature* (Kirkby 2011). The results confirmed those of the Svensmark team, which had used an electron particle beam to ionize the air instead of relying on natural and generated cosmic rays at CERN.[10] Thus, it is now clear that cosmic rays do induce nucleation of aerosol particles that, in turn, lead to cloud formation. In other words, when the strength of the sun's magnetic field declines (the sun becomes less active), the Earth is bombarded with more cosmic rays that result in aerosol production and cloud formation that cools the Earth. Clearly, the sun plays an overwhelmingly important role in climatic changes on Earth.

The experimental results enable us to give an alternative explanation for the rise in temperatures during the latter part of the 1900s. The explanation is twofold. First, the cosmic rays that reach the solar system fluctuate in intensity, as they depend on what is happening elsewhere in the universe – they depend on past cosmological events such as exploding stars and collisions that took place somewhere in the universe and took unknown light years to reach our solar system.

Second, the cosmic rays that bombard the Earth also depend on the activities of the sun. The sun throws off a non-stop stream of charged particles. This stream of particles is referred to as the solar wind and consists primarily of protons (because the sun is primarily hydrogen) plus positively charged atoms of quite a few other elements and a sufficient number of negatively charged electrons to keep the gas neutral. The solar wind also 'drags' the sun's magnetic field along as additional protection (Svensmark and Calder 2007; Svensmark and Friis-Christensen 1997). The ability of the sun to protect the Earth from cosmic rays depends on the strength of the solar wind and the sun's magnetic field.

The strength of the solar wind depends on solar activity, which is only weakly related to sunspot counts as the intensity of solar flares is more important than actual numbers. Further, although cosmic rays are generally stronger when there are fewer sunspots, there is no simple connection: "Effects on cosmic rays can lead or lag behind the rise and fall in sunspot counts by a year or so. And the influx of penetrating cosmic rays at the solar maximum around 2000 was cut to roughly the same extent as it was around 1979, when sunspots were much more numerous" (Svensmark and Calder 2007, pp. 222–223).

How powerful is this solar wind-cosmic ray interaction? Carslaw et al. (2002) found that the intensity of cosmic rays varies by about 15 % over a solar cycle, which in turn is associated with variation in low cloud cover over a solar cycle of

---

[10] See http://calderup.wordpress.com/2011/08/24/cern-experiment-confirms-cosmic-ray-action/ (viewed August 31, 2011). Indeed, the CLOUD results seem to show that Svensmark underestimated the nucleation rate (see http://www.wanliss.com/?p=1546 viewed October 4, 2011). Svensmark felt that the CLOUD team was moving too slowly in testing his hypothesis, and conducted his own tests with the results published in May 2011 (Enghoff et al. 2011). Meanwhile, there was some controversy concerning the *Nature* article as the key finding was hidden in the supplemental material and reference to the earlier experimental result was excluded, contrary to scientific protocol.

about 1.7 %. This change in cloud cover corresponds to a change in the planet's radiation budget of about 1 W/m^2, which is highly significant considering the radiative forcing from anthropogenic emissions of CO_2 is estimated to be 1.4 W/m^2.

Nir J. Shaviv and Jan Veizer (2003) found that between two-thirds and three-fourths of the variance in the earth's temperature over the past 500 million years may be attributable to cosmic ray flux. They argue that, once this is taken into account, a doubling of the air's CO_2 concentration could account for only about a 0.5 °C increase in global temperature, and not the 2 °C or more projected by climate modelers.

Finally, the solar magnetic field doubled during the twentieth Century, but it is not clear whether it will continue to get stronger in the future. No one knows. However, a strong solar magnetic field implies fewer cosmic rays penetrate to reach the Earth, thereby reducing cloud formation and leading to rising temperatures.

5.2.3 Solar-Arctic Connection

Some scientists believe that small changes in solar radiation entering the Earth's atmosphere are amplified by positive feedbacks involving the transfer of energy between the equator and the Arctic via wind patterns and oceans. Bond et al. (2001) envisioned solar variability provoking changes in North Atlantic deep water formation that alter the thermohaline circulation of the global ocean.

Soon (2009) demonstrates the plausibility of a three-part mechanism whereby variation in total solar irradiance affects Arctic temperatures by modulating the thermohaline circulation, the Inter-Tropical Convergence Zone rainbelt and tropical Atlantic ocean conditions, and the intensity of the wind-driven subtropical and subpolar gyre circulation – the ring-like system of ocean currents rotating clockwise in the Northern Hemisphere and counterclockwise in the Southern Hemisphere. Soon tested this 'TSI-Arctic thermal-salinity-cryospheric coupling mechanism' by showing the predicted 5-to-20-year delayed effects of total solar irradiance variability on the peak Meridional Overturning Circulation flow rate centered near 30°–35°N, and sea surface temperature for the tropical Atlantic. He found very close fits on multidecadal to centennial timescales.

5.2.4 Planetary Motion

Some scientists contend that most or all of the warming of the latter part of the twentieth century can be explained by natural gravitational and magnetic oscillations of the solar system induced by the planets' movements through space. These oscillations modulate solar variations and/or other extraterrestrial influences that then drive climate change. The Serbian astrophysicist Milutin Milankovitch (1941) was the first to suggest that planetary motion could affect climate on a multi-millennial timescale. More recent discoveries have enabled scientists accurately to measure these effects.

The Earth's orbit around the sun takes the form of an ellipse, not a circle, with the Earth passing farther away from the sun at one end of the orbit than at the other end.[11] The closest approach of the planet to the sun is called the 'perihelion' and the farthest is called the 'aphelion.' The perihelion now occurs in January, making northern hemisphere winters slightly milder. The change in the timing of the perihelion is known as the precession of the equinoxes, and it occurs every 22,000 years. The shape or 'eccentricity' of the Earth's orbit also varies on cycles of 100,000 and 400,000 years due to the tug of other planets, specifically the largest planets of Jupiter and Saturn. It shifts from a short broad ellipse that keeps the Earth closer to the sun, to a long flat ellipse that allows it to move farther from the sun and back again. The Earth also spins around an axis that tilts lower and then higher during a 41,000-year cycle. More tilt roughly means warmer northern hemisphere summers and colder winters; less tilt means cooler summers and milder winters.

The coincidence of these cycles is known to lead, with the help of climatic feedbacks from such things as water vapor, to the cooling and warming periods we recognize from historical data as ice ages and interglacial periods. Scientists now know that the critical sweep of rotation (known as precession) of the Earth's orbit means that, in about 11,000 years from now, the northern midwinter will fall in July instead of January, and the continental glaciers may return.

Could variation in the planet's movement through space account for climate change on a decadal scale as well as a millennial scale? The planets affect the Earth's climate via two possible mechanisms (Scafetta 2010). First, the varying tidal gravitational and magnetic forces of the planets on the sun, in particular those of Jupiter and Saturn, modulate solar activity and then solar variations modulate the terrestrial climate (as discussed above). Second, the varying gravitational and magnetic fields generated by the movement of Jupiter and Saturn modulate some terrestrial orbital parameters, such as the spinning of the Earth or the 'length of the day' (LOD), which then drives the ocean oscillations and, consequently, the climate.

Scafetta (2009, 2010) tested this theory using the sun's movement relative to the center of mass of the solar system as a proxy for all the known and unknown cycles involving natural oscillations of the solar system. His model closely tracked alternating periods of warming and cooling since 1860: peak warming in 1880 repeats in 1940 and again in 2000, while the smaller 1900 peak repeats in 1960. According to Scafetta (2010), some 60 % of the observed 0.5 °C increase in global average temperatures since 1970 is the result of natural, cosmological factors. He projects cooling between 2000 and sometime in the 2030s.

As noted in Chap. 4, climate models are unable to recreate past temperature variation without extensive tweaking of the models' parameters. Scafetta's model, on the other hand, explains most of the warming of the twentieth century. Here is where nature has the final say: the climate models forecast rapid warming over the next two decades, while Scafetta forecasts primarily cooling for the period to at least 2030.

[11] See National Oceanic and Atmospheric Administration, 'Astronautical theory of climate change' at http://www.ncdc.noaa.gov/paleo/milankovitch.html (viewed August 24, 2010).

None of this suggests that Soon's or Scafetta's or de Jager's or any other model fully explains the impact of solar irradiance forcing on climate. Nor should any of the other models considered in this section be viewed as a total explanation of climatic changes. Rather, these and other models indicate that there are well thought out, scientifically-sound explanations of climate forcings that pose a formidable challenge to the view that human emissions of greenhouse gases are the principal or only driver of global warming. Indeed, the peer-reviewed literature is replete with alternative explanations to the IPCC consensus that humans are culpable in causing the global warming observed since the mid 1800s, or to blame for major climate events such as hurricanes, drought or floods.

The dance of cosmic rays, the solar wind, the sun's magnetic field and the physics of cloud formation all impact the Earth's climate, with sunspots playing a minor role on some occasions and a more important one at other times (depending on their intensity). This dance is totally unpredictable, but likely has a major impact on the Earth's climate and the extent or lack of global warming in the future. Certainly, according to this theory, the role of CO_2 as a driver of climate change is much smaller, and so too are human activities that emit greenhouse gases.

5.3 Climate Science, Climategate and the IPCC

Two events in late 2009 are of particular importance for economic policy related to climate change. Both have been mentioned previously. The first event was the climategate revelations, while the second was the failure of governments at COP 15 in Copenhagen to reach a new, far-reaching climate accord to replace the Kyoto Protocol. As a result of the former, there appears to be a renewed energy among scientists and commentators, many of whom have been referred to as 'skeptics' or, more derogatorily, 'deniers', to point out problems with the IPCC's 2007 Fourth Assessment Report and with the science more generally. From the point of view of science, this should be viewed positively, but many climate scientists and environmentalists see these two events as major obstacles to be overcome. That is, rather than letting the scientific process decide whether global warming is a serious enough problem to warrant spending untold sums of money and slowing the economic development of many countries (thereby adding to the misery of millions of poor people), the tactic has been to paint the views of opponents as unscientific, conduct ad hominem attacks and fortify the defences in the name of science.[12]

[12] Ad hominem attacks are typical: while discussing the proxy temperature record and global warming with a Canadian climate scientist, the names of Ross McKitrick and Roy Spencer came up. As the discussion concluded, my colleague quickly pointed out that McKitrick was a fellow of the Fraser Institute, "a right-wing institute with a poor reputation" (of which I was unaware), while Spencer was castigated for his religious beliefs. I wondered what membership in a 'conservative' organization and religious belief had to do with the results they had published in reputable international scientific journals.

Climategate had been referred in previous chapters. What is clear from the climategate emails is that there was a deliberate attempt to prevent the publication of evidence contrary to the views of the Climate Research Unit at East Anglia University and their collaborators in the United States and elsewhere. This included attempts to blacklist journals, bully and/or remove journal editors, influence members of the editorial boards of journals, provide referee reports whose sole purpose was to deny publication as opposed to fair adjudication of the scientific merit, and promote publication of papers that supported the view that recent years were warmer than previous years.

One example is highly illustrative of the tactics employed, while, at the same time, questioning the science used to argue that global warming is already unprecedented. The Russian Institute of Economic Analysis in Moscow has long questioned the CRU's result that Siberian temperatures showed a warming trend as a result of anthropogenic warming Delingpole (2009). The Russians felt the CRU was deliberately tampering with the raw data, which showed no trend, to obtain an upward trend for Russian and thereby global temperatures. The Russians appear to have been vindicated: In an email sent by Phil Jones, Director of the CRU, to Michael Mann on Wednesday, March 31, 2004, he describes efforts to prevent publication of papers showing that Siberian temperatures were not increasing:

Mike,

> ... Jan [Esper] doesn't always take in what is in the literature even though he purports to read it. He's now looking at homogenization techniques for temperature to check the Siberian temperature data. We keep telling him the decline is also in N. Europe, N. America (where we use all the recently homogenized Canadian data). The decline may be slightly larger in Siberia, but it is elsewhere as well. Also Siberia is one of the worst places to look at homogeneity, as the stations aren't that close together (as they are in Fennoscandia and most of Canada) and also the temperature varies an awful lot from year to year.
>
> Recently rejected two papers (one for JGR [*J of Geophysical Research*] and for GRL [*Geophysical Research Letters*]) from people saying CRU has it wrong over Siberia. Went to town in both reviews, hopefully successfully. If either appears I will be very surprised, but you never know with GRL.
>
> Cheers
> Phil[13]

Not only were the Russians upset with the way the CRU had reconstructed Siberian temperatures, there was controversy over New Zealand temperatures. The New Zealand Climate Science Coalition wanted to know why New Zealand's National Institute of Water and Atmospheric Research shows a warming trend

[13] http://di2.nu/foia/1080742144.txt (viewed January 7, 2012). A searchable database for the 2009 and 2011 climategate emails is available at (viewed December 7, 2011): http://foia2011.org/. Also at (February 13, 2012): http://blogs.telegraph.co.uk/news/willheaven/100020210/climategate-why-the-russians-experts-might-not-have-our-best-interests-at-heart/

when raw data indicate otherwise. The reason relates to the homogenization of data to get rid of non-climate factors, although not everyone appears satisfied with this.[14]

5.3.1 Neutrality of IPCC Assessors

The purpose of the IPCC reports that are produced every 6 or so years is to gauge the state of climate science research – the ongoing advance in knowledge about climate change. The authors of the various chapters are to assess the research that has been undertaken in the previous 6-year period (or longer), adjudicate whatever disagreements appear in the literature, and make an informed judgement about the science that politicians can use for guiding policy. The IPCC reports are meant to be policy or advocacy neutral. While the various chapters in the IPCC assessment reports are supposed to be written by the 'experts' in the fields covered by the chapters, it turns out that this is not the case. Experts are those who are acknowledged to know the most about the field in question, having a PhD in the area and having authored scientific papers on the subject. However, Donna Laframboise (2011) provides damaging evidence indicating that many lead authors (those responsible for writing the main sections of the IPCC reports) had little or no expertise (often having no more than a Masters degree in the area and with no scientific publications) and/or had ties to the environmental movement. In other cases, Laframboise found that climate scientists were little more than computer modelers lacking experience in the scientific methods related to hypothesis testing. Further, in most cases where lead authors were experts in the area, they turned out to be the very same scientists who had written the articles they now had to adjudicate.

An example of this has already been discussed in Chap. 2 (Sect. 2.3). In the IPCC WGI (2007) chapter on "Observations: Surface and Atmospheric Climate Change", the co-ordinating lead authors (assessors) Kevin Trenberth and Phil Jones respond in dismissive fashion to two articles that show warming is strongly correlated with non-climate factors and thus disagrees with their own research: when spatial considerations are taken into account "the correlation of warming with industrial and socioeconomic development ceases to be statistically significant. In addition, observed warming has been, and transient greenhouse-induced warming is expected to be, greater over land than over the oceans ..., owing to the smaller thermal capacity of the land" (p. 244). The IPCC authors provide no evidence that, when spatial factors are taken into account, the non-climate factors cease to be statistically significant. They have carried out no tests, but simply state this to be the case, contrary to scientific protocol. Further, they argue that the research with which they disagree

[14] The controversy has been dubbed 'Kiwigate'. See 'Kiwigate is a Carbon Copy of Climategate' by John O'Sullivan, April 26, 2010 at Suite101.com: http://climatology.suite101.com/article.cfm/kiwi-scientists-copy-data-trashing-technique-of-climategate#ixzz0mJkLtKzW (viewed May 4, 2010).

includes sea surface temperatures (SSTs) and not only land temperatures, but this is blatantly untrue (McKitrick 2010).

Another example is provided by A.W. Montford (2010). He uses the climategate emails to demonstrate that the IPCC's own rules were flouted when a lead author of the paleoclimate chapter in the IPCC's Fourth Assessment Report (IPCC WGI 2007, pp. 431–497) relied on an unpublished paper provided by the authors to rebut critics of the hockey stick view of past climate. Not only was the paper not provided to reviewers of the chapter in question, but the paper in question, plus an accompanying one by another set of authors, was accepted for publication after the IPCC deadline (Montford 2010, pp. 424–434).[15] Further, an IPCC review editor, who was to adjudicate disputes and ensure that both sides of a debate were fairly represented, not only "believed the Hockey Stick used a biased methodology and gave results that were not statistically significant" (p. 447), but then helped the authors of the paleoclimate chapter craft a response to critics of the hockey stick and, finally, signing off the chapter as 'a reasonable assessment of the evidence' (pp. 447–448).

There is a general failure in the IPCC assessment reports to include scientists with differing viewpoints. Of course, the Intergovernmental Panel on Climate Change is aware of this problem, and seeks to rectify it through a review process – scientists representing all points of view are invited to submit comments on earlier drafts of IPCC chapters. Although this is good practice in principle, contentious comments are generally ignored or dismissed off hand with a response similar to the following: 'this has now been dealt with.' The original reviewer is not told how it was 'dealt with' as this statement could mean everything from 'we have changed our view and rewritten the report accordingly' to 'we have noted your concerns in a footnote' to 'we have thought about this but not changed anything.' Another common response is that 'appropriate revisions and editing made', but nothing has actually been done.[16]

As noted in Chaps. 2 and 3, the data and methods (computer code) used to homogenize raw weather data or construct temperature proxies are for the most part unavailable to researchers who wish to check the validity of the temperature reconstructions. The IPCC does not, it would seem, demand that the data and methods that underlie the various chapters be available to other researchers. The assessment of the IPCC authors and reviewers must be done sans the data and methods employed in the research to be evaluated. Given that decisions affecting the lives of billions of people are based on the outcomes, it would seem that the data and computer codes used in any of the studies reviewed by the IPCC, the vast majority of which was publicly funded, should be made widely accessible to any researcher. Further, researchers should be required to help researchers duplicate their results where computer code is poorly documented.

---

[15] As a contributing author to Chapter 9 of Working Group III (IPCC WGIII 2007, pp. 541–584), the current author was told that research published after the IPCC deadline could not be brought to bear on the discussion and conclusions regarding the role of forestry in mitigating climate change, even though the recent research had important implications for some of the conclusions.

[16] In the next subsection, we discuss the imminent disappearance of the Himalayan glaciers. Reviewer criticisms of this conclusion were simply ignored, or not addressed; see (viewed June 10, 2010) http://www.dailymail.co.uk/news/article-1245636/Glacier-scientists-says-knew-data-verified.html.

5.3.2 *The IPCC 'Gates'*

The main objection that has been raised against climate-skeptical articles or papers that appear on line (the so-called 'gray literature'), or commissioned reports such as that by Wegman et al. (2006), is that they are not peer reviewed. Peer review is considered the gold standard. However, it is well known that peer review does not guarantee the validity of scientific findings, because the reviewers frequently fail to identify problems with the research. This can occur for a variety of reasons: the reviewer may not have the time to conduct a thorough review or even have the expertise to judge the quality of the work. One reason for this is that an individual often accepts to conduct a review as a necessary condition for submitting papers to the same journal. Even if the reviewer has the expertise, he or she may not have access to the data and computer code; indeed, this is the most common state of affairs and few reviewers request access to the data, although a few may stipulate that the author(s) provide a web link to the data in the body of their paper. Whether a link is provided is at the discretion of the editor. The basic point is that reviewers rarely if ever attempt to replicate the analysis.

In some cases, especially where the review process is single blind so that the referee knows who the authors are although the authors do not know the identity of the referee,[17] the reviewer knows who wrote the manuscript, and this generally biases the report one way or another. In other cases, the referee has previously seen the paper when one of the authors presented it at one or other professional meeting. Sometimes presentation at a particular workshop or professional forum is a prerequisite for a favorable review and/or acceptance at a particular journal.

Although peer review is helpful in getting authors rethink their research and crafting better arguments, it is not a necessary or sufficient condition to guarantee quality. Likewise, research that is not peer reviewed, whether appearing in some published/printed form or only as an electronic document on the internet, might still make a scientific contribution. Indeed, with a proliferation of journals, particularly eJournals, and the ability of individual researchers or groups of researchers to make their research readily available on the web, the line between peer-reviewed and non-reviewed research is increasingly blurred. A good search engine will enable a researcher to find relevant research in his or her area regardless of whether it was peer reviewed. In this world, research will be deemed worthy by those who cite it, and build upon it.

The IPCC prides itself on the quality of the science upon which its conclusions are based. However, in addition to the problems noted above, some one-third of the citations employed by the IPCC are to non-reviewed studies, reports, newspaper stories, student papers, and so on. As a result, several anomalies have come to light.

[17] An open review process is one where the authors also know the referee(s), while a double-blind process is one where the names of the authors are not revealed to the reviewers and vice versa. In some cases, and this varies from field to field, the reviewer can guess the names of the author(s), or Google the title or abstract and find an earlier version of the paper that identifies the author(s).

While any one of these irregularities can be justified as an honest mistake, they do point out that the process of reviewing, assessing and writing the various chapters that make up this scientific report is less than ideal. A process that was supposed to be bullet proof has found that human foibles are unavoidable. Consider some of the most prominent.

Perhaps the error that received the most attention concerned the retreat of the Himalayan glaciers (Cogley et al. 2010). The glaciers are reported by the IPCC as most likely to disappear by 2035.

> Glaciers in the Himalaya are receding faster than in any other part of the world … and, if the present rate continues, the likelihood of them disappearing by the year 2035 and perhaps sooner is very high if the Earth keeps warming at the current rate. Its total area will likely shrink from the present 500,000 to 100,000 km^2 by the year 2035 (WWF 2005). The receding and thinning of Himalayan glaciers can be attributed primarily to the global warming due to increase in anthropogenic emission of greenhouse gases. The relatively high population density near these glaciers and consequent deforestation and land-use changes have also adversely affected these glaciers (IPCC WGIII 2007, p. 493).

Notice that the reference is to an unpublished World Wildlife Fund report from 2005.[18] What is interesting is that, despite retaining the March 2005 date, the WWF report now contains a 'Correction' page that reads as follows:

> On page 29 of the following report WWF included the following statement: "In 1999, a report by the Working Group on Himalayan Glaciology (WGHG) of the International Commission for Snow and Ice (ICSI) stated: 'glaciers in the Himalayas are receding faster than in any other part of the world and, if the present rate continues, the livelihood [sic] of them disappearing by the year 2035 is very high.'" This statement was used in good faith but it is now clear [as a result of much publicity] that this was erroneous and should be disregarded. The essence of this quote is also used on page 3 in the Executive summary where it states: The New Scientist magazine carried the article "Flooded Out – Retreating glaciers spell disaster for valley communities" in their 5 June 1999 issue. It quoted Professor Syed Hasnain, then Chairman of the International Commission for Snow and Ice's (ICSI) Working Group on Himalayan Glaciology, who said most of the glaciers in the Himalayan region 'will vanish within 40 years as a result of global warming.' This statement should also be disregarded as being unsound. WWF regret any confusion this may have caused.

This 'confusion' became known as 'Himalayagate' because it rested on magazine interviews given by the glaciologist Syed Hasnain.[19] The IPCC clearly failed to take into account published research on the Himalayan glaciers (Cogley et al. 2010).

A closer look at the IPCC quote suggests that the authors are anything but neutral in their assessment of the science. They report the imminent demise of the massive Himalayan glaciers with what might be considered a certain amount of glee, attributing it entirely to human causes – both anthropogenic global warming and overpopulation in the countries surrounding the Himalayan Mountains! Indeed, Murari Lal, a co-ordinating lead author of the IPCC's Chap. 10 on Asia in which the error about

---

[18] WWF (WorldWildlife Fund) (2005). Note that the original reference in the IPCC report has 79 pages, not the 80 it now has.

[19] Rose (2010).

the melting of the Himalayan glaciers appears, acknowledged that the statement did not "rest on peer-reviewed scientific research, ... [but] we thought that if we can highlight it, it will impact policy-makers and politicians and encourage them to take some concrete action."[20]

More recent findings reported by *The Economist* (November 20, 2010, p. 92) indicate that black carbon (soot) is covering the Himalayan glaciers, reducing their ability to reflect light by 2–5 %. The soot originates primarily in India and increases the amount of melting by 12–34 %. The worry is not that this will cause the glaciers to disappear sometime in the future, but that the increased melting will increase the potential for downstream flooding.

In what has been referred to as 'icegate', the IPCC WGII (2007, p. 86) uses evidence from mountain climbers for their assertion that glaciers in the tropical Andes, European Alps and Africa are disappearing at unprecedented rates. Although there might be truth to this, the objection is that such reports do not constitute scientific evidence.

'Amazongate' refers to the IPCC's projection that 40 % of the Amazon rainforest will disappear by 2050: "in the worst-case scenario, by 2050 the projected deforestation trend will eliminate 40% of the current 540 Mha of Amazon forests, releasing approximately ... 109 tonnes/ha of carbon to the atmosphere" (IPCC WGII 2007, p. 594). This conclusion is based on simulation models by Soares-Filho et al. (2005), but has been criticized because the deforestation results from logging and burning, which are non-climate factors, and not temperature changes.[21]

The problem here is that the IPCC blurs its role of assessing climate science and climate impacts with the broader issue of government policies relating to the environment. Economic research by van Kooten and Bulte (2000), van Soest et al. (2002), van Kooten and Folmer (2004), Folmer and van Kooten (2007), and others concludes that tropical deforestation is often the result of explicit or implicit government policy to reduce the forest cover to a more socially desirable level. Thus, even though deforestation does increase emissions of CO_2, the debate about tropical deforestation has more to do with economic and social issues than climate. Governments might permit exploitation of forests because they provide economic rents, forest-related jobs and, eventually, a higher-valued agricultural use of land. This is often the case in southeastern Asia. In Brazil and some other countries, deforestation is permitted as part of a policy to promote economic development in regions away from large urban areas, thereby reducing the population pressure on these urban areas.

---

[20] Same David Rose source as previous note. An Indian report (see Bagla 2009) indicating that Himalayan glaciers had not exhibited abnormal annual retreat had, in the meantime, been criticized as "voodoo science" by Raj Pachauri, the IPCC Chairman (same David Rose source).

[21] See Landsbaum (2010). The IPCC reference should probably be to Soares-Filho et al. (2006) rather than Soares-Filho et al. (2005), as the latter refers to a paper found in *Estud. Avançados* (volume 19, pp. 137–152).

'Dutchgate' came about because the Fourth Assessment Report placed much of the Netherlands below sea level: "The Netherlands is an example of a country highly susceptible to both sea-level rise and river flooding because 55% of its territory is below sea level where 60% of its population lives and 65% of its Gross National Product (GNP) is produced" (IPCC WGII 2007, p. 547). In fact, only 20% of the Netherlands might be considered below sea level, and the country is likely at greater threat from flooding from rivers upstream than from sea level rise per se. To militate against such flooding, the government is looking to contract with landowners to encourage them to accept flooding of their fields in order to protect developed residential, commercial and industrial areas from flooding. However, higher sea levels will make it more difficult to remove flood waters.

'Africagate' refers to the erroneous notion that rising temperatures will be a disaster for African agriculture. In the executive summary to the chapter on Africa in Working Group II's report, entitled *Impacts, Adaptation and Vulnerability*, the authors write:

> A number of countries in Africa already face semi-arid conditions that make agriculture challenging, and climate change will be likely to reduce the length of growing season as well as force large regions of marginal agriculture out of production. Projected reductions in yield in some countries could be as much as 50% by 2020, and crop net revenues could fall by as much as 90% by 2100, with small-scale farmers being the most affected. This would adversely affect food security in the continent (IPCC WGII 2007, p. 435).

The executive summary is based on the following assessment of the science found in the main text of the chapter:

> In other countries, additional risks that could be exacerbated by climate change include greater erosion, deficiencies in yields from rain-fed agriculture of up to 50% during the 2000–2020 period, and reductions in crop growth period (Agoumi 2003). A recent study on South African agricultural impacts, based on three scenarios, indicates that crop net revenues will be likely to fall by as much as 90% by 2100, with small-scale farmers being the most severely affected (IPCC WGII 2007, p. 448).

The report by Agoumi (2003) focuses only on North African countries and is by no means peer reviewed; rather, it is a short highly speculative document and should not have been used to justify the above statements. (No reference is provided to the South African study.) But the greater error is that the authors of the chapter took what was definitely local results pertaining only to rain-fed agriculture and incorrectly attributed the findings to the entire continent, which has the appearance of advocacy. Results certainly do not take into account the ability of farmers and landowners to adapt to changing market conditions brought about by climate change (see, e.g., Adams et al. 1990; Darwin et al. 1995; Mendelsohn et al. 2000; Schimmelpfenning et al. 1996).

There are other instances where the IPCC is faulted for failing to base its findings on scientific studies,[22] but some of the criticism is specious at best, as illustrated for the case of 'reefgate.' Reefgate refers to the link between temperatures and coral degradation based on studies by advocacy groups. There is a ring of truth to the

[22] See reference to Mark Landsbaum in previous footnote.

criticism that some results lack scientific proof, simply because studies attempt to extrapolate information about the effect of climate change on coral reefs from observations at particular sites and/or certain weather events. Extrapolation beyond the range of available observations is always tricky. Thus, the following conclusion from the Fourth Assessment Report needs to be taken with a grain of salt: "Recent risk analysis of coral reefs suggests that between 14 and 30 % of the reefs in Asia are projected to be lost during the next 2–10 years and 10–30 years, respectively (14% and 18% for global), unless the stresses are removed and relatively large areas are protected" (p. 485). Clearly, nothing near 6 % of the globe's coral reefs and 12 % of Asia's have disappeared in the 4 years since the Report was written, nor will 14 % of the globe's and 30 % of Asia's coral reefs disappear by 2017. While various scientific studies are cited as evidence of the eminent degradation of coral reefs, the qualifier that "unless the stresses are removed and relatively large areas are protected" should be noted: the stresses are not specifically linked to global warming, although that might be the implication.

One example illustrates the uncertainty that remains. Hoegh-Guldberg et al. (2007) expect higher temperatures due to climate change and increasing ocean acidification due to rising atmospheric CO_2 to lead to a dramatic decline in coral calcification, thus posing a major future threat to the globe's coral reefs and oceans. Upon testing this theory for temperate coral reefs (Mediterranean *zooxanthellate* coral), Rodolfo-Metalpa et al. (2010) find that "an increase in CO_2 [such as forecast for 2100], alone or in combination with elevated temperature, had no significant effect on photosynthesis, photosynthetic efficiency and calcification." However, a 3 °C rise in winter temperatures increased gross photosynthesis by 72 %, as well as daytime calcification rates. At least for temperate coral, climate change was not a threat to coral calcification or the eventual demise of coral reefs. It appears that a rapid change (rise) in temperature will 'bleach' out the coral's symbiotic algae, as occurred during the 1998 *El Niño* event, but that higher temperatures per se do not impact production of coral reefs; after all, coral reefs are found in the Persian Gulf where water temperatures reach 35 °C (Ridley (2010).

With respect to 'reefgate', it appears a study attributable to Greenpeace (Hoegh-Guldberg et al. 2000) is used to justify the high costs that would be attributable to climate change impacts on coral reefs (IPCC WGII 2007, p. 336). An examination of this study indicates that none of the methods that economists use for estimating economic costs and benefits was employed. These methods are described in the next chapter.

5.4 Discussion

The objective of this chapter was not to propose an alternative to the model that says carbon dioxide and other greenhouse gases emitted by human activities are the principal cause of climate change. Rather, we simply point out that the current IPCC view of the greenhouse effect is not without its critics and that these critics are highly respected scientists who have published their ideas in high-quality, peer-reviewed journals. Further, scientists have published alternative theories that explain how temperatures change, and, importantly, observational data often supports these theories. Thus, they cannot easily be discarded.

What is perhaps most disturbing is that the IPCC process is itself flawed and climate scientists have behaved inappropriately in defending their view of the world. And that is the point: behavior that belittles alternative explanations of climate change, seeks to interfere with the publication of contrary views, and relies on ad hominem attacks is anti-scientific behavior. It is not interested in discovering the truth about the future course of the globe's climate, at least as far as it is predictable. In this regard, computer models cannot take the place of observations and empirical analysis; they can only provide additional insights.

In the end, however, it does not matter whether climate change is happening as described by the IPCC or not. There is simply no way to stop the burning of fossil fuels and the addition of carbon dioxide into the atmosphere. Fossil fuels are too abundant and cheap, and reducing their use to the point where it will matter (at least if one accepts the anthropogenic global warming story) will simply be politically infeasible, which explains why emission-reduction targets are set to occur after the current politicians are likely to have left office (see Chap. 8). As discussed in Chap. 10, there are no technological fixes on the horizon that will enable countries to de-carbonize their economies. The costs will be too onerous and, therefore, a realistic approach to policy needs to be taken. Perhaps the alternative views expressed in this chapter will become more accepted as this realization sets in.

References

Adams, R. M., Rosenzweig, C., Peart, R. M., Richie, J. T., McCarl, B. A., Glyer, J. D., Curry, R. B., Jones, J. W., Boote, K. J., & Allen, L. H., Jr. (1990). Global climate change and US agriculture. *Nature, 345*(6272), 219–224.

Agoumi, A. (2003). Vulnerability of North African countries to climatic changes: Adaptation and implementation strategies for climatic change. *Developing perspectives on climate change: Issues and analysis from developing countries and countries with economies in transition*. Retrieved June 10, 2010, http://www.cckn.net/pdf/north_africa.pdf

Bagla, P. (2009, November 13). No sign yet of Himalayan meltdown, Indian report finds. *Science, 326*(5955), 924–925.

Bond, G., Kromer, B., Beer, J., Muscheler, R., Evans, M. N., Showers, W., Hoffmann, S., Lotti-Bond, R., Hajdas, I., & Bonani, G. (2001, December 7). Persistent solar influence on North Atlantic climate during the Holocene. *Science, 294*(5549), 2130–2136.

Braun, H., Christl, M., Rahmstorf, S., Ganopolski, A., Mangini, A., Kubatzki, C., Roth, K., & Kromer, B. (2005, November 10). Possible solar origin of the 1,470-year glacial climate cycle demonstrated in a coupled model. *Nature, 438*(7065), 208–211.

Burgermeister, J. (2007). Missing carbon mystery: Case solved?. *Nature Reports: Climate Change, 3*, 36–37. doi:10.1038/climate.2007.35.

Carslaw, K. S., Harrison, R. G., & Kirby, J. (2002). Cosmic rays, clouds, and climate. *Science, 298*(5599), 1732–1737.

Cogley, J. G., Kargel, J. S., Kaser, G., & van der Veen, C. J. (2010, January 29). Tracking the source of glacier misinformation. *Science, 327*(5965), 522.

Darwin, R., Tsigas, M., Lewandrowski, J., & Raneses, A. (1995, June). *World agriculture and climate change: Economic adaptations* (AE Report No. 703, p. 86). Washington, DC: U.S. Department of Agriculture, Economic Research Service.

de Jager, C. (2008). Solar activity and its influence on climate. *Netherlands Journal of Geosciences, 87*(3), 207–213.

De Laat, A. T. J., & Maurellis, A. N. (2004). Industrial CO_2 emissions as a proxy for anthropogenic influence on lower tropospheric temperature trends. *Geophysical Research Letters, 31*, L05204. doi:10.1029/2003GL019024.

Delingpole, J. (2009, December 16). Climategate goes SERIAL: Now the Russians confirm that UK climate scientists manipulated data to exaggerate global warming. *The Telegraph*. Viewed June 8, 2010, from, http://blogs.telegraph.co.uk/news/jamesdelingpole/100020126/climategate--goes-serial-now-the-russians-confirm-that-uk-climate-scientists-manipulated-data-to-exaggerate-global-warming/

Duhau, S., & de Jager, C. (2010, June). The forthcoming grand minimum of solar activity. *Journal of Cosmology, 8*, 1983–1999.

Easterbrook, D. J. (2008, December). *Solar influence on recurring global, decadal, climate cycles recorded by glacial fluctuations, ice cores, sea surface temperatures, and historic measurements over the past millennium*. Abstract of American Geophysical Union annual meeting, San Francisco. Viewed September 1, 2010, from, http://adsabs.harvard.edu/abs/2008AGUFMGC21A0725E

Enghoff, M. B., Pedersen, J. O. P., Uggerhøj, U. I., Paling, S. M., & Svensmark, H. (2011). Aerosol nucleation induced by a high energy particle beam. *Geophysical Research Letters, 38*(L09805), 4. doi:doi:10.1029/2011GL047036.

Fagan, B. M. (2000). *The little ice age: how climate made history 1300–1850*. New York: Basic Books.

Folmer, H., & van Kooten, G. C. (2007). Deforestation. In B. Lomborg (Ed.), *Solutions for the world's biggest problems: Costs and benefits*. Cambridge: Cambridge University Press.

Ghosh, P., & Brand, W. A. (2003). Stable isotope ratio mass spectrometry in global climate change research. *International Journal of Mass Spectrometry, 228*(1), 1–33.

Hoegh-Guldberg, O., Mumby, O. J., Hooten, A. J., Steneck, R. S., Greenfield, P., Gomez, E., Harvell, C. D., Sale, P. F., Edwards, A. J., Caldeira, K., Knowlton, N., Eakin, C. M., Iglesias-Prieto, R., Muthigo, N., Bradbury, R. H., Dubi, A., & Hatziolos, M. E. (2007). Coral reefs under rapid climate change and ocean acidification. *Science, 318*(5857), 1737–1742.

Hoegh-Guldberg, O., Hoegh-Guldberg, H., Cesar, H., & Timmerman, A. (2000). Pacific in peril: Biological, economic and social impacts of climate change on Pacific coral reefs. *Greenpeace* (p. 72). Viewed June 6, 2010. http://www.greenpeace.org/raw/content/australia/resources/reports/climate-change/coral-bleaching-pacific-in-pe.pdf

Idso, C., & Singer, S. F. (2009). *Climate change reconsidered: 2009 report of the Nongovernmental International Panel on Climate Change (NIPCC)*. Chicago: The Heartland Institute.

IPCC WGI. (2007). *Climate change 2007: The physical science basis – working group I contribution to the fourth assessment report of the Intergovernmental Panel on Climate Change*. Cambridge: Cambridge University Press.

IPCC WGII. (2007). *Climate change 2007: Impacts, adaptation and vulnerability – working group II contribution to the fourth assessment report of the Intergovernmental Panel on Climate Change*. Cambridge: Cambridge University Press.

IPCC WGIII. (2007). *Climate change 2007: Climate change mitigation – working group III contribution to the fourth assessment report of the Intergovernmental Panel on Climate Change*. Cambridge: Cambridge University Press.

Kirkby, J. (2011, 25 August). Role of sulphuric acid, ammonia and galactic cosmic rays in atmospheric aerosol nucleation. *Nature, 476*, 429–433. doi:10.1038/nature10343.

Knorr, W. (2009). Is the airborne fraction of anthropogenic CO_2 emissions increasing? *Geophysical Research Letters, 36*, L21710. doi:10.1029/2009GL040613.

Laframboise, D. (2011). *The delinquent teenager who was mistaken for the world's top climate expert*. Toronto: Ivy Avenue Press.

Lamb, H. H. (1995). *Climate, history and the modern world* (2nd ed.). New York: Taylor & Francis.

Landsbaum, M. (2010, February 12). What to say to a global warming alarmist. *Orange County Register*. Viewed June 10, 2010, from http://www.ocregister.com/common/printer/view.php?db=ocregister&id=234092

Lindzen, R. S., & Choi, Y.-S. (2009). On the determination of climate feedbacks from ERBE data. *Geophysical Research Letters, 36*, L16705. doi:10.1029/2009GL039628.

Lindzen, R. S., & Choi, Y.-S. (2011, February 12). On the observational determination of climate sensitivity and its implications. *Asia-Pacific Journal of Atmospheric Sciences, 47*(4), 377–390. doi:10.1007/s13143-011-0023-x.

Lindzen, R. S., Chou, M.-D., & Hou, A. Y. (2001). Does the earth have an adaptive infrared iris? *Bulletin of the American Meteorological Society, 82*(3), 417–432.

Loehle, C. (2004). Climate change: Detection and attribution of trends from long-term geologic data. *Ecological Modeling, 171*(4), 433–450.

Loehle, C. (2009). Cooling of the global oceans since 2003. *Energy & Environment, 20*(1&2), 101–104.

Manning, A. C., Keeling, R. F., Katz, L. E., Paplawsky, W. J., & McEvoy, E. M. (2003). Interpreting seasonal cycles of atmospheric oxygen and carbon dioxide concentrations at American Samoa Observatory. *Geophysical Research Letters, 30*(6), 1333. doi:10.1029/2001GL014312.

Matsui, T., & Pielke, R. A., Sr. (2006). Measurement-based estimation of the spatial gradient of aerosol radiative forcing. *Geophysical Research Letters, 33*, L11813. doi:10.1029/2006GL025974.

McKitrick, R. (2010). Circling the bandwagons: My adventures correcting the IPCC. *SPPI reprint series*. Retrieved January 19, 2011, http://scienceandpublicpolicy.org/images/stories/papers/reprint/Circling_the_Bandwagons_Correcting_the_IPCC.pdf

Mendelsohn, R., Morrison, W., Schlesinger, M. E., & Andronova, N. G. (2000). Country-specific market impacts of climate change. *Climatic Change, 45*(3–4), 553–569.

Milankovitch, M. (1941). *Kanon der Erdbestrahlungen und seine Anwendung auf das Eiszeiten problem. Canon of insolation and the Ice Age problem* (Israel Program for Scientific Translations, Trans. 1969 ed.). Springfield: U.S. Dept. of Commerce, Clearinghouse for Federal Scientific and Technical Information.

Montford, A. W. (2010). *The hockey stick illusion: Climate and the corruption of science*. London: Stacey International.

National Research Council. (2005). *Radiative forcing of climate change: Expanding the concept and addressing uncertainties*. Committee on Radiative Forcing Effects on Climate Change, Climate Research Committee, Board on Atmospheric Sciences and Climate, Division on Earth and Life Studies. Washington, DC: The National Academic Press.

Niyogi, D., Chang, J.-I., Saxena, V. K., Holt, T., Alapaty, K., Booker, F., Chen, F., Davis, K. J., Holben, B., Matsui, T., Meyers, T., Oechel, W. C., Pielke, R. A., Sr., Wells, R., Wilson, K., & Xue, Y. (2004). Direct observations of the effects of aerosol loading on net ecosystems CO_2 exchanges over different landscapes. *Geophysical Research Letters, 31*, L20506. doi:10.1029/2004GL020915.

O'Dowd, C. D., Jimenez, J. L., Bahreini, R., Flagan, R. C., Seinfeld, J. H., Hämeri, K., Pirjola, L., Kulmala, M., Jennings, S. G., & Hoffmann, T. (2002, June 6). Marine aerosol formation from biogenic iodine emissions. *Nature, 417*(6889), 632–636.

Oreskes, N. (2004, December 3). The scientific consensus on climate change. *Science, 306*(5702), 1686.

Pielke, R. A., Sr., Beven, K., Brasseur, G., Calvert, J., Chyahine, M., Dickerson, R. R., Entekhabi, D., Foufoula-Georgiou, E., Guptak, H., Gupta, B., Krajewski, W., Krider, E. P., Lau, W. K. M., McDonnell, J., Rossow, W., Schaake, J., Smith, J., Sorooshian, S., & Wood, E. (2009). Climate change: the need to consider human forcings besides greenhouse gases. *EOS, Transactions American Geophysical Union, 90*(45), 413. doi:10.1029/2009EO450008.

Ridley, M. (2010, June 15). Threat from ocean acidification greatly exaggerated. *The Rational Optimist*. Viewed June 19, 2010, from http://www.thegwpf.org/the-observatory/1106-matt-ridley-threat-from-ocean-acidification-greatly-exaggerated.html

Rodolfo-Metalpa, R., Martin, S., Ferrier-Pagès, C., & Gattuso, J.-P. (2010). Response of the temperate coral *Cladocora caespitosa* to mid- and long-term exposure to pCO_2 and temperature levels projected for the year 2100 AD. *Biogeosciences, 7*, 289–300.
Rose, D. (2010, January 24). *Glacier scientist: I knew data hadn't been verified*. Viewed June 10, 2010, from http://www.dailymail.co.uk/news/article-1245636/Glacier-scientists-says-knew-data-verified.html
Scafetta, N. (2009). Empirical analysis of the solar contribution to global mean air surface temperature change. *Journal of Atmospheric and Solar-Terrestrial Physics, 71*(17–18), 1916–1923. doi:10.1016/j.jastp. 2009.07.007.
Scafetta, N. (2010, March 18). Climate change and its causes: A discussion about some key issues. *Science and Public Policy Institute (SPPI) Paper.* Retrieved August 24, 2010, http://scienceandpublicpolicy.org/images/stories/papers/originals/climate_change_cause.pdf
Scafetta, N., & West, B. J. (2007). Phenomenological reconstructions of the solar signature in the northern hemisphere surface temperature records since 1600. *Journal of Geophysical Research, 112*, D24S03. doi:10.1029/2007JD008437.
Scafetta, N., & West, B. J. (2008). Is climate sensitive to solar variability? *Physics Today, 61*(3), 50–51.
Schimmelpfenning, D., Lewandrowski, J., Reilly, J., Tsigas, M., & Parry, I. (1996). *Agricultural adaptation to climate change: Issues of long-run sustainability* (Agric. Econ. Report 740, p. 57). Washington, DC: USDA Economic Research Service.
Sciare, J., Mihalopoulos, N., & Dentener, J. F. (2000). Interannual variability of atmospheric dimethylsulfide in the Southern Indian Ocean. *Journal of Geophysical Research, 105*(D21), 26369–26377. doi:10.1029/2000JD900236.
Shaviv, N., & Veizer, J. (2003). Celestial driver of Phanerozoic climate? *GSA Today, 13*(7), 4–10.
Soares-Filho, B. S., Nepstad, D. C., Curran, L. M., Cerqueira, G. C., Garcia, R. A., Ramos, C. A., Voll, E., McDonald, A., Lefebvre, P., Schlesinger, M. E., & McGrath, D. (2005). Cenários de desmatamento para a Amazônia. *Estudos Avançados, 19*(54), 137–152. doi:10.1590/S0103-40142005000200008.
Soares-Filho, B. S., Nepstad, D. C., Curran, L. M., Cerqueira, G. C., Garcia, R. A., Ramos, C. A., Voll, E., McDonald, A., Lefebvre, P., & Schlesinger, P. (2006). Modeling conservation in the Amazon Basin. *Nature, 440*(7083), 520–523.
Soon, W. W.-H. (2005). Variable solar irradiance as a plausible agent for multidecadal variations in the arctic-wide surface air temperature record of the past 130 years. *Geophysical Research Letters, 32*, L1672. doi:10.1029/2005GL023429.
Soon, W. W.-H. (2009). Solar arctic-mediated climate variation on multidecadal to centennial timescales: empirical evidence, mechanistic explanation, and testable consequences. *Physical Geography, 30*(2), 144–148.
Spencer, R. W., Braswell, W. D., Christy, J. R., & Hnilo, J. (2007). Cloud and radiation budget changes associated with tropical intraseasonal oscillations. *Geophysical Research Letters, 34*, L15707. doi:10.1029/2007GL029698.
Sud, Y. C., Walker, G. K., & Lau, K.-M. (1999). Mechanisms regulating deep moist convection and sea-surface temperatures in the tropics. *Geophysical Research Letters, 26*(8), 1019–1022.
Svensmark, H., & Calder, N. (2007). *The chilling stars: A new theory of climate change*. Cambridge: Icon Books Ltd.
Svensmark, H., & Friis-Christensen, E. (1997). Variation of cosmic ray flux and global cloud coverage – A missing link in solar-climate relationships. *Journal of Atmospheric and Solar-Terrestrial Physics, 59*(11), 1225–1232.
Travis, D. J., Carleton, A. M., Johnson, J. S., & DeGrand, J. Q. (2007). U.S. jet contrail frequency changes: Influences of jet aircraft flight activity and atmospheric conditions. *International Journal of Climatology, 27*(5), 621–632. doi:10.1002/joc.1418.
van Kooten, G. C., & Bulte, E. H. (2000). *The economics of nature: Managing biological assets*. Oxford: Blackwell Publishers.
van Kooten, G. C., & Folmer, H. (2004). *Land and forest economics*. Cheltenham: Edward Elgar.
van Soest, D. P., Bulte, E. H., Angelsen, A., & van Kooten, G. C. (2002). Opening Pandora's box? Technological change and deforestation. *Environment & Development Economics, 7*(2), 73–84.

Weaver, A. (2008). *Keeping our cool: Canada in a warming world.* Toronto: Viking Canada/ Penguin Group.
Wegman, E. J., Scott, D. W., & Said, Y. H. (2006). *Ad hoc committee report on the 'Hockey Stick' global climate reconstruction.* Retrieved October 16, 2009, http://republicans.energycommerce.house.gov/108/home/07142006_Wegman_Report.pdf
Wingenter, O. W., Haase, K. B., Zeigler, M., Blake, D. R., Rowland, F. S., Sive, B. C., Paulino, A., Thyrhaug, R., Larsen, A., Schultz, K., Meyerhöfer, M., & Riebesell, U. (2007). Unexpected consequences of increasing CO_2 and ocean acidity on marine production of DMS and CH_2ClI: Potential climate impacts. *Geophysical Research Letters, 34*, L05710. doi:10.1029/2006GL028139.
WWF (WorldWildlife Fund). (2005). *An overview of glaciers, glacier retreat, and subsequent impacts in Nepal, India and China* (80 pp.). WWF Nepal Programme. Viewed June 8, 2010, from http://assets.panda.org/downloads/glacierssummary.pdf

Chapter 6
How Economists Measure Wellbeing: Social Cost-Benefit Analysis

A simple man believes every word he hears; a clever man understands the need for proof

(Proverbs 14:15)

Yet, in holding scientific research and discovery in respect, as we should, we must also be alert to the equal and opposite danger that public policy could itself become the captive of a scientific-technological elite.

– Dwight D. Eisenhower, Farewell Address, January 17, 1961

There is much confusion about how economists measure economic wellbeing – what economists refer to as 'welfare.' Governments usually measure the effectiveness of their policies by the number of jobs created. Employment estimates are inflated by taking into account the indirect jobs created upstream and downstream as a result of the public expenditures. Upstream jobs are created, for example, to satisfy demand for inputs required by a public project, while downstream ones arise as a result of spending by those employed by the public project. Thus, when the Canadian government provides a university professor with research funds, a crucial reporting requirement relates to the training of graduate students and the employment of technicians and support staff, numbers that are then inflated to account for the indirect jobs associated with the public spending on research. While important, jobs should not be confused with the true benefits of the research. The number of people paid by the research grant, like the numbers employed as a result of any public spending program, are simply one measure of the inputs required to achieve the program's targets, whether a research outcome, greater production of energy from renewables or improved health care benefits.

The job creation metric completely neglects alternative uses of public funds – the opportunity cost of funds. The money used to create jobs could have been spent in other ways that would also have resulted in expanded employment, and jobs created by government might well have crowded out private sector jobs. Indeed,

G.C. van Kooten, *Climate Change, Climate Science and Economics: Prospects for an Alternative Energy Future*, DOI 10.1007/978-94-007-4988-7_6,

had the funds been returned to taxpayers to spend as they saw fit, jobs would have been created, perhaps even more than those created by the public works project. These forgone jobs need to be taken into account in determining the true level of job creation; indeed, if the government expenditures are directed into the wrong areas, the number of jobs actually lost might exceed those created. Thus, it is important to take into account the opportunity cost of funds spent.

Employment is not even the correct measure of societal wellbeing, and job creation might even reduce overall social welfare. Jobs could be redistributed from current residents to immigrants who have specialized skills not available to current residents. Jobs could be lost in one sector, but created in another. In many cases, public programs and policies do little more than transfer jobs and/or income from one group to another. Unless 'wealth' is actually created, there is no benefit to society, and there is a loss (wealth destruction) if economic costs exceed benefits. This raises the question: How do economists measure costs and benefits, or changes in society's overall wealth? In this chapter, we review methods economists use to measure costs and benefits, particularly as these relate to climate change.

We begin in the next section by focusing on private cost-benefit analysis, or financial analysis, because economic agents will not generally take into account the greater good of society. We then focus, in Sect. 6.2, on the economic surplus measures used in social cost-benefit analysis. Given that environmental and other amenity values are important in the context of global warming, but environmental goods and services are not traded in markets, in Sect. 6.3 we consider how non-market amenity values can be measured. Because costs are incurred and benefits accrue at different points in time, it is necessary to weight costs and benefits according to when they occur so that costs and benefits can be brought to a common point in time (whether today or some future date). Without this weighting, it is not possible to compare costs and benefits, or one project with another. The weighting scheme is referred to as discounting and the weights are discount rates. This is the subject of Sect. 6.4. Finally, in Sect. 6.5 we consider extreme events and how to account for them in cost-benefit analysis.

6.1 Financial Analysis

Consider the perspective of the private firm. If a supplier of power to an electrical grid is considering the construction of an additional thermal power plant, for example, the costs of the project equal the up-front construction costs related to land, labor and materials; annual operating (fuel and other), maintenance and (routine) replacement (OM&R) costs; estimates of the costs of unscheduled breakdowns and the risks imposed by changes in fuel prices (and other input costs) over time; costs of meeting environmental regulations; and any costs related to the eventual mothballing of the facility. All costs are discounted depending on when they are incurred. Benefits are provided by the discounted stream of expected revenues from sales of electricity to the system operator (or directly to households and industry if the system operator is also the operator of the plant), plus any 'salvage' value at the end of the facility's useful life. As long as financial benefits over the lifetime of the project exceed costs, the private investor determines the investment to be feasible. That is, the rate at which the power producer weights the streams of costs and revenues is the rate of return that she hopes to earn on the investment, and equals the rate of return should the funds be invested elsewhere in the economy – the opportunity cost of the funds. Thus, if the weighted stream of benefits exceeds that of costs, the project earns a higher rate of return on the investment than could be earned elsewhere.

Financial analysis, or private cost-benefit analysis (CBA), excludes spillovers (also known as externalities) unless the authority specifically requires the firm to pay for access to unpriced natural resources, to pay compensation to those 'harmed' by the firm's activities, to pay an environmental tax, to purchase 'pollution rights', and/or to post a bond to offset society's potential future need to mitigate environmental damage caused by the firm's activities. These costs would be included by the firm in its financial analysis of a project. Further, a financial analysis uses market prices for natural resources, labor, land and other inputs instead of the (shadow) value that these resources have to society. Regardless of these limitations, it is important that public projects are valued from the perspective of private firms. For example, if the government wants to implement a given project and the financial performance of the project is attractive from a private perspective and it imposes little or no external costs on other economic agents, it is likely wise just to let the private sector pursue the project – to provide the good or service in question.

Projects are usually ranked on the basis of financial criteria such as net present value (NPV), the benefit-cost ratio (BCR), internal rate of return (IRR), and/or modified internal rate of return (MIRR).

6.1.1 Net Present Value (NPV)

For ranking projects on the basis of NPV, the following assumptions are needed (Zerbe and Dively 1994):

1. the discount rate is given and usually taken as the market interest rate;
2. capital is always readily available;
3. the interest rate for borrowing is the same as the interest rate for lending;
4. cash flow projections include all relevant costs and benefits, and taxes; and
5. projects are mutually exclusive (so that they can be evaluated separately). Any combination of projects should be considered as a separate option.

If these assumptions are valid, the NPV is the sum of the discounted benefits minus the sum of the discounted costs of the project over the project lifetime:

$$\mathrm{NPV} = \sum_{t=0}^{T} \frac{B_t - C_t}{(1+r_t)^t}, \tag{6.1}$$

where B_t represents the benefits derived from the project in period t, C_t the costs in period t, T is the lifespan of the project and r_t is the interest rate in period t.

The interest rate or discount rate is generally assumed to remain constant in each period because it may be difficult to forecast future values of the rate.

If we are evaluating a single project and NPV is greater than zero, the project is worth undertaking as it increases net wealth. If we are evaluating several projects, the one with the highest NPV should generally be chosen, although that will depend on factors unique to each project. For example, some projects may be riskier than others, or projects have different life spans (in which case one might wish to annualize the net discounted benefits of each project in order to make the comparison).

6.1.2 Benefit-Cost Ratio (BCR)

This is the ratio of the discounted total benefits from a project divided by the discounted total costs of the project:

$$\mathrm{BCR} = \frac{\sum_{t=0}^{T} \frac{B_t}{(1+r_t)^t}}{\sum_{t=0}^{T} \frac{C_t}{(1+r_t)^t}}. \tag{6.2}$$

If the BCR for a single project is greater than 1, the project increases real wealth.

When comparing different projects, the problem of scaling appears. For example, a project with total benefits of $1 million may generate a greater increase in real wealth than a project with total benefits of $100, but the ratio of benefits to costs may not be as high. Thus, projects must have an equal outlay basis if they are to be compared. This is why in the case of choosing among several or many projects it is desirable to examine and rank projects on the basis of both the NPV and BCR criteria.

6.1.3 Payback Period

Given that costs are usually 'front-loaded', with only costs incurred in the first several periods while benefits do not accrue until after construction is completed, the payback period is the point in time when a project's time-weighted total benefits exceed its time-weighted total costs. At that time, the project has 'paid back' its initial investment. The major problem with the payback method is that it ignores cash flows, including potentially negative ones (e.g., costs of clean up), that occur beyond the payback period. If the payback period is the only financial criterion taken into account, it is possible to accept a project that has a negative NPV. Nevertheless, the payback period is a useful indicator for firms that are unsure about future cash-flows and their position in the market. Obviously, firms prefer projects with a shorter payback period.

6.1.4 Internal and Modified Rates of Return: IRR & MIRR

The IRR is a popular criterion for private project appraisal. The IRR is the discount rate for which the NPV is zero – where the project's discounted benefits exactly balance discounted costs. In Eq. (6.1), it is found by setting NPV = 0 and solving for r (which assumes r does not change over time). The project with the largest IRR is generally preferred, subject to the proviso that the IRR exceeds the interest rate. Despite its popularity, the IRR criterion needs to be used with caution. First, for complex cash flows, there might be more than one IRR associated with a project. Second, the IRR approach assumes that the project can both borrow and lend at the internal rate of return. In other words, excess funds generated by the project can be invested externally at the IRR. This is certainly not the case.

The modified IRR (MIRR) is the average annual rate of return that will be earned on an investment if the cash flows are reinvested at the firm's cost of capital. Therefore, MIRR more accurately reflects the profitability of an investment than does IRR. To determine the MIRR, it is necessary to solve the following equation:

$$K_0(1+\text{MIRR})^T = \text{FV}_{\text{cash flow}}, \tag{6.3}$$

where K_0 is the capital investment (effectively calculated at time zero) and $\text{FV}_{\text{cash flow}}$ is the future (as opposed to present) value of the cash flow estimated using the interest rate that reflects the firm's cost of capital.

6.1.5 Informal Analysis

Depending on the manager or owner, and on the size of the project (the sums of money involved in the investment), a private company may decide to conduct an in-depth project evaluation, or it might eschew any formal analysis relying instead on the intuition of the manager or owner. But even intuition can be regarded as a form of project evaluation, and certainly 'paper and pencil' (or 'back-of-the-envelope') calculations would qualify. As the size of an investment project increases, formal analysis using tools such as those discussed above are more prevalent, although, again, there is nothing to prevent managers from relying solely on intuition and rough calculations.

Informal analysis is less likely for projects under consideration by government ministries and international quasi-governmental organizations, for example, although intuition and 'rough analysis' cannot be ruled out entirely in some cases (e.g., decisions sometimes announced by politicians in a media scrum). However, just because a government body conducts formal project evaluations does not mean that the criteria it uses differ much from those used in the private sector. Many government agencies are concerned only with the impact of decisions on their 'bottom line', and are much less concerned about the impact of their decisions on society more generally. The reason is that many government agencies, such as the US

Bureau of Land Management, US Forest Service, and Canada's Ministry of Native Affairs and Northern Development, operate under a broad mandate but in practice are concerned primarily about their own survival and influence. The same is true of international agencies such as the International Monetary Fund, World Bank and United Nations Environment Program. As a result, the evaluation of projects and policies is very much from the perspective of the agency – from a private perspective – rather than from the perspective of society as a whole. This is partly justified by the argument that the agency serves a particular clientele, while it is the job of politicians to ensure that the wellbeing of others in society is represented.

Social cost-benefit analysis is much broader in scope than private cost-benefit analysis because it takes into account the effect that projects have on all facets of society – on all citizens. However, the private perspective is not ignored in social CBA. In many cases, the private decision is adequate, and there is no need for public intervention. The only reason why the public authority would be involved in private investment decisions is if there are important externalities or spillovers, or if the private sector has no incentive to provide the good or service. If spillovers are small, the transaction costs of rectifying them might be too great to warrant intervention. If the spillover/externality is sufficiently large, or public provision is required, then criteria of social cost-benefit analysis are needed to evaluate government policies and public projects.

6.2 Measuring Social Costs and Benefits

Greenhouse gas emissions constitute the ultimate externality, and government intervention is required to rectify the problem and potentially reduce emissions to a socially optimal level. Intervention might take the form of regulations that require manufacturers to employ best available technology, electricity system operators to rely on renewable energy for some proportion of their power generation, and car producers to meet fuel efficiency standards for their automobile fleet. Regulations that require a certain proportion of biodiesel to be sold at the pump might be effective in encouraging biodiesel production, but such regulations impose no costs to the public purse. Alternatively, some investments in technologies that reduce CO_2 emissions and are considered worthwhile undertaking from a public standpoint might not proceed without subsidies or direct involvement by the authority. For example, the government might consider providing a subsidy to wind energy producers to encourage substitution of wind for fossil fuels in power generation, thereby reducing CO_2 emissions. In either event, such interventions must pass a social cost-benefit test, where a benefit of the action or policy is the reduction in CO_2 emissions.

There are alternatives to regulations and specific emission-reduction projects, although these are likely more in the realm of macroeconomic policy. Carbon taxes and carbon emission trading are two instruments that governments can use to reduce CO_2 emissions. These will be considered in Chap. 8. Here we are interested specifically in social cost-benefit analysis related to specific projects. The reason is that social cost-benefit analysis implicitly assumes that the policy or project has little impact elsewhere in the economy. If this is not the case, then general equilibrium analysis is a more appropriate tool to employ because general equilibrium models take into account how changes in one market affect prices and output in all other markets.

This highlights one of the main problems with estimates of the damages from global warming. General equilibrium models tend to be static, at least from the perspective of the long-term nature of the climate change problem; such models are difficult enough to calibrate over the short run, let alone attempting to calibrate them for future scenarios. As a result, economists rely on dynamic integrated assessment models (IAMs) that seek to maximize the well being of citizens, as represented by a social welfare or representative utility function, over a period of perhaps 100 years. (IAMs were considered briefly in Chap. 4 and are discussed in more detail in Chap. 7).

There are several oddities that should be noted. First, as noted in Chap. 4, the climate models themselves are driven by emission scenarios that are derived from economic models, many of which have elements that are similar to integrated assessment models. Second, integrated assessment models assume that damages are a function of temperature; that is, a relationship between temperature and damages is explicitly assumed, whether it is true or not. Third, the models assume a rate of technological change, although there is no way to predict where and how technology might change. Fourth, given that IAMs must project human and physical (perhaps even biophysical) relationships some 50–100 years or more into the future, the relationships in the model are either identities that must necessarily hold or relations whose functional form comes from experience in the theoretical realm and a parameterization based on what can best be described as ad hoc calibrations. Calibration amounts to nothing more than answering the following question in the affirmative: Are the results in the realm of the possible? Do the results seem reasonable?

Finally, as debate regarding the work by Nicholas Stern (2007) and his colleagues in the UK government shows (Chap. 7), the rate used to discount utility or wellbeing over the period in question is extremely important.

In this section, we examine three issues related to social cost-benefit analysis (CBA). First, we consider what constitutes valid measures of wellbeing, of costs and benefits. The short answer is that economists measure costs and benefits as surpluses; the longer answer requires some elaboration, which is done in the next section. Second, we discuss the methods used to measure the costs and benefits of amenities that are not directly traded in markets, such as spectacular views, nature, open spaces and recreation. Finally, we turn to the issue of discount rates.

6.2.1 Benefits and Costs as Rent and Surplus

Social cost-benefit analysis does not ignore financial costs and benefits, but it does proceed differently than private evaluation of costs and benefits. As discussed in Sect. 6.4 below, it employs a social rather than a private rate of discount, with the former generally lower than the latter. Further, social CBA considers opportunity costs (shadow prices) of resources as opposed to market prices. For example, market wage rates might be higher than social rates because of market impediments that

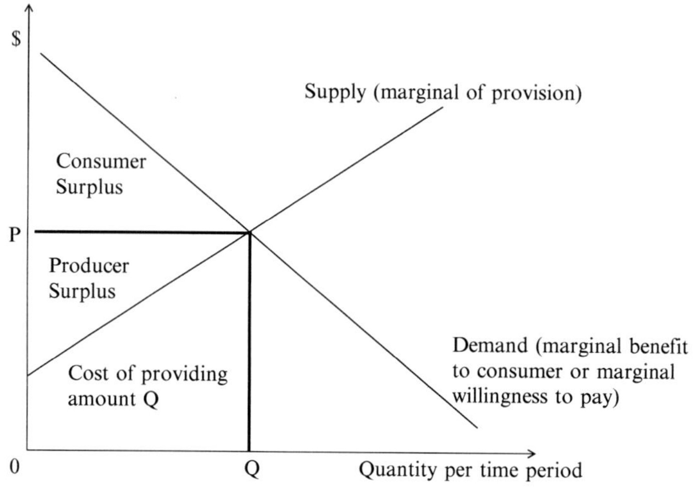

Fig. 6.1 Consumer and producer surplus

cause the wage rate to exceed the marginal value product – the value of additional output that the next unit of labor produces. In other words, the amount that labor is paid at the margin exceeds the value of what it produces. In that case, the economist recommends either that the wage rate be lowered (its shadow value is less than what is actually paid) or that less labor be hired as this will raise its marginal productivity, thereby increasing marginal value product. Where there exists a large pool of unemployed workers, the shadow price of labor is approximately zero.

In economics, costs and benefits constitute a surplus that is either lost (cost) or gained (benefit). There are four types of economic surplus.

1. *Consumer surplus* is the difference between the value that consumers place on goods and services – their willingness to pay – and the actual expenditure to obtain those goods and services. In essence, it is the difference between the total benefit that consumers derive (maximum willingness to pay) and what they pay. It can be measured by the area below the marginal benefit (demand) function and above price. It is illustrated in Fig. 6.1.
   Consumer surplus is not always directly measurable. Consider the case where a project does not affect consumer surplus in the market you expect. For example, it is unlikely that decisions concerning the harvest or protection of a single commercial forest landscape, or the development of a wind energy project, will affect the prices of timber products or power. Thus, the direct consumer surplus associated with such a project is unlikely to change; indeed, unless the project lowers price, the consumer is not going to gain surplus from the project. In that case, consumer surplus becomes relevant only in some other market, but not the market for lumber or energy. If, in addition to the market for lumber or energy, there is a demand for an environmental amenity that is somehow impacted by the

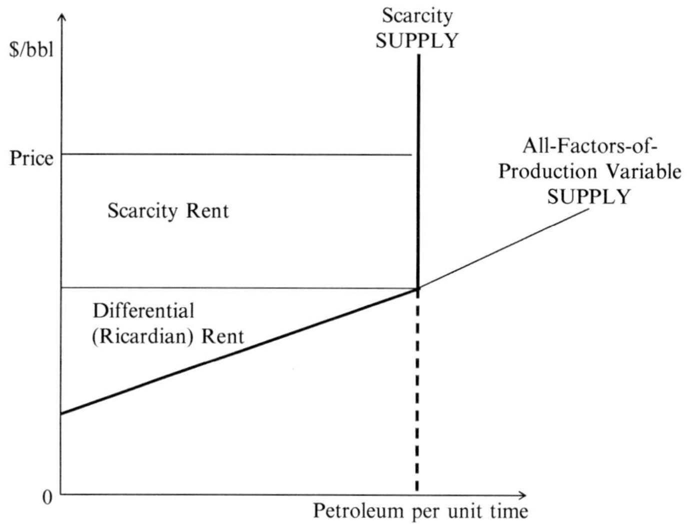

Fig. 6.2 Resource rent and its components

logging decision or energy project, then there may be surplus that needs to be taken into account in evaluating the logging or energy project. This would be an indirect cost or benefit associated with the project, which is discussed below as the fourth type of surplus.

2. *Producer surplus* or *quasi rent* constitutes the difference between total revenue and total variable cost. It can also be measured by the area below price and above the marginal cost (supply) function, as indicated in Fig. 6.1.[1] While constituting a true welfare benefit, producer surplus constitutes a rent accruing to fixed factors of production and entrepreneurship. That is, the supply curve in Fig. 6.1 is a short-run supply function, which means that returns to the fixed factors of production must come from producer surplus. Hence, attempts to tax this rent will adversely affect firms' investment decisions.
3. *Resource rent* accrues to natural resources and consists of two components that are often indistinguishable from each other in practice, and difficult to separate from the second type of surplus – the quasi rent (van Kooten and Folmer 2004). We illustrate the concept of resource rent with the aid of Fig. 6.2, noting in particular that the supply curve in this figure differs from that in Fig. 6.1.
   The first component of resource rent is *differential* (or *Ricardian*) *rent* that arises because of inherent or natural advantages of one location relative to another.

[1] Of course, the supply/marginal cost function is much flatter before the project is built than afterwards. Once the project is built, the construction cost is ignored in the determination of quasi-rent, as bygones are bygones.

Consider oil production. The price is such that the marginal oil sands producer earns at least an internal rate of return higher than the market interest rate. In comparison, Middle East producers earn a huge windfall, which constitutes a differential rent. Likewise, a woodlot located near a transportation corridor (highway, water transport) or a sawmill earns a windfall compared to one with the same amount of commercial timber volume and harvest cost structure, but located farther from the transportation corridor or sawmill.

Second, there is a *scarcity rent* that results simply from oil scarcity or a limit to the number of stands with commercial timber. That is, if the oil sands or timber producer, despite being the highest cost producer, earns a windfall over and above what could be earned elsewhere in the economy, there is a scarcity rent because price exceeds the marginal cost of production.

Resource rent is the sum of the differential and scarcity rents, and must be considered as a benefit in decisions about whether to harvest a forest, develop an energy project, or invest in a biofuels refinery. Interestingly, it is possible for government to tax resource rents without adversely affecting private investment decisions. However, because measurement of resource rents is difficult, government must be careful in taxing such rents lest quasi rents be taxed instead.

4. Finally, the *indirect surplus* refers to benefits or costs that accrue in markets for substitute and/or complementary goods and services. However, indirect benefits occur only if price exceeds marginal cost in one of the affected markets. Whenever price exceeds marginal cost, for example, this implies society values the good or amenity more than it costs to provide it. Hence, if the demand function in a related market shifts outward, more of the good or amenity is purchased, leading to a benefit; the opposite is true if demand shifts inward. If price equals marginal cost in each of the markets for substitutes and complements, there are no indirect effects (Harberger 1971, 1972).

We illustrate the concept using Fig. 6.3. Suppose the marginal cost of providing an environmental amenity is given by MC, but the amount of the amenity provided is less than what is socially desirable – provision is restricted to E_R while the optimal amount that should be provided is E*. At E_R, citizens' marginal willingness to pay (MWTP) for the amenity is $MWTP_1$, while the cost of providing an additional unit of the amenity is only c. The total cost of providing E_R is h, while total benefits amount to the area under D_1 up to E_R, or area $(a+d+f+g+h)$. The net benefit is area $(a+d+f+g)$.

Now suppose that logging a forest in one jurisdiction shifts the demand for the amenity in Fig. 6.3 outwards, from D_1 to D_2. Because the market is out of equilibrium since marginal willingness to pay (price) exceeds marginal cost, the social costs and benefits of logging timber in one region must take into account the indirect surpluses generated in the market for environmental amenities. Now the total benefit (total willingness to pay), given by the area under the demand function, is $(a+b+d+e+f+g+h)$ and the total cost of providing E_R is still h. Thus, the net increase in surplus is given by area $(b+e)$. To determine this benefit, it is necessary to employ one of the non-market

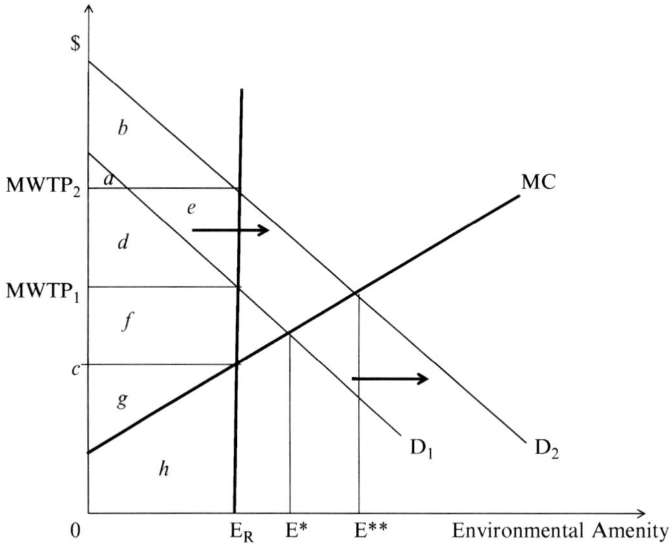

Fig. 6.3 Indirect surplus gain due to increase in timber harvests in other jurisdiction

valuation techniques described in Sect. 6.3. Notice also that the socially desirable level of the environmental amenity has also increased to E^{**}.

It is important to note that environmental spillovers, such as global greenhouse gas emissions, fall into the last category. Since markets are absent, price cannot possibly equal marginal cost. Therefore, it is necessary to determine the costs (benefits) in those markets using a non-market valuation method (see Sect. 6.3). It is also important to recognize that environmental damage is measured as a loss to consumers akin to consumer surplus.[2]

The cost of environmental damage is measured as lost surplus, which becomes a benefit (the damages avoided) of a project that reduces the environmental 'bad' (atmospheric CO_2 concentration). When all of the changes in surpluses resulting from a project are appropriately summed, the discounted net social benefit must exceed the project's capital cost.

Notice that the criteria for judging whether one project is preferred or somehow better than another from society's perspective is the same as that used under private CBA. That is, Eqs. (6.1), (6.2) and (6.3) remain valid. What differs between the private and social perspective is what one measures and includes as costs and benefits, and the discount rate that one employs (which is considered further in Sect. 6.4 below).

[2] Consumer surplus is not the theoretically correct measure in the case of non-market environmental amenities; rather, the correct measures are compensating and equivalent surplus (variation). A clear discussion is found in van Kooten and Folmer (2004, pp.13–25).

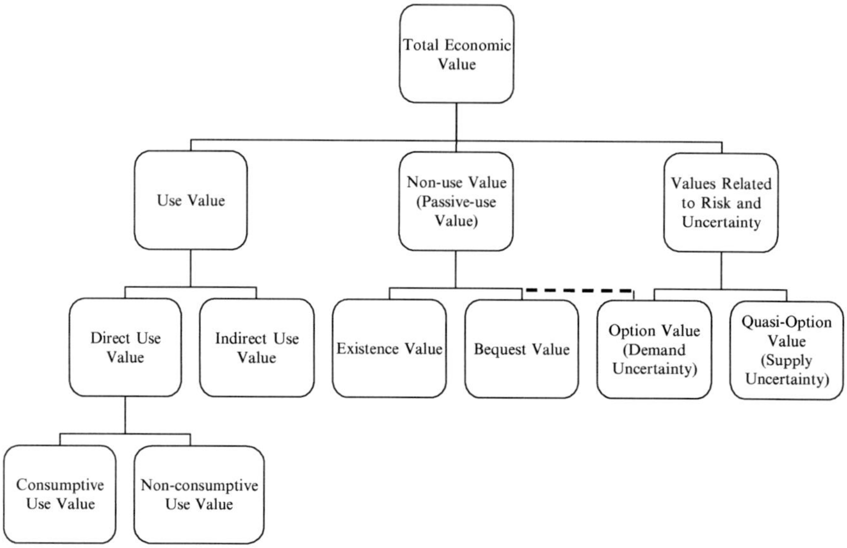

Fig. 6.4 Components of total economic value

6.2.2 *Total Economic Value*

Another way to look at social CBA is via the concept of total economic value (TEV), which is the sum of direct use values, indirect use values, non-use values, and the values associated with remaining flexible in the face of risk and uncertainty (e.g., see Pearce and Warford 1993; van der Heide 2005). A summary of the various types of values that comprise total economic value is provided in Fig. 6.4 (which is adapted from van der Heide 2005). In the figure, it is clear that many of the values that economists attribute to natural resources are ignored in private valuations, and even in the evaluation of public projects. In particular, the focus is generally on the far left branch of the figure, namely, on consumptive, direct use values. From Fig. 6.4, total economic value is given by:

$$\text{TEV} = \text{Total use value} + \text{ total non} - \text{use value} + \text{ value of remaining flexible},$$

where the value of remaining flexible is related to risk and uncertainty. All values are discounted so that they are in present value terms.

Consider the example of a policy regulating biofuel content in gasoline that causes wetlands, native rangeland and/or forested areas to be converted to crop production. Let E_t refer to the net environmental benefits that these lands provide in their original state at time *t*. These benefits include ecosystem services of wetlands in reducing soil salinity and seepage of nitrogen from adjacent cropped lands into ground and surface water, benefits of wildlife habitat and so forth. Of these environmental benefits, ecosystem services may be the most difficult to measure, while other benefits are easier to measure. For example, non-market valuation surveys and other evaluation techniques can be used to determine the values that recreationists place on wildlife viewing, hiking, hunting of waterfowl and ungulates, and so on; but the benefits of reduced soil salinity and nitrogen seepage can only be measured using a great deal of detective work and sophisticated theory and estimation techniques.

In the context of Fig. 6.4, *E* can be thought of as the various use values that the wetland, native grassland and forested areas provide; it consists of values related to consumptive use (hunting, grazing services), non-consumptive use (wildlife viewing, hiking) and indirect use (ecosystem services such as waste assimilation, water quality control). Then the cost-benefit rule for implementing a biofuels regulation that adversely affects marginal land currently in its natural state is:

$$\sum_{t=0}^{T} \frac{B_t - C_t - E_t}{(1+r)^t} > 0, \tag{6.4}$$

where B_t are the benefits from the policy in each period *t*, C_t are the OM&R plus capital costs of investments brought about by the regulation, and *r* is the social rate of discount. Benefits in this case would include the value of reduced CO_2 emissions brought about by the policy. The time horizon is *T*, which is the expected life of the project. In period *T*, there may be salvage benefits and/or environmental or other clean-up costs.

The variable *E* is treated as a cost separate from *C* in order to emphasize that the environmental costs are different from the commercial operating costs of the policy to regulate biofuel content in gasoline, with the latter borne by the energy provider but not the former. Depending on the project or policy, the environmental costs might also include costs associated with the transport and storage of hazardous wastes, potential radiation from and terrorist threats to a nuclear power facility, and the loss of visual amenities when a landscape is converted from its more natural state to the monoculture of energy crops (say corn). While one expects *E* to be positive because it measures lost environmental benefits, there might be situations when it is negative and not a cost to society (e.g., tree planting on denuded land with biomass used to reduce CO_2 emissions from fossil fuels).

In the context of the conversion of wetland, native grassland and forest to crop production, there are two further considerations. First, even in a deterministic world with no uncertainty about the potential future loss of these natural areas, they have existence and bequest value. People attribute value to the knowledge that these natural areas exist and can be passed to the next generation, even though they themselves do not visit or intend to visit them. In Fig. 6.4, we refer to such value as non-use value.

Second, however, there is likely to be uncertainty both with regard to supply and demand. Demand uncertainty is related to people's concern about the future availability of environmental services that may be threatened by the loss of wetlands due to the policy that converts the natural area to crop production. It results because future income and preferences are uncertain, so that individuals might value the environmental amenity more in the future. Option value (*OV*) is the amount a person would be willing to pay for an environmental amenity, over and above its current value, to maintain the option of having that environmental asset available in the future (Graham-Tomasi 1995; Ready 1995). Option value is usually measured in conjunction with existence and bequest value (as indicated by the dashed line in Fig. 6.4); indeed, non-market valuation techniques generally elicit all three at the same time making it difficult to separate them, although this can be done in survey methods by asking questions that specifically focus on separating option value into its various components.

Supply uncertainty is related to irreversibility, and its measurement is known as quasi-option value (*QOV*) (Graham-Tomasi 1995). The idea behind *QOV* is that, as the prospect of receiving better information in the future improves, the incentive to remain flexible and take advantage of this information also increases. Having access to better information results in greater revision of one's initial beliefs, so it is 'greater variability of beliefs' rather than 'improved information' that leads one to choose greater flexibility over potentially irreversible development (say, as a result of cropping marginal agricultural land). Thus, *QOV* is always positive.

The problem with *QOV* is that it is difficult to measure in practice, so its use in cost-benefit analysis is limited.[3] Rather, the concept provides support for the notion of a safe minimum standard of conservation, which suggests that an irreversible development should be delayed unless the costs of doing so are prohibitive. This concept is discussed in more detail in Sect. 6.5.

The cost-benefit model is extended to account for all of these costs and benefits. The decision rule to allow the conversion of 'natural' land, which currently serves as habitat for waterfowl and ungulates, to energy-crop production is now:

$$\sum_{t=0}^{T} \frac{B_t - C_t - E_t}{(1+r)t} - (TNUV + OV + QOV) > 0, \quad (6.5)$$

where *TNUV* refers to total non-use value, and the remaining terms in parentheses refer to the existence value of the marginal land and the benefits of keeping the land in its current state and remaining flexible as opposed to cropping the land. This formulation takes into account all social benefits and social costs associated with the proposed project.

6.2.3 Total (Average) Value Versus Marginal Value

Several caveats remain. What is neglected in the foregoing framework is the impact that the existence of alternative sites for producing energy crops and the availability of alternative amenities have on non-market (environmental) values. For example,

[3] For marginal agricultural land that provides wildlife habitat benefits and visual amenities, *OV* and *TNUV* (total non-use value) are measured using a contingent valuation device (see next section), while *QOV* can be determined using stochastic dynamic programming, for example, as demonstrated by Bulte et al. (2002) for the case of forest protection in Costa Rica.

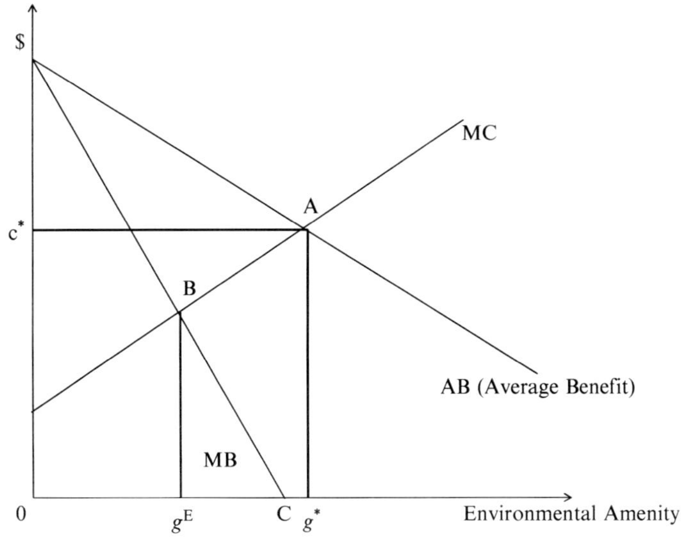

Fig. 6.5 Marginal versus average benefits of decision making

what someone is willing to pay for an option to visit a particular wetlands area is sensitive to the availability of similar sites in other locations. If there is an abundance of wetlands, one expects option value to be small; if there are few, option value is much larger. Hence, it is not the total or average non-market/environmental value that is of importance, but the marginal value. Too often the focus is on total as opposed to marginal value.

Making decisions on the basis of average or total value leads to loss of economic welfare, as illustrated with the aid of Fig. 6.5. In the figure, the curve labelled *AB* represents the average benefits from the environmental amenity (not to be confused with the demand function for the amenity), and is determined as the total area under the marginal benefit (demand) curve, labelled *MB*, divided by the levels of the amenity. The marginal cost (*MC*) of providing the environmental amenity increases as more of the amenity is provided; for example, if the costs of providing wetlands equal the foregone net returns from cropping, it is necessary to 'convert' increasingly higher quality cropland into wetlands, which increases the per hectare costs of providing the next amount of wetlands. A decision based on average or total value would lead to the provision of g^* amount of the amenity (determined from point *A*), while the correct amount to provide as determined by economic efficiency considerations is g^E. The social cost of providing the last unit of the amenity is given by c^*, but the marginal benefit to society of this unit is zero. The total loss in economic well being from providing too much of the amenity (the cost to society) is therefore given by area $ABCg^*$.[4]

[4] This is the difference between the area under *MC* (total costs) and that under *MB* (total benefits) between g^E and g^*. It is the net social cost (negative benefit) of providing g^* of the environmental amenity.

This thinking cuts both ways. Suppose, rather than an environmental amenity, it is output of energy crops that is the object. If a decision is made on the basis of average and not marginal returns, the last acre planted to energy crops would cost more to plant and harvest than it yields in revenue.

Finally, the dynamics of wildlife and the agriculture-nature ecosystem will affect both the value of the agricultural crop and the environmental service benefits. If wetlands can be recreated on cropped land after a short period of time, so that the former attributes of the nature are regained, planting energy crops is not irreversible and quasi-option value is negligible. If it takes a very long period of time to recover the wetlands, the development of cropland may essentially be irreversible, but the benefits of planting energy crops and converting marginal agricultural lands may still exceed costs and be worthwhile undertaking.

There is a conundrum here because the irreversibility of wetlands conversion to production of energy crops needs to be balanced against the potential irreversibility caused by climate change that the energy crops seek to mitigate. This issue is considered further in Sect. 6.5.

6.2.4 Conclusion

Social cost-benefit analysis assumes that everything of interest to the decision maker can somehow be measured in monetary terms. Nevertheless, there remain some things of importance to society that simply cannot be included in the money metric. Since these items are only important if they are somehow (directly or indirectly) affected by the project, these 'intangibles' must be evaluated or judged against the money metric. If the focus is on employment (which is not a true surplus) then any gain in employment that a policy or project brings about needs to be evaluated in terms of the net social loss, preferably measured in terms of the forgone opportunities per job created. If the focus is on CO_2 emissions, a project that reduces the amount of CO_2 in the atmosphere needs to be evaluated with respect to the change in a society's 'surpluses' (economic wellbeing broadly defined). Society might accept a project that removes carbon dioxide from the atmosphere at a cost of \$25 per tonne of CO_2 (t CO_2), but not at a cost of \$250/t CO_2.

6.3 Valuing Amenities: Non-market Valuation

Indirect costs and benefits occur when projects have, respectively, negative or positive spillovers (externalities) that are not taken into account in private decisions about resource use. Interestingly, externalities are just as often ignored by public decision makers, who are supposed to look after the wellbeing of all citizens in society but tend to focus on the clientele they serve. An externality occurs, for example, when surface water used for secondary or enhanced recovery in oil wells is not priced to take into account the value of water in other uses. Surface water injected into oil wells reduces stream flow, thereby affecting water recreation activities (e.g., swimming, boating), fish and other wildlife habitat, irrigators, and downstream generation of hydroelectricity. Likewise, farmers may not pay the true marginal cost of the water they use because losses to recreational users, the hydro facility and so on are neglected. Carbon dioxide emissions that result in climate change are a significant externality because costs are imposed on global society, but no individual agent or country has the incentive to reduce CO_2 emissions. The problem here is measuring the externality effects.

In the example of enhanced oil recovery using water, the surplus lost to agriculture and the electrical grid can be measured, with some effort, using market data, but the loss to water recreationists and the negative effects on aquatic species cannot easily be determined. These losses can be measured using a variety of non-market valuation methods that are now generally accepted and, in some countries, even mandated.

It is possible to distinguish approaches for measuring the value of non-market amenities according to whether changes in the environmental amenity in question leave traces in markets, whether market information can be used to estimate indirect surplus values.[5] Choice-based models employ information about a related activity (as opposed to the environmental amenity itself) to provide estimates about the amenity value. In particular, it may be possible to estimate a *cost function* or and *expenditure function* that includes both market goods and the environmental amenity as variables, and from it draw inferences about the demand for the amenity. Theoretically, if it is possible to estimate a cost function (in the case of production processes) or an expenditure function (in the case of consumers), so-called duality theory can then be used to derive the input or output demand functions, respectively. Since the price of the environmental amenity is effectively zero in most cases, the entire area under the relevant demand function between the amenity's with-and-without-project levels will constitute the surplus measure of benefit or cost (depending on whether the amenity increases or decreases). The best known of these methods are *hedonic pricing* and the *travel cost* methods, but they also include the *damage functions*. Each of these is briefly described below.

In many situations, however, market information cannot be relied upon to derive a cost or expenditure function because the environmental amenity is strongly separable in individuals' utility functions.[6] That is, increments or decrements in the environmental amenity are valued by individuals because it affects their wellbeing (utility), but such changes do not affect how they allocate their budgets. For example, suppose

[5] The term environmental amenity is used in a generic sense to refer to any good or service that is unpriced or priced well below its marginal cost of provision, whether that is wildlife habitat, water/air quality, wilderness areas, recreation sites, visual landscapes, risk of exposure to radiation, et cetera. All of these have value because individuals would be willing to pay something to have more of it or require compensation to put up with it. Of course, this presumes that the individual has some property right over the externality.

[6] A function $U(x_1, x_2, \ldots, x_n)$ is strongly separable if $U(x_1, x_2, \ldots, x_n) = U_1(x_1) + U_2(x_2) + \ldots + U_n(x_n)$. In this case, the marginal utility of x_i unaffected by changes in x_j, so $\partial^2 U/\partial x_j \partial x_j = 0$. This does not imply, however, that the price of x_j has no affect on x_i; that is, $\partial x_i/\partial p_j \neq 0$.

a forest that can be viewed from the road is now clearcut. For the person who travels this road, utility has gone down – she has been negatively impacted by the loss of the visual landscape and would likely be willing to pay some amount to have prevented the clearcut. Nonetheless, since she does not pay, she does not change the way in which she allocates her spending on market goods and services. To determine the value of her loss, we would need to ask her directly about the value she placed on the forest versus the clearcut. We require a survey instrument to elicit directly her *willingness-to-pay* (WTP) for the scenic amenity or her *willingness-to-accept* (WTA) compensation to forgo the amenity (put up with the clearcut), with the latter sometimes referred to as the *compensation demanded.*

Notice that WTP and WTA are alternative measures of consumer surplus, something discussed in more detail below. Here we simply point out that, since this approach requires individuals to respond to hypothetical questions, it is referred to as the *contingent valuation method* (CVM) if actual values are requested, or the contingent behavior method if a behavioral response is desired. Alternative approaches in this genre include contingent ranking, choice experiments (or *stated preferences*), which require respondents to state their preference between situations (much like in marketing surveys), conjoint analysis and other techniques that are briefly discussed below.

6.3.1 Cost Function Approach

The cost function approach to the measurement of environmental values relies on the estimation of a relationship between the output of some market traded commodity and the environmental amenity. For example, the output of an energy crop, such as corn for ethanol or canola for biodiesel, might be adversely impacted by soil salinity. By estimating what is known as a damage function, it is possible to determine the effect that different levels of soil salinity have on yields. Using this relationship and the price of the energy crop, one can estimate the costs that different levels of soil salinity impose. If salinity is related to certain land use practices, the spillover costs of such practices can be determined. Thus, increased salinity may be the result of cropping marginal land that, in turn, is brought about by regulations requiring greater use of biofuels. The damage function approach could be used to value one component of the environmental cost.

Another example of a damage function relates to soil conservation. Agricultural economists have estimated relations between soil depth and crop yield similar to that illustrated in Fig. 6.6. The damage function intercepts the vertical axis above zero because crops can grow in subsoil. Notice also that a drop in soil depth from D_0 to D_1 leads to a loss of y_0 to y_1, with the damage obtained by multiplying the crop loss by its price. If there is less soil on the site, similar soil erosion leads to a much greater loss in yield, as indicated by the downward arrow.

Finally, technology can mask the adverse impacts of soil erosion, making soil conservation appear less attractive, as indicated by the increase in yield from y_0 to y_2

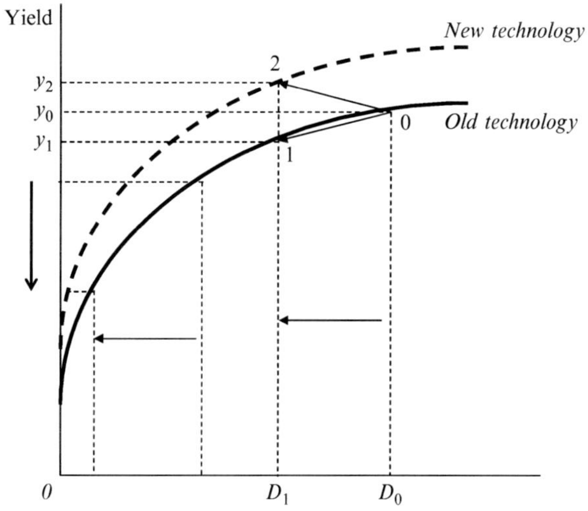

Fig. 6.6 Damage Function between Soil Depth and Crop Yield

when soil depth declines from D_0 to D_1 because technological change has shifted the relationship between soil depth and crop yield upwards. Rather, the true loss in yield is measured by the difference between y_2 and y_1. While this is a simple example of a damage function, it illustrates the difficulty of measuring environmental damages. In Chap. 7, we replace soil depth with temperature and crop yield with a variety of goods or services that are traded in markets.

Also falling into the category of non-market valuation are the costs of averting damages. Whenever people take action to avoid the adverse effects of spillovers (e.g., pollution in a big city, risk of exposure to radiation), the costs of such actions provide information about the value of the spillover. For example, if the municipal drinking water supply contains dissolved minerals or is contaminated with nitrogen, purchases of bottled water can be used to provide one estimate of the benefits of improving water quality, although it would be difficult to separate purchases of water for that purpose from those of convenience, the trendiness of bottled water and so on. Purchases solely to avoid the poor water quality provided by the municipality are an averting expenditure.

6.3.2 *Expenditure Function*

6.3.2.1 Hedonic Pricing

Hedonic pricing relies on market evidence related to property values to determine the value that people assign to improvements in access to public and quasi-public goods (e.g., police and fire protection, local parks) and environmental quality. It is assumed that individuals choose the amount of public goods and environmental quality they want by the choices they make concerning residential purchases. People choose to live in areas that have cleaner air or less crime, they choose to live near airports or along highways, and they choose to live on quiet or on busy streets. The choice is determined by what they are willing and able to pay for housing. Hedonic pricing exploits these choices by estimating implicit prices for house characteristics that differentiate closely related housing classes. In this way, it is possible to estimate demand curves for such characteristics or public goods as air quality and noise. The hedonic technique requires that the following three methodological questions are answered in the affirmative:

1. Do environmental variables systematically affect land prices?
2. Is knowledge of this relationship sufficient to predict changes in land prices from changes in air pollution levels, say?
3. Do changes in land prices accurately measure the underlying welfare changes?

If any of these is not answered in the affirmative, the methodology cannot be applied.

Hedonic pricing is a two-stage procedure (Freeman 1995; Smith 1997): In the first stage, the hedonic or implicit price function is obtained by regressing various house characteristics (such as lot and house size, number of bedrooms and bedrooms, etc.), neighborhood factors (e.g., nearness to schools, parks, fire hall) and environmental characteristics (e.g., air quality) on the property's price. The implicit price of any characteristic is found by differentiating the hedonic price function with respect to that characteristic.

In the second stage, then, the implicit price is regressed on income, quantity of the characteristic and other (instrumental) variables. This constitutes the inverse demand function. The area under the demand function between the current and proposed levels of the characteristic constitutes a measure of the (consumer) surplus associated with the proposed change.

Empirical studies that have used the hedonic pricing method to determine the effect of aircraft and traffic noise on housing prices find that there is a measurable effect. For aircraft noise, a one-unit change in the measure of noise (as related to human hearing and discomfort) resulted in housing prices that were 0.5–2.0% lower, while traffic noise reduced house prices by 0.1–0.7 % per decibel (Lesser et al. 1997, p. 281).

6.3.2.2 Recreation Demand and the Travel Cost Method

To assess benefits from recreation, the travel cost method emerged as perhaps the first technique for valuing non-market benefits (Clawson 1959; Thrice and Wood 1958). The travel cost method is a type of revealed preference model where

1. individuals are observed to incur costs so as to consume commodities related to the environmental amenity of interest, and
2. the commodities consumed are not purchased in a market where prices are determined by supply and demand.

A number of different approaches are available for estimating welfare gains/losses in what is termed the 'travel cost' framework. In general, the travel cost method assumes that costs incurred to travel to a site are identical to an entry fee to the site. This knowledge along with number of visits to a site (and in some variants visits to multiple sites on the same trip) can be used to construct a demand function for the site(s) in question. Again, the area under the demand function yields information about the consumer surplus, which is then used as a measure of benefit or cost.

The hedonic pricing method can also be applied to recreation demand estimation, but the problems involved are complex. Simply, total household expenditures on recreation at a particular site take on the role of property value in the hedonic or implicit price function. Expenditures by a large number of households engaged in recreation at more than one site are regressed on a variety of private and public characteristics of the various sites. Again, by differentiating the hedonic price function with respect to any of the public attributes, an implicit price for that attribute is obtained. In the second stage, the implicit prices for the attribute are regressed on household characteristics, particularly income, and the amount of the attribute available, howsoever measured. The resulting equation is the demand function for the attribute. The area under the demand function can then be used to measure the benefit of a change in the amount of the public good. In practice, it is not easy to implement hedonic travel cost methods.

6.3.3 *Contingent Methods or Direct Approaches*

It is generally thought that the damage function, travel cost and hedonic pricing methods provide reasonable estimates of true values because they rely on market data. Hence, they are best employed to estimate use values (see Fig. 6.4), which relate to the unpriced benefits environmental amenities provide in the production or consumption of some other good or service. For instance, a forest provides ecosystem functions such as flood control, water storage and waste assimilation, as well as recreational and other consumptive and non-consumptive (e.g., wildlife viewing) use benefits.

Measures of non-use or passive-use value, on the other hand, cannot be derived from market data. Non-use values include existence, bequest, altruism and other inherent values that are independent of people's spending on market goods and services. Existence value is the value of simply knowing that an environmental asset exists – people express a willingness to pay simply for the knowledge that the asset exists. Bequest value refers to people's willingness to pay to endow the future generation with the asset, while altruism refers to the benefit that a person places on the benefit another person gets from the environmental asset (and not explicitly identified in Fig. 6.4). Additionally, option value is often indistinguishable from bequest and existence values; it too cannot be derived from market data. Indeed, existence, bequest and option values are together often referred to as preservation value. Preservation values are determined primarily with contingent methods.

Contingent methods are required whenever the amenity to be valued leaves no behavioral trail in the marketplace. Therefore, contingent devices involve asking individuals, in survey or experimental settings, to reveal their personal valuations of increments (or decrements) in unpriced goods – constructing contingent markets. These markets define the good or amenity of interest, the *status quo* level of provision and the offered increment or decrement therein, the institutional structure under which the good is to be provided, the method of payment, and (implicitly or explicitly) the decision rule which determines whether to implement the offered program. Contingent markets are highly structured to confront respondents with a well-defined situation and to elicit a circumstantial choice upon the occurrence of the posited situation. But such markets remain hypothetical, and so too are the choices people make within these markets.

Because the constructed markets used by economists to elicit value are hypothetical, some argue that the values obtained using the methods described below are imperfect, so much so that they are essentially worthless. In most cases, the contingent valuation devices are used to value natural and ecosystem capital, and such capital clearly has value; indeed, natural and ecosystem capital may be of utmost importance to the long-term survival of society (Diamond 2005). Thus, it would be a grave error for decision makers to ignore the non-market services provided by forests, rangelands/grasslands, wetlands, lakes, rivers and riparian zones, and even croplands (Olewiler 2004), whether these services entail carbon storage and sequestration, commercial timber harvests, food production, maintenance of water quality, provision of wildlife habitat/refuge, or recreational and scenic amenities.

6.3.3.1 The Contingent Valuation Method (CVM)

The contingent valuation method was initially proposed nearly 50 years ago in an effort to value non-market amenities (Krutilla 1967). Subsequently, CVM has been approved by the U.S. Department of the Interior for implementing regulations under the Comprehensive Environmental Response, Compensation, and Liability Act (CERCLA) of 1980 and its amendments of 1986. In 1990, the U.S. Oil Pollution Act extended liability to oil spills (as oil was not considered a hazardous waste). A 1989 decision by the District of Columbia Court of Appeals involving CERCLA in the case of *Ohio v. Department of Interior* affirmed the use of CVM and permitted inclusion of non-use values in the assessment of total compensable damages. In the early 1990s, an expert panel led by two Nobel prize-winning economists (Kenneth Arrow and Robert Solow) supported the use of the contingent valuation method for valuing non-market amenities (Arrow et al. 1993). Thus, in the U.S. at least, CVM is used both for determining compensation when firms or individuals damage the environment and in cost-benefit analyses.[7]

[7] In court cases, CVM can be used to estimate compensatory damages, but not the punitive damages that the court might assess.

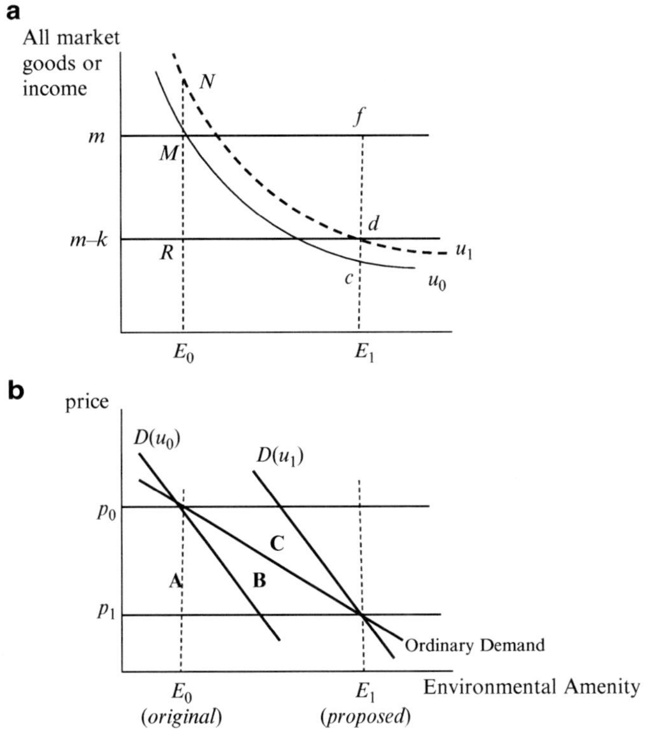

Fig. 6.7 Willingness to pay and willingness to accept compensation as surplus measures in the utility domain (panel a) and price-quantity domain (panel b)

Surveys are used in CVM to elicit information regarding the minimum level of compensation required by an individual to forgo an environmental amenity or public good (compensation demanded) or the maximum amount the individual would be willing to pay to obtain the non-market amenity. These measures are rooted in economic theory and constitute a surplus measure equivalent to consumer surplus as indicated below.

Suppose the current level of an environmental amenity is given by E_0 and we wish to know the benefit of a policy that causes the level to increase to E_1. In Fig. 6.7a, the wellbeing or utility of a respondent to a valuation question is given by u_0 at E_0. The combination of income *m* and amenity E_0 results in a utility of u_0. All combinations of income and the environmental amenity that lie on the u_0 curve lead to the same level of utility. However, if income is reduced to *m*–k from *m* while the level of the environmental amenity is increased from E_0 to E_1, the person's wellbeing increases to u_1. That is, the person is made better off by giving up *k* amount of income to move from point *M* to point *d*, thus gaining E_1–E_0 amount of the amenity. The maximum amount she would be willing to pay (WTP) for the move from *M* to *d* is measured by the distance *cf*; any proposed loss of income less that *cf*, such as amount *k* (=*df*), would be accepted.

Despite the fact that environmental amenities are not traded in a market, we draw three demand curves in Fig. 6.7b. These can be thought of as shadow demand curves that exist in theory but not in practice. Consider first the ordinary demand function. As discussed previously, the benefit of a policy that increases the amount of the environmental amenity is given by area **A**+**B**, which is the consumer surplus. However, since prices do not exist, we cannot estimate such a demand function. The other two demand curves are so-called compensated demand functions because the individual either gives up or gains income in order to remain at the same level of utility as the level of the environmental amenity is varied. As noted above, if a person starts at point M in panel (a) and moves to point d, her income would need to be reduced by amount cf to keep her at u_0; this keeps her on the compensated demand curve $D(u_0)$. The equivalent of cf in panel (a) is area **A** in panel (b) of Fig. 6.7. This is known as the *compensating surplus*.

Notice that in the above analysis the individual is assumed to have a right to E_0 and not E_1. However, if the person had the right to E_1 but was only able to access E_0, we would need to ask her what the minimum amount of compensation she would demand to put up with E_0 rather than the E_1 to which she is entitled. The minimum amount she is willing to accept (WTA) as compensation is given by distance RN in panel (a) and it too constitutes a surplus measure akin to consumer surplus. In this case, the appropriate compensated demand function is $D(u_1)$ and the appropriate surplus measure is given by area **A**+**B**+**C** in panel (b), which equals RN in panel (a). This area is known as the *equivalent surplus*.

In the case of environmental amenities, therefore, there are three measures of surplus from the standpoint of 'consumers' – consumer surplus (CS), compensating surplus (WTP) and equivalent surplus (WTA). These are given in Fig. 6.7b by areas **A**+**B**, **A** and **A**+**B**+**C**, respectively, so that WTP<CS<WTA. In theory, areas **B** and **C** are considered to be very small, so that WTP≈CS≈WTA – the three measures are approximately equal. However, studies consistently find that compensation demanded (WTA) is significantly greater than willingness to pay, so that the initial endowment or one's property right matters a great deal (see Horowitz and McConnell 2002).[8]

In the absence of market data, a contingent valuation approach, whether CVM or some other approach that relies on direct elicitation of value, is needed to determine the surplus from changes in the availability of an environmental amenity. While primarily used to determine non-use values, CVM can also be employed to value market-traded goods and services, which is useful for testing how well responses to hypothetical purchasing questions correspond to actual ones.

An important use of contingent valuation surveys is to determine preservation values for such things as tropical rain forests and wildlife. For example, Kramer and Mercer (1997) found that U.S. residents were willing to make a one-time payment of \$1.9–\$2.8 billion to protect an additional 5 % of the globe's tropical forests.

---

[8] We could just as well examine the case where the 'original' level of the environmental amenity in Figure 6.7 is E_1, and then ask what the associated measures would be. In this case, WTP would be a negative value (indicating that compensation is required), while WTA is positive (indicating the respondent would need to pay). By switching the subscripts in the figure, we then find that WTA<CS<WTP.

Preservation benefits for wildlife were estimated by Canadian economists to be in the neighborhood of $68 million per year for Alberta residents (Phillips et al. 1989), while preservation of old-growth forests is valued at perhaps $150 per household per year (van Kooten 1995) This suggests that ignoring these values in the management of natural resources can lead to substantial misallocation of resources.

6.3.3.2 Choice Experiments or Stated Preferences

Unlike the contingent valuation method, the approach of choice experiments (CE) or stated preferences does not require survey respondents to place a direct monetary value on a contingency (Adamowicz 1995; Adamowicz et al. 1998). Rather, individuals are asked to make pairwise comparisons among environmental alternatives, with the environmental commodity (alternatives) characterized by a variety of attributes. For example, a survey respondent is asked to make pairwise choices between alternative recreational sites or activities, with each distinguished by attributes such as the probability of catching a fish, the type of fish, the amenities available to fishers (e.g., whether or not there are boat rentals), distance to the site, and so on. It is the attributes that are important, and it is these that are eventually assigned monetary value. In order to do so, one of the attributes must constitute a monetary touchstone (or proxy for price). Distance to a recreational site might constitute the proxy for price (as in the travel cost method), but, more generally, one of the attributes will be a (hypothetical) entry fee or an associated tax. Once the values of all attributes are known (using the monetary touchstone and the pairwise rankings), the overall value of the amenity is determined by assuming additivity of the attributes' values. Of course, it is possible that the total value of the amenity is greater than the sum of its components, or vice versa.

While the methodology has been used primarily to value recreational sites, Adamowicz et al. (1998) apply CE to the estimation of non-use values. It is argued that CE avoid the 'yea-saying' problem of dichotomous choice surveys as respondents are not faced with the same 'all-or-nothing' choice, although recent advances in CVM questionnaire design have addressed this issue (Shaikh et al. 2007).

Another advantage of choice experiments over the traditional contingent valuation approach occurs when it comes to the transfer of benefits (e.g., transfer of estimated benefits for water quality improvements in one jurisdiction to those in another). This issue is discussed further below. Further, repeated questioning of the same respondent in CE enables consistency testing that is not possible in CVM where one valuation question is usually asked. CE may also be a means of getting around the embedding problem of CVM. Embedding is used to describe a situation where people state they are willing to pay $40 per year to protect grizzly bears, for example, but they are also willing to pay no more than $40 per year to protect wildlife per se. Of course, if asked to breakdown the latter into the valuation of various species or categories of wildlife, grizzly bears are worth much less than $40. Finally, by allowing some attributes to take on levels both above and below the *status quo* level, CE enables one to estimate both willingness to pay and the compensation demanded.

Conjoint analysis differs from CE because it asks respondents to rank all of the alternatives from highest (best) to lowest (worst). Such a ranking can then be used to infer the importance of the attributes that characterize each alternative within one's preference function. Conjoint measurement is a marketing technique that uses revealed choice among goods with different characteristics (as in hedonic pricing) with a survey that asks people to choose among or rank hypothetical alternatives (contingent ranking) to impute the values of the characteristics. It is used primarily to predict the potential for new products, but efforts are ongoing in the application of this technique to the valuation of non-market commodities in ways that different from CE (Smith 1997).

6.3.4 Benefit Transfer

Use of non-market valuation techniques to obtain surplus data for use in social cost-benefit analysis can be quite expensive and time consuming, especially with regards to administering a survey instrument. The decision maker needs to determine whether the expense is warranted. In this regard, Allen and Loomis (2008) offer some guidance as to when a valuation study should be undertaken or benefit transfers employed.

A further question that arises is: Can one use the values estimated elsewhere and apply them to the situation under consideration? Under certain circumstances, it is possible to avoid large transaction costs associated with the valuation of spillovers and yet provide reasonable values for decision making. That is, the benefits estimated in one jurisdiction might be transferable to other jurisdictions under the right circumstances. Indeed, in her study of the value of natural capital in settled regions of Canada, Olewiler (2004) employs estimates from a variety of sources and jurisdictions. The drawback is that the values are not as precise, but, in many instances, simple knowledge of a range of values is sufficient to take into account non-market costs or benefits. In other cases, it is impossible to determine the appropriate monetary values, in which case a description of the 'with-without' project attributes of the 'externality' under consideration will have to suffice.

Recent initiatives have sought to facilitate the use of benefit transfers. These have relied on meta-regression analysis of data from various studies of the same resource, such as the meta-analysis of wetland services conducted by Woodward and Wui (2001). These and many more studies have subsequently been collected by John Loomis and colleagues at Colorado State University in an effort to provide some notion of the non-market values that can be used for benefit transfer purposes.[9] An example of the types of values available is provided for the case of wetland services in Table 6.1.

---

[9] Information about the Colorado State University benefit transfer project and a toolkit can be found at: http://dare.colostate.edu/tools/benefittransfer.aspx (viewed February 12, 2011). Another effort to collect information for the purposes of benefit transfer is underway at Central Queensland University in Australia under the guidance of John Rolfe and Jill Windle; see 'benefit transfer' at http://resourceeconomics.cqu.edu.au/ (viewed February 12, 2011).

Table 6.1 Value of wetland services for benefit transfer purposes ($ per acre of wetland)

	United States				
	Northeast	Southeast	Inter-mountain	Pacific	Canada
Min	$33	$0.41	$6	$124	$51
Max	$908,492	$6,494	$456	$5,657	$198
Average	$49,873	$448	$80	$1,555	$137
Median	$618	$21	$17	$718	$149

Source: Calculated using data from http://dare.colostate.edu/tools/benefittransfer.aspx

6.4 Discounting and Choice of Discount Rate

Because costs are incurred and benefits accrue at different points in time, cost-benefit analysis relies on discounting financial flows (costs and benefits) to a common date so that they can be compared. Without discounting, for example, it would be possible to advocate spending a large sum today in anticipation of a larger benefit in the future, whether such a benefit came about in several years, 100 or 1,000 years. Clearly, it would be foolish to spend money today so as to obtain a benefit in 1,000 or even 200 years from now. Discounting is required so that rational decisions can be made concerning how we as a society spend and invest scarce resources.

To reiterate, it is necessary to measure and compare the stream of benefits and the stream of costs at a single point in time, whether that is at the beginning or at the end of the time horizon, or at some intermediate point. Further, since individuals prefer to delay pain (costs), while they are eager not to delay pleasure (benefits), it is necessary to weight gains and losses as to when they occur, a procedure known as discounting. Since $1 today is worth more to an individual (or society) than that same dollar received at some future date (say, next year), it is necessary to discount future dollars so that they are worth less today. And it is not only money that is discounted: clearly, it is preferable to remove CO_2 from the atmosphere today rather than next year or 100 years from now – CO_2 removal at a future time is worth less than its removal today. It is the purpose of the discount rate to weight future costs and benefits, no matter whether they are in monetary or physical units. The problem is to choose an appropriate discount rate that reflects society's preferences for current over future 'consumption'. Whether a project is desirable will depend to some extent on the discount rate – the outcome is sensitive to the rate of discount. What, then, is the appropriate rate of discount to use in weighting future costs and benefits? This turns out to be a rather difficult question to answer.

Compared to low interest (discount) rates, high rates encourage savings and investment that lead to higher future incomes. But high interest rates also cause one to focus more on the short run because gains and losses that occur farther in the future are valued less today (as they are discounted more highly). Despite some common sense aspects about interest rates and discounting, the economic literature on this topic is vast and, surprisingly, there is no ready consensus about what discount rate to use when analyzing public policies and projects.

On moral grounds, some advocate the use of a zero discount rate in comparing one generation with another (e.g., Heal 2009). Yet, people behave as if they discount the future because they prefer something today (the sure thing) over tomorrow (because it is unsure) – they exhibit an implicit rate of time preference, so that a future dollar is valued less than a dollar today. Economists get around the dilemma of discounting the value of future generations by arguing that it is wrong to discount the utility or wellbeing of a future generation, but that it is appropriate to discount their consumption. Consumption is related to the ability of the economy to produce goods and services, and growth in consumption is the result of investment in activities that enhance the economy's ability to increase output. Thus, the rate of growth in per capita consumption is sometimes taken as the starting point for determining the discount rate (see below). While consumption goods increase utility, utility goes beyond consumption as it addresses quality of life, and thereby includes environmental goods (e.g., clean air and water), biological diversity, the inter- and intra-generational distribution of income, et cetera.

A major problem in choosing a discount rate is that individuals have different rates of time preference, but even the same individual employs different discount rates. In determining a social rate of discount, not only is it difficult to reconcile the fact that different people use different rates to discount the future (although practically speaking individual rates are equated to the market rate at the margin), but evidence from behavioral economics indicates that people commonly discount future losses at a lower rate than future gains, and that they use higher rates to discount outcomes in the near future than those in the distant future (Knetsch 2000). In one survey, half of respondents were asked for the largest sum of money they would be willing to pay to receive $20 a year from now, while the other half was asked to provide the smallest sum of money they would accept today to give up receiving $20 a year from now. "The rate used to discount the future gain was, on average, about three times higher than the rate used to discount the future loss" (Knetsch 2000, p. 283).

There are other quirks associated with discounting, although these also relate to risk perceptions. People express greater willingness to discount environmental benefits from a government program at a lower rate than the benefits of a program that enhances future consumption of material goods. Individuals express greater willingness to pay to avoid extremely small risks of death from an environmental disaster (e.g., related to construction and operation of a nuclear power plant) than they do to avoid much higher risks of death associated with something with which they are more familiar (e.g., riding on a motorcycle) (see Fischhoff et al. 1981).

6.4.1 How to Discount the Future When Considering Future Generations

A particular controversy about the discount rate relates to the weighting of different generations. This is particularly important for climate change where future generations benefit from current investments in climate mitigation, but also bear the costs of reduced incomes from current investments that lock a future society into an inappropriate technology. Whatever society does today will have an impact on future generations.

Consider the following argument for a low discount rate in comparing across generations. An individual may require a payment of \$1.05 next year in order to forgo receiving \$1 today, which implies a discount rate of 5 %. However, the same individual may be willing to give up \$1 in 20 years' time to obtain \$1.01 in 21 years, implying a discount rate of 1 %. In other words, the discount rate declines as costs and benefits accrue in the more distant future – the discount rate declines as a project's or program's time horizon increases. This is referred to as 'hyperbolic discounting' in contrast to exponential discounting that uses a constant rate of discount (see Dasgupta 2002; Weitzman 1998, 1999). This notion has been used to argue that, when comparing investments that affect future generations, a very low rate of discount should be employed.

The problem with 'hyperbolic discounting' is that, in the above example, when the individual in 20 years' time needs to make the choice between \$1 today and \$1.01 next year, she will choose \$1 today, *ceteris paribus* (assuming her current-period discount rate continues to be 5 %). The use of a declining discount rate leads to time-inconsistent decisions because the mere passage of time causes an individual to modify their choice. However, if the discount rate itself is uncertain because the world is uncertain, then there is always the possibility that "*ex ante* good decisions turn out to be regrettable *ex post*, once nature has revealed herself" (Newell and Pizer 2003, p. 10). The notion of uncertainty about the rate of discount is considered further below.

The long-run rate of growth in per capita consumption is often used as a starting point for calculating the discount rate to use in comparing inter-temporal costs and benefits related to climate change, because it indicates by how much the material wellbeing of the future generation can be expected to rise above that of the current one. To this is added a rate of time preference of 1 or 2 % – the rate that individuals might use in preferring to have something today as opposed to delaying it to a future time. Thus, if the rate of growth in consumption is 1.3 %, then the actual rate of discount might be 2.3 %. The Stern Report (Stern 2007) employed a discount rate of 1.4 %, with the result that future damages (which were already overstated) appeared much larger in current terms than under a more realistic assumption about the discount rate.

To put a technical perspective on the issue, let β be the pure rate of time preference and $C(t)$ the aggregate per capita (global) consumption at time t. Then, following Heal (2009), the discounted present value of per capita consumption over all time is given by

$$\int_0^\infty U(C(t))e^{-\beta t}dt, \tag{6.6}$$

where $U(C)$ is the instantaneous utility of consumption. Let $C'(t)=\mathrm{d}C(t)/\mathrm{d}t$ be the rate of change in consumption, which has generally been positive ($C'(t)>0$). Further, assume $U'=\mathrm{d}U/\mathrm{d}C(t)>0$ and $U''=\mathrm{d}^2U/\mathrm{d}^2C(\mathrm{t})<0$, which tell us the following:

Given that, as consumption rises beyond some threshold (presumed to be low and not included in the mathematical derivations provided here), people will get less enjoyment (utility) out of an extra unit of consumption as consumption rises. Thus, the enjoyment that someone in the future would get from consuming material goods and services would be less as more becomes available to them; on the other hand, if it is assumed that environmental goods are declining over time as a result of climate change or other factors, then utility would actually fall. The consumption discount rate, r, is then given by $e^{-\beta t} U'(C(t))$, which can be written in such a way that the pure rate of time preference is independent of the changes in consumption and the utility function (Heal 2009, p. 277):

$$r = \beta + \varepsilon(t)C'(t). \tag{6.7}$$

where $\varepsilon(t) = -C\ U''/U' > 0$ is the elasticity of the marginal utility of consumption, which tells us how fast the marginal utility of consumption, U', falls over time as consumption rises. In essence, then, there are two discount rates to consider – the pure rate of time preference which is based on an ethical decision and the consumption discount rate which is endogenous.

The change in per capita consumption over time, $C'(t)$, can be determined using historical data, although we have no guarantee that consumption will continue to grow in the future as it has in the past. The choice of other parameters in the above equation is a matter of value judgment. Even the assumption that the rate of growth in per capita consumption is increasing at 1.3 % – that the second term in the above expression is growing at 1.3% – is a value judgment because utility is ignored. Including the consumption elasticity of marginal utility, however, implies that one needs to choose a functional form for utility and that is a value judgment.

Further, Heal (2009) argues that, from an ethical standpoint, the pure rate of time preference is zero, $\beta = 0$, because it deals with cross-generational comparisons. This is only partly true because the pure rate of time preference is as much intra as it is inter generational in context.

Finally, Heal (2009) points out that the above relation is based on a single consumer good or bundle. If there are multiple goods, the above expression needs to be modified, but essentially the same conclusion results. However, if a minimal level of some good is required for survival, such as threshold or minimal level of environmental services, then utility is not defined when provision of that good falls below the critical threshold. Thus, in the case of technological limits to the substitutability between produced goods and natural resources, for example, it is possible for the appropriate discount rate for discounting the costs and benefits of mitigating climate change to be negative.

6.4.2 *What Discount Rate?*

So what discount rate do we use? Consider, first, whether a nominal or real rate of discount is to be employed. While a nominal rate might be used in cases where one wishes to examine cash flows, it is generally preferable not to use a nominal rate of discount because it requires that inflation be taken into account. Since the allocation of investment and consumption over time is based on expectations, adjusting the nominal discount rate by *ex post* inflation is not quite correct. Further, it is not possible to predict inflation over the life of a project/program, which could quite well exceed 100 years. There is already enough uncertainty about the future real rate of interest (see below). In any case, economists generally prefer to use the real discount rate.

It also makes sense as a principle for choosing a discount rate to focus on consumption. Then, the consequences of government program/regulation "should be converted into effects on consumption (versus investment) and then these consumption effects should be discounted using a consumption rate of interest – the rate faced by consumers when they save, rather than businesses when they borrow" (Newell and Pizer 2003). In the United States, the real rate of return on investments by large companies over the period 1926–1990 was about 7%, after taxes, while it was 8% over the period 1926–1998. Given a corporate income tax rate of about 35%, the pre-tax rate of return is thus about 11–12%. Since individuals in the U.S. pay up to 50% in income taxes, the rate of return to individuals as owners of companies is closer to 4%, which can then be considered the consumption rate of interest – the rate at which people trade off spending over time. Interestingly, the U.S. Office of Management and Budget requires the use of 7% for valuing costs and benefits external to the government and 4 % for internal costs and benefits (Newell and Pizer 2003).

Despite this straightforward reasoning for deriving a (social) discount rate from market data, there are several problems that need to be considered. First is the ethical issue of discounting across generations, which was discussed above. Then it is necessary to recognize that the use of 4 % as the consumption rate of interest does not agree with actual behavior in many circumstances. People willingly invest their savings in Treasury bills and guaranteed investment certificates that yield perhaps as little as 2% after taxes (and perhaps even less). Of course, these are riskless investments.

Also, when a government invests in a natural resource project, for example, funds could come from income taxes (displacing an equal amount of consumption) or from increased public-sector borrowing. Funds borrowed by government displace an equal amount of private investment, so it might be appropriate to use the higher rate of 7–8 %. If borrowed funds originate with private savings or if income taxes are used, the lower interest rate is more appropriate. In practice, of course, public funds come from a mix of sources. Thus, it might be appropriate to calculate the discount rate as the opportunity cost of the funds. Suppose that a public investment project costs $100, and that $40 displaces private investment and $60 comes from consumption. If the rate of return to private investments is 10% and the consumption discount rate is 4%, then the opportunity cost of the funds is 6.4% ($=0.40\times 10\%+0.60\times 4\%$). The main difficulty in deriving the opportunity cost rate is that it is not easy to determine where *marginal* funds originate. Further, not all government revenues come from income taxes or domestic borrowing, as governments earn income through charges, tariffs on imported goods, and so on.

Further, society may choose to save more collectively than the sum of all individual savings decisions. The government is considered a trustee for unborn generations, whose wealth will (at least in part) depend on the state of the environment that they inherit, so real consumption (and rates of return on investments) may not grow, and may even decline, when we degrade the environment. Because of risk and uncertainty (giving rise to 'risk premiums'), society's rate of time preference will be lower than that of individuals, as society as a whole is better able to pool risks; certain individual risks are mere transfers at the level of society. While individuals face a real chance of dying, society does not face a similar risk. All in all, these more or less ethical arguments suggest that society's rate of discount is lower than that of individuals making up the society. The social discount rate is likely lower than the opportunity cost of capital rate (real rate of return on investments) or the marginal rate of time preference, but it is not immediately clear how much lower.

Based on the above reasoning, a case can be made for using a very low discount rate to discount consumption by future generations. Again, a 2 % rate of discount might be appropriate. This is a somewhat arbitrary low rate and might be considered to be the social rate of time preference.

Since any rate between about 2 and 8 % appears justifiable, what might constitute the appropriate social rate of discount for use in social CBA? Newell and Pizer (2003) make the case that rates in the lower end of this range should be employed. Their argument rests on an analysis of uncertainty about the future path of interest rates. Using Monte Carlo simulation and historical information on the pattern of inflation-adjusted interest rates, and assuming the stochastic process for interest rates is not mean reverting (does not trend towards a mean in the absence of exogenous shocks), they find that the value of $100 received 400 years in the future is worth many orders of magnitude more today if interest rate uncertainty is taken into account than if a constant discount rate is used (see Table 6.2). While a constant discount rate is to be used in CBA, the results indicate that, because actual discount rates vary in unpredictable fashion (i.e., follow a 'random walk'), the discount rate to be employed should be lower than in the absence of this consideration. Thus, if a 4 % consumption rate of discount is considered appropriate because it is market derived, the true (constant) rate might be 2–3% if uncertainty about future interest rates is taken into account. Indeed, "correctly handling uncertainty lowers the effective discount rate in the future in a way that all generations after a certain horizon are essentially treated the same".

Clearly, there is a strong case to be made for the use of a low discount rate in the evaluation of natural resource and energy projects. Given continued controversy about what might constitute an appropriate rate, one suggestion is to use a rate of 2 % for evaluating policies/projects that affect more than one generation, and then use sensitivity analysis about this rate to determine how choices might be affected if the future is somehow weighted differently.

Finally, consider a declining discount factor approach that partially addresses some of the issues raised above, including hyperbolic discounting.[10] Standard

[10] I am indebted to Brian Scarfe for suggesting this approach.

Table 6.2 Value today of $100 received in 200 and 400 years: comparison of constant versus random walk discounting, selected discount rates

	Constant discounting		Nonmean-reverting random walk	
Discount rate (%)	200 years	400 years	200 years	400 years
2	$1.91	$0.04	$7.81	$3.83
4	$0.04	$0.00	$1.54	$0.66
7	$0.00	$0.00	$0.24	$0.09

Source: Derived from Newell and Pizer (2003)

discounting generates adjacent period weights such that $\frac{w(t+1)}{w(t)}=\frac{1}{1+r}$, which is constant for all adjacent time periods. An alternative system for adjacent period weights can be constructed by first letting $w(t)=b/(b+t)$, where $b>0$ is a parameter. Then $\frac{w(t+1)}{w(t)}=\frac{b+t}{b+t+1}, t>0$. This ratio converges to 1.0 as $t\rightarrow\infty$ (t gets larger).

Suppose one wishes to employ a standard discount rate to future costs and benefits for a period of T years, but one wants to weight years beyond T higher so as to favor future generations, say. That is, for $t>T$, the discount rate falls until it eventually is zero so that a period in the future is weighted the same as a current period. Such a scheme would weight costs and benefits in the years after T higher than if the standard rate were applied throughout the entire time horizon. In that case, one sets

$$\frac{b}{b+T}=\frac{1}{(1+r)^T}, \tag{6.8}$$

where r is the standard discount rate applied to the early period and T is the year when the weighting of the future begins to diverge.

Given r and T, formula (6.8) generates a unique value of the fundamental parameter b of the declining discount factor model. The weight attached to a future year is $w(t)=1/(1+r)$ for $t\leq T$ and $w(t)=b/(b+t)$ for $t>T$. An illustration is provided in Fig. 6.8, where we assume $r=0.05$ and T takes on values of 10 and 30. Notice that the importance (weight) of a future year in the cost-benefit analysis initially declines quite quickly, but after T it declines slower than in the case of the standard discount model. Needless to say, the problem with this approach is it arbitrariness, especially in the choice of T.

6.4.3 Discounting Physical Entities

A second issue related to the use of a zero discount rate involves the weighting of physical things. For example, should physical carbon be discounted according to when it is released to or removed from the atmosphere? Interestingly, some economists object to discounting of physical carbon, although they accept discounting if

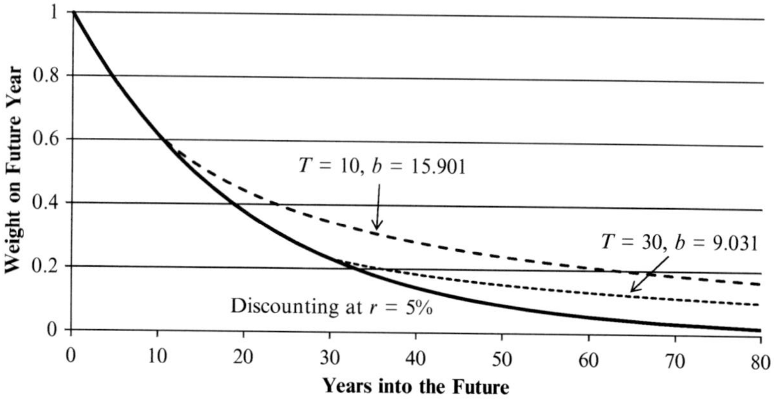

Fig. 6.8 Declining discount factor model, example with r=5 %

the physical carbon is multiplied by an arbitrary constant that converts the carbon into monetary units. Discounting or weighting of physical units is clearly an acceptable practice in economics, as is evident from Ciriacy-Wantrup (1968) and subsequent literature on conservation. One cannot obtain a consistent estimate of the costs of carbon uptake unless both project costs and physical carbon are discounted, even if at different rates of discount.

Suppose a tree-planting project results in the reduction of CO_2–equivalent emissions of 2 tons of carbon (tC) per year in perpetuity (e.g., biomass burning to produce energy previously produced using fossil fuels). In addition, assume the project has a permanent sink component that results in the storage of 5 tC per year for 10 years, after which time the sink component of the project reaches an equilibrium. How much carbon is stored? Suppose the present value of project costs has been calculated and that these are then allocated equally across the years of the project – so that the discounted stream of the equal annual costs is the same as the calculated present value of costs. If costs and carbon uptake are compared on an annual basis, does one use 2 or 7 tC per year? Suppose the discounted project costs amount to $1,000, or annualized costs of $40 if a 4% rate of discount is used. The costs of carbon uptake are then estimated to be either $20/tC if 2 tC is used, or $5.71/tC for 7 tC.

Suppose instead that we divide the present value of project costs (or $1,000) by the sum of all the carbon that eventually gets removed from the atmosphere. Since 7 tC gets taken up annually for the first 10 years, and 2 tC per year thereafter, the total amount of carbon sequestered is infinite, so that the cost of carbon uptake is essentially $0.00/tC. Therefore, an arbitrary planning horizon needs to be chosen. If the planning horizon is 30 years, 110 tC are sequestered and the average cost is calculated to be $9.09/tC; if a 40-year planning horizon is chosen, 130 tC are removed from the atmosphere and the cost is $7.69/tC. Thus, cost estimates are sensitive to the length of the planning horizon, which is not usually made explicit in most studies.

Cost estimates that take into account all carbon sequestered plus the timing of uptake can only be achieved if physical carbon is discounted. Then, using the methods described in the previous section, the total discounted carbon saved via our hypothetical project amounts to 147.81 tC if a discount rate of 2 % is used, and the correct estimate of costs is $6.77/tC. If carbon is discounted at a rate of 4 %, the project results in costs of $10.62/tC.

Finally, what discount rate should be applied to physical carbon? Richards (1997) demonstrates that, if physical carbon is not discounted, this is the same as assuming that damages from rising atmospheric concentrations of CO_2 are increasing at the same rate as the social rate of discount, but there is no reason to think that this might be the case. It also implies that there is no difference between removing a unit of carbon from the atmosphere today, tomorrow or at some future time; logically, then, it does not matter if the carbon is ever removed from the atmosphere. Only if damages rise slower than the growth in atmospheric CO_2 is a positive discount rate on physical carbon appropriate. This issue is addressed again in Chap. 9.

6.4.4 Risk Adjusted Discount Rates

If outcomes are unknown but estimable with some probability, the decision-maker faces risk that is measured by the expected variability in outcomes. If variability of returns from one project is higher than for another project, it is said to be riskier. The variance and standard deviation are measures of variability or spread and, thus, measures of risk. Most decision makers are risk averse, or reluctant to take risks. Given equal expected net returns, a risk-averse individual will choose the project with the 'narrower' distribution of payoffs as there is more certainty on the outcome.

There are ways to account risk in investment projects. A commonly applied method is the use of risk-adjusted discounted returns. The Capital Asset Pricing Model (CAPM) requires that riskier projects have higher rates of return, surely greater than the market rate of return (market rate of interest). Otherwise, no agent would invest in them. The fundamental equation of the CAPM is:

$$r_{\mathrm{i}} = r_{\mathrm{f}} + \beta\,(r_{\mathrm{m}} - r_{\mathrm{f}}), \tag{6.9}$$

where r_{i} is the required return for risky asset i, r_f is the risk-free rate of return, r_m is the market rate of return, and β measures the investment's contribution to risk relative to the market.[11] Returns are assumed to be normally distributed, so β is estimated as the ratio of the covariance of the asset and market returns to the variance of the market return:

$$\beta = \frac{\mathrm{cov}(r_i, r_m)}{\mathrm{var}(r_m)}, \tag{6.10}$$

[11] Note that β here is defined differently than its earlier use in Eqs. (6.6) and (6.7).

*β*s are usually calculated from past behavior of the investment and market returns. If time series data are available on rates of return, β is the regression coefficient that compares the responsiveness of the investment returns with changes in the market returns. Published data on βs can be useful for private and public projects. For example, Merrill Lynch and Value Line publish βs for stocks of a large number of companies. For project evaluation, asset βs instead of stock βs are required, although the latter can be converted into the former by recognizing that the asset value of a firm equals debt plus equity. Thus, the β of an asset is the weighted sum of the stock β plus the debt β.

Consider an example of the use of CAPM in the energy sector (see Zerbe and Dively 1994). Suppose a North American investor is considering construction of a power plant similar to ones operated by others. By checking βs published by Merrill Lynch for other electrical generating companies, some idea of the relevant β for the project can be obtained. The average β for 23 large utilities in the U.S. is 0.45. Assume that the investor has 40 % of her assets as debt and the debt β is zero. Then, the asset β for the project would be 0.27. If the nominal risk-free rate is 9 % and the market rate is 8.8 percentage points higher than this, the required return for the new investment project using the above formula is: $r = 9\ \% + 0.27(8.8\ \%) = 11.4\ \%$. This means that the energy investment is worth undertaking only if its expected NPV is positive when future costs and benefits are discounted at a rate of 11.4 %.

Risk is often relevant when dealing with externalities. For example, the benefits of mitigating global warming depend on so many variables that scientists cannot accurately estimate costs or benefits. Also, it is often the case where the emission reductions resulting from a carbon mitigation project are risky (e.g., carbon sequestration in agricultural soils). Therefore, it is reasonable to think that private investors involved in carbon mitigation investments might require a rate of return that is higher than the risk-free rate.

6.5 Extreme Events and Irreversibility

There are three alternatives for addressing extreme events and the possibility of irreversibility resulting from a decision either 'to do something' or 'not to do something'. Climate change might potentially be considered an extreme event.

1. The first is to determine the cost of the extreme event or irreversibility and the probability of its occurrence, and then include the expected cost in a social CBA. If the probability of the event or its cost, or some combination of the two, is sufficiently high, the expected cost may be such that avoiding the extreme event or irreversibility will be the optimal decision. In other cases, the cost will be small and the social cost-benefit criterion indicates that the project should proceed. In cases where the probability of the extreme event/irreversibility is not known and/or the cost associated with it is vague, Monte Carlo cost-benefit analysis (simulation across the range of probabilities and possible costs) can be used to

determine the probability that the social CBA criterion is violated.[12] As argued below, this approach to extreme events is the most consistent.

2. Economists have long debated another criterion that is invoked only when dealing with extreme events and irreversibility, namely, the notion of a 'safe minimum standard' (SMS) of conservation (van Kooten and Folmer 2004, pp. 219–221). Begin by ignoring the probability that an event occurs, and consider the maximum potential loss (maximum cost) associated with any strategy under some state of nature. We could choose the strategy that minimizes the maximum loss – the min-max strategy. However, such a decision criterion would prevent us from choosing a project whose net benefit to society might be very large simply because there is a tiny risk of an extreme event that imposes large costs. It is also possible that we avoid choosing the 'conservation' strategy because it has a potential loss that is only slightly larger than the loss that would occur by doing nothing. That is, the min-max criterion could lead us to choose in favor of a strategy with high probability of a large loss over an alternative that has an extremely low probability of a slightly greater loss.

   Clearly, the min-max strategy is not in the best interests of society because it fails to take into account event/outcome probabilities and the scale of cost differences. The safe minimum standard of conservation addresses this and other shortcomings via the following decision rule: Choose in favor of the strategy that provides the greatest flexibility and smallest potential loss, unless the social cost of doing so is 'unacceptably large'. This rule places development of natural resources and impacts on the environment beyond routine tradeoffs, and it does not permit deferral of resource development, say, at a cost that is intolerably high. The problem lies with the term 'unacceptably large'. Who decides when the cost is unacceptably large? In some cases, society can readily agree to accept risks that are extremely small but the potential benefits are large. In other cases, it is difficult to make such a decision and it must be made in the political arena, with all of the facts made available to citizens.

3. The criterion that is most commonly applied to situations where there exists the potential for extreme events and/or irreversibility is the 'precautionary principle'. Environmentalists define it as follows: "When an activity raises threats of harm to human health or the environment, precautionary measures should be taken even if some cause and effect relationships are not fully established scientifically".[13] While the European Union has taken the lead in promoting the precautionary principle as a basis for making decisions about the environment, Hahn and Sunstein (2005) and Sunstein (2005) have pointed out the logical inconsistency of the precautionary principle. For example, a decision based on the precautionary

[12] For example, under the social CBA criterion, a project is desirable only if the benefit-cost ratio is greater than 1.0. Monte Carlo cost-benefit analysis might generate 10,000 benefit-cost ratios, of which some proportion are less than 1.0.

[13] Statement adopted by 31 individuals at the Wingspread Conference, Racine, Wisconsin, 23-25 January 1998 (http://www.gdrc.org/u-gov/precaution-3.html as viewed February 25, 2010).

principle would prevent China from building nuclear power plants, even though doing so would reduce health problems associated with pollution from coal-fired power plants, deaths from coal mining, and emissions of CO_2. Yet, if China relied only on nuclear power, a decision to mine coal and use it to generate electricity would be squashed on the basis of the precautionary principle – that electricity generated from coal could lead to adverse environmental consequences and that it is therefore preferable to rely on nuclear power.

If the precautionary principle is to be taken seriously, it would thus provide no direction for and paralyze decision making. By balancing costs against benefits, and perhaps applying the notion of a safe minimum standard, there is at least a foundation for making difficult decisions (see Hahn and Sunstein 2005).

The use of either the safe minimum standard or the precautionary principle implies that one no longer employs social CBA as the decision criterion. In the case of SMS, the social CBA criterion is jettisoned in favor of a somewhat arbitrary criterion whenever there is potential for a decision to bring about an irreversible change. In the case of the precautionary principle, no other criteria are employed unless there is no risk whatsoever to human health or the environment. The chances that this is the case in decisions are rare – wind turbines endanger birds, fossil fuels lead to global warming, hydro dams endanger fish, biomass energy encourages destruction of wildlife habitat as marginal lands are cropped, nuclear power plants might meltdown, and so on.

The economist will almost certainly favor cost-benefit analysis over other criteria for making decisions, even decisions that entail some probably of irreversible loss. The tacit argument is that it is technically feasible to monetize all of the costs and benefits, including spillovers; it is possible to use expert judgments of health and environmental risks; it is possible to account for the ranges of costs associated with spillovers; people's perceptions of risk can be included; and, subsequently, it is possible to calculate the probability that a project results in losses to society, and the distribution of those losses. This information can then be used to determine whether the risks are worth undertaking – whether the benefit associated with accepting the risk (of building a nuclear power plant, say) is 'sufficiently great enough.'

Yet, there is a large element of subjectivity in cost-benefit analysis, particularly as it relates to extreme events. As we will see in Chap. 7, social-cost benefit analysis can be adapted to take account of potential extreme events in several ways. There we find some climate economists recommending a policy ramp (slowly increasing carbon taxes over time) for mitigating climate change, while others recommend immediate and drastic action to control carbon dioxide emissions. The reasons relate to the underlying assumptions employed in cost-benefit analysis to deal with extreme events. The results in the next chapter are briefly discussed from this perspective, keeping in mind that the discounted present value of expected damages avoided by taking action to prevent global warming climate must, in the cost-benefit framework, exceed the costs of acting.

Given uncertain information, economists must decide upon the potential costs of action to mitigate climate change, the potential damages from rising temperatures, the probabilities that damages will occur (although these are supposedly available from climate models), and the discount rate. Costs of mitigation can be low or high; the relationship between temperature increase and damages can be linear, quadratic or exponential; the probability of an extreme event (catastrophic runaway global warming) could be elevated; and the chosen discount rate can make the current value of future damages seem large or small. The policy ramp strategy takes a middle-of-the-road position on these parameters, setting them in such a way that … well a slow policy ramp turns out to be optimal.

The government of the United Kingdom has long been a proponent of immediate action to prevent climate change. A study by the government assumes very large future damages, based primarily on estimates of irreversible ecosystem damages obtained from contingent valuation studies; it also assumes an unusually low discount rate for determining the present value of those damages. Along with presumed low mitigation costs, the forgone conclusion of the UK study is that immediate and drastic action to mitigate climate change is imperative.

Finally, while criticizing the UK study for using a low discount rate, Harvard economist Martin Weitzman argues that the potential for catastrophic damage is understated. In his view, the probability distribution of future damages should reflect high probabilities of extreme events – the distribution of future temperatures should reflect a high probably of extreme future temperatures. The probability distribution should be asymmetric with 'fat tails'. Along with an exponential relation between temperature and damages, the 'fat tails' story leads to extremely large expected damages. Surprisingly, a safe minimum standard type of policy is recommended.

While greater details are provided in the following chapter, each cost-benefit study relies on assumptions about the economic parameters (mitigation costs, temperature-damage relation, probability distribution of temperatures and discount rate) to reach what might be considered a preconceived conclusion. In this regard, there is little difference between the adoption of CBA or some other criterion, including even the precautionary principle, for reaching a decision when confronted by an unknown and unknowable future.

6.6 Discussion

Economists employ four measures of surplus in the evaluation of projects or government programs, including programs to mitigate climate change. While in many natural resource and environmental situations it is difficult to estimate economic surpluses, economists have been able to provide decent enough estimates to facilitate decision making. In the context of climate change, however, the measurement problems are more nuanced. As we will see in the next chapter, uncertainty about the potential damages from climate change in a variety of sectors is unusually large. Such wicked uncertainty makes it difficult to implement a straightforward cost-benefit decision criterion. How does one determine the costs and benefits of mitigating CO_2 emissions to prevent climate change when damages avoided occur decades from now?

To the inherent uncertainty in dealing with climate issues must be added the perhaps more puzzling aspect of discounting when time frames are on the order of many decades or even centuries. As the controversy surrounding a study for the UK government (see Chap. 7) indicates, small differences in the discount rate used in cost-benefit analysis can lead to significantly different policy conclusions. The problem is that the world changes greatly over the course of a half century or more. One hundred years ago, the automobile was only slightly more than a curiosity; today the economies of many industrial nations (and even some developing ones) depend on automobile production, and many countries spent billions of dollars in 2009 to prevent the collapse of their automotive sectors. Electricity, refrigeration, airplanes, radio, television and computers were largely unknown, but today we cannot envision doing without them. How can we predict the potential damages (or benefits) from climate change in 2050 or 2100, much less 2200, without knowing the technical, social and economic changes that will occur on a global scale during this period?

By far the best and most rational cost-benefit analysis of future climate change has been conducted by Bjørn Lomborg (2007). It is the only one of which we are aware that takes into account technical progress in assessing climate change.[14] Lomborg's approach is simple: He indicates that the climate change that has occurred in the past century is about what models predict for the next century, both in terms of global temperature rise and sea level rise. He then compares life at the turn of the twentieth century with that today, showing how well people have adapted, and considers it rational for people likewise to adapt to future changes in climate.

Given the obstacles that confront cost-benefit analysis of climate change mitigation, policymakers have tended to promote mitigation policies on the basis of the precautionary principle. In that case, economists need to examine the costs of various mitigation schemes – to focus on efficiency (minimizing costs as a surplus metric) in the implementation of policy. This is the topic of Chaps. 8, 9, 10, and 11.

References

Adamowicz, W. L. (1995). Alternative valuation techniques: A comparison and movement to a synthesis. In K. G. Willis & J. T. Corkindale (Eds.), *Environmental valuation. New perspectives* (pp. 144–159). Wallingford: CAB International.

Adamowicz, W. L., Boxall, P., Williams, M., & Louviere, J. (1998). Stated preference approaches to measuring passive use values: Choice experiments versus contingent valuation. *American Journal of Agricultural Economics, 80*(1), 64–75.

[14] This is not to suggest that Nordhaus and others ignore technological changes. Integrated assessment models of climate change do include parameters for technical progress, but these translate into rate of change as a function of time or income. They do not attempt to address technologies that change a society's very structure or the utility functions of individuals. The reason is simple: No one can predict what changes lie in the future, even the near future.

Allen, B. P., & Loomis, J. B. (2008). The decision to use benefit transfer or conduct original valuation research for benefit-cost and policy analysis. *Contemporary Economic Policy, 26*(1), 1–12.

Arrow, K., Solow, R., Leamer, E., Portney, P., Randner, R., & Schuman, H. (1993, January 15). Appendix I – Report of the NOAA panel on contingent valuation. *Federal Register, 58*(10), 4602–4614.

Bulte, E. H., van Soest, D. P., van Kooten, G. C., & Schipper, R. (2002). Forest conservation in Costa Rica: Optimal forest stocks under uncertainty and rising nonuse benefits. *American Journal of Agricultural Economics, 84*(1), 150–160.

Ciriacy-Wantrup, S. V. (1968). *Resource conservation: Economics and policies* (3rd ed.). Berkeley: Agricultural Experiment Station, University of California.

Clawson, M. (1959). *Measuring the demand for and value of outdoor recreation. RFF Reprint #10*. Washington, DC: Resources for the future.

Dasgupta, P. (2002, June 24). *Discounting: Public vs private and constant vs hyperbolic*. Paper presented at the 2nd World Congess of Environmental and Resource Economists, Monterey.

Diamond, J. (2005). *Collapse*. New York: Viking.

Fischhoff, B., Lichtenstein, S., Slovic, P., Derby, S. L., & Keeney, R. L. (1981). *Acceptable risk*. Cambridge: Cambridge University Press.

Freeman, A. M., III. (1995). Hedonic pricing models. In D. W. Bromley (Ed.), *The handbook of environmental economics* (pp. 672–786). Oxford: Blackwell Publishers.

Graham-Tomasi, T. (1995). Quasi-option value. In D. W. Bromley (Ed.), *The handbook of environmental economics* (pp. 594–614). Cambridge, MA: Basil Blackwell Publishers.

Hahn, R. W., & Sunstein, C. R. (2005). The precautionary principle as a basis for decision making. *The Economists' Voice* 2(2), Article 8. Retrieved from Berkeley Electronic Press, http://www.bepress.com/ev

Harberger, A. C. (1971). Three basic postulates for applied welfare economics: An interpretive essay. *Journal of Economic Literature, 9*(3), 785–797.

Harberger, A. C. (1972). *Project evaluation: Collected papers*. Chicago: University of Chicago Press.

Heal, G. (2009). The economics of climate change: A post-Stern perspective. *Climatic Change, 96*(3), 275–297.

Horowitz, J. K., & McConnell, K. E. (2002). A review of WTA/WTP studies. *Journal of Environmental Economics and Management, 44*(3), 426–447.

Knetsch, J. L. (2000). Environmental valuations and standard theory: Behavioural findings, context dependence and implications. In H. Folmer & T. Tietenberg (Eds.), *International yearbook of environmental and resource economics 2000/2001* (pp. 267–299). Cheltenham: Edward Elgar.

Kramer, R. A., & Mercer, D. E. (1997). Valuing a global environmental good: U.S. residents' willingness to pay to protect tropical rain forests. *Land Economics, 73*(2), 196–210.

Krutilla, J. V. (1967). Conservation reconsidered. *American Economic Review, 57*(4), 777–786.

Lesser, J., Dodds, D., & Zerbe, R. (1997). *Environmental economics and policy*. New York: Addison-Wesley.

Lomborg, B. (2007). *Cool it. The skeptical environmentalist's guide to global warming*. New York: Knopf.

Newell, R. G., & Pizer, W. A. (2003). Discounting the distant future: How much do uncertain rates increase valuations? *Journal of Environmental Economics and Management, 46*(1), 52–71.

Olewiler, N. (2004). *The value of natural capital in settled areas of Canada* (p. 36). Stonewall: Ducks Unlimited Canada and the Nature Conservancy of Canada.

Pearce, D. W., & Warford, J. J. (1993). *World without end*. New York: Oxford University Press.

Phillips, W. E., Adamowicz, W. L., Asafu-Adjaye, J., & Boxall, P. C. (1989, August). *An economic assessment of the value of wildlife resources to Alberta* (Department of Rural Economy Project Report No. 89-04, pp. 70). Edmonton: University of Alberta.

Ready, R. C. (1995). Environmental evaluation under uncertainty. In D. W. Bromley (Ed.), *The handbook of environmental economics* (pp. 568–593). Cambridge, MA: Basil Blackwell Publishers.

Richards, K. R. (1997). The time value of carbon in bottom-up studies. *Critical Reviews in Environmental Science and Technology, 27*(Supp 1), 279–292.

Shaikh, S., Sun, L., & van Kooten, G. C. (2007). Treating respondent uncertainty in contingent valuation: A comparison of empirical treatments. *Ecological Economics, 62*(1), 115–125.

Smith, V. (1997). Pricing of what is priceless: A status report on nonmarket valuation of environmental resources. In H. Folmer & T. Tietenberg (Eds.), *The international yearbook of environmental and resource economics 1997/1998* (pp. 156–204). Cheltenham/Lyme: Edward Elgar.

Stern, N. (2007). *The economics of climate change: The Stern review*. Cambridge: Cambridge University Press.

Sunstein, C. R. (2005). *Laws of fear: Beyond the precautionary principle*. Cambridge: Cambridge University Press.

Thrice, A. H., & Wood, S. E. (1958). Measurement of recreation benefits. *Land Economics, 34*(3), 195–207.

van der Heide, C. M. (2005). *An economic analysis of nature policy* (Academisch Proefschrift, Tinbergen Institute and Vrije Universiteit, Amsterdam).

van Kooten, G. C. (1995). Economics of protecting wilderness areas and old-growth timber in British Columbia. *The Forestry Chronicle, 71*(1), 51–58.

van Kooten, G. C., & Folmer, H. (2004). *Land and forest economics*. Cheltenham: Edward Elgar.

Weitzman, M. L. (1998). Why the far-distant future should be discounted at its lowest possible rate. *Journal of Environmental Economics and Management, 36*(3), 201–208.

Weitzman, M. L. (1999). *Gamma discounting*. Cambridge, MA: Harvard University Press.

Woodward, R. T., & Wu, Y.-S. (2001). The economic value of wetland services: A meta-analysis. *Ecological Economics, 37*(2), 257–270.

Zerbe, R., & Dively, D. (1994). *Benefit-cost analysis in theory and practice*. New York: HarperCollins.

Chapter 7
Economic Assessment of the Damages Caused by Global Warming

The world will always need economists because economic matters are too important to ignore.

– Robert Nelson in The New Holy Wars (2010), p.91.

In a recent report, entitled *Climate Change 2009*, the United Nations Environment Program (UNEP) attributes every severe weather event to global warming, and thereby implicitly if not explicitly assumes humans are responsible for all weather-related damages (McMullen and Jabbour 2009, pp. 776–777).[1] The subtitle of the report, *Science Compendium*, appears to lend the report scientific authenticity. However, even the UN's Intergovernmental Panel on Climate Change would not attribute particular weather events to climate change, let alone anthropogenic emissions of greenhouse gases. In the latest IPCC report, the authors write: "Determining whether a specific, single extreme event is due to a specific cause, such as increasing greenhouse gases, is difficult, if not impossible, for two reasons: (1) extreme events are usually caused by a combination of factors and (2) a wide range of extreme events is a normal occurrence even in an unchanging climate" (IPCC WGI 2007, p.696).

At the very outset of the report, the UNEP identifies some 60 significant climate anomalies that occurred in various parts of the world during 2007–2009, clearly suggesting that these are related to climate change.[2] By attributing all extreme weather events to global warming, the report overstates the case for the damages that might be expected from climate change. Further, by assuming that severe weather events can be attributed to anthropogenic global warming and then advocating for government

[1] The report can be downloaded from http://www.unep.org/compendium2009/ (viewed February 26, 2011). According to the website, the "UNEP welcomes further constructive comments so that the report evolves as a living document containing the latest peer-reviewed science."

[2] The heat wave that struck Russia and floods that hit Pakistan in the summer of 2010 were part of the same weather system that was, at the time, widely attributed to anthropogenic climate change. However, a NOAA study (Dole et al. 2011) found that this extreme weather event was due to natural causes, and that frequency and intensity of blocking patterns such as happened in this event were not driven primarily by heat. Nor should such events increase with a rise in global temperatures.

G.C. van Kooten, *Climate Change, Climate Science and Economics: Prospects for an Alternative Energy Future*, DOI 10.1007/978-94-007-4988-7_7,

action to forestall climate change, the UNEP is telling us that humans are masters of their own fate – that we can control the weather! This is an arrogant claim to say the least, particularly given that weather events are considered by most to be uncontrollable and unpredictable, and that severe weather events ('acts of God') have frequently altered the course of human history (Durschmied 2000).

In this chapter, we want to provide some perspective regarding the possible damages from global warming. This is important given that the expected damages from global warming constitute the benefits (damages avoided) from taking action to mitigate greenhouse gas emissions.

There are many problems and pitfalls associated with attempts to calculate the damages from climate change. First off, it is necessary to separate damages attributable to the anthropogenic component of global warming and the global warming that is due to natural factors. If 90% of global warming is natural, then damages are essentially unavoidable and the benefits of reducing human emissions of greenhouse gases, of CO_2, are going to be small. However, if 90% or more of global warming can be attributed to human causes, the benefits of avoiding damages will be much higher. The problem is that, as shown in earlier chapters, the contribution of natural versus human factors to global warming remains unknown – a source of speculation.

Given scientific uncertainties, it is not clear whether a particular future event, such as a drought, an unusual storm or an early spring (which negatively affects tree growth in boreal climes, for example), can be attributed to climate change of anthropogenic origin, or whether it is simply a natural occurrence well within the usual vagaries of weather patterns. It may well be a bit of both, and sorting that out is nigh impossible.

The relevant question is: What would the subsequent weather-related damage of an event be if action is taken to prevent global climate change versus what is would be without such action? This is the 'with-without' principle of economic evaluation, and must be the principle that guides any discussion of damages. It is another way of saying that opportunity cost must be taken into account.

Even if we know that human activities are primarily responsible for climate change, it is necessary to determine how much temperatures will rise and what effect this will have on such things as sea level rise, increased weather events (more frequent droughts, hurricanes, tornados, etc.), the impact on biodiversity and ecosystems, the impact on disease, and so on. The scientific uncertainties are enormous, but once the impacts are known, it is necessary to estimate the costs (or benefits) of such changes and balance these damages against the costs of mitigating CO_2 emissions. It is quite possible that there is a socially optimal level of atmospheric CO_2 that is much higher than the current level, a point where the marginal benefits of further reducing CO_2 emissions equal the marginal costs of doing so. Given wicked uncertainty, could we ever find such an optimal point?

Finally, there is the question of tipping points, which has become a cause de celebre among certain economists, particularly the UK's Sir Nicholas Stern (2007) and Harvard University's Martin Weitzman (2009a, b, c). One version of the story is that global warming will cause the boreal tundra to melt, thereby releasing vast amounts of the potent greenhouse gas methane (CH_4). Once that happens, it is argued, run away global warming will take place. Therefore, based on the 'precautionary principle,' it is necessary to take drastic action to control human activities that release greenhouse gases into the atmosphere. It is important to note that arguments based on the precautionary principle (Chap. 6) cut both ways, as William Anderson (2010), a philosopher from Harvard University, points out in a commentary on climategate. If *drastic* action is taken to curb human emissions, this could, for example, lead to political instability that results in a cataclysmic global conflict that must be avoided at all costs; therefore, we should not take drastic action to curb emissions.

Clearly, estimating the damages avoided by mitigating climate change is a much more difficult and uncertain prospect than estimating costs, even though ascertaining costs is a difficult enough task. Since costs are related to what is happening in the economy today, they are supposedly easier to get a handle on. Yet, as will be seen in Chap. 8, where we look at the costs of implementing legislation proposed in the U.S. Congress to reduce CO_2 emissions, such cost estimates are controversial.[3] Determining damages avoided is a significantly more difficult task, and not one that policymakers are willing to take on, with the exception perhaps of Stern (2007). However, even if specific estimates of damages are unavailable or unstated, political decisions taken to address climate change give some indication of the costs that politicians are willing to incur, and thereby some notion of the damages they think might be avoided (even if these relate only to their chances of getting reelected at some future date).

In this chapter, we examine two approaches to damage estimation, commonly known as 'bottom up' and 'top down.' Bottom-up approaches determine the impacts of global warming at the sector, region or country level. For example, a bottom-up study might focus on the effect that climate change has on crop yields in the Canadian prairie provinces, or the potential impact of sea-level rise on New York City, or the effect of increased drought on biodiversity in the U.S. Pacific Northwest, or the impact of reduced precipitation on the Amazon rainforest. Such studies are somewhat speculative in the sense that climate change is about global trends in average temperatures and precipitation and not specific details. But the main drawback of bottom-up estimates of climate-change damages is that they do not take into account adaptation and impacts elsewhere. Thus, while tourists may no longer visit one region, another region may see an increase in tourism. A reduction in crop yields in one region could be more than compensated for by increased yields in another region. Prices change which causes estimates of damage to change, but bottom-up approaches do not and cannot take price changes into account.

Top-down models are less detailed and, in that sense, less accurate in their estimates of damages. Given that climate change is all about global mean temperatures and precipitation, and trends in these averages, crude estimates and indicators of

[3] CO_2 is considered the most important greenhouse gas (GHG) emitted by humans (but see Chap. 4) and other GHGs are generally measured in terms of their CO_2 equivalence, denoted $CO_{2\text{-e}}$. For convenience, here and throughout we simply use CO_2 to refer to carbon dioxide plus other greenhouse gases measured in terms of their CO_2 equivalence.

direction and potential magnitudes of damages are all that can realistically be expected. This is what top-down models provide. Integrated assessment models (IAMs) are a particular type of top-down model and they offer perhaps the best means for estimating global damages. IAMs are generally mathematical programming models that seek to optimize an objective function subject to static and dynamic constraints. The objective is usually to maximize the discounted sum of economic surpluses (generally only producer and consumer surpluses) over time, with constraints on technology, available resources, et cetera.

We begin in this chapter by considering estimates of climate-change damages related to (1) the primary sectors (agriculture and forestry), (2) ecosystems and biodiversity, (3) sea level rise, (4) increased severe weather incidents, (5) health effects, and (6) other impacts (e.g., on tourism). Sea level rise, health and biodiversity in particular are considered to be particularly vulnerable to global warming and thus to lead to large estimated costs if climate change occurs. In agriculture and forestry, sophisticated modeling and statistical methods have been used to determine damages from climate change, but even such estimates remain highly speculative. Estimates of damages in other sectors are crude and unreliable at best. Most of the damage estimates provided here are derived from bottom-up studies, but not all.

In the second half of the chapter, we focus on top-down modeling efforts. These are likely to be more controversial and, from an economist's point of view, a more exciting line of inquiry. Here the research centers less on the actual magnitude of the damage estimates, although that is of importance, but more on choice of an appropriate discount rate (which has a huge effect on the magnitude of damages when brought to a single year) and the potential for catastrophe (infinite damages).

When evaluating various sector- or regional-level estimates of damages, it is well to recall the measurement issues identified in the previous chapter. In many instances, estimates of damages constitute little more than an income transfer from one region to another or from one group to another, and are not true costs of climate change. One should also note that estimates of damages likely have little relation to a particular policy, such as the Kyoto Protocol, which actually did little to prevent global warming. Nor do all authors come to the conclusion that climate change will only lead to large reductions in overall global well being, although there is agreement that some regions will lose while others gain. In this regard, it is important to note that, even if climate change results in benefits (albeit with some probability), a risk-averse society may still be willing to pay some amount to avoid climate change.

7.1 The Climate Damages Landscape

Not surprisingly, sector-, regional- and country-level estimates of climate-change damages come from every quadrant, and are certainly not the purview only of economists. In some cases, damages are estimated using the principles espoused in the previous chapter, estimates based on solid economic science using economic measures of surplus. In other cases, estimates confuse true economic costs and benefits with economic transfers, generally leading to overestimates of damages. In yet other cases, damages (or benefits) are not even measured in monetary terms, as non economists weigh in with their views. This is all for the good, except it must be confusing for policymakers who must balance expenditures on mitigating or adapting to climate change against expenditures on public infrastructure, education, health and other programs of concern to citizens. It is because of these concerns that top-down modeling is generally done by economists.

In this section, we examine what is known about sectoral impacts, particularly focusing on areas of potential vulnerability – where costs might be greatest. In our discussion, we not only look at monetary and even non-monetary estimates of possible damages, but at some of the controversies related to what should be measured and how global warming might impact a particular sector.

7.1.1 Agriculture and Forestry

As noted by the *New York Times* (December 21, 2010), agriculture is likely the sector to be the most impacted and damaged by climate change[4]: "The risks [of rising atmospheric CO_2] include melting ice sheets, rising seas, more droughts and heat waves, more flash floods, worse storms, extinction of many plants and animals, depletion of sea life and – perhaps most important – difficulty in producing an adequate supply of food." Early estimates of potential climate change damages in agriculture employed crop simulation models and assumed a 'dumb' farmer, who would continue to plant the same crops with the same methods as those employed prior to the change in climate conditions. As economists became involved in damage estimation, greater possibilities for adaptation entered into the agricultural production models as 'smart' farmers were assumed to respond to changes in prices and climate.

Early Canadian studies by Louise Arthur and her colleagues at the University of Manitoba (Arthur 1988; Arthur and Abizadeh 1988; Arthur and van Kooten 1992; Mooney and Arthur 1990) suggested that, even if farmers only adopted crops suitable to the changed climate, western Canadian farmers could benefit. For the United States, Adams (1989) and Adams et al. (1990) used crop simulation and economic models to conclude that climate change in that country could lead to an overall increase or decrease in wellbeing, but that such changes were generally small. Indeed, results depended on which of several climate models was employed, but they were unambiguous in finding that the distributional impacts of climate change were the largest and most important aspect (see also Kaiser et al. 1993a, b).

Two methods have been used to determine climate-change damages in the primary sectors, both of which are rooted in economic theory. Each has its advantages and drawbacks. The hedonic approach employs the theory of Ricardian land rents

---

[4] See http://www.nytimes.com/2010/12/22/science/earth/22carbon.html?emc=eta1 (viewed December 22, 2010).

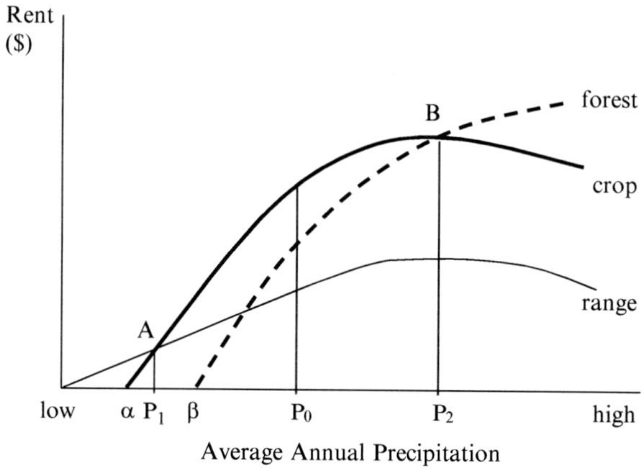

Fig. 7.1 Impact of changing precipitation on land use choices

(discussed in Chap. 6) to develop an econometric (statistical) model for estimating damages. The second approach, which was used by the researchers cited above, employs numerical solutions to a constrained optimization model. Each is discussed in turn.

7.1.1.1 Hedonic Pricing: Ricardian Land Use Models

Ricardian land-use models have become popular for determining the costs or benefits from climate change. Ricardian analysis simply assumes that landowners will use land in its best alternative, thereby maximizing the available rent. Observed land values reflect the fact that land rents diverge as a result of different growing conditions, soil characteristics, nearness to shipping points, and so on. Since producers face nearly the same output price, differences in land prices are the result of differences in the Ricardian or differential rents, which, in turn, are attributable to the various factors affecting production. Thus, marginal agricultural land that is used for extensive grazing has much lower rent than land that is intensively cropped.

The idea is illustrated with the aid of Fig. 7.1, where three land uses are considered and the factor determining differential or Ricardian rent is precipitation. When annual precipitation is low, the land can only support livestock, but, as precipitation increases, the rangeland yields increasingly higher rents because more forage can be produced. When annual precipitation increases beyond P_1 (associated with point A), crop production is the alternative land use that provides the highest rents. To the right of A, landowners will cultivate land and grow crops to achieve higher net returns to the land. Rents to crop production are even higher on land parcels that experience more rainfall. If too much rain falls, however, the land yields a greater rent in forestry – beyond P_2 (point B) rents in crop production are lower than those when the same land is in forestry. Points A and B represent *intensive margins*, precipitation thresholds where land use changes, while α and β represent *extensive margins* for land uses crop production and forestry, respectively, because rent to those land uses falls below zero if precipitation is below those thresholds.

The Ricardian approach assumes that a landowner experiences average annual precipitation of P_0 before climate change occurs. If precipitation under the altered climate falls below P_1, the landowner will stop crop production and rely on forage for livestock; if precipitation increases beyond P_2, she will encourage a forest to establish on the land. It is assumed that, for annual precipitation levels between P_1 and P_2, the owner will adjust use of other inputs to maximize the rent accruing to the land.

A hedonic model estimates farmland values as a function of climate variables, such as growing degree days (number of days during the growing season that temperature exceeds 5°C) and precipitation, and various control variables (soil quality, latitude, nearness to urban area or population density, nearby open spaces, presence of irrigation, etc.). The climate and control variables constitute the explanatory variables or regressors. Once the model parameters have been estimated for a sample of farms for which actual sales data are available, the results are first used to predict the farmland values across the entire study region or country. Then the climate variables are changed to reflect the projected change in climate, and the model parameters are again used to predict the farmland values for the entire region/country. (The control variables and the estimated model parameters remain the same in the current and changed climate states of the world.) The model implicitly assumes that, if landowners face different climate conditions, they will choose the agricultural land use (crop and technique) that maximizes their net returns. The differences between farmland values in the current climate state and the projected future climate regime constitute the costs (if overall values fall) or benefits (if overall farmland values rise) of climate change.

The Ricardian approach is considered by economists to be the most appropriate method for estimating the potential impact of climate on agriculture and even forestry, if forest use of land is taken into account. However, forestlands are often ignored because there is little information on private forestland prices (much is owned by institutional investors) or the forestland is owned by the government and no price data are available. Further, there is no reason to suppose that the estimated parameters will continue to hold under a changed climate regime, which might be the case if growing conditions under a future climate regime are outside observed values (which increases uncertainty of predicted values); it is also difficult to use hedonic models estimated for a current period to project how the same land might be used some 50–100 years later. Ricardian models do not take into account technological and economic changes that might occur, nor can they be expected to do so. But they also fail to take into account the fertilizer impact of CO_2, which is discussed below. Despite these flaws, the Ricardian method is one of the few statistical approaches that can be used to determine potential damages from global warming, and it is solidly rooted in economic theory.

Using econometric analysis of land use values, Mendelsohn et al. (1994) projected a small increase in U.S. GDP as a result of global warming. On the other hand, Schlenker et al. (2006) found that, if agricultural regions were separated into irrigated and dryland areas, the conclusions from econometric modeling would be reversed. Climate change would unambiguously impose net costs upon agriculture in dryland regions of the United States, although some dryland areas in the northern states would gain. It was also believed that "climate change will impose a net economic cost on agriculture in irrigated counties, whether in the form of higher costs for replacement water supply or lower profits due to reduced water supply."

Reinsborough (2003) also used a Ricardian land rent model to analyze the potential impact of global warming, but then for Canada. She found that Canada would benefit marginally as a result of climate change – some $1.5 million per year or less. In sharp contrast, Weber and Hauer (2003) find that Canadian agricultural landowners could gain substantially as a result of climate change. Their Ricardian rent model employed a much finer grid and greater intuition regarding agricultural operations than did Reinsborough. They projected average gains in land values of more than 50% in the short term (to 2040) and upwards of 75% or more in the longer term (to 2060). Canada will clearly benefit from global warming, as most likely would Russia.

7.1.1.2 Mathematical Programming Models

A second class of models uses economic theory to develop a mathematical representation of land-use allocation decisions. An economic objective function is specified and then optimized subject to various economic, social, climate and technical constraints, with the latter two representing the production technology. The objective function might constitute net returns to landowners, the utility (wellbeing) of the citizens in the study region, or, most often, the sum of producers' and consumers' surpluses. The choice of an objective function depends on the purpose of the analysis, the size of the study (whether country, region or worldwide level) and the number of sectors included. The numbers and types of constraints also depend on the size of the model and its purpose (multi-region models with as many as 100,000 or even more constraints are not unusual), but somewhere (usually in the production constraints) climate factors are a driver. Models are calibrated to the current land uses and other conditions (e.g., trade flows) using a method such as positive mathematical programming, which employs economic theory to find calibrating cost functions (Howitt 1995, 2005).[5] Models are solved numerically using a software environment such as GAMS (McCarl et al. 2007).

To determine the costs (or benefits) associated with climate change, the calibrated model is solved with the current climate conditions, and subsequently re-solved with the projected future climate conditions. Differences between the base-case objective function and the future scenario (or counter factual) constitute an estimate of the cost or benefit of climate change.

[5] Weintraub et al. (2007) provide examples of these techniques in the area of natural resources, including an example of agricultural land use in Europe and calibration via positive mathematical programming.

Some numerical constrained optimization models are static, while others are dynamic in the sense that current activities (the land uses chosen today) affect the state of nature in the next period (future possibilities), and thus the choices one can make in the future. This is the idea behind integrated assessment models. Most models of land use in agriculture and forestry are static, although the Forest and Agricultural Sector Optimization Model (FASOM) is an exception (Adams et al. 1996). It optimizes the discounted sum of producers' and consumers' surpluses across forestry and agriculture, determines optimal harvest times of commercial timber, permits reallocation of land between the agricultural and forest sectors over time, and takes into account carbon uptake and release. To keep things manageable, it employs a 10-year time step. The impact of climate change is not modeled, per se, as FASOM is primarily used for policy to determine how carbon penalties and subsidies might affect the allocation of land use within and between the two primary sectors. The primary limitation of FASOM is that it only applies to the forestry and agricultural sectors of the United States, ignoring climate impacts in other countries that may affect U.S. prices.

One variant of static numerical optimization models is the computable general equilibrium model (CGE). A CGE model maximizes a utility or social welfare function subject to equality constraints. Each sector in an economy is somehow represented in the constraint set (even if subsumed within a larger sector) and sometimes in the objective function. The extent to which sector detail is modeled depends on the question to be addressed (purpose of the study) and the extent to which detailed macroeconomic level data are available. The best known work employing CGE models in agriculture has been done at the Economic Research Service of the U.S. Department of Agriculture (Darwin et al. 1995; Schimmelpfenning et al. 1996).

Upon comparing econometric results with those from mathematical programming models, we find that the results of Weber and Hauer, as well as those of Schlenker et al. (2006) for the northern United States, are in line with those reported by Darwin et al. (1995) for Canada. Darwin et al. used a land-use model linked to a computable general equilibrium model to estimate the global welfare impacts of climate change as it affects output in the primary sectors. They found that, if landowners were able to adapt their land uses to maximize net returns (as assumed in the Ricardian analyses), global GDP would increase by 0.2–1.2% depending on the particular climate model's projections employed.

The majority of studies of damages to the agricultural and forestry sectors are for the United States and Canada. The general conclusion is that the U.S. agricultural sector will likely be harmed by climate change but damages may be minor compared to the size of the sector, while Canada's sector will benefit overall (although some regions could be harmed). Clearly, while future research might improve the methods of analysis, scientific and economic uncertainty will make it difficult to obtain more than ballpark estimates of the potential damages from global warming to the agricultural and forestry sectors – and even estimates of gains cannot be ruled out entirely. For example, in a study of the impacts of global warming on individual countries, William Cline (2007) concludes that there could be gains to global agriculture in the short run, but in the longer run the sector's output will decline.

Few economic studies include potential CO_2-fertilization benefits that would cause crops and trees to grow faster. One reason is that there is much debate about the impact of CO_2-fertilization. Thus, Cline (2007) takes it into account and attributes short-term gains in agricultural output to the fertilization effect, but, based on diminishing returns argument and the adverse effect of excessive warming, argues that agricultural yields will decline in the longer run.

7.1.1.3 CO_2 Fertilization

The increase in CO_2 during the twentieth century has contributed to about a 16% increase in cereal crop yields (Idso and Singer 2009). Levitt and Dubner (2009) indicate that there will be a 70% increase in plant growth with a double CO_2 atmosphere.

Research gathered by Michigan State University professor emeritus of horticulture, Sylvan H. Wittwer, indicates that, with a tripling of CO_2, roses, carnations and chrysanthemums experience earlier maturity, have longer stems and larger, longer-lasting, more colorful flowers with yields increasing up to 15%. Yields of rice, wheat, barley, oats and rye increase by upwards of 64%, potatoes and sweet potatoes by as much as 75%, and legumes (including peas, beans and soybeans) by 46%. The effect of carbon dioxide on trees, which cover one-third of Earth's land mass, may be even more dramatic: According to Michigan State's forestry department, some tree species have been found to reach maturity in months instead of years when the seedlings were grown in a triple-CO_2 environment (Sussman 2010, p.66).

7.1.2 Biodiversity and Ecosystems

People are much more interested in protecting mega fauna, such as elephants and tigers, than species such as burrowing beetles (van Kooten and Bulte 2000). The polar bear (*ursus maritimus*) is touted as the most prominent species threatened by global warming. Polar bears are endangered by loss of habitat – by the decline in sea ice in the Arctic. It is argued that the bears need the ice to survive, because it is the platform from which they hunt seals, their primary food source. A picture of a lone polar bear on a small piece of ice floating in a vast sea (although the distance to the nearest ice flow or land cannot be determined from the picture, and such a distance could well be very short) serves to highlight the threat global warming poses.[6]

[6] Al Gore uses a picture of a mother polar bear and her cub on "an interesting ice sculpture carved by waves" that is available at (viewed March 3, 2011): http://www.whoi.edu/beaufortgyre/dispatch2004/dispatch02.html to highlight the plight of the polar bear. The following quote accompanies the photo: "Currently we are traveling towards the 150 deg. longitude line to the next CTD and mooring sights, but had to slow down to a crawl because of an interesting and exciting encounter; a polar bear was sighted swimming off of the ship's port bow. It looked to be a juvenile, but is still considered to be very dangerous. Later on a mother and cub were also spotted on top of an extraordinary ice block." The quote is due to Kris Newhall, who also took the picture and regularly sent

While other species and ecosystems are considered to be threatened by climate change, it is the polar bear that is most widely reported upon. Thus, we examine it in a bit more detail to determine what the potential cost of its demise might be.

Let us begin by employing contingent valuation data from other mega fauna and apply it to the polar bear as a form of crude benefit transfer (discussed in the previous chapter). Information provided in van Kooten and Bulte (2000, p.305) indicates that hunters were willing to pay \$36.58 (1993 U.S. dollars) for a grizzly bear permit (that would permit the killing of one bear). Respondents to contingent valuation surveys indicated that they would be willing to pay between \$15.10 and \$32.94 annually to avoid the complete loss of bald eagles, and \$12.36 to avoid disappearance of bighorn sheep. The largest annual values for preventing loss of any species were associated with the monk seal and humpback whale (\$119.70 and \$117.92, respectively). There are a number of problems with these measures, including that the amount a person is willing to pay to protect a single species is not mutually exclusive – it is affected by the need to pay to protect other species, which is not asked in a survey – and respondents to a contingent valuation survey employ high discount rates. With regard to the former, there might be an imbedding effect – people respond as if they are protecting all wildlife species and not just the species in question. If people are asked about their willingness to pay to protect all wildlife species they might well provide answers that are close to those of a single species. Thus, they might be willing to pay \$120 per year to protect all mega fauna, but, within that category, only \$15, say, to protect humpback whales. Likewise, in responding, people do not envision paying the stated amount in perpetuity, usually assuming a period about the same length of time as a car loan. Thus, a 5% rate of discount would actually translate into an effective discount rate of 22%. On the other hand, if compensation demanded was used as opposed to willingness to pay to avoid a loss, then the amount involved might be some two or more times larger (Horowitz and McConnell 2002).

To find out what people would be willing to pay to avoid the loss of polar bears, one first needs to determine a suitable value. Given that the polar bear might have more in common with the humpback whale or monk seal than the bald eagle, we choose \$120 per year, but then assume payment occurs for only 5 years, in which case each household would be willing to make a one-time lump sum payment of \$600 in 1993 dollars; this amounts to \$900 in 2010 dollars if U.S. inflation between 1993 and 2010 is used. If only North American and European households were willing to pay this amount, and that there are some 200 million households, then ensuring survival of polar bears is worth \$180 billion, an enormous sum. However, this type

dispatches from the Woods Hole Oceanographic Institute's 2004 Beaufort Gyre Expedition – the quote is from dispatch 2, August 7–8, 2004. There is no suggestion that polar bears are in any way threatened, but, rather, suggests that polar bear sightings as a common occurrence. The polar bears in the photo used by Al Gore were within swimming distance of shore. The ice was melting, but it always melts in summer. In the winter of 1973–1974, abnormally heavy ice cover in the eastern Beaufort Sea resulted in a major decline in polar bear numbers (as reported in Armstrong et al. 2008, p.386). Thus, both more ice cover as well as less cover appear to be harmful to polar bears.

of calculation is dangerous because it is likely that people are willing to pay that amount to protect all mega fauna, not just polar bears. Alternatively, one might think of the $180 billion as a one-time payment to protect global biodiversity against loss due to climate change (given that biodiversity and mega fauna are closely related in such valuations). In that case, the sum might seem small. In that case, the *guaranteed survival* of polar bears might only garner a payment of several hundred million dollars, still a significant amount.

Yet, there remain caveats. First, there is a chance that the polar bear will not go extinct, that some members will survive. Polar bears might adapt to conditions much as their counterparts to the south, only to become polar bears again when the globe cools and the Arctic ice returns to pre-anthropogenic global warming conditions.[7] Second, there is a chance that the Arctic ice does not disappear entirely and that sufficient numbers of bears continue to live on. After all, the minimum viable population required to prevent most large mammal species from going extinct may be rather small, even as low as 20–50 members (see van Kooten and Bulte 2000, pp.199–201, p.281). Finally, the polar bear may not only adapt to changing climate but may actually increase in numbers – the forecast that polar bears are disappearing and that the cause is climate change may simply be wrong (see Armstrong et al. 2008).[8]

7.1.2.1 Arctic Sea Ice

Consider the question of sea ice. There have been several attempts by various environmental groups to show that sea ice is now so deficient that it might be possible to reach the North Pole by boat.[9] The truth is that substantial sea ice remains and may even be expanding. This is shown in Fig. 7.2: the area of sea ice declined from 2004, reaching its minimum extent during 2007; thereafter, the extent of sea ice seems to have increased, as indicated by the arrows in the figure. Even so, winter sea ice extent is well within historical averages, with only summer sea ice slightly lower than long term (1950–2002) average.

Historically, the Arctic is characterized by warm periods when there were open seas and the Arctic sea ice did not extend very far to the south. Ships' logs identify ice-free passages during the warm periods of 1690–1710, 1750–1780 and 1918–1940,

[7] For example, polar bears appear to mate with grizzly bears and other large bears.

[8] Armstrong et al. (2008) conduct a forecasting audit of studies used by the U.S. Fish and Wildlife Service to recommend listing of the polar bear as an endangered species under the U.S. Endangered Species Act, which was done in 2008. They find that many principles of evidence-based forecasting are violated and that the grounds for listing the polar bear as endangered are unwarranted from a scientific standpoint.

[9] Paul Driessen and Willie Soon provide interesting accounts of public figures attempting to make journeys to the North Pole in the summer in order to publicize the demise of Arctic ice, only to be turned away by cold and ice after having barely started their journeys. See (viewed June 2, 2010): http://townhall.com/columnists/PaulDriessen/2010/05/01/desperately_looking_for_arctic_warming?page=1.

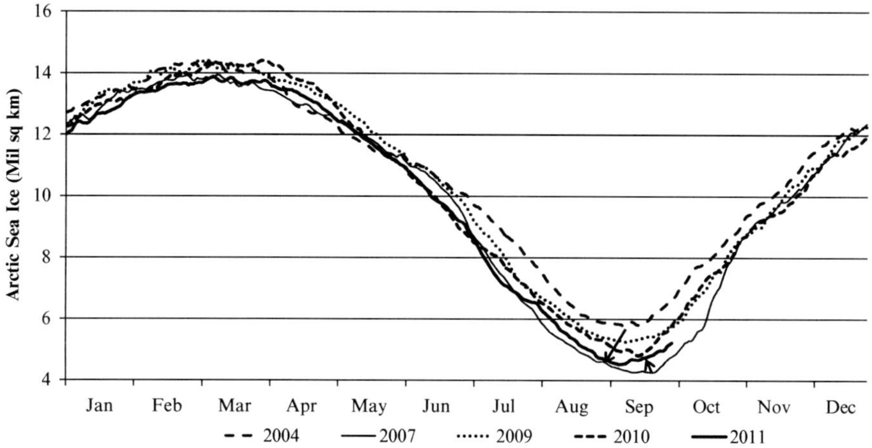

Fig. 7.2 Extent of sea ice in the twenty-first century, selected recent years, lowest year is 2007 (Data were derived from AMSR-E sensor and provided by Japan Aerospace Exploration Agency (JAXA) through the IARC-JAXA Information System (IJIS), and used with permission. See http://www.ijis.iarc.uaf.edu/en/home/seaice_extent.htm)

although each of these warm periods was generally preceded and followed by colder temperatures, severe ice conditions and maximum southward extent of the ice (e.g., during 1630–1660 and 1790–1830). The IPCC WGI (2007, pp.351–352) asserts with "high confidence that sea ice was more extensive in the North Atlantic during the nineteenth century" than now, although this would not be unexpected given that the Little Ice Age ended sometime between 1850 and 1870.

Since there are no empirical measures of the extent of sea ice prior to the age of satellites, one must rely on written accounts, anthropological evidence, ships' records and so on. Clearly, there must have been little ice in the Davis Strait west of Greenland as the Vikings established colonies at Godthab (Nuuk), known as the Western Settlement, sometime at the beginning of the eleventh century or somewhat earlier (Diamond 2005). The Swedish explorer Oscar Nordkvist reported that the Bering Sea region was nearly ice free in the summer of 1822, while Francis McClintock (captain of the 'Fox') reported that Barrow Strait (north of Somerset Island or northwest of Baffin Island) was free of ice in the summer of 1860, but had been completely frozen up at the same time in 1854. Even the famous explorer Roald Amundsen noted in 1903, during the first year of his 3-year crossing of the Northwest Passage, that ice conditions were "unusually favorable."[10] In a highly speculative treatise of Chinese navigation, Gavin Menzies (2002, pp.343–357) com-

[10] Accounts reported in "(Desperately) Looking for Arctic warming," Townhall, May 1, 2010 http://townhall.com/columnists/PaulDriessen/2010/05/01/desperately_looking_for_arctic_warming?page=1 (viewed June 2, 2010).

ments on the likelihood that the entire North may have been sufficiently free of ice circa 1422 to enable Chinese explorers to map the coast of Siberia.[11] Sussman (2010, p.113) provides a photograph of the submarine, U.S.S. *Skate*, on the surface in ice-free water at the North Pole in March 1959.[12] Russian scientists meantime continue to argue that the Arctic is getting colder and not warmer.[13]

The point of these observations is that the extent of Arctic sea ice has fluctuated over time. Research by the Norwegian Torgny Vinje (see Vinje and Kvambekk 1991) suggests that 1979 and 1981 may have been particularly bad years for sea ice, with 1976 comparatively ice free. The researchers point out that ice conditions are primarily driven by winds and ocean currents, that ice can accumulate at the rate of 20 cm/day resulting in a ice thickness of 6 m in a month, and that open water ('polynyas') can be observed on the leeward side of islands in winter 10–30% of the time in the summer-to-winter ice forming zones. Vinje and Kvambekk (1991) make three very relevant observations: First, the average area covered in ice in the Barents Sea in April during the years 1973–1976 was "about 700,000 km^2 and about 1,150,000 km^2 in 1969 and 1979, revealing a variation of as much as 400,000–500,000 km^2 in the annual maximum extension over a period of four years" (p.61). That is, the extent of sea ice in the Barents Sea in April was found to vary by as much as 65% within 4 years. While the dominant ice flow in the Arctic is the Transpolar Ice Drift Stream, which can bring 4,000–5,000 km^2 of ice (equivalent to the annual water discharge of the Amazon) into the Barents Sea from the Greenland Sea, although this clearly cannot account for the differences in sea ice coverage.

Second, Vinje and Kvambekk (1991) found a downward trend in sea ice area over the 23-year period 1966–1988, as measured in late August. This trend amounted to a loss of average sea ice area of 5,400 km^2/year. Clearly, this trend amounts to no more than 4–5% of the total change in sea ice that can easily occur over 4 years – a 23-year record is too short in this case to derive definitive empirically-based conclusions.

Finally, the researchers point out that ice of various ages gets mixed up as a result of wind, wave, tidal and ocean current factors. They conclude that much of the theory and science of ice formation still needs to be sorted out, an observation that remains valid some two decades later.

Debates about the reasons for changes in Arctic ice rage on. In a recent paper, Wood and Overland (2010) attempt to explain why the Arctic ice sheet was noticeably diminished during the period 1918–1940 (as noted above). They conclude that the "early climatic fluctuation is best interpreted as a large but random climate

---

[11] Menzies cites, among others, Needham (1954).

[12] Also http://wattsupwiththat.com/2009/04/26/ice-at-the-north-pole-in-1958-not-so-thick/ (viewed March 8, 2012).

[13] This claim is made by Oleg Pokrovsky of the A.I Voeiko Main Geophysical Observatory; see (viewed June 7, 2010): http://www.upi.com/Science_News/2010/04/23/Scientist-says-Arctic-getting-colder/UPI-94431272034113/. See also "Challenging the basis of Kyoto Protocol" by Vladimir Radyuhin (*The Hindu*, July 10, 2008). Found at (June 10, 2010): http://www.thehindu.com/2008/07/10/stories/2008071055521000.htm.

excursion imposed on top of the steadily rising global mean temperature associated with *anthropogenic forcing* [italics added]." But it could just as easily be concluded that the early warming is best interpreted as a large but random climate excursion imposed on top of the steadily rising global mean temperature associated with Earth's natural recovery from the global chill of the Little Ice Age. Further, there is no reason not to conclude the same about the most recent Arctic warming, because, for example, White et al. (2010), in an analysis of past rates of climate change in the Arctic, conclude that: "thus far, human influence does not stand out relative to other, *natural* causes of climate change [italics added]."[14]

Recent observations of the decline and subsequent increase in the extent of Arctic sea ice, both historically and more recently (as evident in Fig. 7.2), are not unprecedented and cannot at this time be attributed to anthropogenic warming or some other known cause. The matter still needs to be resolved from a scientific point of view.

7.1.2.2 Polar Bear Populations

Current available data relating to polar bear populations are best considered inconclusive in terms of scientists' ability to state that polar bears are threatened by human-caused global warming. It can even be argued that polar bears are not in decline, but that their numbers may even be growing. There are some 20,000–25,000 polar bears in the world, with some 60% found in Canada.[15] This compares with some 5,000 polar bears in the 1950s.[16] Polar bears are divided into 19 subpopulations. According to a 2009 meeting in Copenhagen of the Polar Bear Specialist Group V of the International Union for the Conservation of Nature (IUCN), eight subpopulations are considered to be in decline, three are stable, one is increasing and there is insufficient evidence to determine trends for the remainder. Canada's Western Hudson Bay population has dropped 22% since the early 1980s. Since the number of subpopulations considered to be in decline has grown from five (at the group's 2005 Seattle meeting) to eight in 2009, some argue that polar bears are in decline as a result of global warming. However, the decline cannot be linked directly to global warming and certainly not to human emissions of greenhouse gases.

---

[14] See http://www.nipccreport.org/articles/2010/dec/8dec2010a2.html (viewed December 13, 2010) for additional details.

[15] http://www.polarbearsinternational.org/bear-facts/ (viewed February 25, 2010).

[16] http://www.polarbearsinternational.org/ask-the-experts/population/ (viewed February 25, 2010). Note that both the reference here and that in the previous note are from a polar bear lobby group website. Because of improved monitoring and greater efforts at enumerating bear populations, the more recent figures are more accurate than the historical one, but this is not to suggest that populations in the 1950s were actually four to five times greater than the estimates made by biologists at the time.

The reason why polar bears are in decline, if at all, is hunting. Over a 5-year period to 2005, an average annual total of 809 bears were hunted; this covered 15 of the 19 regions for which information was available.[17] Hunting is permitted by government officials in the various jurisdictions where polar bears are found. For example, with the permission of the Renewable Wildlife Economic Development Office of Canada's Northwest Territories, big-game hunters can purchase a non-resident hunting license and permit to kill a polar bear for $800 (plus 7% tax), although they will need to fly to Inuvik or Tuktoyaktuk and pay unspecified guiding, outfitting and trophy export fees; the overall cost could exceed $35,000 to bag a polar bear.[18] Therefore, if one feels that polar bears are threatened, it is much more efficient to stop hunting than to reduce CO_2 emissions as hunting is a vastly greater threat to polar bears than global warming.

Similar stories can be told of other species. Human hunting of polar bears, bald eagles (see 'Impacts on human health' below), whales and other animals have proven a great threat to the survivability of many mega wildlife populations. Human development of the habitat of elephants, tigers, bison and other mega fauna have contributed to the demise of many species, and will likely continue to do so in the future (van Kooten and Bulte 2000). The introduction of invasive species has also posed an enormous threat to many indigenous species, even causing some to disappear because they cannot compete. These three factors are a greater threat to wildlife populations than global warming. Yet, many wildlife species are extremely resilient, surviving and sometimes even flourishing when temperatures warm. In the case of polar bears, for example, one must ask: How did this species survive previous episodes when there was little Arctic ice, such as during the Medieval Warm Period?

Undoubtedly, some species will not survive under some of the global warming scenarios that are envisioned, but it is not clear as to which climate outcomes will lead to the greatest loss of species. Further, it is not clear to what extent ecosystems will migrate, or simply disappear, or how quickly changes will take place. If the pace of ecosystem change is slow, many species will be able to survive, migrating with the ecosystem itself or adapting to new conditions. From the point of view of economic analysis, the ideal is to know which species are most in danger of extinction as a result of climate change and the value that global society attaches to their survival. This would require knowing the probabilities attached to various outcomes, the probabilities of each species' demise or survival under each climate scenario, and households' willingness to pay for each of the combinations of various outcomes relative to one another, which would also depend on how their incomes and other choice sets are impacted by warming. The point is this: It is impossible to determine the damages that global warming will impose on ecosystems and biodiversity. Attempting to do this as an exercise might be good fun, but it cannot

[17] http://pbsg.npolar.no/en/status/status-table.html (viewed March 1, 2010).

[18] http://www.polarbearhunting.net/ and http://www.independent.co.uk/environment/nature/bag-a-polar-bear-for-35000-the-new-threat-to-the-species-1649547.html (viewed March 1, 2010).

lead to realistic estimates of climate-induced damages. Therefore, as indicated later in this chapter, economists employ much simpler damage functions in integrated assessment models.

7.1.3 Sea Level Rise

Several years ago, Sierra Club held a press conference in Victoria to draw attention to the perils of global warming. They showed that much of the city would be flooded if global warming was allowed to continue unabated. Scaremongering, they suggested that sea level would rise by some 100 m or more, which would change the map of Victoria dramatically.[19] Various studies have found evidence one way or another for changes in sea level (depending on location of the measurements and the time interval chosen), and there is no doubt that sea levels are rising. Indeed, sea levels have been rising ever since the last major ice age, but not as a result of anthropogenic emissions of CO_2 and other greenhouse gases. Thus, it needs to be demonstrated that sea levels are now rising faster than historically and that this is attributable solely to human activities. At this stage, one can only conclude that the science is highly uncertain, making economic estimates of potential damages from rising sea level even more so.

What are the facts? During the twentieth century, sea level rose by some 1.7 ± 0.5 millimeters (mm) per year, or about 17 centimeters (cm) over 100 years. There is evidence suggesting that sea levels have been rising even faster in the past several decades. From 1961 to 2003, average sea levels rose by 1.8 ± 0.5 mm annually, but they rose by 3.1 ± 0.7 mm/year for the sub-period 1993–2003 (IPCC WGI 2007, p.387). This difference needs to be placed in proper perspective, however, because the latter measures are based on satellite altimetry observations, whereas the earlier measures are based on tidal gauges. Further, historical evidence indicates that rates of change in sea level vary considerably from one decade to the next, so it is impossible to determine whether the latest observed rates of increase are due to decadal variability or indicative of a longer-term trend, as noted by the IPCC report.

Determining the causes of past sea level rise and what might cause it in the future is not an easy task. Three factors affect changes in sea level. First, as the ocean warms, it expands, causing the sea level to rise. According to the IPCC WGII (2007, p.317), sea surface temperatures (SST) might increase by upwards of 3°C. Second, when continental glaciers melt, there is an increase in runoff into oceans and the sea level will rise accordingly. Melting of Arctic ice, for example, does not cause sea levels to rise because the ice floats on top of the water and, when it melts, contributes nothing to reduce or raise sea levels, as the effect of floating ice is already included in the current sea level.

Unlike the previous ones, the third factor could lead to a reduction in sea level. As global warming occurs and the oceans themselves get warmer, there is greater evaporation and, as a consequence, greater precipitation. This causes a buildup of

[19] See http://wattsupwiththat.com/2012/07/24/back-to-the-future-paradise-lost-or-paradise-regained/ (viewed July 24, 2012).

glaciers and a reduction in sea levels. It is not clear what influence each factor has had on past sea levels and what it will have on future sea levels. Data for the period 1961–2003 attribute 0.42 mm/year (23.3%) to thermal expansion of the oceans and 0.69 mm/year (38.3%) due to loss of mass from glaciers, ice caps, and the Greenland and Antarctic ice sheets; for the period 1993–2003, 1.6 mm (51.6%) was considered to be due to thermal expansion of the oceans and 1.2 mm (38.7%) due to ice melt (IPCC WGI 2007, p.419). Thus, upwards of 40 % or more of the observed increase in sea level rise cannot be explained, indicating that something else must be going on.

In addition to these factors, the distribution of water between oceans causes some areas to experience a higher increase in sea levels than other areas, with some even experiencing decline. The cause for this 'redistribution' of water is attributed to various factors including the Pacific Decadal Oscillation (PDO), the atmospheric-driven North Atlantic Oscillation (NAO), the *El Niño-Southern Oscillation (ENSO)*, which occurs on average every 5 years but varies from 3 to 7 years, and ocean currents (IPCC WGI 2007, pp.416–417).

Church and White (2006), for example, found that there was a slight acceleration in sea level rise of 0.013 ± 0.006 mm/year^2 throughout the period 1870–2004 (i.e., each year the rate of increase in sea level rise would increase by 0.013 mm). If this rate of increase continued to 2100, sea levels would rise by 28–34 cm. In a more recent study, Siddall et al. (2009) use a temperature-sea level model based on 22,000 years of data to predict sea-level rises of 7–82 cm by the end of the twenty-first century for respective increases in temperatures of 1.1 and 6.4°C. However, in a retraction (Siddall et al. 2010), they point out that, as a result of unforeseen errors related to the size of their time step for the twentieth and twenty-first centuries and failure to account for the rise in temperatures consequent upon coming out of the Little Ice Age, the projected increases in sea level for the period to 2100 are overstated (although their simulations for the remaining periods remain valid).

Sea levels are forecast by the IPCC WGI (2007, p.750) to rise by 18–59 cm, or by somewhat more in a worst case scenario by 2100. This translates into an increase of 1.8–5.9 mm/year, implying that the rise in sea level in the next century will be similar to that experienced in the past century to as much as 3.3 times higher than that of the past century, depending on whether average global temperatures rise by a projected 1.2°C or 4.0°C. Certainly, despite fears to the contrary, the projected increase in sea level is manageable. For example, during the 1960s, the city of Hamburg in Germany experienced an increase in storm surges of more than half a meter as a result of a narrowing of the Elbe River.[20] The city easily countered this by building dykes – a simple solution.

Where do the fears of unprecedented sea-level rise originate? These originate with the possible collapse of the largest mass of ice in the world – the Western Antarctic Ice Sheet (WAIS). Doomsday scenarios postulate the sudden collapse of the WAIS, which would lead to increases in sea level measured in meters rather than centimeters, although the projection would be an increase in sea levels of about 5 m

[20] See Evers et al. (2010).

(IPCC WGI 2007, pp.776–777). However, there is no evidence that the WAIS collapsed in the past 420,000 years, despite temperatures that were significantly higher than any experienced in human history, and there is no scientific basis to fear that a collapse of this ice sheet is imminent as a result of projected global warming (Idso and Singer 2009). Turner et al. (2009) find that, despite rising mean global temperatures and rising atmospheric CO_2, "the Antartica sea ice extent stubbornly continued to just keep on growing." Likewise, the IPCC WGI (2007, pp.818–819) indicates that there is little likelihood that the WAIS will collapse sometime during the twenty-first century.

The other major concern is the Greenland Ice Sheet. In this case, the IPCC indicates that the melting caused by higher temperatures will exceed additions due to increased snowfall, perhaps with the rate of net melting increasing over time, but this will occur slowly over the next several centuries. The expected increase in global sea levels due to total melting of the Greenland Ice Sheet is about 7 m (IPCC WGI) (2007, pp.818–819).

There is a great deal of uncertainty regarding the potential extent of sea level rise because of the dynamical behavior of ice sheets. In essence, too little is known about the processes working inside the large ice caps, and the impact of increased precipitation resulting from warmer temperatures, for scientists to make definitive statements about sea-level rise. Thus, increases in sea levels experienced in the past plus observed rates of increase are the best predictors of future sea-level rise. These suggest that potential future increases in sea levels will be manageable. Even unprecedented loss of ice sheets over the next century is very unlikely to raise sea levels by more than a few meters, and certainly not the 100 m envisioned by some environmental groups.

The best study of sea level rise was conducted by the PALSEA (PALeo SEA level) working group (Abe-Ouchi et al. 2009). Given that the sea-level rise predicted for the twenty-first century is considered one of the greatest potential threats from climate change, with the absolute worst-case scenarios varying between 0.59 and 1.4 m, PALSEA asks whether runaway sea level rise is likely. Based on information about sea level rise at the concluding years of the last glacial period, Abe-Ouchi et al. (2009) conclude that, if climate models are correct in their temperature projections, sea levels will rise quickly in the early part of the twenty-first century, but then level off to a much smaller increase. According to these experts, sea levels are certainly not expected to rise exponentially, as suggested by many climate change alarmists (including Greenpeace).[21]

Finally, as was the case with temperature data (Chaps. 2 and 3), there is evidence that contradicts the notion that sea levels are rising, or at least rising as quickly as indicated.[22] For example, a study of sea level trends on 12 Pacific Islands found that cyclones and tsunamis induced false readings that should have been ignored when

---

[21] See also http://thegwpf.org/science-news/1837-no-cause-for-alarm-over-sea-level-or-ice-sheets.html (viewed November 14, 2010).

[22] There is much debate about sea levels in Australia and the South Pacific. Some is clearly rhetoric, but some is also based on empirical evidence. See http://www.warwickhughes.com/blog/?p=283 and http://www.bom.gov.au/pacificsealevel/ (June 1, 2010). The latter is a good source of data.

Table 7.1 Effect of adjusting trends in sea level rise in 12 Pacific Islands

Island State	Claimed sea level trend (mm/year)	Years with zero trend
Cook Islands	+4.3	1994–2006
Federated States of Micronesia	–	2003–2007
Fiji	+3.5	2000–2007
Kiribati	+5.1	1993–2008
Marshall Islands	+4.4	1994–2008
Nauru	+6.0	1993–2008
Papau New Guinea	+7.0	1995–2008
Samoa	+6.3	1996–2008
Solomon Islands	+6.1	1999–2008
Tonga	+8.7	1997–2008
Tuvalu	+6	1994–2008
Vanuatu	+3.6	1995–2008

Source: Gray (2009)

calculating a trend. Further, these extreme weather events also disrupted the leveling of equipment. As a result, readings from years characterized by cyclones and tsunamis, and until the equipment could be tested and recalibrated, affected the calculation of trends. When the effects of extreme weather are taken into account, the measured rise in sea levels disappears, as illustrated in Table 7.1.

7.1.4 Extreme Weather Events

It is impossible to attribute extreme weather events to global warming. The Fourth Assessment Report of the Intergovernmental Panel on Climate Change argues as follows: "Single extreme events cannot be simply and directly attributed to anthropogenic climate change, as there is always a finite chance that the event in question might have occurred naturally. However, when a pattern of extreme weather persists for some time, it may be classed as an extreme climate event, perhaps associated with anomalies in SSTs (such as an *El Niño*)" (IPCC WGI 2007, p.310). Notice that the IPCC leaves room to interpret extreme weather events as attributable to global warming, presumably the result of human emission of greenhouse gases. Yet, rather than making a statement that rules out a link between climate change and extreme weather events, the IPCC prefers to leave open to interpretation the potential that any single extreme weather event is part of a pattern that could be attributed to global warming.[23]

[23] In the Third Assessment Report of 2001, the IPCC points out that there is no evidence of increased storm events (IPCC WGI 2001, pp.162–163, p.664) and, despite progress in climate modeling, current GCM models are *not* up to predicting increased future storm or weather events (IPCC WGI 2001, pp.573–575). It is unlikely that much has changed in the intervening 6 years to convince scientists otherwise (although see below).

Table 7.2 Saffir–Simpson Hurricane scale

Category	Wind speed (km/h)	Storm surge (m)	Assumed probabilities before warming	Assumed probabilities after warming
5	≥250	>5.5	0.020	0.025
4	210–249	≥4.0 and <5.5	0.060	0.065
3	178–209	≥2.7 and <4.0	0.070	0.073
2	154–177	≥1.8 and <2.7	0.100	0.102
1	119–153	≥1.2 and <1.8	0.150	0.150
Tropical storm	63–117	≥0 and <1.2	0.200	0.195
Tropical depression	0–62	0	0.400	0.390

Source: http://www.nhc.noaa.gov/sshws_table.shtml?large (viewed 13 February 2012)

Few reasonable scientists would attribute individual weather events, whether extreme or not, to global warming simply because it is impossible, as the IPCC recognizes, to determine whether the event would not have occurred had there been no change in climate whatsoever. Consider the probability that a particular weather event occurs, say a category 4 hurricane weather event (see Table 7.2). Suppose that, on average, 40 storms develop in a particular hurricane season (June 1 through November 30) in the North Atlantic and that the current probability of a storm of a particular intensity is given in the second to last column of Table 7.2. Thus, only 40% of storms develop into hurricanes, many of which are unlikely to make landfall. Using these assumptions, in any given hurricane season, one expects there might be one storm that reaches category 5, between five and six that attain a category 3 or 4 rating, four category 2 hurricanes, six category 1 hurricanes, and some 24 other tropical storms or depressions. This is more than found in the usual hurricane season, and, of course, it is important to remember that a large proportion may never make landfall.

The effect of rising temperatures on the number of tropical storms and hurricanes, and their intensity, is unknown. (We consider this in more detail below.) Suppose, however, that rising temperatures cause the above probability distribution to shift slightly so that more intense hurricanes appear more frequently. Further, assume that the probability of a storm occurring increases so that, rather than an average of 40 storms per year, now 45 storms are expected. In that case, in any given hurricane season, one expects there might be one storm that reaches category 5, between six and seven that attain a category 3 or 4 rating, between four and five category 2 hurricanes, between six and seven category 1 hurricanes, and some 26–27 other tropical storms or depressions. Yes, there is a slight increase in nearly every category of weather event, but it would require many years of observations to determine whether any given weather event had been drawn from one of the following three probability distributions: (1) the original distribution where the mean number of storms events per year was 40; (2) the original distribution but with a mean of 45 annual storm events; or (3) the after-warming probability distributions, with an annual average of 45 storm events and distribution of storm events given by the last

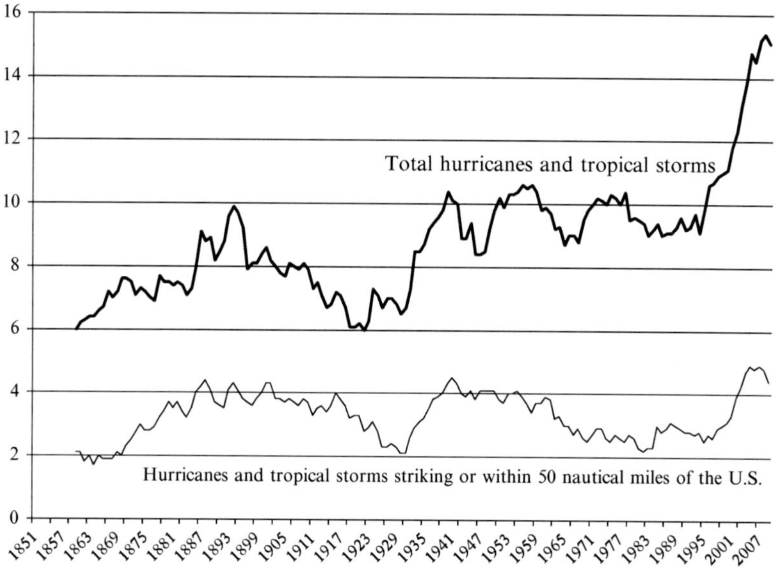

Fig. 7.3 Total hurricanes plus tropical storms in the Atlantic, and Atlantic storms and hurricanes affecting the United States, 10-year moving averages, 1851–2009

column of Table 7.2. It is a nearly impossible to attribute any single extreme weather event and even a sequence of events to anthropogenic climate change; there is simply insufficient information about how global warming impacts the above types of probability distributions.

A 10-year moving average of the number of hurricanes and tropical storms arising in the North Atlantic Ocean from 1851 through 2009 is provided in Fig. 7.3.[24] Also provided in the same figure is a 10-year moving average of the Atlantic tropical storms and hurricanes coming within 50 nautical miles of the U.S. coast or actually striking the U.S. Interestingly, the number of Atlantic storms rose between 1850 and about 1900, falling back to the earlier number by about 1915. Numbers rose during the 1920s and 1930s, only to level off until 1995, when storms appeared to increase rapidly for some 10 years, and then begin to fall to the end of the record (2009); the number of storms in the Atlantic appears to track increases in global temperatures for at least part of the record, as can be seen by comparing the dark line in Fig. 7.3 with the 10-year moving average of global temperatures in Fig. 2.5. A similar pattern is observed for the number of storms striking the U.S. or, at least, coming within 50 nautical miles of the coast, and thus having some, perhaps only

[24] Source of data for Figs. 7.3, 7.4, and 7.5: NOAA Coastal Services Center at (viewed June 23, 2010): http://csc-s-maps-q.csc.noaa.gov/hurricanes/download.jsp. Also see Davis et al. (1984) for explanation of data.

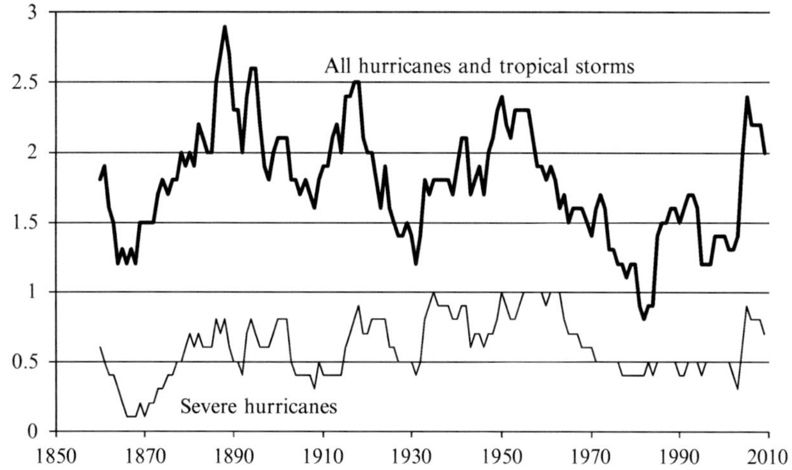

Fig. 7.4 Total hurricanes plus tropical storms (*dark line*), and Category 3, 4 and 5 hurricanes (*thin line*), making landfall in the U.S., 10-year moving averages, 1851–2009

minor, impact. This variable is indicated by the thin line in Fig. 7.3. For storms affecting the U.S., however, it is much more difficult to discern a trend that might be related to climate change. The reason for the difference might be attributable to the fact that the reported increase in Atlantic storms is the result of better measurement methods, including the use of satellites, rather than more storms per se.

Hurricane Katrina was downgraded to a category 3 by the time it struck and devastated much of New Orleans on August 29, 2005 (killing 1,833 people and causing more than $100 billion in damages). It was seen as a harbinger of more frequent and fiercer storms to come, all to be attributed to anthropogenic climate change. In Fig. 7.4, we provide a plot of all tropical storms and hurricanes that actually made landfall in the United States, and a plot of only category 3, 4 and 5 hurricanes to make landfall; both plots are based on 10-year moving averages. From the figure, it is clear that hurricanes affected the U.S. more frequently in the period 1890–1970 than thereafter. In the earlier period, there was an average of 1.84 hurricanes per year, while the average after 1970 was 1.56 (and 1.70 for the 20 years 1990 through 2009).

The number of really severe hurricanes (Category 4 and 5) declined from an average of one every 5 years during 1890–1970 to one every 12 years thereafter. Clearly, there is no discernable trend over the more than 150 years of data, and it is impossible to attribute hurricane events, such as Katrina to anthropogenic global warming. Of course, based on Fig. 7.3, more of the tropical storms and hurricanes could have struck Caribbean islands, Mexico or countries of Central and South America, but historical data regarding such events are not available.

We turn now to cyclones in the eastern and central Pacific Ocean for additional information on trends in storminess. In Fig. 7.5, we plot the annual numbers of

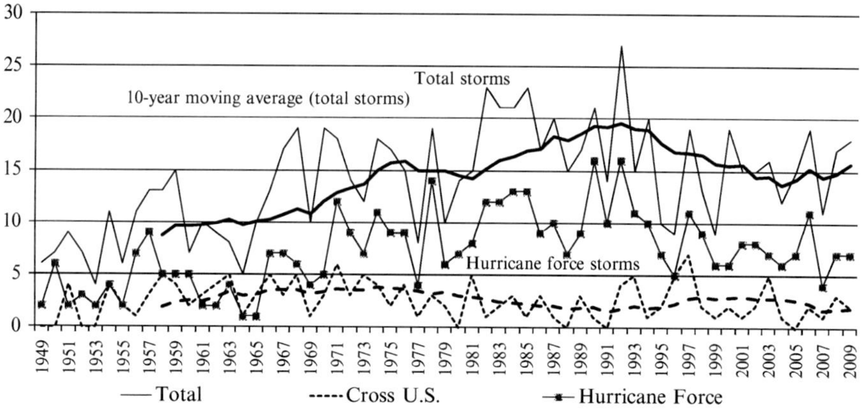

Fig. 7.5 Cyclones and tropical storms, Eastern and Central Pacific Ocean: total storms, hurricane force storms and storms striking or coming within 50 Nautical Miles of the United States, 1949–2009 (dark solid and dashed lines indicate 10-year moving averages)

tropical storms and cyclones appearing in the eastern and central Pacific Ocean during the period 1949 through 2009.[25] Included in the figure are the total numbers of storms (tropical storms plus cyclones of hurricane force), hurricane force storms, and storms striking the U.S. or coming within 50 nautical miles of the U.S. (including Hawaii). The number of total storms and hurricane force cyclones peaked in 1992, falling somewhat since then, while there is no discernable trend in the annual numbers affecting the U.S. over the past 50 years.

To determine the impact of global temperatures on storm events, we regress the number of storms in each year on the annual global HadCRUT3 temperature series (which runs from 1850 through 2009) and on year, where year is used to capture a secular trend independent of temperature (e.g., from better observations of off-shore storms). The results are provided in Table 7.3 for two regression models. If Atlantic and Pacific storms are considered to be related because of common factors, such as an El Niño, this is taken into account by estimating the two storm equations simultaneously, assuming that the error terms are correlated (see Greene 2008). In that case, we can only use 61 of the 159 annual observations for the North Atlantic because there are only 61 annual observations of storms in the Pacific. We also employ an independent, single-equation linear regression model for each of the Atlantic and Pacific storms. Notice that, when the equations are estimated simultaneously, the estimated coefficients do not change, but their estimated standard errors (provided in parentheses) are smaller, indicating a higher level of confidence in the estimated value (as evidenced by a lower probability in the square brackets).

The results indicate that storms in the North Atlantic are positively correlated with higher temperatures, as measured by the HadCRUT3 temperature data. The estimated

[25] Source: http://csc-s-maps-q.csc.noaa.gov/hurricanes/download.jsp (viewed June 23, 2010). Earlier data are not available.

Table 7.3 Effect of temperature on tropical storm events in the North Atlantic Ocean and Eastern and Central Pacific Ocean, seemingly unrelated regression and ordinary least squares regression[a]

	North Atlantic		Eastern & Central Pacific	
	OLS	SUR	OLS	SUR
Intercept	−23.159	21.956	−591.657	−591.657
	(18.670)	(107.313)	(122.918)	(119.858)
	[0.217]	[0.838]	[0.000]	[0.000]
Temperature	4.411	6.731	−14.191	−14.191
	(1.701)	(4.105)	(4.702)	(4.585)
	[0.010]	[0.101]	[0.004]	[0.002]
Year	0.017	−0.006	0.307	0.307
	(0.010)	(0.054)	(0.062)	(0.061)
	[0.077]	[0.915]	[0.000]	[0.000]
RMSE	3.410	3.641	4.171	4.067
R^2	0.2244	0.1418	0.3497	0.3497
χ^2 or F[b]	22.57	10.08	15.59	32.80
	[0.000]	[0.007]	[0.000]	[0.000]
Observations	159	61	61	61

Notes: [a]Two models are estimated: linear or ordinary least squares (OLS) regression, and seemingly unrelated regression (SUR), where the error terms in the two equations are assumed to be correlated. Regression was conducted using Stata 10 using the 'surge' and 'regress' functions. An explanation of the regression models can be found in Greene (2008), for example. The standard error of the estimated coefficient is provided in round parentheses and the associated probability in square brackets

[b]The Chi-square statistic (χ^2) measures overall goodness of fit of the estimated model for the SUR regressions, while the F-statistic does the same for the OLS regressions

coefficient for the period 1851 through 2009 is statistically significant at the 1% level of significance, but the statistical significance of the coefficient drops to a significance probability of more than 10% for the period 1949–2009. At the same time, the estimated coefficient falls, indicating that the effect of increasing temperature is smaller for the period 1949–2009 than for the entire period 1851–2009. This is surprising when one compares this result to Fig. 7.4, but it comes about because of the secular time trend, which is more pronounced in the latter period as opposed to the former. Further, even though the model is appropriate (as determined by the goodness of fit statistics), it explains less than 15% of the variation in storm activity for the period 1949–2009 and less than 23% for the period 1851–2009. Clearly, factors other than temperature are affecting storm formation in the North Atlantic.

If we turn to events in the Eastern and Central Pacific Ocean, we find that, for the period for which data are available (1949–2009), storm events are inversely correlated with temperature once adjusted for secular trends. The model with temperature and trend as regressors explains nearly 35% of the variation in storm activity. Further, the inverse effect of temperature on storm activity in the Pacific is highly statistically significant in both the OLS and SUR models.

When we regress storms that affected the United States on temperature and trend, we find no statistically significant relation whatsoever, which is why these results are not reported. The same is true if we look only at category 3, 4 and 5 hurricanes impacting the U.S. That is, neither temperature nor year could explain storm activity affecting the U.S. Atlantic coast. This is also evident from Fig. 7.4.

In conclusion, despite our finding that rising global temperatures appear to affect storm activity in the North Atlantic, there is no similar evidence for this from the Pacific Ocean, and there is no evidence that the number of storms impacting the U.S. has increased as a result of climate change. Further, the regression analyses indicate that other factors not considered here are more important determinants of storm activity. Overall, however, one must conclude that there is no convincing evidence that extreme weather events are impacted by climate change. More information must be gathered before any such conclusion can be reached, and that might well take another 50 years of observations.

Nonetheless, two studies recently examined the impact of human activities on the incidence of extreme precipitation events using output from climate models (Min et al. 2011; Pall et al. 2011). Using Hadley Centre, grid-point data on 1-day and 5-day precipitation accumulation events for 49 years (1951–1999) and precipitation outputs from an ensemble of climate models for the same gridpoints and years, Min et al. (2011) find that the model outputs track actual precipitation extremes rather closely when climate model outputs are based on CO_2 than when CO_2 is absent from such simulations. The authors conclude that this shows human activities are responsible for extreme weather events.

Pall et al. (2011) consider the probability that floods in England and Wales during autumn 2000 were the result of anthropogenic climate change. The authors used forecasts from the Hadley climate model to obtain temperature and precipitation forecasts; these forecasts were fed into a precipitation-runoff model to simulate daily river runoff and the potential and magnitude of floods. The climate model was run using sea surface temperatures, atmospheric greenhouse gas levels and sea ice levels found in year 2000 and again with conditions as they existed (or were presumed to exist) in 1900. In each case, the climate model was run several thousand times for a full year, with runs differing according to their starting values. The authors concluded: "The precise magnitude of the anthropogenic contribution remains uncertain, but in nine out of ten cases … results indicate that twentieth century anthropogenic greenhouse gas emissions increased the risk of floods occurring in England and Wales in autumn 2000 by more than 20%, and in two out of three cases by more than 90%."

Although interesting, this research is certainly not conclusive for several reasons. Climate models have been calibrated so that they can replicate the recent past. As noted in Chap. 5, this makes such models less suitable for predicting the future but, because they can replicate the past, it is not surprising that they provide a decent tracking of past precipitation. Earlier in this chapter we indicated that sea ice varies considerably and obtaining information on its extent in the early 1900s is fraught with uncertainty. The same holds with sea surface temperatures, which are influenced by climate events such as El Niño. Thus, it is anyone's guess as to what sea ice and temperature were in 1900. Finally, climate model replications of the past cannot substitute for actual observations.

To determine the effects of human activities on the risks of high precipitation events and flooding, it is necessary to use observed and not simulated data. Records from the UK Met Office show no upward trend in UK rainfall between 1961 and 2004. While autumn 2000 rainfall was unusual, it was exceeded in 1930, while 1768 and 1872 were wetter than 2010. Nor do historical UK precipitation data provide evidence of an upward trend.[26] Likewise, Chu et al. (2010) found that extreme precipitation events in Hawaii were declining, while Xie et al. (2010) found that there was no trend in the size of hail stones in China. Larger hail stone sizes are indicative of extreme weather events. Meanwhile, Czymzik et al. (2010) used data from lake bed sediments to determine that flood events in Germany actually declined over a 450 year period; indeed, the worst flooding occurred during cold periods and not warm periods. Frustrated by the lack of evidence for a human responsibility in bringing about climate disasters, Bouwer (2011) pleads for the use of models rather than actual evidence: "Lacking significant impact from anthropogenic warming so far, the best way to assess the potential influence of climate change on disaster losses may be to analyze future projections rather than historical data."

7.1.5 Impacts on Health

Health is potentially an area where there might be significant damages from global warming. Global warming poses a threat to human health primarily because of the projected spread of malaria and other tropical diseases. However, tropical diseases are a problem of economic development and preventative health, not of rising global temperatures. The West Nile virus has spread into cold regions (including Canada), while malaria killed thousands of Russians even at the Arctic Circle in the 1920s. Both Canada and the United States experienced malaria as late as the 1950s, while the last cases in The Netherlands occurred in the 1970s (Spielman and D'Antonio 2001, pp.116–137).[27]

Malaria was eradicated in northern countries not because the mosquitoes carrying the disease could no longer breed in those countries, but because countries were able to treat people with malaria using quinine, drain swamps where malaria-carrying mosquitoes bred, and spray chemicals in areas with the highest malarial incidence (Spielman and D'Antonio 2001). Indeed, malaria has been in recession in many locations. That is, despite rising global temperatures over the past century or more, the range of malaria shrunk because of economic development and disease control (Gething et al. 2010).

---

[26] See Booker (2011).

[27] As late as 1914, some 1% of the population of Mississippi died as a result of malaria and the disease (parasite) was only eradicated in the U.S. South after an effective anti-malaria campaign by the U.S. government (Spielman and D'Antonio 2001).

Mosquitoes and malaria know no boundaries, even without global warming. Malaria remains a deadly disease, infecting annually nearly 500 million people while killing between two and three million, mainly children. In developing countries, mosquitoes carrying malaria are best controlled using a chemical discovered in 1939 – *dichloro-diphenyl-trichloroethane*, commonly known as DDT.[28] DDT spraying began in earnest in 1958 and, despite its success in having eradicated malaria in developed countries by about 1967, was subsequently banned worldwide in 1973 because of the dangers it posed, having been discovered in breast milk and thought to be linked to declining numbers of bald eagles.[29] Although a persistent organic pesticide, the main problem with DDT may have been its indiscriminant use.

In developing countries, indiscriminant use without care to prevent re-introduction of mosquitoes from non-treated areas reduced the effectiveness of spraying programs and caused mosquitoes to develop some resistance to the chemical. DDT use has now been permitted since 2000, but lobbying against its use continues, making it difficult to implement effective programs. The chemical can be used effectively if applied to the walls of homes and mosquito nets, but it may take a long time to bring the parasite under control, let alone eradicate it – and then only if countries work in concert.

Malaria has little if anything to do with climate change. Rather, as noted above, it is a problem of development and health care. This is not to say that global warming will have no impact on malaria, but it will be difficult to discern the effect of climate change on the disease in relation to other factors, as non-climate factors have profoundly confounded the relationship between geographical climate and malarial outbreak; indeed, the relationship between climate and malarial endemicity

[28] Dr. Paul Hermann Müller discovered the insecticide qualities of and patented DDT, for which, in 1948, he won a Nobel Prize in Medicine (Fisher et al. 2003). DDT was effective against malaria and yellow fever. The 1973 ban on DDT is considered by some to have led directly to the death of more than 90 million people, if one assumes some 2.5 million people per year die of malaria. This is the legacy some have attributed to the 1962 best-selling book, *Silent Spring*, by Rachel Carson who was not a scientist but an excellent writer.

[29] The link between DDT and declining populations of bald eagles has subsequently been proven false. First, it appears that bald eagles were nearly hunted to extinction, with concern expressed as early as 1921. A total ban on hunting was put in place in the U.S. with the Bald Eagle Protection Act (1940). As a consequence populations began to rise, with Marvin (1964) reporting that overall bird populations had risen by 25% between 1941 and 1960, including the particularly vulnerable robin despite many years of DDT spraying. Various studies indicated that DDT did not have an adverse impact on bald eagles, nor was there evidence that thin egg shells were correlated with DDT (see Coon et al. 1970; Reichel et al. 1969). Cromartie et al. (1974) found that, in a sample of 37 bald eagles found dead in 1971–1972, 13 had been shot and 13 had died of insecticide poisoning (dieldrin and thallium), but no deaths were directly attributable to DDE (*Dichloro-diphenyl-dichloro-ethylene*), which is a form of DDT. Likewise, Belisle et al. (1972) found that, in a sample of 39 bald eagles found dead in 1969–1970, only one died from DDE, 18 had been shot and 6 died from dieldrin, an alternative insecticide to DDT. See also E.J. Gordon and S. Milloy, "100 Things You Should Know about DDT" at http://www.junkscience.com/ddtfaq.html (viewed May 25, 2010). Yet, a Google search of DDT and bald eagles finds that the majority of websites (mainly associated with environmental groups) maintain that DDT almost led to the demise of the eagle, when the scientific literature attributes the cause to legal and later illegal hunting.

has effectively been decoupled (Gething et al. 2010; Spielman and D'Antonio 2001). This makes economic estimates based on coupled climate-biology models of potential damages from malaria in a warmer world extremely speculative, and certainly orders of magnitude higher than warranted by empirical observations (Gething et al. 2010). Similar comments can be made about dengue fever and other diseases and parasites that are also spread by mosquitoes.

While respiratory ailments could increase as a result of global warming, new threats to health, such as severe acute respiratory syndrome (SARS), do not necessarily need to be related to global warming – they are more a matter of globalization than climate change. Many diseases and pests (such as West Nile virus) are spreading regardless of climate. It is also not at all clear that current tropical diseases and pests will increase their range as a result of global climate change, except maybe in the developing countries, which cannot cope because they lack public health infrastructure – they are poor and the remedy is to increase their incomes, not to rely on mitigation to prevent higher incidence of disease.

The question here is whether funds currently meant to mitigate greenhouse gas emissions can be better spent in Africa, say, improving the public health care infrastructure and fighting AIDS, improving access to quality drinking water, or simply providing DDT to apply to walls and netting to reduce incidence of malaria. Indeed, the Copenhagen Consensus concerning the world's biggest problems found that AIDS and water quality improvements, and several other issues facing global society, were greater problems than climate change (Lomborg 2007b). If the principal objective of climate change policy is to help poor countries and future generations, a better strategy might be to direct funds spent on mitigation by industrial countries to improve incomes in developing countries. This is the position taken by Lomborg (2007a, b) and others.

In terms of empirical studies of the costs to health of climate change, Moore (1998) estimates that an average global temperature increase of 4.5°C will yield some \$30–\$100 billion in health benefits (not losses) to U.S. residents. Goklany (2008, 2009) also reports that global warming actually reduces mortality rates as fewer people will die from exposure to cold temperatures. It seems that people are better able to cope with warmer temperatures than colder ones.

The summer of 2003 was a particularly warm one. In Europe, many deaths were attributed to the heat. However, as elsewhere, people whose death is attributable to heat are generally the elderly and weak, who are more than likely to have died from other causes in the following months. This is not to deny the value of their lives, only that exceptionally warm weather may simply have been a factor triggering a mortality that was inevitable within the next several months. This hypothesis can be verified empirically by comparing incidents of death among various age categories before, during and after a period of exceptional heat. The same is true of cold periods. By controlling for access to central heating, air conditioning, age, health status, and other non-climate factors, it is possible to determine the effect that climate (unusual cold or heat) has on mortality.

In Tables 7.4 and 7.5, we provide data on deaths from various weather related events. Prior to 1989, droughts were by far the most important contributing factor to

Table 7.4 Global deaths and death rates for various types of weather events, 1900–1989 and 1990–2006

	Death rates per year		Death rates per year (per million people)	
	1900–1989	1990–2006	1900–1989	1990–2006
Droughts	130,042	185	57.99	0.03
Floods	75,212	7,637	31.95	1.29
Windstorms	10,856	13,650	3.96	2.45
Slides	469	868	0.16	0.15
Waves/surges	128	207	0.06	0.03
Extreme temperature	110	5,671	0.03	0.91
Wildfires	21	47	0.01	0.01
TOTAL	216,839	28,266	94.16	4.87

Source: Goklany (2007)

Table 7.5 U.S. Deaths due to weather-related events, 1979–2002

	Cumulative deaths	Deaths per year	% of annual, all-cause deaths
Extreme cold	16,313	680	0.031
Extreme heat	8,589	358	0.016
Flood	2,395	100	0.005
Lightning	1,512	63	0.003
Tornado	1,321	55	0.003
Hurricane	460	19	0.001
Sum	30,590	1,275	0.058
Total deaths, all causes, average		2,189,000	100.0

Source: Goklany (2007)

mortality, mainly because poor countries were least able to cope with drought. Floods were the second most important weather-related cause of death, followed by windstorms, again because developing nations are least able to prevent such natural disasters. Extreme temperatures ranked sixth out of seven weather-related events as a contributor to mortality. For the period 1990 through 2006, death rates from all weather-related causes dropped dramatically as nations learned how to cope with severe weather events and as a result of relief efforts by rich countries. Annual mortality from drought and floods fell significantly, while it rose for the other five weather events; however, annual death rates fell for all categories, with the exception of extreme temperature events. Global average annual death rates fell by some 7–99 %, but rose by several 100 % in the case of extreme weather events.

While one might draw the conclusion that extreme temperature events refer to heat waves, such as the one in Europe in 2003, it turns out that more people die from extreme cold than heat, as indicated in Table 7.5; almost twice as many people in the U.S. died from extreme cold than died from extreme heat over the period 1979–2002. Further, based on U.S. data, weather-related mortality is extremely low, with severe weather events accounting for less than 0.06% of U.S. deaths during 1979–2002.

7.1.6 Other Economic Effects

Given that Arizona and Nevada have been the fastest growing states in the United States, it appears that people express a preference for living in warm (even hot) and dry climates. Empirical measures of the values of these amenities are generally lacking. Maddison and Bigano (2003) found that, in Italy, higher summer temperatures are regarded negatively as are lower January temperatures and higher January precipitation. Rehdanz and Maddison (2009) use a hedonic pricing model to determine that Germans prefer warmer and drier winters; however, they could find no statistically significant gain or loss to Germans from IPCC-projected changes in climate. Lise and Tol (2002) found that people have a preference for warmer climates, as evidenced by their choices regarding vacation destinations. One can only conclude that there exists no firm information about the economic effects of climate on health and amenity values – any conclusions are speculative at best. Hamilton and Tol (2007) use an econometric simulation model to show that, under climate change, tourism in Ireland and the UK would shift northwards, while in Germany it would shift towards the south. Initially, the UK and Ireland would lose some international tourists but gain domestic ones, but as climate change continues there would be a growth in international tourists as northern Europe warms.

Some researchers have regressed countries' GDP levels on their mean temperature and a variety of control variables, including the latitudes of capital cities (as a control variable to account for differences in development opportunities between countries). For example, Choiniere and Horowitz (2000) regress per capita GDP on average temperature for 1980, 1985 and 1990 using a double-logarithmic functional form. Their conclusion is that the effect of temperature has become more pronounced over time, not less. That is, they find that the world might become more vulnerable to changes in climate over time. Using a similar approach, Horowitz (2001) further reports that a 3 °F (1.67°C) increase in temperature leads to a 4.6% decline in global GNP.

There are several problems with this analysis. First, as the authors themselves point out, average temperature taken at the capital city of a country may not be representative of the average annual temperature for the country as a whole. Second, and perhaps more important, the authors neglect the fact that temperatures in a given year, or even over a decade, may not be representative of the actual temperatures that the country/region has historically experienced and might experience in the future. Climate is not the same as weather, nor is climate variability the same as weather variability. Average temperature in any given year may be an anomaly, as may the change in average annual temperatures over any 5- or 10-year interval. Further, for many regions, it is not the average annual temperature that is most important. Regions that experience large differences between summer and winter temperatures incur higher costs from such things as increased road and other infrastructural repairs plus heating/cooling needs. Finally, there is no causal mechanism, precipitation is ignored, and amenity values are neglected. In particular, citizens in some countries might simply desire warmer and drier weather.

7.2 Economic Modeling of Climate Change Damages

With some exceptions, economists take the view that meteorological, atmospheric and ocean science are outside their realm of expertise, and they accept without qualification the science of climate change – that human emissions of greenhouse gases cause global warming and that, if we want to stop warming, we need to control such emissions. Economists then attempt to balance the costs and benefits of climate change, focusing on what an optimal economic response might look like. William Nordhaus of Yale University has led the way by developing an integrated assessment model to guide policy makers (Nordhaus 1991, 1994, 2008). He summarizes his position and that of most economists as follows:

> "Global warming is a serious, perhaps even a grave, societal issue [and] there can be little scientific doubt that the world has embarked on a major series of geophysical changes that are unprecedented in the past few thousand years. . … A careful look at the issues reveals that there is at present no obvious answer as to how fast nations should move to slow climate change. Neither extreme – either do nothing or stop global warming in its tracks – is a sensible course of action. Any well-designed policy must balance the economic costs of actions today with their corresponding future economic and ecological benefits" (Nordhaus 2008).

7.2.1 Integrated Assessment

One method used by economists to determine what policies to pursue in addressing climate change is the use of integrated assessment models (IAMs), which seek to balance costs and benefits of taking action. An integrated assessment model is essentially a constrained mathematical optimization model. The present (discounted) value of social wellbeing is maximized subject to dynamic and static constraints that represent the potential damages, production possibilities, and interactions among markets and world regions. The objective function includes the sum of consumer and producer surpluses (recall Chap. 6), the potential damages from global warming as a function of temperature, and costs of mitigating climate change. Damages are a function of temperatures, which, in turn, are a function of the level of greenhouse gas (CO_{2e}) emissions in each period of the model. The level of CO_{2e} emissions in each period is affected by the technology, which is generally determined outside the model, and a carbon tax that provides incentives to reduce emissions. The level of the carbon tax and its rate of increase are determined endogenously as the global economy seeks to optimize the objective function by reducing emissions and the amount of tax to be paid. Different climate scenarios can be examined by varying assumptions concerning the technology (CO_2 emissions per unit of output), growth in population, and so on (see Chap. 4).

In a series of books and articles, Nordhaus developed two IAMs that have been used by many economists to examine the costs of climate change and the optimal level of mitigation (Nordhaus 1991, 1994, 2008) – DICE (Dynamic Integrated model of Climate and the Economy) and RICE (Regional dynamic Integrated model of Climate and the Economy). To obtain some idea of what this involves, consider the following outline of the structure of DICE (Nordhaus and Boyer 2000). The objective is as follows:

$$\text{Maximize} \quad \sum_{t} \{N(t)\log[c(t)]\delta(t)\}, \tag{7.1}$$

where $N(t)$ is the global population at time t, $c(t)$ is per capita consumption over period t, and $\delta(t)$ is the social rate of time preference or discount factor for period t. Since the time step is 10 years, the discount rate represents a 10-year rate; further, the discount rate is assumed to differ from one period to the next, although it could also be kept constant or fall over time (see Chap. 6). Population is also modeled to grow over time, although it too can be held constant or adjusted so that it falls over time, or rises and then falls. These are assumptions in the model and each is represented by one or more equations (that are not shown here).

A Cobb-Douglas or double-logarithmic production function is assumed, but it is adjusted by climate factors. It takes the following form in DICE:

$$Q(t) = \frac{1}{1+D(t)}\left[1-b_1(t)\mu(t)^{b_2}\right]A(t)K(t)^{\gamma}\,N(t)^{1-\gamma}. \tag{7.2}$$

where $Q(t)$ is output (global GDP), $A(t)$ is total factor productivity or technology in period t, $K(t)$ is the total capital stock at time t, $D(t)$ are climate-related damages as a fraction of net output, $\mu(t)$ is the industrial emission control rate, $b_1(t)$ is the coefficient on the control rate in the abatement cost function (which changes over time), b_2 is the exponent on the control rate (a parameter that is fixed over time), and γ is the elasticity of output with respect to capital. A key equation is the damage function:

$$D(t) = \theta_1 \mathrm{T}(t) + \theta_2 \mathrm{T}(t)^2 \tag{7.3}$$

where damages increase as a quadratic function of global mean temperature, $\mathrm{T}(t)$, at time t and θ_1 and θ_2 are parameters of the damage function. The parameters are calibrated ('guessed at') rather than statistically estimated because data are lacking.

Human emissions of CO_2, denoted $E(t)$, are a function of output, the base-case ratios of industrial emissions to output, $\sigma(t)$, and the industrial emissions control rate:

$$E(t) = [1-\mu(t)]\sigma(t)A(t)K(t)^{\gamma}\,N(t)^{1-\gamma}. \tag{7.4}$$

Total consumption in a given period, $C(t)$, is determined by output in that period, $Q(t)$, minus investment in maintaining and/or enhancing the capital stock, $I(t)$, and minus the amount paid as carbon taxes:

$$C(t) = \mathrm{Q(t)} - I(t) - \tau(t)E(t), \tag{7.5}$$

where τ(t) is the tax on emissions (\$ per tCO_2) in period t, although (7.5) can be modified so that the term $\tau(t)E(t)$, or total tax paid, is expressed as a cost of purchasing permits. The tax rate or level of emission permits constitute policy variables in the IAM.

Remaining constraints deal with the change in the capital stock from one period to the next (taking into account deterioration of the capital stock and new investment), the rise in temperatures as a function of human emissions, the release of CO_2 from changes in land use, CO_2 emissions from the oceans, previous temperatures, and radiative forcing parameters representing various factors that contribute to the buildup or drawn down of CO_2 in the atmosphere. Including initial conditions, there are more than 30 constraints in the model, although the actual number of constraints is closer to 300 as a result of incrementing time over ten periods. RICE has significantly more constraints because many DICE-equivalent constraints apply to individual regions; there is the need to include region-specific parameters, such as for the production function (7.2) and emissions function (7.3). In addition, aggregation constraints enter into the mix.

In the DICE model, Eq. (7.1) is maximized subject to constraints (7.2), (7.3), (7.4), (7.5) and the other constraints mentioned (Nordhaus and Boyer 2000). The constrained optimization problem is a straightforward nonlinear programming (NLP) problem. Once parameterized, the NLP can be solved numerically (analytic solution is impossible) using a computer software package, such as GAMS (McCarl et al. 2007); it might also be solved in Excel, although this requires add-on software because the standard solver in Excel cannot handle that many nonlinear constraints.[30]

7.2.1.1 Copenhagen Consensus[31]

Many researchers have employed the DICE model in their own work. William Cline of the Institute for International Economics and Center for Global Development in Washington used the DICE-99 model to find the world's optimal CO_2-abatement strategy and the associated optimal path of carbon taxes. Relative to business-as-usual (BAU) CO_2 emissions, the optimal strategy is to reduce emissions immediately by 35–40 % followed by further reductions to nearly 50% of BAU emissions by 2100 and to a peak of 63% by 2200, followed by a tapering off (Cline 2004). The associated optimal carbon tax starts in 2000 at $35 (in 1990 dollars) per ton of CO_2 (t CO_2) rises to $46/t$CO_2$ in 2005, to $67/t$CO_2$ by 2025, to $100/t$CO_2$ by 2050, and to a peak of $355/t$CO_2$ in 2200 before declining. Cline (2004) also investigates the Kyoto Protocol (assuming it remains in place in perpetuity) and a value-at-risk

[30] An updated description of the Nordhaus and Boyer (2000) versions of DICE and RICE can be found at http://www.econ.yale.edu/~nordhaus/homepage/dicemodels.htm (viewed March 3, 2010). Both GAMS and spreadsheet versions of the models can be downloaded from this website.

[31] The Copenhagen consensus referred to here should not be confused with the climate conference that was held in Copenhagen in late 2009. At the latter, nations failed to reach an agreement on reducing emissions of CO_2 and other greenhouse gases, which was generally considered a 'disaster' in the environmental community.

Table 7.6 Cost-benefit analysis of three emission reduction scenarios, 300-year time horizon, 1.5% discount rate, 1990 US dollars

	Scenario		
Item	Optimal carbon tax	Kyoto protocol	Value-at-risk carbon tax
Benefits ($\times 10^{12}$)	\$271	\$166	\$1,749
Costs ($\times 10^{12}$)	\$128	\$94	\$458
Benefit-cost ratio	2.12	1.77	3.82
Annualized benefits ($\times 10^{12}$)	\$0.90	\$0.55	\$5.83
Annualized costs ($\times 10^{12}$)	\$0.43	\$0.31	\$1.53

Source: Cline (2004, p.38)

scenario that identifies the maximum expected loss over the time horizon up to a probability of 95%; that is, the value-at-risk scenario determines the optimal carbon tax required to reduce the chance of a maximum possible loss to 5% or less. A summary of Cline`s results is provided in Table 7.6.

The scenarios and analyses developed by Cline were used in the original 'Copenhagen Consensus' project to rank the world's most pressing problems (Lomborg 2004). The Copenhagen Consensus project, headed by the Danish environmentalist Bjørn Lomborg, consists of reports on the globe's most pressing problems and an assessment and ranking by a panel of experts that, in the 2004 project, consisted of eight top economists, including three Nobel laureates. Originally 32 challenges facing humankind were identified, but these were subsequently reduced to ten that warranted further investigation and became the subject of the Copenhagen Consensus project. Global warming was classified as one of the ten problems to be considered by the expert panel, but the panel ranked it last in terms of urgency and in terms of where governments should direct limited financial and other resources.

The panel's ranking is somewhat surprising because Cline's analysis in Table 7.6 is more along the lines of a later analysis by Nicholas Stern, which employs a low discount rate, assumes high damages, and recommends immediate action because the benefit-cost ratio from taking action is much greater than one. (The Stern analysis is discussed further below.) Yet, communicable diseases (especially HIV/AIDS), access to sanitation and clean water, government corruption, malnutrition and hunger, and trade barriers were considered greater problems whose solution yielded higher benefits than attempts to mitigate climate change.

In a follow-up to the 2004 Copenhagen Consensus that asked a number of experts to rank the world's biggest problems, Bjørn Lomborg edited a book that identifies 23 global issues and provides a cost-benefit analysis of various promising policy solutions (Lomborg 2007b). Readers are asked to make their own prioritizations. Then, in a second 'Consensus' project, Lomborg brings together experts in an effort to prioritize policy options for addressing global warming. The results are discussed in Sect. 7.4.

7.2.1.2 Integrated Assessment: Further Results

Nordhaus subsequently used an updated version of the DICE model to conclude that global society should make an effort to mitigate climate change by reducing CO_2 emissions relative to what they would otherwise be, but not stop warming entirely. Further, controls on emissions should ramp up over time. In particular, based on his later estimates and to prevent temperatures from rising more than 2.3°C, greenhouse gas emissions should be reduced by 15% in the current period (2010–2019) relative to what they would be without any action, by 25% of business as usual emissions in 2050, and by 45% in 2100. This implies that the optimal carbon tax (measured in real 2005 purchasing power US dollars) should rise from $9.50 per ton (t) of CO_2 ($35 per ton of carbon) in 2005 to about $25/t CO_2 in 2050 and $56/t CO_2 in 2100 – or 12¢ per gallon of gasoline in 2005 to nearly 70¢ per gallon by 2100 (Nordhaus 2007b, 2008).

This optimal path for a carbon tax is predicated on unmitigated damages from climate change that amount to nearly 3% of global output in 2100 and 8% by 2200. Future damages from climate change are related in Nordhaus's integrated assessment model to projected temperature increases via Eq. (7.3) that is calibrated to take into account estimates of damages and benefits from global warming found in the literature. For example, agricultural economists had found that (some) warming was actually beneficial for agricultural production (Darwin et al. 1995; Mendelsohn et al. 2000; Mendelsohn et al. 1994; Tol),[32] but such benefits appear to be outweighed by losses elsewhere (Nordhaus 2007a). Three scenarios of projected damages from different calibrations of the power function used by Nordhaus are provided in Table 7.7. It is important to note that these are calibrations and not statistical evidence, so they really amount to nothing more than an assumed relation between temperature increase and economic damages that is based on projections of possible damages made by researchers examining specific sectors such as agriculture. And each of these sectoral analyses has its own sometimes dubious assumptions regarding the relationship between projected climate change and damages, as discussed in Sect. 7.1.

Integrated assessment models are now finding that an optimal climate strategy needs to combine mitigation and adaptation (Prins et al. 2010). For example, Bosello et al. (2010) link adaptation, mitigation and climate change damage in an integrated assessment model of the world economy and the energy and climate systems. They find that an optimal combination of adaptation policies (reactive and anticipatory, plus investment in R&D) and mitigation would see no more than 20% of emissions abated over the period to 2100, while expenditures on adaptation would rise rapidly beginning in 2060. Depending on the discount rate and perceptions of future damage,

[32] Others find similar results, although some find that agriculture will eventually suffer as temperatures continue to rise, with developing countries to be hardest hit. Thus, Cline (2007) finds a small decline in agricultural output in the latter half of this century, with the greatest decline in poor countries. However, any results depend on regional precipitation projections, which are the weakest component of any predictions from climate models.

Table 7.7 Power function between temperature rise and damages

	Damages as proportion of global output		
Temperature rise (°C)	Worst case (%)	Mid case (%)	Best case (%)
0	0.00	0.00	0.00
1	0.32	0.28	0.10
2	1.27	1.14	0.58
2.5	1.98	1.77	1.01
3	2.85	2.55	1.60
4	5.07	4.54	3.28
5	7.93	7.10	5.74
6	11.41	10.22	9.05

Source: Based on unpublished notes by Nordhaus (2007a) made available to explain the methodology. William Nordhaus points out that estimates for temperature increases greater than 3°C are unreliable because of lack of sources for increases that large

the combination of mitigation and adaptation would account for between 44 and 73% of total damages (the remainder simply borne by the economy), while the proportion dealt with by adaptation would vary from a low of 20% (equal to that borne via mitigation) to 53% (with mitigation only addressing 9% of damages).

7.2.2 *Bottom-Up Approach*

In contrast to the approach used by Nordhaus (1994, 2008) and Tol (2002), which rely upon integrated assessment models, Goklany (2008, 2009) measures the impacts of projected global warming on human risks, mortality and ecosystems using a bottom-up approach. Surprisingly, he is one of the few who begins with the IPCC's (2001) emission scenarios, which are the principal driver of climate models' projections of temperature increase (see also Tol 2005). A brief description of four key scenarios is provided in the first 11 rows of Table 7.8. The scenarios indicate the range of possible greenhouse gas emissions for different economic development trajectories (and include assumptions about technological change, land use changes and the energy mix) if nothing is done to mitigate climate change. The final three rows summarize Goklany's (2009) estimates of the associated changes in mortality, changes in populations at risk due to water stress, and losses of coastal wetlands.

The one thing to note about the assumed future emissions scenarios is the projected increase in per capita GDP (measured in 2005 US dollar equivalents); as shown in Chap. 4, these are highly optimistic for all scenarios. Even the scenario leading to the lowest increase in income (scenario A2) and the highest increase in population would have those living in developing countries producing more than $16,000 per person, equivalent to standards currently existing in some eastern European countries. Two scenarios (A1F1 and B1) see those in developing countries with incomes equivalent to those in rich countries today, while those in rich countries will see a doubling of their real incomes.

Table 7.8 Selected emission scenarios and projected impact of climate change on population and ecosystem health

Item	IPCC scenarios			
	A1F1	A2	B2	B1
1. Population in 2085 ($\times 10^9$)	7.9	14.2	10.2	7.9
2. Average global per capita GDP in 2085 ($)[a]	78,600	19,400	29,900	54,700
3. Average per capita GDP in 2100, Industrialized countries ($)[a]	160,300	69,000	81,300	108,800
4. Average per capita GDP in 2100, Developing countries ($)[a]	99,300	16,400	26,900	60,000
5. Technological change	Rapid	Slow	Medium	Medium
6. Energy use	Very high	High	Medium	Low
7. Energy technologies	Fossil fuel intensive	Regionally diverse	"Dynamics as usual"	High efficiency
8. Land-use change	Low-medium	Medium-high	Medium	High
9. Atmospheric CO_2 concentration in 2085 (ppmv)	810	709	561	527
10. Global temperature change in 2085 (°C)	4.0	3.3	2.4	2.1
11. Sea level rise in 2085 (cm)	34	28	25	22
12. Change in total mortality in 2085 compared to baseline[b,c]	−2,064,000	+1,927,000	−1,177,000	−2,266,000
13. Total population at risk due to water stress compared to baseline[c]	299,000	5,648,000	2,746,000	857,000
14. Average net global loss in coastal wetlands by 2085 compared to baseline[c]	13%	9%	9%	10%

Source: Adapted from Goklany (2009)

[a] GDP per capita in 2005 US $, converted from 1990 $ using the US CPI

[b] Mortality due to hunger, malaria and flooding; deaths directly due to climate change increase slightly, but are offset in the A1F1, B2 and B1 scenarios by reduced mortality resulting from improved living standards

[c] The baseline assumes incomes are kept at the 1990 level and there is no climate change

Suppose that there is no adaptation to global warming. Even so, Goklany finds that things will generally improve compared to a situation where there is no climate change and incomes remain at the level they were in 1990. Indeed, the negative impacts of climate change are offset by rising incomes, so much so that the overall climate impact is essentially negligible. Among scenarios, the greatest damages occur for the situation where people are poorest.

Goklany (2009) also reports that net biome productivity will increase as a result of climate change and that less wildlife habitat will generally be converted to cropland as a result of global warming, a finding similar to that of Sohngen et al. (1999). Finally, compared to mitigation efforts through emissions reductions, such as the Kyoto process, Goklany finds that targeted adaptation can yield large benefits – adaptation is an optimal policy response.

Gary Yohe of Wesleyan University prepared the chapter on climate for a follow-up report to the Copenhagen Consensus (Yohe 2007). Again, the DICE model was used to obtain estimates of costs and benefits. However, Yohe points out that non-market values of the damages avoided are not sufficiently taken into account in integrated assessment models. Therefore, in addition to the costs and benefits (damages avoided) included in the integrated assessment model, he calculates the benefits of mitigating global warming by examining the reduction by 2080 in the number of people at risk from hunger, water scarcity and coastal flooding.

Results are provided in Table 7.9. These indicate first off that, for all of the scenarios investigated, the DICE model predicts discounted costs exceeding benefits – that net present value is negative. When the benefits of reducing hunger, water scarcity and coastal flooding are included, the benefit-cost ratio is still below 1.0, except for one scenario, although none are less than 0.96. When other non-market benefits, such as ecosystem services and biodiversity, are taken into account, argues Yohe, discounted benefits will definitely exceed discounted costs by a large amount. Thus, the benefit-cost ratios of taking action to avoid climate change, as presented in Table 7.9, must be considered an absolute lower bound.

Notice that Yohe applies very low carbon taxes in all of his scenarios compared to Cline's optimal taxes (see Table 7.6). This is one reason why the net discounted benefits (net present value) of mitigating climate change turn out to be negative, until one starts to add in non-market damages avoidance. He also uses a very low discount rate in one set of scenarios, which causes benefits in 2080 to be more valuable today. This is why the net present value is much lower in scenarios 2 and 5, those with the low carbon taxes. For the second Copenhagen Consensus, Lomborg 2007b) asks readers to make up their own mind as to how they spend money to address 23 of the world's biggest problems. However, if one looks at the benefit-cost analysis of mitigating climate change, it is clear that other problems are more pressing.

In addition to not taking into account many non-market values, Yohe (2007) correctly points out that risks of tipping points, such as the collapse of the Atlantic Thermohaline Circulation (presumably the result of rapid Greenland glacial melt), are ignored in his analysis. The probability of such an event (or something similar) occurring as a result of human activities is extremely tiny, so that, even if the associated cost is extremely large, the discounted expected value is small but not

Table 7.9 Costs and benefits of mitigating climate change: further results from the DICE model (US$1995)

	Scenarios				
Scenario description	1	2	3	4	5
Climate sensitivity	3°C	1.5°C	5.5°C	3°C	1.5°C
Carbon tax (/tCO_2)	$13.70	$1.40	$20.50	$27.30	$2.20
Tax starts in	2006	2006	2006	2016	2016
Number with less risk of hunger in 2080 ($\times 10^6$)	26	19	48	26	19
No. with less risk of water scarcity in 2080 ($\times 10^6$)	2,070	1,160	2,680	2,070	1,160
No. with less risk of coastal flooding in 2080 ($\times 10^6$)	74	16	76	74	16
High disc. rate (3% social rate of time preference; DICE – effective 5% declining to 4%)[a]					
Discounted costs ($\times 10^{12}$)	$12.73	$0.22	$19.23	$16.14	$0.73
Net present value ($\times 10^{12}$)	–$0.46	–$0.44	–$0.74	–$0.59	–$0.44
Per person cost to reduce risk of hunger	$17,69	$24,444	$15,417	$22,692	$24,444
Per person cost to reduce risk of water scarcity	$222	$379	$266	$285	$204
Per person cost to reduce risk of coastal flooding	$6,216	$27,500	$9,737	$7,973	$27,500
Cost-benefit ratio (lower bound)	0.96	2.00	0.96	0.96	1.91
Low disc. rate (0% social rate of time preference; DICE – effective 2% declining to 1%)[a]					
Discounted costs ($\times 10^{12}$)	$110.81	$2.22	$141.52	$129.43	$2.33
Net present value ($\times 10^{12}$)	–$0.76	–$0.20	–$1.07	–$0.95	–$0.02
Per person cost to reduce risk of hunger	$29,231	$1,111	$22,292	$36,538	$1,111
Per person cost to reduce risk of water scarcity	$367	$17	$385	$459	$9
Per person cost to reduce risk of coastal flooding	$10,270	$1,250	$14,079	$12,838	$1,250
Cost-benefit ratio (lower bound)	0.99	0.99	0.99	0.99	0.99

Source: Adapted from Yohe (2007)

Notes: [a]As indicated in Eq. (7.1), the DICE model employs a logarithmic utility function, which causes the effective real discount rate to fall over time insignificant. In benefit-cost analysis, these costs can easily be accounted for and are unlikely to have a major impact on the overall conclusions, especially if one also uses probabilities to account for the possibility that human activities might only have a small impact on climate (see Chaps. 2 through 4).

7.2.3 Economic Consensus Breaks Down

In addition to the idea of a policy ramp, economists almost unanimously favor market incentives and, in particular, a carbon tax that uses the proceeds to reduce income and other taxes – a revenue-neutral tax scheme. (Whether carbon taxes or cap-and-trade are a better way to deal with global warming is discussed further in Chap. 8.) A carbon tax could theoretically lead to higher wellbeing as the economic distortions caused by other taxes would be reduced – the so-called 'double-dividend' of a green tax; it would also increase employment (see Bovenberg and Goulder 1996). As the work of Nordhaus indicates, the optimal policy would be to impose a carbon tax set low to begin with and then slowly increased over time. One compelling reason for a tax is to avoid getting locked into an emission-reduction technology that might prove inferior to another option yet to be developed. For example, one might not want to lock into the internal combustion engine by promoting and subsidizing production of ethanol and biodiesel, with its production facilities and transportation networks, in case a much better option, such as an electric vehicle capable of going 300 km or more on a single charge, should come along. Doing so might be prohibitively expensive and militate against the very development of such an electric vehicle.[33]

Two unrelated events changed the foregoing consensus among economists that the optimal tax should begin at a low rate and ramp up slowly over time. First was the publication of the Stern Report (Stern 2007).[34] Contrary to all previous economic analyses (e.g., Kennedy 1999, 2002; Nordhaus 1994; Tol; van Kooten 2004), the Stern Report finds that the benefits of severely restricting CO_2 emissions today exceed the costs of doing so; there is no ramping up policy, only the conclusion that immediate severe restrictions on CO_2 emissions are warranted. The reasons are soon apparent, but they are rooted in the cost-benefit approach used in the Report, and particularly regarding the appropriate discount rate to apply in cost-benefit analysis.

---

[33] An overview of the state of electric cars is found in *The Economist* (September 5, 2009, pp.75–77). The main obstacle remains the battery, although new battery-automobile technologies are potentially capable of 200 km on a single charge (although current vehicles such as GM's Chevy Volt can only go about 60 km). Along with infrastructure that permits quick recharging or exchange of batteries, innovations in auto design to take advantage of electric motors, and economic and institutional innovations (e.g., separating ownership of batteries and vehicles), it could well be that the electric motor replaces the internal combustion engine for land transportation.

[34] The report was prepared for the British government by civil servants under the guidance of Sir Nicholas Stern, a well known economist.

Stern does not reject the notion of discounting, because it only makes sense when comparing viable alternatives with different flows of costs and benefits over time, but he relies on a *very low* 1.4% rate of discount, which is determined as the rate of growth in per capita consumption plus 0.1% (Mendelsohn 2006). This implies that distant damages (costs) of global warming are much more highly valued today than had heretofore been assumed (Nordhaus 2007), thereby raising the discounted benefits of acting today.

Further, the Stern Report assumes damages from global warming to be three or more times higher than what has been previously assumed, while costs of mitigating CO_2 emissions are taken to be rather small (Mendelsohn 2006; Nordhaus 2007; Tol 2006). But it is only when the non-market environmental damages from global warming are taken to be extremely large that an argument can be made for immediate drastic action to reduce CO_2 output.[35] Yet, the Stern Report did not immediately change the majority view of economists that society should wait before taking costly action on global warming. Rather, economists widely condemned it as "the greatest application of subjective uncertainty the world has ever seen" (Weitzman 2007, p.718), and an analysis that is not based on "solid science and economics" (Mendelsohn 2006, p.46) and that "can therefore be dismissed as alarmist and incompetent" (Tol 2006, p.980).

Finally, the Stern Report is alarmist. It attributes any and all potential future climate disasters solely to anthropogenic emissions of CO_2. Thus, Stern argues it would be folly not to take action immediately to avert such a potential disaster; when a low discount rate is employed, the present value of extremely large damages occurring some distance into the future is also very large – thus, take immediate action. This is a theme to which we return shortly.

The second event was the global financial crisis that originated with U.S. financial institutions, which had created a variety of suspect financial derivatives that were overlooked or not well understood by most economists and investors. Unregulated financial markets enabled institutions to sell financial derivatives that consisted of very shaky loans (mainly high-risk mortgages that financed 100% or more of the price of a home) combined with sounder assets, thereby hiding the true risks of the asset. In addition, insurance derivatives were created to insure the combined assets, and these insurance derivatives were also sold in financial markets. When loans could not be repaid because house prices stagnated and then fell, the financial derivatives unraveled and a credit crisis ensued.

[35] Non-market values are difficult to measure, and there has been quite a bit of controversy surrounding attempts to assign high values to such things as forest ecosystems, wildlife species, etc. In addition to the problem of budget constraints in the estimation of values (some studies find that people are willing to pay more than their entire income to protect nature), there is much confusion about average versus marginal values. For example, an old-growth forest might have tremendous worth, but harvesting one more hectare of the forest might benefit society, as a single hectare might have little non-market value at the margin, much as the hundredth pair of shoes provided to an individual has no value to the person. See Chap. 6.

The financial crisis affected the real economy because people's wealth and earnings were adversely affected, and it shook the faith of many in the ability of markets to create desired outcomes leading instead to a renewed interest in regulation. This might explain why Jeffery Sachs of Columbia University even praised President Obama for favoring regulation in addressing climate change: "Obama is already setting a new historic course by reorienting the economy from private consumption to public investments. ... Free-market pundits bemoan the evident intention of Obama and team to 'tell us what kind of car to drive.' Yet that is exactly what they intend to do ... and rightly so. Free-market ideology is an anachronism in an era of climate change." The backlash was so severe that Nobel Laureate Robert Lucas of the University of Chicago felt compelled to write an article for *The Economist* (8 August 2009, p.67) defending the 'dismal science' and markets in particular.

7.2.4 Doomsday Climate and 'Fat Tails'

A very different approach to that of Stern (2007) is taken by Martin Weitzman, who first criticized the Stern Report for its highly speculative nature but then set about to provide an alternative defense for taking immediate action on global warming. His approach is not based on low discount rates and optimistic estimates of mitigation costs (Weitzman 2009a, b, c), but is still rooted in cost-benefit analysis. Weitzman considers what happens when there is a high probability of a catastrophic event. Weitzman bases his case on 'fat-tailed' probability density functions that, using his methods (discussed below), provide a 5% probability that average global temperatures rise by more than 10°C and a 1% probability that they increase by more than 20°C. What would be the implications of 10–20°C warming? "At a minimum such temperatures would trigger mass species extinctions and biosphere ecosystem disintegration matching or exceeding the immense planetary die-offs associated in Earth's history with a handful of previous geoenvironmental mega-catastrophes" (Weitzman 2009a, p.5). Thus, Weitzman begins with the view that anthropogenic global warming is not only occurring, but that its implications are catastrophic.

The attributing cause of the current catastrophe is the result, according to Weitzman, of the product of unprecedented greenhouse gas emissions and a critical, climate sensitivity parameter $s^* = s_{2\times CO2}$ that converts atmospheric CO_2 into temperature increases. This was discussed in Chap. 4 in relation to Eq. (4.20). Recall that the climate sensitivity parameter s^* was determined to be equal to 1.2°C, but that climate models project much higher temperature increases as a result of two feedbacks that involve (1) water vapor, clouds and ice-albedo (denoted f_1), and (2) a potentially catastrophic secondary release of greenhouse gases (including CO_2) attributable to the initial warming (denoted f_2). This led to Eq. (4.21) reproduced here:

$$\Delta T = \frac{\Delta \ln(CO_2)}{\ln(2)} \times \frac{s^*}{(1-f_1)(1-f_2)}. \tag{7.6}$$

The values of the parameters discussed in Chap. 4 were as follows: $s^*=1.5°C$ to $s^*=5.5°C$, which are taken from climate models rather than based on the historical value $s^*=1.2°C$; $f_1=0.20$ to $f_1=0.73$; and $f_2=0.042$ to $f_2=0.067$. Weitzman argues that the subsequent scaling multiplier, $s_m=\dfrac{s^*}{(1-f_1)(1-f_2)}$, is highly uncertain, so much so that its probability distribution is necessarily characterized by 'fat tails' that bring about high probabilities of large increases in temperature. How does Weitzman come to this conclusion? He bases this on four exhibits (Weitzman 2009b, c):

1. According to Antarctic ice core data reported by Dieter et al. (2008), current atmospheric concentrations of CO_2 are the highest ever recorded in the past perhaps 850,000 years, and the current rate of increase in atmospheric CO_2 is historically unprecedented. This unprecedented increase can only be attributed to human causes according to Weitzman and others.
2. There are 22 studies reported in Table 9.3 and Box 10.2 of the IPCC's Fourth Assessment Report (IPCC WGI 2007, pp.721–722, pp.798–799). These studies report probability density functions (PDFs) with high probabilities of large temperature rise. Weitzman sums the reported PDFs into a single PDF using what he calls a meta-analysis based on Bayesian model averaging. From the meta-derived single PDF, Weitzman finds that the probability that the temperature increase exceeds 7°C is 5%, or that $P(s^*\geq 7°C)=P(\Delta T\geq 7°C)=0.05$, and $P(\Delta T\geq 10°C)=0.01$. This is the feedback effect of CO_2 warming on water vapor discussed earlier.
3. Next, he assumes that the higher temperatures brought about by increased concentrations of atmospheric CO_2 will cause permafrost and boggy soils to release methane, thereby amplifying global warming beyond even the water vapor feedback. The possibility that this feedback effect takes place is discussed by Scheffer et al. (2006), Matthews and Keith (2007)), and *The Economist* (1 August 2009, p.70).[36] The possibility of such a feedback effect leads Weitzman to increase the value of the climate scaling multiplier s_m so that, based on information from Torn and Harte (2006), the probability that temperatures could rise above 11.5°C is 5% and that they could rise above 22.6°C is 1%, or $P(s_m\geq 11.5°C)=P(\Delta T\geq 11.5°C)=5\%$

[36] These feedbacks ignore others in the climate system. Without a natural greenhouse gas effect, the Earth's surface temperature would be about 18°C; with it, but without any feedbacks, it would be about 60°C; with feedbacks, it is about 15°C. In the natural system, then, feedbacks eliminate 68% of GHG warming – that is, negative feedbacks outweigh positive ones. But the climate models used by the IPCC assume that positive feedbacks outweigh the negative ones – precisely the opposite of what is found in nature. The result of Spencer's and Lindzen's and others' studies (noted in Chap. 4) is generally that climate sensitivity $s_{2\times CO2}$ is about 0.5°C instead of the IPCC's midrange of 3.0°C. That, in turn, greatly diminishes the probabilities of 10 and 20°C warming.

and $P(s_m \geq 22.6°C) = P(\Delta T \geq 22.6°C) = 1\%$. However, recognizing the crude and speculative nature of his calculations, Weitzman rounds this down: there is a 5% probability that the expected increase in temperature exceeds 10°C and a 1% probability that it exceeds 20°C – that is, $P(\Delta T \geq 10°C) = 5\%$ and $P(\Delta T \geq 20°C) = 1\%$.

4. Finally, given the potential for huge increases in temperature, Weitzman argues that economic damage (utility) functions parameterized on the basis of current fluctuations in temperature make no sense. Recall that William Nordhaus uses a quadratic damage function: $D(T) = bT + cT^2$, where T is temperature as before, and a particular parameterization is given in Table 7.7. But Weitzman argues that it should be exponential, so that $D(T) = e^{f(T)}$, where $f(T) = bT + cT^2$, or some other function. Clearly, the exponential damage function results in much higher damages the farther into the future one projects rising temperatures.

Based on these four points, all of which are highly speculative, Weitzman concludes that there is a real possibility that, regardless of the discount rate, the damages from climate change could be infinite – that humans could cease to exist as a species. Integrated assessment models ignore this possibility and, by failing to take it into account, any policy path is not truly optimal.

Weitzman's analysis also entails a methodological issue. The so-called probabilities provided by the 22 studies reported by the IPCC WGI (2007, pp.798–799) are based solely on computer models, beginning with those that develop the emission scenarios and then followed by the climate models that project the associated future climates (see Chap. 4). These are not probabilities in the classical sense – based on repeated observations, as in the case of a fair coin toss yielding a 50% probability that the coin comes up as a 'tail,' or the probability that a driver involved in an accident has no valid driver's license. Weitzman's exercise is nothing more than a means for specifying a prior belief (in the Bayesian sense) that there is a high probability that anthropogenic emissions of CO_2 will trigger dangerously high changes in temperature.[37] Further, as noted in Chap. 6, his 'fat-tail' probability distribution and exponential damage function are an attempt to place a precautionary type principle in a cost-benefit framework.

In some sense, this line of argument is similar to that of Brander and Taylor (1998), who use an economic model to argue that the advanced civilization on Easter Island disappeared because people simply "ate up their natural endowment,"

[37] McKitrick (2011) makes two observations: First, "given the finite number of observations on which to estimate s_m, the Bayesian posterior of the expectation of the marginal rate of substitution between current and discounted future income corresponds to the moment generating function of a t distribution, which does not exist, or on Wietzman's interpretation, is infinite." Second, to prevent the outcome that people would have to pay more than their entire current income to insure against a possible future climate catastrophe, Weitzman must either truncate the distribution of *sm* or assume thin-tailed posterior distributions.

and then speculate that we are doing the same thing on a global scale.[38] This is simply a more modern expression of Malthus's original argument that population growth outpaces growth in food production, thereby leading to inevitable misery for the human race.

Needless to say, if one accepts Weitzman's premises, he makes a reasonable case that something should be done to prevent global warming, invoking something akin to a 'generalized precautionary principle' (see Chap. 6). But he then introduces further value judgments by arguing that "the handful of other conceivable environmental catastrophes are not nearly as critical as climate change" (2009a, p.14), comparing global warming only against genetically modified organisms ('Frankenfoods') and "the possibility of a large asteroid hitting Earth" (2009a, p.14; see also Weitzman 2009b, c).[39] Clearly, there exist a lot of other threats to Earth besides these two, including ones perpetrated by humans (such as nuclear holocaust), and deciding which constitutes the worst threat is not an easy task without a lot more information and a crystal ball.

Yet, despite his conclusions, Weitzman backs away from advocating immediate action to stop CO_2 emissions entirely. He comes to this conclusion partly because his results imply, in the end, that the chance of catastrophe, and thus the optimal insurance policy required to offset it, is quite arbitrary. Thus, he advocates spending on a 'put-a-man-on-the-moon' type of research and development (R&D) program that will lead to a technological solution that will enable humankind to control the climate. This is discussed further in the Sect. 7.3.

Here we simply conclude that Weitzman's economics hinges crucially on two points: (1) human activities contribute to the observed increase in atmospheric CO_2 and humans can devise means to control the level of CO_2 in the atmosphere; and (2) increased atmospheric CO_2 leads to increased global temperatures via an extremely high climate sensitivity parameter (s_m). If either of these suppositions is false, or even if one of them is only partially true, then the economic conclusions disappear.

---

[38] Jared Diamond's (1997, 2005) works set a similar tone as he blames human folly related to the environment for the collapse of various civilizations, including that of Easter Island. In the context of global warming specifically, Brian Fagan takes a similar tack. Although he argues throughout that humans are courting disaster by emitting CO_2 and other GHGs that will bring about unprecedented and dangerous warming, his examples show exactly the opposite. Fagan's (2000) book on the 'little ice age' presents a litany of suffering due to cold weather, while demonstrating the huge benefits that the 'medieval warm period' brought to most societies in his later book (Fagan 2008). Yet, he argues that the medieval warming was actually colder than today and that it resulted in vast population shifts (a typical argument of climate doomsayers) that led to war and suffering. He also argues that the warming period brought about unprecedented droughts, but that cold periods had the same consequence. For an alternative perspective to both the arguments of Diamond and Fagan, particularly as these pertain to the Americas, see Mann (2005).

[39] Interestingly, he avoids the possibility of a nuclear war and a potentially killer virus akin to the Black Plague of Medieval times, both of which are more likely than catastrophic climate change according to the Copenhagen Consensus.

7.2.5 A State-Contingent Pricing Rule

With the aid of integrated assessment models, researchers are able to identify a policy ramp – an escalating carbon tax or an increasingly stringent cap on CO_2 emissions. The policy ramp is smooth but it is only pseudo-optimal, because the key structural parameters and future trends (of population, technology, etc.) need to be specified a priori, once and for all within the integrated assessment model. Using a Bayesian learning approach, Andrew Leach (2007) tested the policy paths derived from the IAMs when there is uncertainty about some of the structural parameters in the model. He found that uncertainty in even one or two structural parameters was sufficient to delay the identification of an expected optimal policy regime for a century or more. That is, in the context of climate change, it takes upwards of hundreds of years to be able to determine an optimal policy strategy because it takes that long to obtain sufficient information about the future damages of climate change and other economic relationships to satisfactorily resolve uncertainty.

The solutions offered by Stern and Weitzman attempt to address this uncertainty in the framework of cost-benefit analysis by making various assumptions about the levels of damages and costs and the discount rate (Stern), or the probability of a catastrophic loss and the possibility of insuring against it (Weitzman). Both require a very large present outlay to mitigate climate change (Stern) or fully insure against the potential damages (Weitzman). It is unlikely that this type of outlay will be politically acceptable, even in rich countries (see Chap. 8), let alone developing ones.

McKitrick (2011) offers a solution that is politically acceptable to both those who feel climate change will lead to a climate catastrophe and those who are unconcerned either because they do not think warming will lead to catastrophe or they disagree with the premise of anthropogenic climate change. The innovation introduced by McKitrick is that he ignores abatement costs and, thus, does not seek to demonstrate that discounted benefits of taking action exceed discounted costs. He is only interested in setting the correct (optimal) price on carbon emissions.

McKitrick derives an optimal carbon tax by minimizes the discounted present value of damages subject to the effect that carbon emissions have on a suitable state variable, namely, temperature. The state-contingent pricing rule that he derives is the following:

$$\tau_t = \gamma^{\frac{e_t}{\overline{e}_t}} s(t), \tag{7.7}$$

where τ_t is an approximation of the optimal tax to be set at time t. The tax is a function of the marginal damage rate γ, the current level of emissions e_t, the moving average of emissions over k periods, denoted by $\overline{e}_t$, where k is the number of periods required for CO_2 to leave the atmosphere (or the half life of CO_2 residency in the atmosphere), and $s(t)$ is the value of the state variable (say, temperature). The actual derivation of the optimal path of the tax involves assuming the discount rate is zero,

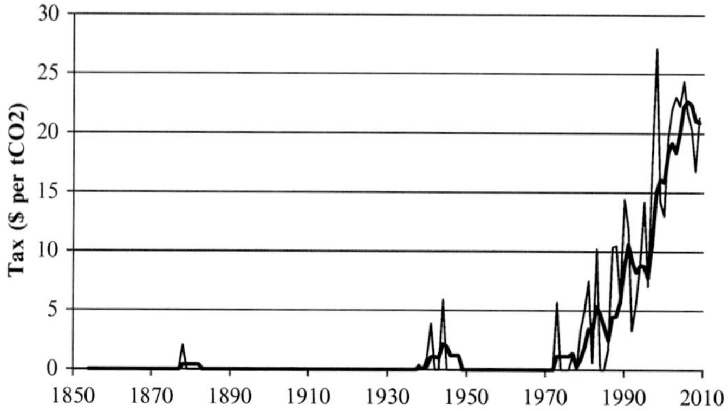

Fig. 7.6 Optimal Tax Rate ($ per tCO_2) to address global warming, annual (*thin line*) and 5-year moving average (*thick line*), 1850–2009

which implies that the rule (7.7) is conservative if temperature (the state variable) is rising.

Tax rule (7.7) is not a prescription for a policy path, but only a rule that links the tax rate to the state of the environment. The only obstacles to implementing the optimal tax are information about the marginal damage rate and the period k for calculating the moving average of emissions. To determine the former, we assume that a tax of $25 per ton of CO_2 was optimal in 2005. Using global fossil fuel emissions data from Oak Ridge National Laboratory in Tennessee,[40] we find that 29,227 Mt of CO_2 were emitted globally in 2005, with average emissions over the preceding 50 years equal to 17,835 Mt of CO_2 (k is assumed to equal 50). The HadCRUT3 temperature anomaly for 2005 was 0.482°C. Solving (7.7) using these values gives $\gamma=31.65$. We then use the value of γ, information on emissions of CO_2 going back to 1801, and the HadCRUT3 temperature anomaly as a state variable to calculate the optimal tax rate that should have been imposed going back to 1850. The optimal tax path is given in Fig. 7.6, where negative tax rates have been assigned a value of zero. Notice that the tax rate exceeds $25/$tCO_2$ on only one occasion, namely, in 1998 when there was a particularly strong El Niño event.

As indicated in Fig. 7.6, if global temperatures rise rapidly, the tax rate will also rise rapidly. In the figure, an average annual global temperature is used as the state variable, but a monthly average could also be employed. The monthly average will surely be more volatile than the annual temperature, which is much more volatile than a 3- or 5-year moving average.

The reason that costs are ignored in setting the tax is that the tax acts as a signal to emitters of carbon dioxide. The market participants will use the information from

[40] Data available at http://cdiac.esd.ornl.gov/trends/emis/em_cont.html (viewed July 22, 2010) and compiled by Tom Boden, Gregg Marland and Bob Andres.

tax trends to make decisions concerning the credibility of the IPCC's (and others') forecasts of climate change. The market will decide on the credibility of the science, because those who decide wrongly (say, by investing heavily in emission reduction equipment) incur costs that make them relatively less competitive. Rather than rely on political or scientific pronouncements, investors will use the market – the trend in tax rates – to guide their decisions, much like commodity and other prices that fluctuate significantly over time currently guide decisions.

Tax rule (7.7) should be acceptable to many more people than the alternative options being proposed by climate scientists and legislators (see Chap. 8). The tax rate appeals to those who fear catastrophic global warming because the tax will escalate rapidly with rising temperatures. It also appeals to those who do not believe in catastrophic anthropogenic climate change because, if their view is correct, the tax will either rise very slowly or not at all, or even fall to zero. Thus, while a majority of citizens are unlikely to support actions that drastically increase energy costs, a majority will be likely to support a tax rule such as (7.7).

7.3 Taking the Debate Beyond CO_2-Emission Reduction Targets

A significant number of economists and policy analysts predicted that the Kyoto Process would fail, because it hopes to achieve greenhouse gas emission-reduction objectives that cannot possibly be attained.[41] The problem is that 80% of the world's peoples live on \$10 per day or less, and 1.5 billion people currently have no access to electricity (Pielke Jr. 2010) – climate policies that prevent economic development will certainly be objectionable to them. Further, as discussed in Chap. 10, non-OECD countries are projected by the OECD and International Energy Agency to account for 93% of the increase in global energy demand between 2007 and 2030, and this will be driven largely by economic growth in China and India. Growth in emissions resulting from the increased consumption of coal by China, India and other Asian countries, let alone growth in consumption of oil and gas, will exceed any possible reduction in emissions that OECD countries could implement. The only conclusions that a realistic observer could possibly come to are that (1) energy prices are

[41] The current author also recognized the inability of the Kyoto Process to achieve its objectives, which are modest compared to those that European Union and United States are currently crafting (see Chap. 8). This is evident from the sub-title of the author's climate economics book (van Kooten 2004): *Why International Accords Fail*. For example, Canada's position as it approached the Kyoto discussions was not to give ground on Canadian emissions, because earlier meetings between the federal and provincial ministers of the environment had concluded Canada was in no position to reduce emissions. Yet, during late-night negotiations, Canada abruptly agreed to a 6% reduction from 1990 emission levels by the 2008–2012 commitment period, while knowing full well it could not possibly achieve any reductions whatsoever. Subsequent growth in Canadian emissions has borne this out. This is discussed further in van Kooten (2004).

currently too high as too many of the earth's citizens are unable to afford to purchase the energy they need to attain even modest standards of living, and (2) addressing climate change by targeting CO_2 emissions is a futile project.

It would appear that current climate policies are now dead in the water as a result of the failure of COP15 at Copenhagen in December 2009 (and COP16 in Cancun, Mexico in December 2010), and the release of the climategate emails in November 2009. As a result, Prins et al. (2010), Pielke (2010), Levitt and Dubner (2009), and many others are making a strong case that a different approach is required. So where does this begin and what form does it take?

7.3.1 Climate Policy

The climate agenda as found in the Kyoto process, for example, is based on what Prins et al. (2010) describe as a 'deficit model' of science: The scientific expert provides the ignorant public and its representatives with the requisite knowledge to remedy their deficit. The public implicitly trusts the superior knowledge and qualifications of the scientists and thereby allows scientists to set forth the actions needed to solve the problem. This model of science works well when the problem is straightforward, such as forecasting where a hurricane might make landfall and instructing people in the path of the hurricane to get out of the way. It works when the valve in someone's heart stops functioning and the prescribed action is to replace the valve with an artificial one. It fails in the case of weapons of mass destruction, for example, not because the experts are unable to destroy the ability of a rogue nation from building nuclear weapons, but because doing so involves value judgments.

The case of climate change is more like the latter example, where knowledge that a country has dangerous weapons does not constitute scientific grounds for concluding that the country will deploy them. The reason has to do with wicked uncertainty. Wicked uncertainty occurs when the problem is too complex and/or too uncertain to resolve by focusing on a single object and the outcomes from taking action, including doing nothing, are unknowable. Climate change is not a conventional environmental problem that can simply be solved by reducing CO_2 emissions! As Prins et al. (2010) point out, it is a problem of economic development, population, technological progress, income differentials, urban planning, agriculture and forestry, lifestyles, and much more. It is, among other things, an economic, energy, development and land-use problem. Given the failure of policies to spur economic growth in poor countries, how can we expect an easy global fix to the climate problem? The Kyoto-IPCC process is a failure because it relied on the deficit model – climate scientists defining the problem and then recommending political solutions to solve it.

Recent efforts by social scientists, economists and others are now moving away from the naïve approach that still seems to dominate policymaking (see Chap. 8). The new approach is still evolving, but the focus is holistic rather than single minded. It takes the view that policies should not be implemented to punish people, as this considers emitting CO_2 to be a sin, but, rather, that policies should be attractive to people because the policies will provide immediate benefits. Two elements of this approach can be identified.

First, there are things which ought to be done regardless of their impact on climate change mitigation, although mitigation is an indirect benefit. One of these is to reduce air pollution – black carbon (soot), which comes from the burning of diesel fuel, cooking stoves (many people still rely on wood stoves for cooking), forest fires, and so on. Soot is thought to be responsible for nearly half of the ice melt in Arctic regions, for example, because the soot particles land on ice and absorb the sun's energy, thereby causing the ice to melt (Prins et al. 2010). Likewise, non-CO_2 greenhouse gases make a significant contribution to anthropogenic warming. By reducing hydrofluorocarbons (HFCs), perfluorocarbons (PFCs), ozone, methane and so on, health benefits can be realized, as well as mitigation benefits.

Land use changes also impact the climate. While tropical deforestation is perhaps the best known example (because it releases CO_2 into the atmosphere as trees are usually burned as land is converted to agriculture), other land use changes can also have a large impact on local and global climates. For example, more of the sun's energy is absorbed as land is paved or converted to development. Planting trees can alleviate some of the adverse temperature and even moisture impacts, while the use of alternative water-permeable surface materials (e.g., clay-gravel driveways as opposed to concrete or pavement) might prevent future flash floods. Agricultural programs and subsidies that promote cropping, including policies to increase production of biofuels, have an adverse impact on forest and grasslands, thereby reducing the carbon stored in terrestrial ecosystems (see Chap. 9) and the ecosystem benefits that are provided. The social benefits of retaining lands in their 'natural' state often exceed the costs of converting them to the subsidized use.

Second, cap and trade cannot succeed as the EU's Emissions Trading System (ETS) has proven (e.g., see Prins et al. 2010). As discussed in Chap. 8, cap-and-trade schemes are likely less desirable than taxes as a policy instrument in the case of climate change. Further, the way in which they have been implemented (including the EU ETS) is not true cap and trade; schemes permit the use of 'outside' credits (CDM, terrestrial offsets, etc.), which leads to corruption. Nor can a carbon tax do the job, partly because of political acceptability issues. Both taxes and credit trading will make energy too expensive if either is implemented on a global scale. If implemented only in some rich countries, prices for fossil fuels will fall elsewhere encouraging greater consumption and reducing the benefits associated with the original tax or emissions trading policy – that is, a leakage will occur that is likely to be quite large. Yet, a global tax is not what is needed as the majority of the globe's citizens need more energy not less and fossil fuels are already too expensive.

What can be done? We need to come up with cheaper, non-carbon energy sources. The costs of alternatives to fossil fuels must be cheaper than the cost of using coal; otherwise, 80% of the world's population will have the incentive to use coal-fired energy. Nuclear power is one possibility. While there are likely other energy options, the task of determining these is onerous. What is required is a research effort similar to that of putting a man on the moon, as advocated by Weitzman. However, climate policy should not scare people into reducing fossil-fuel energy through massive sin taxes, or cap and trade (which amounts to the same thing); nor should society provide massive subsidies for alternative fuels, such as biofuels that may increase rather than reduce greenhouse gas emissions (see Chap. 10), which might lock us into undesirable technologies. Rather, Prins et al. (2010) recommend a carbon tax that is set at a low rate, sufficient to fund an R&D project of the type required, but not so high that it results in adverse or unanticipated consequences, ones we might later regret.

A focus on research and development, and demonstration and adoption of the new technologies that arise is important because it is the only way to de-carbonize economies. This is illustrated by the simple mathematical relation between economic development and energy use.

7.3.2 *Simple Mathematics of Emissions Reduction: The Kaya Identity*

Thinking about a new approach to climate policy might begin with something that is similar to the well known macroeconomic income identity, where income equals the sum of consumption, investment, government expenditure and net exports. The energy-equivalent relation is known as the Kaya identity[42]:

$$C = N \times \frac{Y}{N} \times \frac{E}{Y} \times \frac{C}{E}, \tag{7.8}$$

where C refers to carbon emissions (measured in terms of CO_2), N is population, Y is gross domestic product (GDP), and E is total energy consumption or use. This identity can be applied to the globe, a nation or a region. The first term on the right hand side of the identity is population, the second term is per capita GDP, the third term is the *energy intensity of the economy* and the final term is the *carbon intensity of energy*. An indirect approach to climate mitigation is to reduce the energy intensity of economies and the carbon intensity of energy.

According to the Kaya identity (7.8), there are only a limited number of ways to reduce emissions of carbon dioxide:

- Manage population;
- Limit the generation of wealth (reduce GDP);
- Generate the same or a higher level of GDP with less energy;
- Generate energy with less CO_2 emissions; or
- Some combination of the first four factors.

[42] The Kaya identity is named after Japanese economist Yoichi Kaya (Kaya and Yokobori 1997).

Dramatically reducing population is something that is outside the policy envelop – it is simply not acceptable, although Paul Erlich, James Lovelock, Peter Singer and others have advocated dramatic reductions in population to forestall climate change and other environmental disasters that these writers attribute to humans.[43] Further, climate policies must not cost too much, or they must be done in a way that leads to economic growth. If they cost too much or prevent economic growth, particularly the economic development of poor countries, they will simply not be politically acceptable.

To examine the remaining options, one can rewrite the Kaya identity as:

$$\text{Emissions} = \left[N \times \frac{Y}{N}\right] \times \left[\frac{E}{Y} \times \frac{C}{E}\right] = Y \times \frac{C}{Y} = \text{GDP} \times \text{Technology}, \qquad (7.9)$$

where technology (C/Y) is simply the ratio of CO_2 emissions to GDP. In 2006, 29.12 Gt CO_2 were emitted globally while global GDP amounted to $47.267 trillion, so that the technology or emissions to GDP ratio was 0.62 tCO_2 per $1000 GDP. From 1980 to 2006, the world's C/Y ratio fell from 0.92 to 0.62 tCO_2 per $1,000 GDP. This is seen in Fig. 7.7 where the emissions intensities of selected countries are also provided. In making these calculations, a measure of purchasing power parity (PPP) GDP is used.

The carbon intensity of an economy depends on the GDP value that one employs. In Fig. 7.7a, we provide PPP GDP in constant 1990 Geary Khamis dollars (GK$), but in Figs. 7.7b, c, we employ PPP GDP measured in constant 2000 US$.[44] Notice that, for the United Kingdom, the GK$ measure has carbon intensity falling from 0.85 in 1980 to 0.42 in 2006, while carbon intensity falls from 0.63 to 0.31 in US$ terms. In both cases, however, the proportional decline in carbon intensity is the same.

The global carbon intensity has been falling at a rate of about 0.012 tCO_2 per year. This average rate of decline is about the same for rich and poor countries, but they vary considerably from one country to another. For example, rich countries that have recently shed aluminum production (Japan) or replaced coal-fired power plants with nuclear ones (most notably France) have experienced faster rates of improvement in carbon intensity than other countries.

Once the most carbon-intensive industries have been moved offshore and the least costly transitions to 'green' power have been implemented, it becomes increasingly difficult for a country to increase the rate at which carbon intensity of the economy declines. Indeed, many rich countries have reduced their domestic

---

[43] Possibly the most depressing book on the subject is by David Benatar (2006), who ultimately concludes that a human population of zero is ideal. He says he would commit suicide except that he is needed to make others aware of the harm that humans cause to the earth. See Wanliss (2010) for an alternative perspective.

[44] PPP adjusts country-level GDP for cost of living rather than using current exchange rates. GK$ is explained at: http://en.wikipedia.org/wiki/Geary%E2%80%93Khamis_dollar (viewed May 16, 2011). The data for Fig. 7.7a are available from Pielke Jr (2009). Data on carbon intensities based on 2000 US$ are available from the International Energy Agency at: http://eia.doe.gov/iea/carbon.html (viewed May 19, 2011).

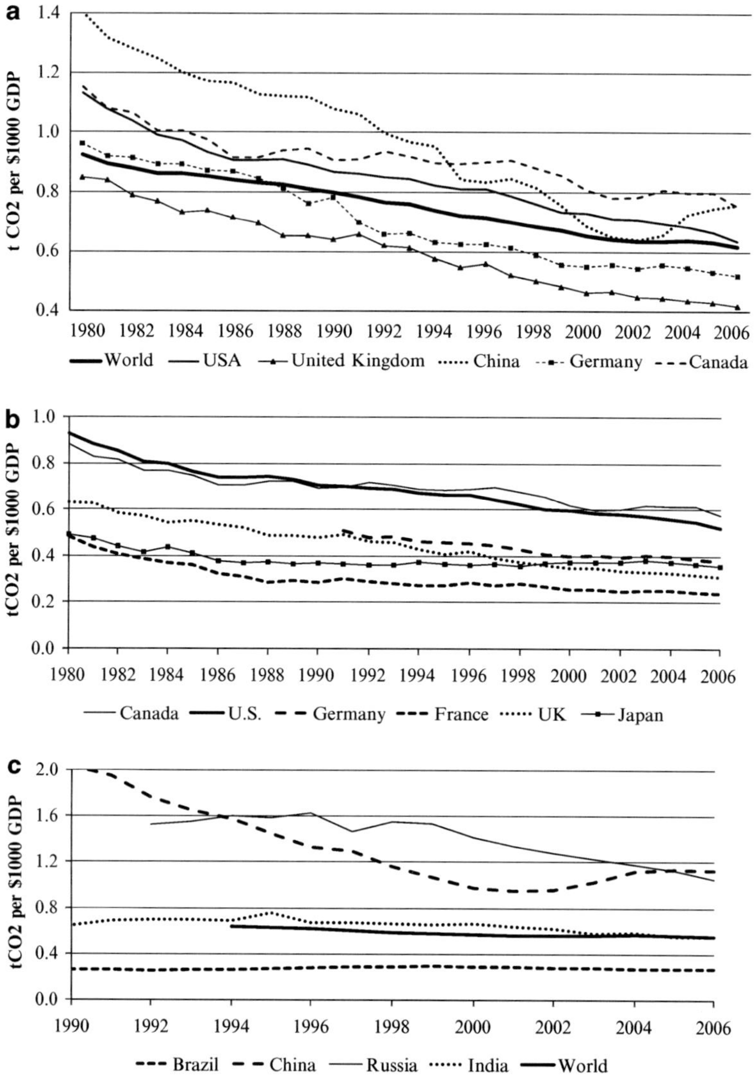

Fig. 7.7 (**a**) Carbon dioxide emissions per $1000 GDP measured in 1990 GK$, global and selected countries, 1980–2006. (**b**) Carbon dioxide emissions per $1000 GDP measured in 2000 US$, selected OECD countries, 1980–2006. (**c**) Carbon dioxide emissions per $1000 GDP measured in 2000 US$, global and the BRIC countries, 1990–2006

CO_2 emissions since 1990 by shifting production of consumer goods to developing nations (Peters et al. 2011). As a result, since 1990 the CO_2 emissions embodied in goods exported from poor to rich countries increased from 400 million tons to 1.6 billion tons, or by 8% per year.

If we consider the so-called BRIC countries (Brazil, Russia, India and China), we find that, while China and Russia have improved upon their carbon emissions

Table 7.10 Power generating facilities offsetting 10% of Canada's annual CO_2 emissions: any single option or combination is needed

Number of units	Capacity per unit (MW)	Type of generating facility
21	1,000	Nuclear power plants
22	600	Hydroelectric dams[a]
4,650	10	Solar photovoltaic plants
18,500	2.5	Wind turbines

Source: Roger Pielke presentation at Pacific Institute for Climate Studies' Vancouver workshop on BC's Future Grid, June 15 (2010)

Notes: [a] This is the size of the Site C facility proposed on the Peace River in northeastern British Columbia, a project that is unlikely to go ahead as a result of environmental, local, native and other lobby groups

intensity, Brazil and India have not made any discernible improvements on this front (Fig. 7.7c). Russian emission intensity may have improved because of recession and the subsequent closure of inefficient manufacturing facilities associated with the downfall of the Soviet Union in 1989. Chinese emissions per \$1000 GDP have improved greatly from 1980 to the present, but have taken an upturn since 2001. Brazil and India are likely to see emissions per \$1000 GDP rise before they fall because their current C/Y is much lower than that of even rich countries that are characterized by large service sectors. Hence, one might expect an intensification of emissions as a result of rapid industrialization (including construction of infrastructure) that generally accompanies economic growth, as witnessed in China.

The United Kingdom has perhaps the most draconian climate legislation of any government: climate legislation passed in December 2008 requires the UK to reduce greenhouse gas emissions by 34% by 2022. The UK also has one of the lowest C/Y ratios (using the 1990 GK\$ measure), with $C/Y=0.42$ in 2006; but France has the lowest carbon intensity index among rich countries, with $C/Y=0.30$ in 2006. The reason for the low rates in these countries is in large part due to their success in moving manufacturing offshore and, in the case of France, heavy reliance on nuclear energy. France took 20 years to move from $C/Y=0.42$ to $C/Y=0.30$, but not as a result of a concerted effort to reduce CO_2 emissions. Roger Pielke Jr. (2009, 2010) estimated that, to meet its climate policy targets, the UK will need to get to $C/Y=0.30$ in 5 years. This would require, for example, the immediate construction of 40 nuclear power plants, each with a capacity of 1,100 megawatts (MW).

The reason why it is unrealistic to achieve stabilization of atmospheric CO_2 at 450 ppmv, or any other lower target, is that developing countries are going to increase their emissions of CO_2 rapidly over the next 50 years. China and India will account for more emissions of CO_2 than Europe and North America, and growth in emissions in these countries will swamp anything developed countries will do to reduce their own emissions. For example, the incremental increase in Chinese CO_2 emissions during 2½ months equals the total of Canada's annual emissions.

The enormity of the task is further illustrated in Table 7.10 by examining what is needed to offset 10% of Canada's CO_2 emissions. It would take about 20 new nuclear power plants or 20 large-scale hydroelectric dams, or huge investments in solar and/or wind farms. None of these are likely to be constructed in the near future, at least not on this scale. The reason has to do with environmental and local opposition to any of these options, including opposition to the construction of transmission facilities that might be needed if electricity is generated in more remote locations and needs to be transmitted to developed areas. Inevitably some residents will be affected and thus mount campaigns opposing construction, which will require years of negotiation and litigation to resolve.

7.4 Discussion: Beyond Carbon Dioxide

The past century witnessed a tremendous reduction in mortality because of improvements in general health, and that was accompanied by large population increases as life expectancy increased and infant mortality declined. At the same time, air and water quality in the developed countries improved significantly as citizens demanded environmental improvements. These improvements are the direct result of rising per capita incomes. As we saw in Chap. 4 (Sect. 4.1), incomes in developing countries are projected by the IPCC to increase substantially over the next 50–90 years, so much so that poverty will effectively be eliminated. This implies that even the poorest countries will have the resources needed to adapt quite easily to climate change.

The higher per capita incomes assumed to be associated with projected climate change also makes it difficult to justify mitigation. Consider the with-and-without effects of mitigation. Because energy is the most important driver of development, reducing the ability of poor people to access cheap energy serves only to keep them poor. Mitigation policies slow the economic growth of developing (and developed) countries, but will prevent the increased mortality associated with climate change. Without mitigation, on the other hand, real per capita incomes of poor people will rise to such an extent that poverty is eliminated. Based on historical evidence, this will lead to large reductions in mortality from almost all causes. As indicated in this chapter, because the Earth's poorest are projected to have high per capita incomes in the absence of action to mitigate climate change, the overall benefits of allowing climate change to occur might well exceed those of taking action and keeping countries in poverty. This provides one explanation why developing countries are not keen on taking action to reduce carbon dioxide emissions if this in any way slows economic growth. Another possible explanation relates to the potentially flawed focus on carbon dioxide.

In their book *Super Freakonomics*, Steven Levitt and Stephen Dubner argue that the focus of climate change mitigation should not be on carbon dioxide. They argue that, if the objective is to mitigate global warming, then this should be done in the socially most efficient manner – it should be done at least cost – and that might not entail controls on CO_2 emissions. After all, carbon dioxide is necessary to plant growth and increasing atmospheric CO_2 has helped drive the green revolution (as noted in section 7.1 above).

One low-cost solution to the problem of global warming has been suggested by the Dutch Nobel Prize winning chemist, Paul Crutzen. He argues that, since it is difficult to get people to reduce greenhouse gas emissions sufficiently to mitigate global warming, it is simpler to inject sulfur into the stratosphere, as this could rapidly reduce temperatures (Crutzen 2006). Levitt and Dubner (2009) point out that it is potentially possible to spray sulfur dioxide into the stratosphere using a specialized 'hose' in the sense of a garden hose and suggest that such a device could be located near Fort McMurray, Alberta – Canada's oil sands where sulfur and energy are readily available and the location is suited to having SO_2 circulate throughout Earth's entire upper atmosphere, thereby reducing warming. The costs would be several hundred million dollars annually compared to the trillions of dollars to achieve the same mitigation benefits from reduced CO_2 emissions, while the negative impact of SO_2 would be relatively small.

There are other geo-engineering solutions that permit us to continue using the Earth's plentiful fossil fuels and emitting greenhouse gases. These do not focus on carbon dioxide but on other factors that affect climate. Indeed, Martin Weitzman's assumption that a technological solution exists was highly influenced by Scott Barrett's argument that there are economically inexpensive, geo-engineering solutions to the problem of global warming (Barrett 2008, 2009). David Keith of the University of Calgary is an enthusiastic proponent of an engineered climate who is cited by Barrett, Weitzman and others. In personal conversations, he assures listeners that one day, when the planet gets hot enough, we will simply use carbon capture and storage technologies to take CO_2 out of the atmosphere on a large scale and thereby cool the globe; alternatively, we can release CO_2 to warm the globe.

Finally, recall that Prins et al. (2010) advocated a small carbon tax that would be used to fund research and development into solutions to the problem of global warming. R&D was also the favored option at the most recent Copenhagen Consensus, which focused solely on climate change. Five economists, including three Nobel laureates, were asked to rank 15 options for addressing climate change. These are listed in Table 7.11 along with the economists's individual rankings, where we use a 15-point scale with the individual's best option given a score of 15 and the lowest ranked alternative a score of 1. The overall ranking of the alternatives is provided in the final column, and differs slightly from that presented in Lomborg (2010, pp.381–382). Included in the table is the type of solution each of the options represents – whether reliance on engineering, R&D, technology transfer, adaptation, terrestrial carbon sinks, or cutting emissions of anthropogenic emissions of carbon dioxide or methane.

Interestingly, the high-profile panel of economists consistently ranked reduction of greenhouse gas emissions at or near the bottom, with carbon taxes and, thereby, emissions trading as the worst possible of all options for addressing climate change. Adaptation was ranked 5th, only behind research and development into cloud whitening, energy (see Chaps. 10 and 11 below), carbon storage and stratospheric aerosol insertion. Adaptation was ranked ahead of the approach that is advocated by environmentalists and seriously considered by policymakers, namely carbon taxes or emissions trading. Ranked in the middle were two other technology options

Table 7.11 Addressing climate change: 2010 Copenhagen consensus results

Rank and solution	Solution category	Individual ranking values					Rank score
1. Cloud whitening research	Engineering	15	15	13	14	8	65
2. Energy R&D	Technology	13	12	11	13	13	62
3. Carbon storage research	Technology	12	7	15	15	12	61
4. Stratospheric aerosol R&D	Engineering	14	13	14	11	7	59
5. Planning for adaptation	Adaptation	11	14	9	7	15	56
6. Research in air capture	Engineering	10	6	12	12	9	49
7. Technology transfers	Tech transfer	9	10	10	8	11	48
8. Expand & protect forests	Forestry	5	11	8	9	14	47
9. Stoves in poor nations	↓ black carbon	2	9	4	6	10	31
10. Methane reduction	Cut methane	4	8	7	4	5	28
11. Diesel vehicle emissions	↓ black carbon	3	4	5	5	6	23
12. $20 OECD carbon tax	Cut CO_2	8	5	6	1	1	21
13. $0.50 global CO_2 tax	Cut CO_2	6	3	2	10	2	23
14. $3 global CO_2 tax	Cut CO_2	7	2	3	3	4	19
15. $68 global CO_2 tax	Cut CO_2	1	1	1	2	3	8

Source: Adapted from Lomborg (2010) using author's calculations. Copyright © 2010 Copenhagen Consensus Center. Reprinted with the permission of Cambridge University Press

(including transfer of technology), forest ecosystem sinks (discussed further in Chap. 9), reduction in methane emissions as methane is a potent greenhouse gas, and means for reducing black carbon output from stoves in poor nations and diesel vehicles. With the exception of research into air capture, the other options were considered to be fair or poor.

Clearly, decision makers need to reconsider their focus when it comes to climate change. Despite efforts by environmentalists to impose controls on fossil fuel emissions and, seemingly, to implement global tax and/or emission trading schemes, if global warming is truly the problem that global society needs to solve (and it might not be), it is necessary to adopt policies that are most effective in addressing climate change. The most effective policies may not be ones that seek to reduce fossil fuel emissions.

References

Abe-Ouchi, A., Andersen, M., Antonioli, F., Bamber, J., Bard, E., Clark, J., Clark, P., Deschamps, P., Dutton, A., Elliot, M., Gallup, C., Gomez, N., Gregory, J., Huybers, P., Kawamura, K., Kelly, M., Lambeck, K., Lowell, T., Mitrovica, J., Otto-Bleisner, B., Richards, D., Siddall, M., Stanford, J., Stirling, C., Stocker, T., Thomas, A., Thompson, W., Torbjorn, T., Vazquez Riveiros, N., Waelbroeck, C., Yokoyama, Y., & Yu, S. (2009). The sea-level conundrum: Case studies from palaeo-archives. *Journal of Quaternary Science, 25*, 19–25.

Adams, R. M. (1989). Global climate change and agriculture: An economic perspective. *American Journal of Agricultural Economics, 71*(5), 1272–1279.

Adams, R. M., Rosenzweig, C., Peart, R. M., Richie, J. T., McCarl, B. A., Glyer, J. D., Curry, R. B., Jones, J. W., Boote, K. J., & Allen, L. H., Jr. (1990). Global climate change and US agriculture. *Nature, 345*(6272), 219–224.

Adams, D. M., Alig, R. J., Callaway, J. M., McCarl, B. A., & Winnett, S. M. (1996). *The forest and agricultural sector optimization model (FASOM): Model structure and policy applications* (p. 60). Portland: U.S. Department of Agriculture, Pacific Northwest Research Station.

Anderson, W., (2010). Some Like it Warm, *First Things*, February. Viewed July 23, 2012 http://www.firstthings.com/article/2010/01/some-like-it-warm.

Armstrong, J. S., Green, K. C., & Soon, W. (2008). Polar bear population forecasts: A public-policy forecasting audit. *Interfaces, 38*(5), 382–405.

Arthur, L. M. (1988). *The implications of climate change for agriculture in the prairie provinces Canadian Climate Centre CCD88-01* (pp. 1–13). Downsview: Atmospheric Environment Service, Environment Canada.

Arthur, L. M., & Abizadeh, F. (1988). Potential effects of climate change on agriculture in the prairie region on Canada. *Western Journal of Agricultural Economics, 13*(2), 215–224.

Arthur, L. M., & van Kooten, G. C. (1992). Climate change impacts on agribusiness sectors of a prairie economy. *Prairie Forum, 17*(Spring), 97–109.

Barrett, S. (2008). The incredible economics of geoengineering. *Environmental and Resource Economics, 39*(1), 45–54.

Barrett, S. (2009). The coming global climate-technology revolution. *Journal of Economic Perspectives, 23*(2), 53–75.

Belisle, A. A., Reichel, W. L., Locke, L. N., Lamont, T. G., Mulhern, B. M., Prouty, R. M., DeWolf, R. B., & Cromartie, E. (1972). Residues of organochlorine pesticides, polychlorinated biphenyls, and mercury and autopsy data for bald eagles, 1969 and 1970. *Pesticides Monitoring Journal, 6*(3), 133–138.

Benatar, D. (2006). *Better not to have been. The harm of coming into existence*. Oxford: Oxford University Press.

Booker, C. (2011, March 7). *Telegraph*. At viewed March 7, 2011: http://www.telegraph.co.uk/comment/columnists/christopherbooker/8349545/Unscientific-hype-about-the-flooding-risks-from-climate-change-will-cost-us-all-dear.html

Bosello, F., Carraro, C., & de Cian, E. (2010). *Climate policy and the optimal balance between mitigation, adaptation and unavoided damage* (Sustainable Development Paper SD 32–2010, pp. 32): Fondazione Eni Enrico Mattei (FEEM).

Bouwer, L. M. (2011). Have disaster losses increased due to anthropogenic climate change? *Bulletin of the American Meteorological Society, 92*(1), 39–46.

Bovenberg, A. L., & Goulder, L. H. (1996). Optimal environmental taxation in the presence of other taxes: General-equilibrium analysis. *American Economic Review, 86*(4), 985–1000.

Brander, J. A., & Taylor, M. S. (1998). The simple economics of Easter Island: A Ricardo-Malthus model of renewable resource use. *American Economic Review, 88*(1), 119–138.

Choiniere, C. J., & Horowitz, J. K. (2000). *A production function approach to the GDP-temperature relationship (November 2000)*. doi:10.2139/ssrn.267352. Retrieved from http://ssrn.com/abstract=267352.

Chu, P.-S., Chen, Y. R., & Schroeder, T. A. (2010). Changes in precipitation extremes in the Hawaiian Islands in a warming climate. *Journal of Climate, 23*(18), 4881–4900.

Church, J. A., & White, N. J. (2006). A 20th century acceleration in global sea-level rise. *Geophysical Research Letters, 13*, L01602. doi:10.1029/2005GL024826.

Cline, W. R. (2004). Climate change. In B. Lomborg (Ed.), *Global crises, global solutions*. Cambridge: Cambridge University Press.

Cline, W. R. (2007). *Global warming and agriculture: Impact estimates by country*. Washington, DC: Center for Global Development and Peterson Institute for International Economics.

Coon, N. C., Locke, L. N., Cromartie, E., & Reichel, W. L. (1970). Causes of bald eagle mortality, 1960–1965. *Journal of Wildlife Diseases, 6*(January), 72–76.

Cromartie, E., Reichel, W. L., Locke, L. N., Belisle, A. A., Kaiser, T. E., Lamont, T. G., Mulhern, B. M., Prouty, R. M., & Swineford, D. M. (1974). Residues of organochlorine pesticides and polychlorinated biphenyls and autopsy data for bald eagles, 1971–1972. *Pesticides Monitoring Journal, 9*(1), 11–14.

Crutzen, P. J. (2006). Albedo enhancement by stratospheric sulfur injections: A contribution to resolve a policy dilemma? *Climatic Change, 77*(3–4), 211–219.

Czymzik, M., Dulski, P., Plessen, B., von Grafenstein, U., Naumann, R., & Brauer, A. (2010). A 450 year record of Spring-Summer flood layers in annually laminated sediments from Lake Ammersee (Southern Germany). *Water Resources Research, 46*, W11528. doi:10.1029/2009WR008360.

Darwin, R., Tsigas, M., Lewandrowski, J., & Raneses, A. (1995, June). *World agriculture and climate change: Economic adaptations* (AE Report No. 703, pp. 86). Washington, DC: U.S. Department of Agriculture, Economic Research Service.

Davis, M. A. S., Brown, G. M., & Leftwich, J., P. W. (1984, September). *A tropical cyclone data tape for the eastern and central North Pacific Basins, 1949–1983: Contents, limitations, and uses* (NOAA Technical Memorandum NWS NHC 25, 17 pp). National Hurricane Center, National Weather Service, NOAA. Miami, FL. Retrieved from http://www.nhc.noaa.gov/pdf/NWS-NHC-1984-25.pdf

Diamond, J. (1997). *Guns, germs and steel*. New York: W.W. Norton.

Diamond, J. (2005). *Collapse*. New York: Viking.

Dieter, L., Le Floch, M., Bereiter, B., Blunier, T., Barnola, J. M., Siegenthaler, U., Raynaud, D., Jouzel, J., Fischer, H., Kawamura, K., & Stocker, T. F. (2008, May 15). High-resolution carbon dioxide concentration record 650,000-800,000 years before present. *Nature, 453*(7193), 379–382.

Dole, R., Hoerling, M., Perlwitz, J., Eischeid, J., Pegion, P., Zhang, T., Quan, X.-W., Xu, T., & Murray, D. (2011). Was there a basis for anticipation the 2010 Russian heat wave? *Geophysical Research Letters, 38*, L06702. doi:10.1029/2010GL046582.

Durschmied, E. (2000). *The weather factor. How nature has changed history*. New York: Arcade Publishing.

Evers, M., Stampf, O., & Traufetter, G. (2010). *Climate catastrophe: A superstorm for global warming research*. Spiegel Online, April 4, 2010 (viewed April 6, 2010) at: http://www.spiegel.de/international/world/0,1518,druck-686697,00.html

Fagan, B. M. (2000). *The little ice age. How climate made history 1300–1850*. New York: Basic Books.

Fagan, B. M. (2008). *The great warming. Climate change and the rise and fall of civilizations*. New York: Bloomsbury Press.

Fisher, A., Walker, M., & Powell, P. (2003). *DDT and DDE: Sources of exposure and how to avoid them* (NAES #52031334, 6 pp). University of Nevada, Cooperative Extension. Reno, NV. Retrieved from http://www.unce.unr.edu/publications/ files/nr/2003/SP0316.pdf

Gething, P. W., Smith, D. L., Patil, A. P., Tatem, A. J., Snow, R. W., & Hay, S. I. (2010, May 20). Climate change and global malaria recession. *Nature, 465*(7296), 342–345. doi: 10.1038/nature09098

Goklany, I. M. (2007, November). *Death and death rates due to extreme weather events. Global and U.S. trends, 1900–2006* (pp. 15). London: International Policy Network.

Goklany, I. M. (2008, February 5). *What to do about climate change* (Policy Analysis Paper No. 609, pp. 27). Washington, DC: CATO Institute.

Goklany, I. M. (2009). Is climate change the 'defining challenge of our age'? *Energy & Environment, 20*(3), 279–302.

Gray, V. R. (2009, September 2). *South Pacific Sea Level: A Reassessment* (Science and Public Policy Institute (SPPI) Report, 18pp). Wellington, New Zealand.

Greene, W. H. (2008). *Econometric analysis* (6th ed.). Upper Saddle River: Pearson-Prentice Hall.

Hamilton, J. M., & Tol, R. S. J. (2007). The impact of climate change on tourism in Germany, the UK and Ireland: A simulation study. *Regional Environmental Change, 7*(3), 161–172.

Horowitz, J. K. (2001, February 14). *The income-temperature relationship in a cross-section of countries and its implications for global warming* (University of Maryland Working Paper No. 01–02). Retrieved from http://ssrn.com/abstract=260990

Horowitz, J. K., & McConnell, K. E. (2002). A review of WTA/WTP studies. *Journal of Environmental Economics and Management, 44*(3), 426–447.
Howitt, R. E. (1995). Positive mathematical programming. *American Journal of Agricultural Economics, 77*(2), 329–342.
Howitt, R. E. (2005). *Agricultural and environmental policy models: Calibration, estimation and optimization*. Manuscript (209 pp). Agricultural & Resource Economics. Davis: University of California at Davis.
Idso, C., & Singer, S. F. (2009). *Climate change reconsidered: 2009 Report of the Nongovernmental International Panel on Climate Change (NIPCC)*. Chicago: The Heartland Institute.
IPCC WGI. (2001). *Climate change 2001. The Scientific basis. Working Group I contribution to the third assessment report of the Intergovernmental Panel on Climate Change*. Cambridge: Cambridge University Press.
IPCC WGI. (2007). *Climate change 2007. The physical science basis. Working Group I contribution to the fourth assessment report of the Intergovernmental Panel on Climate Change*. Cambridge: Cambridge University Press.
IPCC WGII. (2007). *Climate Change 2007. Impacts, adaptation and vulnerability. Working Group II contribution to the fourth assessment report of the Intergovernmental Panel on Climate Change*. Cambridge: Cambridge University Press.
Kaiser, H. M., Riha, S. J., Wilks, D. S., Rossiter, D. G., & Sampath, R. (1993). Farm-level analysis of economic and agronomic impacts of gradual climate warming. *American Journal of Agricultural Economics, 75*(2), 387–398.
Kaiser, H. M., Riha, S. J., Wilks, D. S., & Sampath, R. (1993). Adaptation of global climate change at the farm level. In H. M. Kaiser & T. E. Drennen (Eds.), *Agricultural dimensions of global climate change* (pp. 136–152). Delray Beach: St. Lucie Press.
Kaya, Y., & Yokobori, K. (Eds.). (1997). *Environment, energy and economy: Strategies for sustainability*. Tokyo: United Nations University Press.
Kennedy, P. W. (1999). Learning about environmental damage: Implications for emissions trading. *Canadian Journal of Economics, 32*(5), 1313–1327.
Kennedy, P. W. (2002). Optimal early action on greenhouse has emissions. *Canadian Journal of Economics, 35*(1), 16–35.
Leach, A. J. (2007). The climate change learning curve. *Journal of Economic Dynamics and Control, 31*(5), 1728–1752.
Levitt, S. D., & Dubner, S. J. (2009). *Super freakonomics: Global cooling, patriotic prostitutes, and why suicide bombers should buy life insurance*. New York: Harper Collins.
Lise, W., & Tol, R. S. J. (2002). Impact of climate change on tourist demand. *Climatic Change, 55*(4), 429–449. doi:10.1023/A:1020728021446.
Lomborg, B. (Ed.). (2004). *Global crises, global solutions*. Cambridge: Cambridge University Press.
Lomborg, B. (2007a). *Cool it. The skeptical environmentalist's guide to global warming*. New York: Knopf.
Lomborg, B. (Ed.). (2007b). *Solutions for the world's biggest problems: Costs and benefits*. Cambridge: Cambridge University Press.
Lomborg, B. (Ed.). (2010). *Smart solutions to climate change. Comparing costs and benefits*. Cambridge: Cambridge University Press.
Maddison, D., & Bigano, A. (2003). The amenity value of Italian culture. *Journal of Environmental Economics and Management, 45*(2), 319–332.
Mann, C. C. (2005). *1491 New revelations of the Americas before Columbus*. New York: Random House.
Marvin, P. H. (1964). Birds on the rise. *Bulletin of the Entomological Society of America, 10*(3), 194–196.
Matthews, H. D., & Keith, D. W. (2007). Carbon-cycle feedbacks increase the likelihood of a warmer future. *Geophysical Research Letters, 34*, L09702. doi:10.1029/2006GL028685.
McCarl, B. A., Meeraus, A., van der Eijk, P., Bussieck, M., Dirkse, S., & Steacy, P. (2007). *McCarl GAMS user guide. Version 22.5*. College Station: Department of Agricultural Economics, Texas A&M University.

McKitrick, R. (2011). A simple state-contingent pricing rule for complex intertemporal externalities. *Energy Economics, 33*(1), 111–120.

McMullen, C. P., & Jabbour, J. (Eds.). (2009). *Climate change 2009. Science compendium*. Nairobi: United Nations Environment Programme.

Mendelsohn, R. (2006). A critique of the Stern report. *Regulation, 29*(4), 42–46.

Mendelsohn, R., Nordhaus, W. D., & Shaw, D. (1994). The impact of global warming on agriculture: A Ricardian approach. *American Economic Review, 84*(4), 753–771.

Mendelsohn, R., Morrison, W., Schlesinger, M. E., & Andronova, N. G. (2000). Country-specific market impacts of climate change. *Climatic Change, 45*(3–4), 553–569.

Menzies, G. (2002). *The year China discovered the world*. London: Bantam Books.

Min, S.-K., Zhang, X., Zwiers, F. W., & Hegerl, G. C. (2011, 17 February). Human contribution to more-intense precipitation extremes. *Nature, 470*(7334), 378–381. doi: 10.1038/nature09763

Mooney, S., & Arthur, L. M. (1990). Impacts of 2xCO2 on Manitoba agriculture. *Canadian Journal of Agricultural Economics, 38*(4), 685–694.

Moore, T. G. (1998). Health and amenity effects of global warming. *Economic Inquiry, 36*(3), 471–488.

Needham, J. (1954). *Science and civilization in China*. Cambridge: Cambridge University Press.

Nordhaus, W. D. (1991). To slow or not to slow: The economics of the greenhouse effect. *Economic Journal, 101*(407), 920–937.

Nordhaus, W. D. (1994). *Managing the global commons: The economics of climate change*. Cambridge, MA: MIT Press.

Nordhaus, W. D. (2007a). Accompanying notes and documentation on development of DICE-2007 model: Notes on DICE-2007.delta.v8 as of September 21, 2007.

Nordhaus, W. D. (2007b). A review of the "Stern review on the economics of climate change". *Journal of Economic Literature, 45*(3), 686–702.

Nordhaus, W. D. (2008). *A question of balance. Weighing the options of global warming policies*. New Haven/London: Yale University Press.

Nordhaus, W. D., & Boyer, J. (2000). *Warming the world. Economic models of global warming*. Cambridge, MA: The MIT Press.

Pall, P., Aina, T., Stone, D. A., Stott, P. A., Nozawa, T., Hilberts, A. G. J., Lohmann, D., & Allen, M. R. (2011, 17 February). Anthropogenic greenhouse gas contribution to flood risk in England and Wales in autumn 2000. *Nature, 470*(7334), 382.

Peters, G. P., Minx, J. C., Weber, C. L., & Edenhofer, O. (2011). Growth in emission transfers via international trade from 1990 to 2008. *PNAS Early Edition*. Retrieved May 5, from http://www.pnas.org/content/early/2011/04/19/1006388108

Pielke, R. A., Jr. (2009). The British climate change act: A critical evaluation and proposed alternative approach. *Environmental Research Letters, 4*(2), 024010. doi:10.1088/1748-9326/4/2/024010.

Pielke, R. A., Jr. (2010). *The climate fix: What scientists and politicians won't tell you about global warming*. New York: Basic Books.

Prins, G., Galiana, I., Green, C., Grundmann, R., Hulme, M., Korhola, A., Laird, F., Nordhaus, T., Pielke, R. A. Jr., Rayner, S., Sarewitz, D., Shellenberger, M., Stehr, N., & Tezuka, H. (2010, May). *The Hartwell paper: A new direction for climate policy after the crash of 2009* (pp. 42). London: Institute for Science, Innovation & Society, University of Oxford; LSE Mackinder Programme, London School of Economics and Political Science.

Rehdanz, K., & Maddsion, D. (2009). The amenity value of climate to households in Germany. *Oxford Economic Papers, 61*(1), 150–167.

Reichel, W. L., Cromartie, E., Lamont, T. G., Mulhern, B. M., & Prouty, R. M. (1969). Pesticide residues in eagles. *Pesticides Monitoring Journal, 3*(3), 142–144.

Reinsborough, M. J. (2003). A Ricardian model of climate change in Canada. *Canadian Journal of Economics, 36*(1), 21–40.

Scheffer, M., Brovkin, V., & Cox, P. M. (2006). Positive feedback between global warming and atmospheric CO2 concentration inferred from past climate change. *Geophysical Research Letters, 33*, L10702. doi:10.1029/2005GL025044.

Schimmelpfenning, D., Lewandrowski, J., Reilly, J., Tsigas, M., & Parry, I. (1996). *Agricultural adaptation to climate change: Issues of long-run sustainability* (Agric. Econ. Report 740, pp. 57). Washington, DC: USDA Economic Research Service.

Schlenker, W., Hanemann, W. H., & Fisher, A. C. (2006). The impact of global warming on U.S. agriculture: An econometric analysis of optimal growing conditions. *The Review of Economics and Statistics, 88*(1), 113–125.

Siddall, M., Stocker, T. F., & Clark, P. U. (2009). Constraints on future sea-level rise from past sea-level change. *Nature Geoscience, 2*(8), 571–575.

Siddall, M., Stocker, T. F., & Clark, P. U. (2010). Retraction: Constraints on future sea-level rise from past sea-level change. *Nature Geoscience, 3*(2), 217. doi:10:1038/ngeo780.

Sohngen, B., Mendelsohn, R., & Sedjo, R. (1999). Forest management, conservation and global timber markets. *American Journal of Agricultural Economics, 81*(1), 1–13.

Spielman, A., & D'Antonio, M. (2001). *Mosquito. A natural history of our most persistent and deadliest foe*. London: Faber and Faber.

Stern, N. (2007). *The economics of climate change: The Stern review*. Cambridge: Cambridge University Press.

Sussman, B. (2010). *Climategate. A veteran meteorologist exposes the global warming scam*. New York: WND Books.

Tol, R. S. J. (2002). Estimates of damage costs of climate change. *Environmental and Resource Economics, 21*(1), 47–73.

Tol, R. S. J. (2005). The marginal damage costs of carbon dioxide emissions: An assessment of the uncertainties. *Energy Policy, 33*(16), 2064–2074.

Tol, R. S. J. (2006). The Stern review of the economics of climate change: A comment. *Energy & Environment, 17*(6), 977–981.

Torn, M. S., & Harte, J. (2006). Missing feedbacks, asymmetric uncertainties, and the underestimation of future warming. *Geophysical Research Letters, 33*, L10703. doi:10.1029/2005GL025540.

Turner, J., Comiso, J. C., Marshall, G. J., Lachlan-Cope, T. A., Bracegirdle, T., Maksym, T., Meredith, M. P., Wang, Z., & Orr, A. (2009). Non-annular atmospheric circulation change induced by stratospheric ozone depletion and its role in the recent increase of Antarctic Sea ice extent. *Geophysical Research Letters, 36*, L08502. doi:10.1029/2009GL037524.

van Kooten, G. C. (2004). *Climate change economics. Why international accords fail*. Cheltenham/Northampton: Edward Elgar.

van Kooten, G. C., & Bulte, E. H. (2000). *The economics of nature: Managing biological assets*. Oxford: Blackwell Publishers.

Vinje, T., & Kvambekk, Å. S. (1991). Barents Sea drift ice characteristics. *Polar Research, 10*(1), 59–68.

Wanliss, J. (2010). *Green dragon: Dominion not death*. Burke: Cornwall Alliance.

Weber, M., & Hauer, G. (2003). A regional analysis of climate change impacts on Canadian agriculture. *Canadian Public Policy – Analyse de Politiques, 29*(2), 163–180.

Weintraub, A., Romero, C., Bjorndal, T., Epstein, R., & Miranda, J. (Eds.). (2007). *Handbook of operations research in natural resources*. New York: Springer.

Weitzman, M. L. (2007). A review of the Stern review on the economics of climate change. *Journal of Economic Literature, 45*(3), 703–724.

Weitzman, M. L. (2009a). On modeling and interpreting the economics of catastrophic climate change. *Review of Economic and Statistics, 91*(1), 1–19.

Weitzman, M. L. (2009b, May 17). *Reactions to the Nordhaus critique* (18 pp). Retrieved July 24, 2009, from http://www.economics.harvard.edu/faculty/ weitzman/papers_weitzman

Weitzman, M. L. (2009c, February 19). *Some basic economics of extreme climate change* (29 pp). Retrieved July 24, 2009, from http://www.economics.harvard.edu/faculty/ weitzman/papers_weitzman

White, J. W. C., Alley, R. B., Brigham-Grette, J., Fitzpatrick, J. J., Jennings, A. E., Johnsen, S. J., Miller, G. H., Nerem, R. S., & Polyak, L. (2010). Past rates of climate change in the Arctic. *Quaternary Science Reviews, 29*(15–16), 1716–1727.

Wood, K. R., & Overland, J. E. (2010). Early 20th century arctic warming in retrospect. *International Journal of Climatology, 30*(9), 1269–1279.

Xie, B., Zhang, Q., & Wang, Y. (2010). Observed characteristics of hail size in four regions in China during 1980–2005. *Journal of Climate, 23*(18), 4973–4982.

Yohe, G. (2007). Climate change. In B. Lomborg (Ed.), *Solutions for the world's biggest problems. Costs and benefits*. Cambridge: Cambridge University Press.

Chapter 8
Implementing Policy

I do what every sensible person does, when someone convinces me I am wrong, I change my mind. What do you do sir?

– John Maynard Keynes to a U.S. Senator during Senate hearings.

Concern about global warming led the World Meteorological Organization and the United Nations Environment Program jointly to establish the Intergovernmental Panel on Climate Change (IPCC) in 1988. Its first assessment report came out in 1990 and attributed recent increases in average global temperatures to a buildup of greenhouse gases in the atmosphere, suggesting that this was principally due to human activities and especially the burning of fossil fuels. This led 174 nations to the sign the UN Framework Convention on Climate Change (UNFCCC) at the so-called 'Earth Summit' in Rio de Janeiro in June 1992. The UNFCCC committed signatories to stabilize atmospheric concentrations of CO_2 and other greenhouse gases. At the second Conference of the Parties (COP) to the UNFCCC, which was held in Paris, nations endorsed the IPCC's second assessment report. Then, at COP3 held at Kyoto, Japan, in December 1997, industrialized nations agreed to reduce by 2008–2012 their collective emissions of CO_2 and equivalent greenhouse gases, together known as CO_{2e} (or just CO_2), to an average of 5.2% below what they were in 1990. Subsequent meetings of the parties hammered out how nations might meet their targets, with a focus on the types of activities or offsets that could be used in lieu of emissions reductions (as it became clear that nations would have difficulty simply reducing emissions from fossil fuel burning). At COP7 held at Marrakech in Morocco, a final agreement was reached concerning the particular offsets that would be eligible (see Chap. 9), and the IPCC's third assessment report was endorsed. Subsequent meetings of the parties have since sought to find an agreement to limit future emissions of CO_2.

Suppose therefore that the desire is to control global emissions of greenhouse gases, most particularly CO_2. That is, despite the uncertainty of the science suppose that, for whatever reason, politicians have decided to do something to control CO_2

G.C. van Kooten, *Climate Change, Climate Science and Economics: Prospects for an Alternative Energy Future*, DOI 10.1007/978-94-007-4988-7_8,

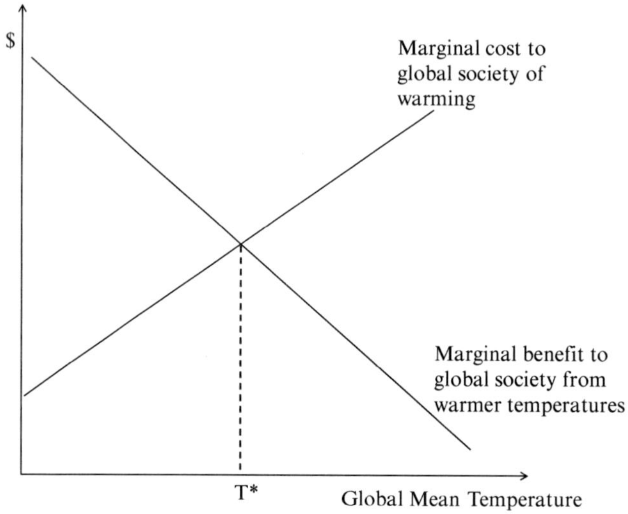

Fig. 8.1 Determining an optimal global mean temperature

emissions. Fine, but to what extent? That is the first question that needs to be answered. If one truly believes that by controlling the amount of CO_2 in the atmosphere one also controls global climate, then it is necessary to ask: "What is the optimal concentration of CO_2 in the atmosphere?" For the true believer, this amounts to asking: "What is the optimal global mean temperature?" The correct answer is illustrated with the aid of Fig. 8.1. The optimal global mean temperature is given by T*, where the marginal cost to global society of further temperature increase exactly equals the marginal benefit to society from warmer temperatures. The temperature in question might be higher or lower than the current global mean temperature.

The problem is that, as discussed in Chaps. 2 and 3, we do not know what the global mean temperature is, or how to measure it. Further, even if we could accurately measure global mean temperatures, we have no clue as to what the marginal costs and marginal benefits at various levels of the global mean temperature might be, let alone what T* should be. Finally, we are not even sure how effective any policy to reduce CO_2 might be in terms of its impact on global climate. As argued in Chap. 5, there are alternative explanations for climate change, while the current explanation that it is caused by human emissions of greenhouse gases remains in dispute. Given that everything about the science of climate change remains uncertain, policy cannot hope to attain T* or any other global mean temperature for that matter. Policymakers are stabbing in the dark. All they can do is set targets for reducing CO_2 emissions and then design policy instruments that hopefully achieve those targets. And, given that climate change is an issue affecting the global commons, persuade other countries to adopt similar targets.

The 1997 Kyoto Protocol, for instance, called for industrial countries to reduce their collective emissions by some 5.2% below what they were in 1990 by 2008–2012 (van Kooten 2004). This target has not been met and, even if it had been, would have made no perceptible change in the future climate, reducing the increase in projected average global temperature expected in 2100 by only 0.08°C (van Kooten 2004, p. 53).

In terms of climate change, policymakers appear to be continuing along the Kyoto path – setting domestic emission reduction targets and cajoling other countries to follow along. Once CO_2-emission reduction targets have been determined, there are essentially three economic instruments available to attain those targets – regulation, taxes/subsidies or cap-and-trade. How these work is discussed in Sect. 8.2. In Sect. 8.3, we examine policies that have been put forward or already implemented, focusing in particular on the costs of these policies. Some of this discussion is left to Chaps. 10 and 11, where we look at energy policies. We begin, however, with a discussion of the role of government.

8.1 Government and Finance

8.1.1 Market Failure and the Role of Government

As we saw in Chap. 6, market failure occurs whenever the marginal social benefit (given by market price) exceeds marginal social cost (marginal cost of provision). If social benefits exceed social costs at the margin, this implies that more of the good, service or amenity ought to be provided. This imbalance occurs with persistent monopoly or when the private sector is unwilling to provide enough of the amenity, as is the case with wildlife habitat, police and fire protection, national defense, and so on. In the latter case, the public authority (government) can subsidize provision of the good, service or amenity by the private sector or provide it publicly. The choice depends on the ability to monitor economic agents, whether these are working in the private sector or as public providers. If quality is easy to measure (so contracting and monitoring are straightforward), then private sector provision might be preferred; if quality is difficult to measure and monitor, public provision is preferable. Public goods are a special case: National defense, police protection and wildlife protection constitute examples of public goods, because no private agent has the incentive to provide them but, once provided, no one can be excluded.

Another form of market failure occurs when an economic agent fails to account for the good or bad effects of her decisions on others – a case referred to as an externality or spillover. Nitrogen pollution of waterways as a result of agricultural operations constitutes an externality – the farmer's activities impose a cost on downstream water users.

Governments are needed to correct market failure through charges (e.g., carbon taxes), subsidies (as noted above), regulation (of pollution, say), or even public provision. However, not all externalities need to be 'fixed' and there is a limit as to how much national defense, income transfer to the poor, health care benefits, wildlife protection and so on that can be publicly provided. It depends on the costs and benefits to society; it might simply be too expensive to prevent all pollution, provide everyone with unpriced health care, et cetera. After all, government revenues must come from the productive sector of the economy and, even leaving aside the question of compensation for takings, there is a limit as to how much this sector can contribute – there is a limit to the available rents that can be captured. Governments must take care not to remove the incentives required to produce the surpluses that can fund programs.

8.1.2 Models of Government

Too often it is assumed that governments act only in the best interests of society. This notion has been referred to as the 'helping hand' model of government (Shleifer and Vishny 1998): The main purpose of government is to produce public goods and correct market failures. However, 'policy failure' is a frequent outcome. Policy or government failure occurs when state intervention to produce a public good, or correct market failure, causes an even greater misallocation of resources.

Policy failure is also a central theme of a second model of government, the 'grabbing hand' model (Shleifer and Vishny 1998). Governments consist of individuals who are not always interested in the wellbeing of society but, rather, act in their own self interests, pursuing their own agendas rather than the common good. Governments consist of bureaucracies that take on a life of their own, with individuals inside these state agencies working to protect their own turf as much or more so than seeking the welfare of the citizens who pay their wages.

A special case of the grabbing hand is corruption. Corruption is said to occur when government officials (elected or otherwise) do a favor in exchange for votes or acquiescence, and/or sell a government service or commodity (e.g., passport, work or building permit, etc.) in return for a 'bribe.' A bribe consists not only of an under-the-table payment of money, but might take various subtle forms, including gifts, dinner, implicit acceptance of abuse, and so on. The flip side of corruption is lobbying by private agents to get government to do something that creates rents and/or distributes rents in the direction of the lobbyist. Rent seeking occurs, for example, when the U.S. Coalition for Fair Lumber Imports in the United States lobbies to put a duty on imports of softwood lumber from Canada. Other examples include lobbying to circumvent zoning regulations by obtaining a variance, and financial institutions arguing in favor of a CO_2 cap-and-trade system (carbon markets) that sees them earning billions of dollars in transaction fees. But lobbying can also come from inside government bureaucracies, where civil servants are protecting or expanding their own turf or power.

An alternative to the helping hand and grabbing hand models of government is the 'invisible hand' model. This paradigm argues for a reduced role of the state. Market failures are considered to be small, and certainly much less of a problem than the problem of policy failure. By minimizing the role of the state many problems will resolve themselves, or so it is thought. The Nobel laureate Ronald Coase's famous insights about externality were less directed at the role of government in correcting them as to the role of government in defining property rights and then relying on the courts to resolve remaining externality.

Nonetheless, whether or not there should be a minimal role for the state is a moot point. It is simply unrealistic to think that the role of government in modern societies can be rolled back. State intervention in the economy is here to stay, although its limits have increasingly been recognized (Fukuyama 1992; Hart et al. 1997; La Porta et al. 1999; Landes 1998). And one of the greatest casualties of larger government has been individual freedom.

Despite differing models of government, the helping hand model is the one that dominates and it forms the basis of cost-benefit analysis – a tool instrumental to policymaking that can also serve as a check on state intervention but is rarely used in that capacity.

8.1.3 Takings

The Fifth Amendment to the U.S. Constitution (November 1791) states: "… nor shall private property be taken for public use, without just compensation." This amendment is frequently called the 'takings clause.' Takings occur, for example, when governments expropriate property to make way for a new road. Such takings are known as a *titular taking* – literally a taking of title to the property – and are accepted as long as the owner is provided with fair market value, or *direct compensation*. Taxes to pay for armed forces or police protection provide compensation in the form of security, while taxes used to build roads, sewers and so on similarly provide indirect benefits. Taxing the better off to provide for the less fortunate also provides *in-kind compensation* in the form of social stability (and satisfying altruistic motives). Social stability or even a social sense of fairness might justify universal medical coverage.

While a social safety net provides in-kind compensation to those who pay the bill, there is no in-kind compensation when social programs encourage abuse, the military or police are large and pose a threat to innocent citizens of one's own or another country, constitute an income transfer from poor to rich (or even rich to rich), or impose one group's idea of what is best for society upon others with an alternative view of what is best. While there are many ingenious arguments for taking things from individuals, political philosophers question whether or not many government actions justified under the takings clause are indeed constitutional (Epstein 1985).

As an example of takings resulting from environmental activism, suppose you had purchased a beachfront property with the hopes of building a house at some future date. Houses are built on the lots on either side of the property, but, after your purchase, the government passes a law preventing further development to protect some endangered species. The new law constitutes a *regulatory taking*. Is such a taking fair? Under the U.S. Constitution, such a taking might require compensation depending upon the circumstances, although courts have been slow in recognizing these forms of compensable takings for two reasons. First, the bureaucracy's ability to fight lengthy legal cases is better than that of citizens, with litigants sometimes passing away before cases are concluded. Second, definitions of 'property' and 'compensation' are not always clear.

In an actual case, the state of South Carolina was ordered by the U.S. Supreme Court in 1992 to buy Mr. David Lucas' beachfront lot at market price if it wanted to prevent anyone building on the property. After gaining title to the property, however, South Carolina sold the lot to a developer. There are few citizens who would argue against providing compensation in the aforementioned case because they can envisage it happening to them. It is personal. It might be an entirely different matter if the beachfront lot had been owned by a large corporation, but, then, such a corporation would also be in a better position to litigate in the courts for compensation.

In Canada, the concept of private property is similar to that in the United States, but private property is not explicitly protected in the Constitution (although constitutional proposals during 1992 included a clause pertaining to private property). *Expropriation* of private property (*condemnation* of property for public purpose) is permitted with or without compensation, and such laws vary from one province to another. Each province has its own legislation concerning compensation in the case of government *expropriation* of private property rights, but the general principle of compensation for takings is well known in Canada. However, rights with respect to regulatory takings are not as clear, as argued, for example, by Richard Schwindt and Steve Globerman (Schwindt 1992; Schwindt and Globerman 1996). Further, the Peace, Order and Good Government provision of Canada's Constitution can be used by the federal government to *take* private property from individuals without compensation (van Kooten and Arthur 1997; van Kooten and Scott 1995). Although not explicitly referred to as takings, other democratic countries have some provision in constitutional law to prevent the government from taking property from citizens without compensation.

8.1.4 Governance

Regardless of the model of government that one might favor, governance is of crucial importance. Economics is the science of allocating scarce or limited resources in a way that leads to the greatest wellbeing of society, or, in a dynamic sense, that leads to economic development which makes citizens better off in the future. Economists have developed cost-benefit analysis to decide how this can be done, but it is a sophisticated tool with many nuances. However, cost-benefit analysis and the making of economically efficient decisions cannot occur unless the proper governance structures are in place. In particular, a jurisdiction's institutional environment and the level of social capital are important drivers that ensure economic efficiency and economic progress.

The institutional environment consists of formal rules (constitutions, laws and property rights) and informal rules (sanctions, taboos, customs, traditions, and norms or codes of conduct) that structure political, economic and social interactions. Informal constraints are commonly referred to as social capital, which the Nobel laureate Elinor Ostrom defines as "the shared knowledge, understandings, norms, rules, and expectations about patterns of interactions that groups of individuals bring to a recurrent activity" (Ostrom 2000, p. 176). Trust is perhaps the most important element of social capital, and it affects the costs of transacting: If one's confidence in an enforcement agency falters, one does not trust people to fulfill their agreements and agreements are not entered into. There is an element of trust in any transaction where one has to decide (make a choice) before being able to observe the action of the other party to the transaction. One has to assume that the other person is not acting with guile, thus not keeping hidden information about themselves that can be used to their advantage at the expense of the other party to the transaction. Trust is the catalyst that makes an economy function efficiently.

While one line of economic research (located in the New Institutional Economics) deals with governance, another line of research pioneered by Oliver Williamson of Harvard University focuses on transaction costs. Transaction costs relate to the organization of a transaction – time and effort searching for a solution, brokerage fees, advertising costs and so on. Transaction costs increase the costs of government actions, for example, policies that address externality. Indeed, transaction costs could be sufficiently large that it is optimal not to try to correct an externality – transaction costs can prevent a policy from achieving its objective. That is, if the transaction costs exceed the benefits from correcting an externality or having government supply a good or service, it is in society's interest not to correct the perceived market failure or provide the good/service.

The true economic test of whether a public policy or project is worthwhile or economically efficient is this: if those who benefit from the policy are able to compensate the losers and still be better off, the program is worth undertaking. Efficient outcomes do not, in principle, require that compensation be paid, but not requiring gainers to compensate losers will, in practice, give them an incentive to overstate the true value of their gains. For example, environmentalists have no economic incentive to limit their demands because they have no requirement to compensate those harmed (e.g., the poor who must pay a carbon tax to heat their homes, a landowner who cannot develop land because of endangered species habitat). Governments may be tempted to pursue programs or adopt policies only because they are able to shift the burden of their implementation onto private individuals who have no power to mitigate the costs or prevent 'wipeouts.' If governments had to pay compensation in all circumstances, they would be more likely to avoid policies that impose large costs on ordinary citizen but few benefits. Therefore, such outcomes are likely to be more efficient, and efficient outcomes are desired because they utilize less of society's scarce resources.

8.1.5 Financing Government and Public Projects

Finally, we might consider how governments finance projects because, when all is said and done, financial constraints will affect the ability of governments to pursue all of the policies and projects that it would like to implement. In other words, financing constraints will limit what the government does and require it to make tradeoffs. This is the main lesson of Francis Fukuyama's (1992) book, *The End of History and the Last Man*: A balance is needed between the extent (size) of the public sector and the size of the private sector. One might add to this that a balance is needed between the primary sectors that produce wealth and all other sectors of the economy, whether the wealth generating sector is in private or public hands. (History has shown, of course, that leaving the wealth generating sector in the public domain leads to inefficiency as agents pursue their own goals rather than those of the public authority.)

The government can pay for its operations, including new projects and policies in a number of ways:

- out of an existing budget surplus (if there is one);
- borrowing on financial markets;
- higher charges for services provided by the public sector (e.g., higher transportation fees, fees for doctor services at hospitals, etc.) and fees for use of natural resources (e.g., collecting a higher proportion of resource rents);
- increased taxes on incomes, properties, consumption, businesses, et cetera; and/or
- printing money.

The latter option is open only to a federal government not in a currency union.

Any decision to raise funds is a political one – no matter how funds are raised, there will be a consequence. Unless there is compensation, any public expenditure involves an income transfer, and one consequence is that there will be people (or companies) that engage in rent seeking behavior to ensure that they are the beneficiaries of income transfers. Likewise, there will be opposition to whatever revenue generating mechanism is employed, whether higher charges (say, for transportation or health-care services) or taxes, or borrowing. Taxes and charges will be opposed because they increase the costs of those who must pay the tax or charge; they may also be opposed because people have an ideological predilection against charges or taxes. But there is also another problem with charges and taxes: there is no guarantee that revenues actually increase.

If corporate income and other taxes are raised, firms may leave for another jurisdiction where taxes or charges are lower. This could reduce overall economic activity, lowering tax revenues from all sources while increasing expenditures on social welfare and so on. Firms do have some, perhaps limited, ability to move to 'tax havens.' This led in the early part of the new millennium to a reduction not only in corporate but also income taxes in Europe as eastern European countries that recently joined the EU attempted to attract investment and highly skilled labor by reducing taxes and simplifying the tax system. Thus, it could turn out that the elasticity of revenue with respect to the tax or charge is such that revenue falls with increased taxes/charges. A number of years ago, when the province of Saskatchewan attempted to raise revenues via a large hike in the provincial sales tax, it found that sales tax revenue actually declined.

Corporate and personal income taxes, along with consumption taxes (provincial and federal sales taxes) remain the major revenue source for most governments. Governments must, however, balance these various tax sources to avoid a flight of investment and skilled labor, and added costs when the economy under-performs as a result. The same applies to charges for public services. Indeed, it may turn out that the costs of public provision of such things as transportation services (bus, ferry, train) or health services could be lowered by letting the private sector provide them. The private sector has a greater incentive to reduce costs, so only quality control is an issue (see above).

Finally, the government can borrow money to finance its programs, but this places a burden on future generations. Further, too much borrowing can increase the government's 'cost of borrowing' – the interest that it needs to pay on the funds it borrows for the project or program in question plus what it pays on all other outstanding debt. If new borrowing triggers a change in the government's credit rating (and even governments can default on loans and thus are a credit risk), the cost of the project might increase significantly as it must pay more on its total debt as that debt is renewed.

A government must be careful as to how it raises funds as there are tradeoffs. In the context of austerity measures needed to address high debts, *The Economist* (April 3, 2010, p. 76) points out that:

> Fiscal adjustments that rely on spending cuts are more sustainable and friendlier to growth than those that rely on tax hikes. Studies show that cutting public-sector wages and [income] transfers is better than cutting public investment. Many cuts, from raising pension ages to slashing farm subsidies, have a double benefit: they boost growth both by improving public finances and by encouraging people to work harder or promoting more efficient allocation of resources.

Raising taxes may harm growth, while some fiscal measures can address budget deficits while still promoting the growth required to get a country out of debt. However, if it is absolutely necessary to rely on taxes, the ones that do the least harm to growth are taxes on consumption and immobile assets, such as land. Green taxes may also make sense, but one has to be careful as these might also harm the poor more than is desired.

8.2 Economic Instruments to Reduce CO_2 Emissions

Four instruments are available to policymakers for reducing greenhouse gas emissions: (1) regulation, (2) a carbon tax, (3) 'cap and trade', and (4) subsidies (the flip side of a tax). Cap and trade requires the authority to set a cap on allowable emissions from large industrial emitters followed by trade to allocate emission permits (also referred to as 'allowances') in an optimal fashion. A weaker version of cap and trade allows emission reduction offsets to substitute for the allowances created by the cap. Offsets can come from biological sinks (e.g., carbon sequestered in trees) in the same jurisdiction, activities that reduce emissions in non-covered jurisdictions such as developing countries (e.g., investments that make burning coal at an existing power plant more efficient), et cetera. This is referred to as 'credit trading' (without reference to cap and trade). However, this weaker version of emissions trading is also more susceptible to corruption and the admittance of dubious, even illegal, credits for sale in legitimate markets.

Taxes and emissions trading (and to a lesser degree subsidies) are market instruments, while regulation is a form of command and control. While all instruments can have the desired effect of meeting a target level of emissions, regulation and cap and trade are often considered more effective than taxes because the authority may set the tax too low to achieve the target – there is no guarantee that the target emission reductions are reached. Economists generally distrust subsidies because of its political and distorting effects (as we will see in the case of biofuels). Economists also dislike regulation because it is less efficient than a price (tax) or quantity (cap and trade) instrument. Indeed, it is quite easy to demonstrate that the regulatory approach lacks incentives and is more costly than a market approach.

8.2.1 Regulation Versus Market Instruments

The problem with regulation is that, while in principle enabling a country to achieve an emissions reduction target, the intervention often leads to policy or government failure, and much higher costs than are incurred under a market instrument (tax or tradable permits). Policy failure is considered further in Sect. 8.3, while the proposition that regulation is more costly than a market instrument is demonstrated with the aid of Fig. 8.2.

Assume that there are two firms in society, *A* and *B*, each with a different marginal cost curve for reducing CO_2 emissions, as indicated by MC_A and MC_B in Fig. 8.2. Firm *A* has lower emission reduction costs than firm *B*. Suppose that the CO_2-emissions reduction target is given by 0*E*, although there is no guarantee that 0*E* is somehow socially optimal or even desirable. The marginal social cost is determined as the horizontal sum of the individual marginal cost functions, and is denoted by MC_A+MC_B.

With regulation, the authority will generally choose to make each firm reduce emissions by the same amount, or by $0K=½\,0E$. The cost to society of this form of regulation is given by area 0*MGK* (the area depicting firm *A*'s costs) plus area 0*NJK* (firm *B*'s costs). If, on the other hand, firms had to pay a penalty (tax) of *P*, firm *A* would reduce emissions by an amount equal to 0*A* and firm *B* would reduce emissions by 0*B*. Notice that $0A+0B=0E$. The cost to society in this case would be area 0*MLA* (firm *A*) plus 0*NIB* (firm *B*), which is identical to area 0*MTZE* – the area under the social marginal cost curve, MC_A+MC_B. This cost is lower than that

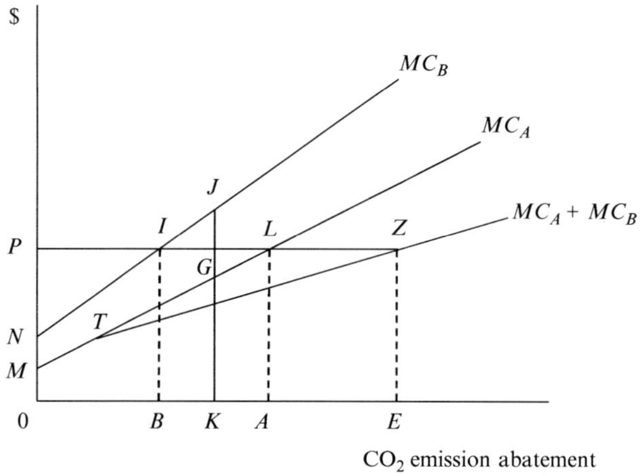

Fig. 8.2 Market incentives versus regulation to reduce CO_2 emissions

incurred under regulation, where each firm must attain the same standard. The firm with the higher cost of reducing CO_2 emissions, firm *B*, would reduce emissions less than would the more efficient firm *A*. Thus, a carbon tax is economically efficient.

Suppose instead that a cap-and-trade scheme is employed. Each firm is granted the right to emit some amount of CO_2 (usually based on historical emissions), with the total amount of rights issued sufficient to ensure that emissions are reduced by amount 0*E* in Fig. 8.2. The emission rights (permits) can then be bought and sold. Assume both firms are given identical emission rights. A firm will wish to purchase emission permits if the cost of doing so is less than its marginal abatement cost, and is willing to sell permits as long as it receives at least the marginal abatement cost. Given the marginal costs in the figure, the possibility of emissions trading exists, with permits trading at price *P*. At that price, the high-abatement cost firm *B* would like to emit more CO_2 than allowed, while *A* is willing to reduce emissions by more than is required by selling excess emission rights to *B*. In effect, *P* represents the cost of emitting CO_2 – for firm *A* it is an opportunity cost because it can either choose to emit more CO_2 or sell the credits from emitting less; for firm *B*, it represents the least cost of pursuing activities that release CO_2. Just as firms had the option of paying the tax or reducing emissions, under a permit system firms have the option of paying for rights to release CO_2 or reducing emissions. In either case, firms choose the least cost option. Thus, a cap-and-trade system is also economically efficient because lower cost firms will reduce emissions, selling permits to firms with higher emission reduction costs.

Both a CO_2 tax and a cap-and-trade scheme for CO_2 emissions satisfy the equi-marginal criterion – that the marginal value of emissions is equalized across sources (Field and Olewiler 2002). This is clear from Fig. 8.2 where $MC_A = P = MC_B$, whether *P* is a tax or the price of an emissions permit.

Alternatively, consider Fig. 8.3. Permits are issued so that an amount CO_2^* is emitted. Given the marginal benefit from emitting CO_2, which is identical to the

Fig. 8.3 Controlling CO_2 emissions using a permit or tax

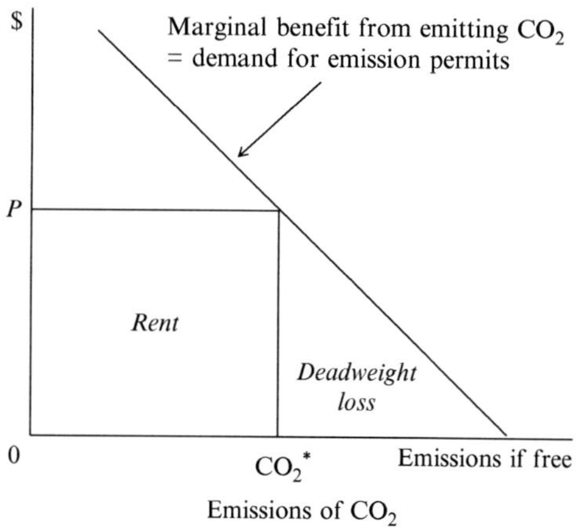

demand for emission permits, the value of an emission permit is *P*, which is determined by the intersection of demand and supply, where supply is simply a vertical line at CO_2^*. The price *P* in the figure is identical to that in Fig. 8.2.

Economists generally like economic incentives because firms have an incentive to adopt technical changes that lower the costs of reducing emissions, because they can then sell permits or avoid buying them, or avoid paying the tax. Further, market instruments provide incentives to change products, processes and so on, as marginal costs and benefits change over time. Because firms are always trying to avoid the tax, or avoid paying for emission rights, they tend to respond quickly to technological change.

8.2.2 Prices Versus Quantities

Auctioned permits and carbon taxes are thought to be identical – opposite sides of the same coin in the sense that auctioned permits target quantity while taxes target price. Consider again Fig. 8.3. The carbon tax determines the level of emissions; if the number of permits to be auctioned is the same as this level of emissions, the auction price should equal the tax. The state can choose the tax level (price) or the number of emission permits to auction (quantity), but if all is known the outcome will be the same – CO_2* emission permits trade at a price *P* (which would equal the tax).

When abatement costs and/or benefits are uncertain (i.e., the demand function in Fig. 8.3 is unknown), picking a carbon tax can lead to the 'wrong' level of emissions reduction, while choosing a quantity can result in a mistake about the forecasted price that firms will have to pay for auctioned permits. Such errors have social costs. If the marginal cost of abatement is steep while the demand curve for permits is

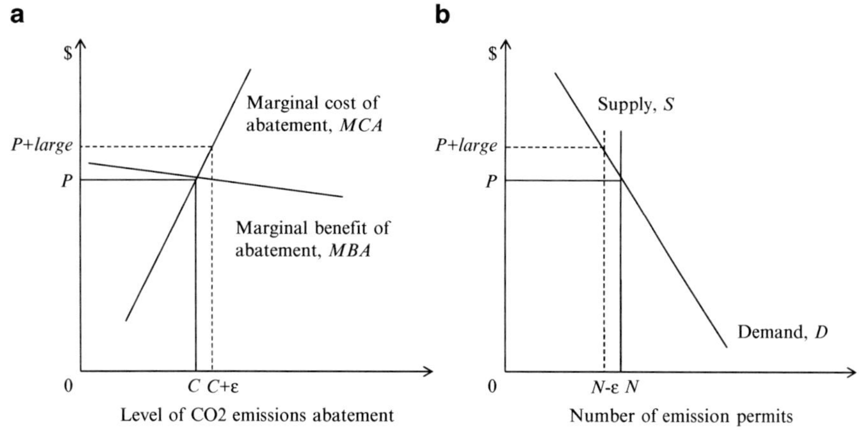

Fig. 8.4 Uncertainty about the costs and benefits of emissions abatement: determining optimal emission abatement and a market for emissions permits

relatively flat, then a small increase in the number of permits that are issued can have a large impact on their price (Pizer 1997; Weitzman 1974, 2002).

This is illustrated with the aid of Fig. 8.4. In panel (a), the level of CO_2-emissions abatement is shown on the horizontal axis. The marginal cost of abatement (*MCA*) is steep compared to the marginal benefits of abatement (*MBA*). Suppose that the optimal level of abatement is *C* but, lacking information on *MCA* and *MBA*, the authority sets emissions abatement slightly higher at $C+\varepsilon$. The result is a relatively large increase in marginal cost. The associated market for emission permits is provided in panel (b) of the figure. The steep *MCA* curve in panel (a) translates into a steep demand for permits in panel (b), while the authority sets the supply of emission permits as indicated by the vertical supply lines in panel (b). If the authority overshoots the target emission abatement by ε, it will issue too few emission permits – amount $N-\varepsilon$ if there is a one-to-one correspondence between emissions and permits. Despite tightening the number of permits by only a small amount ε, the price of permits will increase by a large amount.

If there is uncertainty about the marginal costs and benefits (damages avoided) of abating climate change, the choice of a price-based or quantity-based instrument will depend on which type of uncertainty is most prevalent. Since there appears little that can be done to halt climate change and since damages are likely to increase slowly over time, the choice of a price or quantity instrument will depend on the marginal abatement cost function.

It is clear that, the steeper the marginal abatement cost function, the more beneficial it will be, *ceteris paribus*, to use a price (tax) rather than quantity instrument. More specifically, if the marginal benefit curve is relatively flat compared to the marginal cost curve, a tax is preferred to a quota. Contrariwise, the flatter the marginal abatement cost function, the better it is to employ a quantity instrument (cap and trade).

The success of the grandfathered SO_2 trading scheme in reducing pollution among (primarily) electrical power producers in the U.S. Northeast is often touted as the approach that CO_2 abatement needs to follow. However, one cannot expect the same results with a CO_2 trading system because the marginal benefit curve from controlling CO_2 emissions is relatively flat compared to the associated marginal cost curve, in which case a tax and not quota should be used.

> The prospects for welfare gains through quotas or non-auctioned permits are much dimmer in the CO_2 context. Quotas or non-auctioned permits have a greater chance at yielding welfare gains in the SO_2 case because the central estimates for marginal environmental benefits are relatively high (compared with marginal abatement costs) in this case. In contrast, central estimates for marginal environmental benefits from CO_2 reductions are fairly low relative to marginal abatement cost (Parry et al. 1999, p. 55).

While the above arguments favor a tax over cap and trade in the context of climate change, there is a way in which emission trading can avoid the high costs of overshooting targets, as indicated in Fig. 8.3. This is done by enabling flexibility in the issuance of emission permits. If the authority sees that prices of permits are too high, it can release more permits into the market, effectively reducing the emissions abatement target. The problems with this approach are that environmental groups might object while emitters needing to purchase emission permits could delay purchases knowing that the authority will release more permits into the market. In essence, this approach is not as clear cut as a tax, and is more open to manipulation and rent seeking by especially large emitters.

8.2.3 *Income Re-distributional Effects*

Regardless of how emissions are curtailed, doing so creates a wedge between the marginal costs of providing emission permits (which is effectively zero) and the price at which they sell in the market. This wedge is a form of scarcity rent (see Chap. 6), with the total unearned rent equal to the restricted level of emissions multiplied by their price (Fig. 8.3). The rent represents the capitalized value of the right to emit CO_2, which had previously been free. With a tax or auction scheme, the government captures the rent, but, in the case of grandfathered emission rights, it is captured by extant emitters. Those lucky enough to receive tradable emission permits experience a windfall. As a result, governments will be subject to tremendous lobbying pressure in their decision regarding the allocation of permits. If international permit trading is implemented, countries that have done the most to reduce emissions in the past may lose relative to ones that made no similar efforts; firms that are high-energy users may benefit relative to those firms that invested in energy savings technology. This has been the case with the European trading system.

Yet, emission trading is often preferred by government because firms' total compliance costs are generally lower than with a tax as some or all emission permits are grandfathered, while economic efficiency benefits are the same as under a tax. This makes a cap-and-trade permit scheme politically acceptable to large industrial emitters, which are better able to lobby government to gain the rents at the expense of citizens more broadly. That is, only the distribution of income differs: The cost to firms under a tax is given by the entire rent area in Fig. 8.3, while, if permits are solely grandfathered, firms gain the entire rent area. A tax, on the other hand, results in a transfer of income from firms to taxpayers that is absent under cap and trade unless permits are auctioned.

A permit scheme also provides a government with flexibility in the way it assigns the burden of climate change mitigation. If permits are not auctioned, older less-efficient firms can be assigned a disproportionate share of the permits to enable them to better withstand the adverse impact of the measures. A greater share of the permits could also be assigned to the hardest hit regions of a country. The assignment of permits across firms and/or regions enables politicians to re-allocate income and thereby make climate change mitigation more palatable, but it can also lead to further misallocation of resources as firms spend resources seeking a favorable assignment of permits.

If carbon taxes are used to capture the unearned rent, then the pertinent question is: What becomes of the tax revenue? If permits were auctioned, one would likewise want to know what happens to the revenue. First off, both a tax and a tradable permit scheme increase the costs of consumption goods, thereby reducing the supply of labor and increasing the marginal cost of abatement. This 'tax-interaction' effect is the result of pre-existing taxes on labor, which are known to be distorting. The state can use tax revenue to offset the negative impact of restricting CO_2 emissions – the economic inefficiencies resulting from a misallocation of resources (the deadweight loss in Fig. 8.3) and the tax-interaction effect – by reducing distorting taxes (and associated economic inefficiencies or deadweight losses) elsewhere in the economy. So-called 'revenue recycling' has the potential to reduce substantially the costs of carbon abatement policies. This is known as the 'double-dividend' because the environmental improvement also leads to increased efficiency elsewhere in the economy. However, if the tax revenue is recycled in lump-sum fashion (e.g., every citizen is provided the same tax rebate) and not used to reduce distortionary taxes elsewhere in the economy, the carbon tax actually exacerbates inefficiencies raising the marginal cost of climate change abatement (see Bovenberg and Goulder 1996). As a corollary, a cap-and-trade scheme with grandfathered permits is more costly than a carbon tax or a cap-and-trade auction scheme because the government collects no revenues.

8.2.4 Permits Versus Credits: Cap and No Cap Trading

As alluded to in the introduction to this section, it is important to distinguish between cap-and-trade schemes and emissions trading schemes where there is no real cap on emissions. In Figs. 8.2, 8.3 and 8.4, the respective quantities E, CO_2^* and C represent actual targets or caps beyond which the emitters affected by the cap-and-trade scheme could not go – emissions are capped and the only way a firm could emit more is to purchase more allowances or permits. By demanding more permits, the price of permits would need to increase to clear the market, so that when one firm bought permits another sold them by incurring costs to reduce its emissions and thereby free up permits for sale.

Because no one knows how expensive it is to implement true cap and trade, governments have permitted purchase of carbon offsets. In essence, large emitters are required to curb their emissions but, in lieu of doing so, can purchase 'credits' from a wide variety of sources. Emission permits can be purchased from other industrial emitters that exceed their emission reduction targets, but they can also be purchased elsewhere. Most countries allow firms to purchase emission reduction credits created by activities that sequester carbon in the domestic forestry or agricultural sector (see Chap. 9), or similar credits earned in another country. Firms can also purchase emission permits in developing countries by investing in forest planting, efficiency-improvement, renewable energy and other projects under Kyoto's Clean Development Mechanism (CDM). While CDM credits generally need to be certified by the United Nations, there is no guarantee that these projects actually reduce emissions to the extent that is claimed and monitoring is spotty due to its high costs. The potential for corruption is great and some governments require little in the way of ensuring that credits thus earned are even credible.[1] However, the important point is that firms can meet their emission reduction targets while not needing to reduce emissions; indeed, a firm's CO_2 emissions could even increase.

It is clear that the objective of CDM is a worthy one: Given that climate change is a global problem, it does not matter where CO_2 emissions are reduced (or CO_2 is removed from the atmosphere). But it does matter that this be done at least cost. The CDM facilitates reducing emissions at least cost, as do forest sequestration projects. In theory, the idea is a good one. In practice, there are huge transaction costs related to measuring, monitoring and other factors, and there are thousands of schemes to sell CO_2-offset credits, the majority of which do not come close to addressing climate change mitigation (as discussed in Chap. 9).

Rather than allow firms to purchase questionable CO_2-offset credits in other countries, governments can implement domestic credit trading that focuses on particular sources, such as large industrial emitters, or provide subsidies for particular projects such as sink activities (e.g., tree planting). Thus, for example, British Columbia is seeking to increase electricity production from wood biomass using a variety of incentive mechanisms, including the potential to perhaps trade any emission reductions earned should a trading system be set up at some future date.

[1] A study by CDM Watch, an environmental NGO, found that the most lucrative means of creating certified emission reduction credits under the CDM is via the destruction of hydrofluorocarbon-23 (HFC-23), a potent greenhouse gas. Evidence indicates that chemical plants, located in China and India, inflate their production of HFC-23 simply to sell credits under CDM. Indeed, projects to destroy HFCs dominate CDM emission reduction projects. Chemical plants in developing countries benefit from lucrative payments, while large industrial emitters in Europe purchase offset credits at prices below those traded in the EU's carbon market. See "Firms abusing Kyoto carbon trading scheme: watchdog" (viewed June 18, 2010): http://uk.reuters.com/article/idUKTRE65C1FZ20100613

As with CDM projects, such schemes do reduce emissions relative to what they would otherwise be.

Credit trading schemes have generally performed poorly, however, because of high transaction costs related to oversight and uncertainty, as pointed out by Tietenberg et al. (2003) in their review of greenhouse gas emissions trading. Governments can of course set rules that determine which credits can and cannot be allowed. Although this has the potential to remove some of the corruption and rent-seeking behavior one sees in countries with poor-quality governance structures, there remain problems. In many jurisdictions, burning biomass for electricity is regarded as carbon neutral, so that credits are earned for the fossil fuel CO_2 emissions avoided. However, CO_2 is also released when biomass is burned, even more so on an energy-equivalent basis than with fossil fuels, while growing trees may take as long as 80 years to remove this CO_2 from the atmosphere. This is hardly a carbon neutral activity, but is made so by government fiat.

8.2.5 The Real World of Carbon Trading and Carbon Markets

Thus far there have only been two carbon markets of note in the real world. One was the voluntary Chicago Climate Exchange (CCX), which went bankrupt at the end of 2010. This was not unexpected at the time given that, for a long period prior to shutting down, carbon credits on the CCX traded at \$0.10/tCO_2 or less. The other is the European Union's Emissions Trading System (ETS) that is meant to help the EU achieve its carbon dioxide emissions reduction targets. Other initiatives, such as the Western Climate Initiative and a scheme involving northeastern U.S. states, have been proposed but have never been formalized. As noted below, the U.S. House of Representatives and Senate have also attempted to start emission trading systems, but none is yet in place.

The ETS grandfathers emissions to countries that, in turn, allocate them among their large emitters. When the ETS was first set up, the large industrial emitters signaled their emissions to their governments, which then requested an allocation of permits. Some (mainly northern) European countries requested an allocation of permits that closely reflected the actual CO_2 output of their large emitters. The large emitters then had to reduce their emissions or purchase permits on the ETS exchange from emitters whose permits exceeded their emissions. However, many countries had been allocated significantly more permits than their large emitters required to meet their emissions targets. Eventually these 'excess' permits found their way onto the market causing the price to drop and the trading system to collapse in 2007, as indicated in Fig. 8.5. The market then had to be re-set by reallocating permits, which was done in April 2008 (Fig. 8.5). An indication of the value of trades in the ETS market for the period 2005–2007 and then the period since early 2008 is indicated in the figure; in December 2011, permits traded at between some €12 (approximately \$16) and just under €7 (\$9)/tCO_2, down from a maximum of just under €29 (\$38)/tCO_2 in mid 2008.

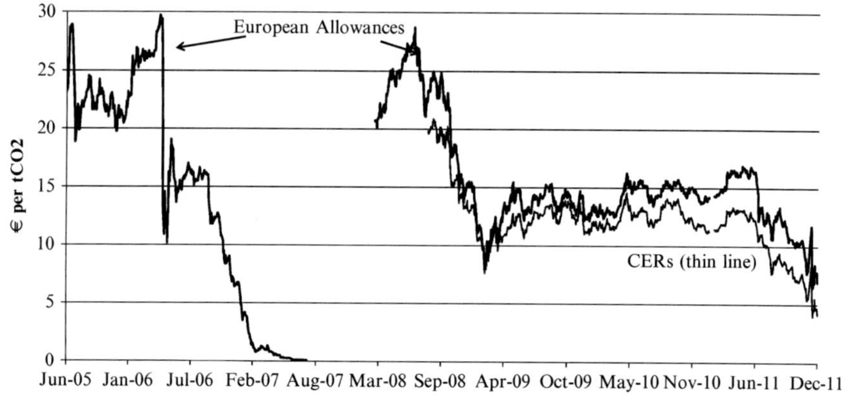

Fig. 8.5 Trading outcomes, European emissions trading system, European allowances (*thick line*) and certified emission reductions (*thin line*), June 2005–December 2011 (Source: http://www.bluenext.fr/statistics/downloads.html (viewed August 29, 2011)) Copyright ©: BlueNext

Currently, there are several other developments that are important to any discussion of real-world carbon markets and the ETS in particular. First, as a result of the 1997 Kyoto protocol, certified emission reductions (CERs) can be created through investments in clean technology (e.g., an investment that enhances the efficiency of a coal-fired power plant thereby lowering CO_2 emissions) in developing countries through the Clean Development Mechanism (CDM), and in countries of the former Soviet Union through Joint Implementation (JI). CERs can be used in lieu of actual emission reduction permits. That is, European countries, or large emitters in the EU, can invest in projects that create CERs or purchase CERs on the European market as CERs can be substituted for actual emissions reductions, although there are some limits on the extent to which CERs can be used in lieu of EU allowances, which explains why their price is some three-quarters or less of that of allowances (see Fig. 8.5). Nonetheless, the use of CERs constitutes a means by which the cap-and-trade provision of ETS can be circumvented. The upshot is this: while carbon trading systems are supposedly characterized by a limit on emissions, in practice politicians will look for a relief valve that keeps the price of permits (allowances) sufficiently low so that the damage to the economy is kept in check.

A second development relates to forestry projects. As described in more detail in Chap. 9, tree planting projects that increase the amount of carbon stored in forest ecosystems can be used in lieu of CO_2 emissions reduction. Thus, an emitter can participate in a domestic tree planting project rather than purchase emission permits. But the emitter can also obtain CER credits from tree planting projects in developing countries under CDM, although the number of such projects has been limited to date because of the inherent difficulties associated with the process of certifying such projects (see Chap. 9).

An even more difficult subject relates to the avoidance of forest degradation. Since deforestation is a major contributor to global carbon dioxide emissions, activities that Reduce Emissions from Deforestation and forest Degradation

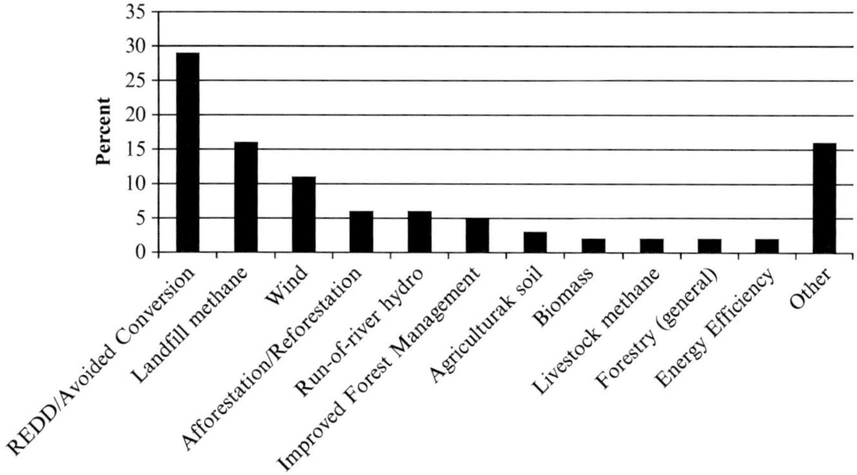

Fig. 8.6 Over the counter (OTC) sales of voluntary carbon offsets by origin, 2010

(REDD) are touted as another means for earning CER credits. The obvious questions concerning REDD are the following: How does one determine that a REDD project is additional – that the forest would not have been degraded in any event? And, if a forest is protected from logging, how do we know that this act does not lead to greater logging elsewhere in the same country, or in some other country – that a leakage occurs?

But the international community is willing to go even further. As a result of negotiations at COP16 in Cancun in December 2010, the narrow role of REDD has been expanded to include sustainable management of forests, forest conservation and the enhancement of forest carbon stocks, collectively known as REDD+. In this way, the UNFCCC and the UN Convention on Biological Diversity (CBD), both of which were signed at the 1992 Earth Summit in Rio de Janeiro, are interlinked. Increasingly, therefore, climate negotiators appear willing to accept REDD+activities as potential carbon offset credits to the extent that these activities also enhance biodiversity. It would appear, therefore, that REDD and REDD+projects in developing countries could well be used to create CERs that could then be sold in carbon markets, although this still needs approval under the CDM. If REDD+projects are accepted for international certification, what does that say about efforts to address global warming? After all, by accepting REDD+, climate negotiators have implicitly accepted that CO_2 and biodiversity targets are similar – that the climate change agenda is about environmental degradation and not climate change per se.

There is then a third development that is of import here. In recent years, there has been remarkable growth in voluntary carbon markets. A number of private companies have developed as certifiers of CERs and of voluntary emission reductions (VERs). In the voluntary market, REDD+ and other forestry activities play a large role, as indicated in Fig. 8.6. There are a number of REDD+standards, including the 'Gold Standard' (GS), the Climate, Community and Biodiversity Alliance's

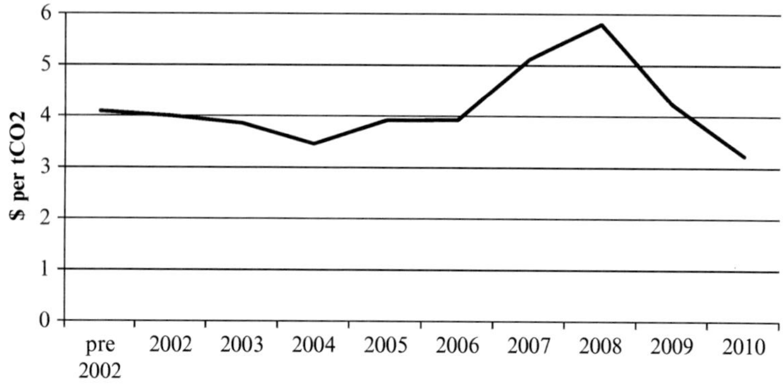

Fig. 8.7 Average prices of voluntary carbon offsets, pre-2002–2010

CCB certification, and the Verified Carbon Standard (VCS). Various sponsors grant the certifying agencies – these environmental nongovernmental organizations (NGOs) – their legitimacy. For example, core sponsors of the Gold Standard include the German Federal Ministry for the Environment, Nature Conservation and Nuclear Safety, WWF International (located in Zeist, The Netherlands), the European Climate Foundation (The Hague), and Merrill Lynch Commodities (Europe) Limited in London; the GS standard is endorsed by Renewable Energy and Energy Efficiency Partnership (REEP) in Vienna, MyClimate, and 'astmosfair' (Kreuzberg, Germany), among others.

The market for VERs amounted to $424 million in 2010, with trades averaging $3.24/t$CO_2$ in 2010, down from a high of $5.81/t$CO_2$ in 2008 (Fig. 8.7). Currently, the voluntary market is small compared to a total global carbon market estimated to be worth $142 billion in 2010. There is evidence, however, that VERs are sold not only in the voluntary but also in mandatory markets, most notably the EU's ETS (e.g., Peters-Stanley et al. 2011). Thus, while CER credits created by REDD+ activities are not currently available for sale in international markets, voluntary REDD+ credits may increasingly be marketed in global carbon markets. It would appear, therefore, that legal markets facilitate the laundering of VER credits, aided and abetted by environmental NGOs, government and financial intermediaries.

Voluntary trading markets are a concern. First, rent seeking on both sides of the market has created a vibrant market in emissions offsets that has little to do with the problem of global warming, but everything to do with the pursuit of (short-run) profits and objectives unrelated to climate change. Large industrial emitters, companies wishing to appear 'green,' and governments and their agencies wishing to demonstrate a commitment to climate change mitigation look to purchase emission offset credits at lowest cost (where cost might include the costs of meeting other objectives).[2] Sellers of emission reduction credits constitute various private companies and environmental NGOs that are willing to supply emission offset credits even if their legitimacy is questionable, because they can thereby earn large windfalls or funds to finance objectives that may be unrelated to climate change. Finally, there are the financial intermediaries that earn money from each transaction. Given that the global carbon market is projected to be in the range of \$1–2 trillion in the future, the potential revenue accruing to financial intermediaries, which earn a percentage on every transaction, is enormous. Grandfathering of allowances ensures the support of industry.

There is the notion that, by freely giving allowances to large emitters (e.g., power companies), there will be little immediate impact on output prices. This is misleading because allowances will have a market value. Allowances (or permits) have market value because they are traded, so a company will consider its 'freely-allocated' allowances to be an asset whose cost must be covered by revenues – they have an opportunity cost (Woerdman et al. 2009). It is similar to a person who owns a house but no longer has a mortgage payment. The homeowner cannot ignore the fact that the house has value, and that, despite having no outlay to live in the house, there are other options. The homeowner could sell the house, put the money in the bank and use the interest to rent another dwelling; the owner could choose to rent the house to another and use the proceeds to rent accommodation elsewhere (perhaps for less money). The upshot is that the prices of goods and services requiring inputs of energy will rise, whether a carbon tax or a permitting system (even one using grandfathered allowances) is employed.

The price distortion caused by the rent-seeking activities of economic agents is primarily driven by how the market is organized. The carbon market has become so complex that parties exploit the market wherever possible, and the scope of rent seeking is usually proportional to the degree of market complexity (Helm 2010). By supplying the voluntary market with questionable credits, such as those based on REDD+ activities, the price mechanism equilibrating the demand and supply of offset credits gets distorted, because non-emission reducing offset credits replace valid emissions reduction. Instead of dealing only with the sale and purchase of permits to emit CO_2, the market mechanism has to deal with emission reduction credits from sources that have nothing to do with CO_2 emissions from fossil fuel burning. REDD+credits derive from protection of biodiversity on private and public forestland and do not contribute explicitly to reductions in CO_2 emissions. By allowing these 'illegitimate' offsets, the carbon market gets distorted, with the price

[2] In British Columbia, for example, the activities of all government departments and agencies are required to be 'carbon neutral.' The agencies must purchase carbon offsets from a public corporation, Pacific Carbon Trust (PCT), for \$25 per tCO_2, but PCT purchases offset credits in voluntary markets. The costs imposed on hospitals, school boards, public universities, et cetera, are over and above the province's carbon tax. Except for the possibly large transaction costs and costs of purchasing VER credits, government grants to the public agencies come back as revenues. Without a proper accounting, it is not clear by how much CO_2 emissions are actually reduced and at what cost. Creators of VERs are likely the main beneficiaries, and not the environment.

of carbon below what it would otherwise be; it lowers incentives to conserve energy and invest in R&D that spurs clean energy innovation (Bosetti and Rose 2011). Thus, REDD+ credits and voluntary markets facilitate activities that enhance preservation of biodiversity enter the global carbon market without really contributing to a net carbon reduction.

Then there is the further problem of illegitimate and questionable VER credits entering legal markets. Consider the several adverse impacts when states allow the sale of REDD+ credits on the European market, for example. First, there is the distorting effect on prices. Because European firms can purchase lower cost emission permits, they can avoid other, more expensive efforts to do something about their carbon dioxide emissions. Second, although this situation enables states and private firms to meet their targets, it fails to address emission reduction obligations. Given that credits earned via carbon sequestration in terrestrial ecosystems were only meant to be a bridge to provide time for an economy or firm to develop and invest in emission-reducing technologies, the sale of such credits has turned out to be an impediment to the implementation of new technology (as carbon prices are lower than necessary), while creating a larger gap between actual emissions and emission targets in the future (van Kooten 2009) and doing little if anything to mitigate climate change.

8.2.6 *Mixing Market Instruments*

British Columbia already has carbon taxes in place, but it still intends to participate in carbon trading schemes. Europe has in place an emissions trading scheme for large industrial emitters, but it also uses carbon taxes. Simultaneous use of multiple instruments to target greenhouse gas emissions leads to inefficiency because one ends up with two prices for carbon, one determined by the tax and the other by emissions trading. This is illustrated with the aid of Fig. 8.8. The sector covered by emissions trading is depicted in panel (a), while the sector faced with a tax is depicted in panel (b). Assume that the marginal costs of using CO_2 in production are the same in the two sectors, while demand in the sector covered by emissions trading is denoted by D_T and demand in the sector that is only taxed is D_{NT}. In the absence of a tax, CO_2 output in the latter sector is C^E, while it is C^* if the tax is in place; likewise, in the trade sector, CO_2 output in the absence of emissions trading would be T^E, while with cap and trade it is T^*.

With a tax affecting only one sector of the economy, the tax revenue is given by the shaded area in panel (b). Now suppose an emissions cap is put into effect in the rest of the economy, and that emission permits are granted freely to firms on the basis of past emissions. Then, there will be a windfall to these firms equivalent to area *adeP* in panel (a), with the value of permits equal to *P–MC*. Rather than only retain the tax in one sector of the economy the government needs to impose the tax on both sectors. This would allow the authority to recover some of the windfall, namely, the amount given by area *abcd*, but a windfall equal to area *bceP* remains.

The problem with a carbon tax that affects only some sectors but not others is that it results in inefficiency. For example, a tax on natural gas consumption but not

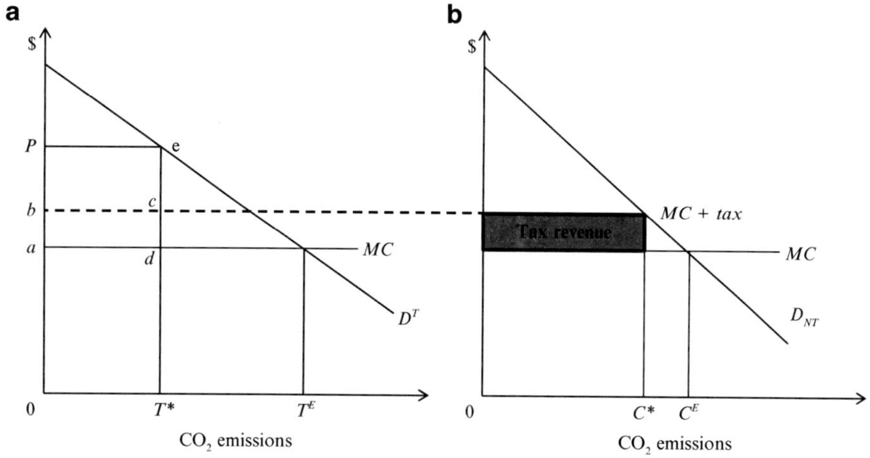

Fig. 8.8 Combining taxes and emissions trading. (a) Emissions trading sector, (b) non-trade sector

on electricity encourages greater use of electricity for space heating rather than the use of more efficient natural gas. Worse still, if natural gas is plentiful and as utilities convert more and more generating capacity to natural gas from coal, this distortion can result in a long-term lock-in as builders invest in electric as opposed to natural gas space heating.

Implementing emissions trading in conjunction with a tax does not eliminate the distortions, although it might reduce them somewhat. As is evident from Fig. 8.8, the cost of reducing CO_2 emissions is higher in the trade sector – equal to P in panel (a) – than in the no-trade sector, where in panel (b) it equals $MC+tax$. The equi-marginal criterion requires that marginal costs be the same in both sectors, which would entail shifting the burden of emissions reduction from the trade to no-trade sector.

8.2.7 Discussion

Governments appear to favor credit trading over taxes. A carbon tax is a straightforward instrument that can be adjusted to the severity of climate change damages, with revenues used to improve economic performance elsewhere in the economy (resulting in a 'double dividend') and used to fund research and development (R&D) for a 'put-a-man-on-the-moon' type of effort, as suggested by Martin Weitzman. Indeed, Ross McKitrick's (2011) optimal carbon tax rule [recall Eq. (7.7)] is likely the best approach for implementing a tax.

In McKitrick's state-contingent pricing rule (7.7), or the optimal tax rate rule, the proposed state variable to target for the climate externality is temperature. In Fig. 7.5, the historical path for an optimal tax was based on the HadCRUT3 temperature product. Annual average temperatures since 1850 and emissions since 1801 were used to construct the graph. Use of the annual HadCRUT3 temperature series might result in a tax that responds too slowly to runaway global warming. Thus, several years ago, McKitrick (2007) had proposed a tax rule based on actual temperatures in the tropical troposphere, which is where an early and strong signal of anthropogenic warming, which is not affected by solar activity, is predicted to occur. Climate models forecast temperatures in the tropical troposphere to respond more quickly than temperatures elsewhere (e.g., IPCC WGI 2007, Fig. 9.1). It would be simple to base the tax rate on temperature data from satellites measured at about 10–15 km above the earth in the region around the equator between 20°S latitude and 20°N latitude – the region predicted to be the weather harbinger for global warming.

According to McKitrick (2007), if the tax is set at 20 times the 3-year moving average of the mean of the tropical troposphere temperature anomalies, it would amount in 2005 to about US$4.70/t$CO_2$. If projections of global warming are correct, the tax would rise aggressively by between $2 and 9 per decade depending on the climate scenario, reaching some $200/t$CO_2$ by the end of this century; the market would certainly reward those who anticipated the large increase in temperatures (McKitrick 2011). As discussed in Sect. 7.2, if global warming is truly a dire threat, the rising tax will bring about the desired changes in anthropogenic emissions and/or the R&D needed to remove CO_2 from the atmosphere. The point is that the tax scheme proposed by McKitrick is unambiguous, it is not dependent on controversies surrounding temperature increases projected from climate models and economic analyses mired in similar assumptions, and can easily be adopted on a global scale.

To provide some indication regarding the effect of a carbon tax, one can calculate the extent to which consumers have to pay more for fuel using the carbon content of various fuels and the tax rate. An example is provided in Table 8.1. Not surprisingly, these indicate that electricity rates might be most impacted by carbon taxes, especially in regions that rely most on coal-fired power and use average as opposed to marginal costs for setting rates. Indeed, prices of coal will increase by between about 60% for a low-level carbon tax to more than seven fold for a carbon tax of $100/t$CO_2$. Oil prices are impacted to a lesser degree and many consumers will not really notice taxes of $10–30/t$CO_2$ because oil prices tend to fluctuate within these ranges. The same is true of natural gas prices, with gas prices for power generation impacted to a lesser extent. Consumers as citizens can be expected to complain about taxes (or carbon trading) if these reach towards $100/t$CO_2$.

Not all economists favor a tax, however, because they feel it would not guarantee adequate emissions reductions. Instead, they prefer quantitative controls and emissions trading, arguing that, if costs become too prohibitive (price of a permit to emit CO_2 becomes too high), the authority can always release more permits into the market. If carbon trading is the instrument of choice, the majority of economists and environmentalists prefer that the government auctions off permits, using the revenues to reduce income taxes and other distorting taxes. But economists are not wedded to the idea of auctions because, other than the revenue benefits to government and the potential for a 'double-dividend', the outcomes are the same whether

Table 8.1 The effect of carbon taxes on various fuels, United States, 2010

	Coal	Oil	Natural gas
CO_2 emissions[a]	2.735 tCO_2/t coal	0.427 tCO_2/barrel	1.925 $tCO_2/m^3 \times 10^3$
Average price[b]	$45.50/t coal	$70.69/barrel	$\$423.78/m^3 \times 10^3$
Carbon tax per unit of fuel			
$10/$tCO_2$	$27.35	$4.27	$19.25
$30/$tCO_2$	$82.05	$12.81	$57.75
$100/$tCO_2$	$273.50	$42.70	$192.50
% increase in price of fuel from carbon tax			
$10/$tCO_2$ (%)	60.1	6.0	4.5
$30/$tCO_2$ (%)	180.3	18.1	13.6
$100/$tCO_2$ (%)	601.1	60.4	45.4
Carbon tax as % of tax-adjusted fuel price			
$10/$tCO_2$ (%)	37.5	5.7	4.3
$30/$tCO_2$ (%)	64.3	15.3	12.0
$100/$tCO_2$ (%)	85.7	37.7	31.2

Notes: [a]Source: http://cdiac.ornl.gov/pns/convert.html (viewed May 26, 2010)
[b]Source: U.S. Energy Information Administration http://www.eia.doe.gov/ (viewed May 26, 2010). Coal price is average price for U.S. utilities in generating electricity in 2008; oil is the world price of crude at Texas gulf, late May 2010; and natural gas price is for U.S. residential customers, May 2010

permits are auctioned or simply given to existing emitters on the basis of past emissions (i.e., grandfathered), or allocated in some other fashion – emission targets are met at least cost to society. Large industrial emitters are certainly more amenable to a scheme that grandfathers permits as opposed to one that requires them to pay for permits; they also prefer it to a carbon tax. However, environmentalists are for the most part against the grandfathering of emission permits as they see this as a reward to polluters for their polluting activities.

There are other problems with emissions trading that are sometimes overlooked, particularly in economic theory. Emissions trading is fraught with political maneuvering, corruption, questionable offset credits, high monitoring costs because of the variety of offsets that are already appearing in carbon markets, lack of revenue recycling (no double dividend), and difficulties in bringing all countries into the scheme. Political maneuvering is already evident in U.S. legislation (discussed in the next section), for example, because some proposals delay much of the pain to 2020 and later, well beyond the next round of elections, and large emitters have been granted an enormous windfall in the form of free credits. In essence, therefore, large industrial emitters can tax energy consumers in lieu of government doing so, while large financial firms will reap huge benefits as intermediaries in the buying and selling of emission permits on financial markets. Again, it is little wonder that large firms not only favor cap-and-trade schemes, but actually promote and lobby for them. Not surprisingly, large industrial emitters and oil companies have backed away from funding climate research that contradicts the mainstream view – with emissions trading there is no financial incentive to contradict claims of anthropogenic warming.

Whether global warming is occurring or not, large companies are better served by emissions trading that would be hard to stop even if the temperatures in the tropical troposphere were to indicate that a more prudent approach would be better.

8.3 Government Failure: Estimating Policy Costs

In an effort to get serious about climate change, the leaders of the largest eight countries (G8) meeting in L'Aquila, Italy, agreed on July 8, 2009 to limit the increase in global average temperature to no more than 2°C above pre-industrial levels. To attain this, they set "the goal of achieving at least a 50% reduction of global emissions by 2050, [with] … developed countries reducing emissions of greenhouse gases in aggregate by 80% or more by 2050 compared to 1990 or more recent years."[3] The U.S. House of Representatives passed the *American Clean Energy and Security Act* (also known as Waxman-Markey after its sponsors) by a vote of 219–212 on June 26, 2009. The Act requires large emitters of greenhouse gases to reduce their aggregate CO_2 and equivalent emissions by 3% below 2005 levels in 2012, 17% below 2005 levels in 2020, 42% in 2030, and 83% in 2050.

One aspect of the Waxman-Markey bill is a cap-and-trade scheme that would require firms to submit permits that allow them to emit CO_2 and other greenhouse gases (measured in CO_2 equivalents). Only large industrial emitters (with emissions exceeding 25,000 ton of CO_{2e} per year) are affected, of which there are some 7,400. The program includes all electrical utilities and producers or importers of liquid fossil fuels beginning in 2012; all industrial facilities that manufacture products or burn fossil fuels are to be included beginning in 2014. Covered firms would receive 4.627 billion (10^9) allowances in 2012 and as few as 1.035 billion in 2050, with each allowance permitting one metric ton of CO_2 emissions. Interestingly, 29.6% of allowances will be auctioned off in the first 2 years, 2012–2013, thereby raising $846 billion in revenue. The proportion of allowances auctioned off actually falls to less than 18% over the period to 2020, rising to 18.4% by 2022 and then gradually to about 70% by 2031, where it would remain.[4] In the first few decades, therefore, significant allowances would be grandfathered.

Grandfathering of allowances ensures the support of industry, although there is the notion that, by freely giving allowances to large emitters such as power companies, there will be little immediate impact on output prices. This is misleading as

[3] Paragraph 65, 'Responsible Leadership for a Sustainable Future' Declaration, G8 Summit, July 2009. Available at (viewed July 22, 2009): www.g8italia2009.it/static/G8…/G8_Declaration_08_07_09_final,0.pdf

[4] This information is based on a report by the Congressional Budget Office and Congressional Joint Committee on Taxation, as reported by Amanda DeBard (CBO: House climate bill to raise $973B, Washington Post Monday, June 8, 2009, and viewed June 11, 2009): http://washingtontimes.com/news/2009/jun/08/cbo-house-climate-bill-raise-973b/. See also Congressional Budget Office (2009a).

already discussed in Sect. 8.2. The large industrial emitter could take its 'free' asset, sell it, and invest the proceeds in a technology that reduces CO_2 emissions (which is the idea behind allowances to begin with) or invest it elsewhere. Whether allowances are auctioned or given away (grandfathered), their cost will be reflected in final output prices. Thus, all citizens will face higher energy costs and higher costs for anything that involves the use of energy in its production and marketing.

As noted earlier, economists do not really care whether emission permits are auctioned or given away – the desired outcome is met at least cost. The only difference relates to the distribution of income and, from a theoretical perspective, that distribution can be adjusted by lump sum transfers, although the potential double-dividend benefit is lost and where such income transfers do occur they may be suspect.[5] However, large industrial firms love climate mitigation schemes, such as Waxman-Markey, that give them free access to emission allowances – the financial gains to such firms can be enormous, with taxpayers and consumers footing the bill. Financial institutions such as Morgan Stanley, Goldman Sachs and JP Morgan-Chase recognize the opportunities afforded by carbon trading; after all, carbon is forecast to become the largest commodity traded in the world, with a trading value estimated to reach \$3 trillion ($\$3 \times 10^{12}$) by 2020.[6] No wonder large financial institutions lobby governments to employ permit trading instead of carbon taxes – this has the makings to be the next crisis with huge amounts of money to be made before the bubble bursts.

Unfortunately, politicians are not content with a simple cap-and-trade scheme. In addition to providing large industrial emitters with a potential windfall, politicians like to introduce subsidies, regulations and provisions that lead to inefficiency – that actually increase the costs of meeting CO_2 emission reduction targets. Thus, the *Clean Energy and Security Act* also comes laden with regulations and provisions that make achieving targets much more expensive than would be the case with a carbon tax or even emissions trading, lock the economy into potentially inefficient investments, and make it much less likely that targets will be achievable. For example, there are mandated biofuel targets, with subsidies to farmers for planting energy crops and processors for building ethanol production plants. Agricultural economists have long opposed ethanol subsidies because they raise food prices (which harm the least well off in society), intensify crop production (increasing chemical use and machinery operations), distort land use by converting grassland into crop production and forestland into agriculture, reduce the performance of automobiles consuming gasoline with ethanol, provide only questionable climate mitigation

[5] The government of British Columbia sent each taxpayer a small remittance in the mail when it first implemented a carbon tax. The idea was to gain political acceptance, demonstrate that the tax would be revenue neutral, and, given the progressive nature of income taxes, favor the poor. The amount involved was small from the perspective of an individual taxpayer while the costs of implementing such a one-time payment were substantial.

[6] See M. Carr, China, Greenpeace Challenge Kyoto Carbon Trading (Update1). June 19: www.bloomberg.com/apps/news?pid=20601080&sid=aLM4otYnvXHQ (viewed August 31, 2009).

benefits, and lock society into ethanol production facilities that will produce ethanol for many years to come (Crutzen et al. 2008; Klein and LeRoy 2007; Morriss et al. 2009, pp. 79–89; Searchinger et al., 2008, 2009; see also Chap. 10).

8.3.1 Potential Costs of Reducing CO_2 Emissions: Evidence from the U.S.

The *American Clean Energy and Security Act* comes with a steep price tag. According to the Congressional Budget Office (CBO) and based on estimates that greenhouse gas emission permits would start to trade at $15 per ton of CO_2 (tCO_2) in 2010 and increase to $26 in 2019, each household will have to pay upwards of $175 per year so that firms can purchase emission allowances (Congressional Budget Office 2009b). The CBO's estimated costs of allowances are low if the EU's trading system is a guide, since permits have already traded in the ETS for more than $30 per tCO_2, although they more recently traded closer to $20 per tCO2 (see Fig. 8.5). The annual budgetary cost to U.S. taxpayers of Waxman-Markey is expected to rise from $52 billion in 2012 to over $800 billion by 2020 (Congressional Budget Office 2009a).

The U.S. Environmental Protection Agency (2009) predicts that CO_2 allowances will trade between $13 and $26 per tCO_2 in 2012 and $17–$33/tCO_2 by 2020 – again below historical trades in the EU's ETS. Consumer expenditures are then forecast to decline by no more than 0.19% by 2020 compared to no restriction on emissions, and by only a maximum of 0.39% by 2030. This amounts to, at most, an annual cost of $140 per household. Meanwhile, the EPA projects an increase in consumption expenditures of 18–19% between 2010 and 2020. These figures are quite optimistic. Given the large increase in consumption over a decade, it would seem that one can only reduce CO_2 emissions by the extent required under Waxman-Markey by purchasing emission reductions from other nations.

The cost estimates provided by the CBO and EPA are misleading, however, because they fail to take into account costs to the economy as a whole. These are difficult to calculate, but several studies have provided some very rough calculations. The more realistic forecasts come from two private sources. First, McKibbin et al. (2009) of the Brookings Institute estimate the costs to consumers of a cap-and-trade scheme that seeks to reduce CO_2 emissions by upwards of 49%, and not the 83% of the 2050 Waxman-Markey target. They estimate that cap-and-trade will lead to a loss in personal consumption of $1–$2 trillion (about $3,225–$6,450 per person) in present value terms. Of course, more stringent carbon targets would produce even higher costs; the authors suggest that an additional 8% cut in CO_2 emissions will increase costs by 45%. U.S. GDP would be lower by 2.5% in 2050 with cap and trade and unemployment would be 0.5% higher in the first decade compared to the without cap-and-trade baseline.

The Heritage Foundation (Beach et al., 2009a) estimates an average annual GDP loss of $393 billion, reaching a high of $662 billion in 2035.[7] Over the period 2012–2035, the accumulated GDP loss is estimated to be $9.4 trillion (in 2009 dollars). It also finds that there will be 1.1 million fewer jobs compared with the baseline assumptions, and that by 2035 there could be 2.5 million fewer jobs. In addition, electricity rates are projected to rise by 90%, gasoline prices by 74% and residential natural gas prices by 55%. (These values can be compared with those in Table 8.1 above.) The average U.S. household's energy costs are expected to rise by over $1,200 per year (Beach et al. 2009b).

None of the studies cited above provides a full economic accounting of costs and benefits. No study attempts to determine the true costs to the U.S. economy using a general equilibrium model that would take into account changes in prices and the economic effects of increased government regulations and subsidies for biofuels, wind energy and so on. Subsidies and regulations could increase costs significantly. However, one would not expect joblessness to continue for long as, in a well-functioning economy where wages can adjust, wages would fall and more people would be employed. Studies also ignore environmental costs and benefits – costs would increase if lands are converted from forest to cropland, for example, while there might be benefits from reduced consumption of certain automotive fuels. Again, calculating all of these costs and benefits is no easy task.

Before climate legislation could be made the law of the land, the Senate would also have to pass a climate bill, and the House and Senate bills would then need to be reconciled. A climate bill by Senator John Kerry, Chair of the Foreign Relations Committee, and Senator Barbara Boxer, Chair of the Environment and Public Works Committee, was considered by the U.S. Senate in Spring 2010. The bill was known as the *Clean Energy Jobs and American Power Act*, dropping any pretense about climate, although it was meant as a climate bill. It proposed an even more radical reduction in CO_2 emissions, a reduction of 20% by 2020 rather than the 17% under Waxman-Markey. Consequently, it would be even more costly. It was touted as a bill that will provide jobs, a subject addressed in the next subsection.

The Kerry-Boxer bill was subsequently re-introduced as the *American Power Act* by Senators Kerry, Joseph Lieberman and Lindsey Graham, although the latter subsequently dropped out as a co-sponsor. The Kerry-Lieberman bill adds to the Waxman-Markey cap and trade a $12 per ton tax on carbon emissions produced by large emitters. The tax floor would be allowed to rise at the rate of inflation plus 3%, while a tax ceiling of $25 per ton would be indexed to inflation plus 5% (so the tax could potentially reach $175/tCO_2$). State-level schemes to reduce CO_2 emissions would be eliminated, with states compensated for the loss of revenue this might entail. Emission reduction targets were the same as those under Waxman-Markey.

Additional components of the Kerry-Lieberman bill included cash subsidies to stimulate production of biofuels and additions to electrical-generating capacity

[7] As a reference point, U.S. GDP was $13,312.2 billion in 2008 (measured in 2005 dollars).

from wind, solar and other renewable sources, loan guarantees to builders of nuclear power plants, exemptions from emissions targets for favored sectors such as steel and agriculture, tax credits and subsidies to the transport industry, and inducements for technological innovation related to energy. The bill also made off-shore drilling for oil much harder, clearly in response to the Gulf oil spill in the summer of 2010 resulting from an accident and fire at a British Petroleum drilling platform.

Subsequently, the Democratic Senate Leader, Harry Reid, withdrew the Kerry-Lieberman bill and proposed one that did not have cap-and-trade provisions. The reasons concerned public opposition to cap and trade, which was correctly viewed as having the same impact on consumers as a carbon tax, and Democratic concerns about their potential success at mid-term elections in November 2010. Prior to the election, however, a last-ditch effort was made on September 21 to implement climate legislation when Democratic Senator Jeff Bingham introduced a bill (S.3813) to create a national 'Renewable Electricity Standard' (RES).[8] It requires that, by 2021, 15% of the electricity sold by an electric utility be generated from wind or certain 'other' renewable energy sources (presumably solar, wave, geothermal or tidal, and not hydro); up to four of the 15% points could, theoretically, be achieved by actions that improve energy efficiency, although these are tightly defined. The bill would create a new agency within the Department of Energy to oversee and enforce the new federal demands.

As a result of Republican gains in the November 2010 mid-term elections and a shift in attention from climate change issues to the large U.S. debt and deficit, Congress has not yet passed climate legislation; this is perhaps fortunate because many bills, such as Kerry-Lieberman, are a labyrinth of incentives, exceptions, subsidies and qualifications that are clear sign of policy failure.

Although the Congress failed to pass climate legislation, the environmental lobby did not despair. The U.S. Supreme Court had ruled in 2007 that carbon dioxide was a pollutant and, thus, that the U.S. Environmental Protection Agency (EPA) would be required to regulate CO_2 emissions. Given that climate legislation was not forthcoming, the EPA indicated that it would begin regulating CO_2. To counter this, in March 2011, the House Energy and Commerce Committee recently passed the *Energy Tax Prevention Act*, H.R. 910, which would prevent the EPA from regulating greenhouse gas emissions. Clearly, the policy debate is far from over, at least in the United States.

8.3.2 Benefits of Green Job Creation

Employment is a controversial element of any government program as politicians are wont to promote job creation as the most essential component of any legislation they have a hand in. So-called green (or environmentally friendly) jobs in particular

[8] See http://www.masterresource.org/2010/10/bingamans-national-res/ (viewed October 11, 2010).

have been touted by proponents of action to reduce reliance on fossil fuels. However, as argued in Chap. 6, employment benefits of projects constitute an income transfer as opposed to a benefit, unless it can be shown that the shadow price of labor is effectively zero, which occurs when there is structurally high unemployment (as opposed to unemployment that would fall with wages and mobility).

An in-depth study by Morriss et al. (2009) explores what is meant by green jobs, indicates that special interest groups have overstated the numbers of green jobs various clean-energy (and other positive environmental) initiatives have created, and questions whether environmental expenditures, such as subsidies to ethanol producers, wind and solar energy, increase jobs overall. In this regard, a Spanish study by Álvarez et al. (2009) found that, for every green job created (in producing renewable energy) at a cost of €571,138, 2.2 jobs were lost elsewhere in the economy. Similarly, the claim that electricity from renewable energy creates more jobs per kilowatt hour than traditional power generation simply implies "that renewable energy is more costly in labor terms than alternatives – hardly a virtue to anyone asked to pay for the energy produced" (Morriss et al. 2009, p. 44).

8.3.3 Politics and the Response to Climate Change

It is also helpful to consider the benefits of spending money on emissions reduction and whether citizens are prepared to pay for climate mitigation efforts. The benefits of climate change mitigation brought about by U.S. actions are miniscule. It translates into a reduction of perhaps 0.20°C in the projected 2100 temperature increase, but then only if the drastic House and Senate targets discussed above are fully implemented. The reduction in projected temperature will be only slightly more if all rich nations follow suit. This assumes climate sensitivity (warming caused by doubled CO_2 with feedbacks) at the midrange estimate of the IPCC, though more recent studies point to an increase of only one-sixth that amount (Lindzen and Choi 2009; Schwartz 2007; Spencer 2008; Spencer and Braswell 2008; Spencer et al. 2007), which would entail an insignificant temperature reduction of 0.03°C instead. The problem is that developing countries, particularly China and India, are not about to curb their economic growth simply because rich countries are concerned about an environmental problem that ranks at the bottom of their list of environmental and development priorities.[9] For example, with AIDS killing more than two million people annually in Africa, and worldwide more than four million children dying of respiratory infections, diarrhea and malaria each year, global warming is mainly a concern of the rich (Lomborg 2004).

[9] From U.S. Senate hearings on 7 July 2009, it is clear that the *American Clean Energy and Security Act* will have no effect on climate unless China and India reduce their CO_2 emissions. See (viewed July 9, 2009): http://epw.senate.gov/public/index.cfm?FuseAction=Minority.PressReleases&ContentRecord_id=564ed42f-802a-23ad-4570-3399477b1393

Fig. 8.9 Respondents' willingness to pay to mitigate climate change

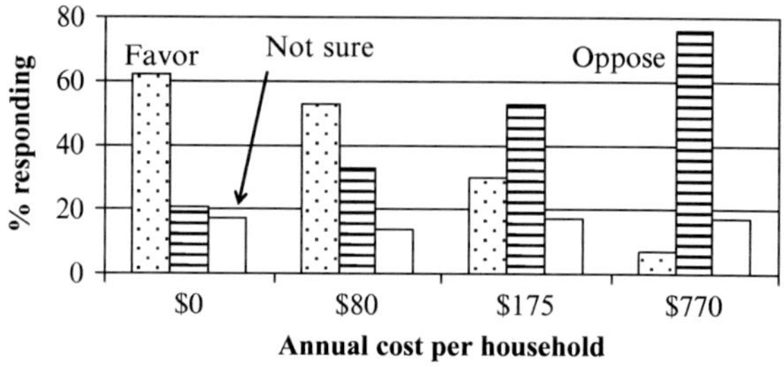

Next consider citizens' willingness to pay and a poll conducted by YouGuvPolimetrix for *The Economist*.[10] Forty-one percent of those polled considered climate change to be a 'very serious' problem with 28% considering it to be 'somewhat serious'. For comparison, 57% of respondents thought it was a 'very serious' problem that many Americans do not have health insurance, while a further 27% rated this to be 'somewhat serious.' When asked to choose between passing health care legislation or legislation to address global warming, 61% chose health care reform ahead of global warming, with only 16% considering climate change to be the more important problem; the remaining 23% were 'not sure'. Finally, Americans tended to favor legislation to reduce CO_2 emissions as long as it did not cost very much; when costs reached the Congressional Budget Office's (2009b) estimate of $175 per household per year, the majority of respondents to the poll were opposed (Fig. 8.9). Needless to say, costs of mitigating climate change are very likely going to be vastly greater than this.

That costs are likely to be high is evident from Table 8.1, where even a 'modest' carbon tax of $30 per tCO_2 will increase energy costs by 13.6–180.3% depending on the fuel source. Suppose the price of gasoline is $0.75 per liter and someone drives 10,000 km annually with a vehicle that averages 9 liters per 100 km. Before the carbon tax, the driver would pay $675 annually for fuel; after the tax, she would be required to pay $797 (about 18% more according to Table 8.1), or $122 more per year. Add to this the increased cost of electricity and purchases of other goods that have increased in price because of the carbon tax, and one quickly exceeds the annual cost where the majority of household would switch from being in favor of climate mitigation to being against. It is something politicians recognize, especially ones who are up for reelection.

The PEW Center in Washington, DC has traced American opinions regarding people's priorities related to health, employment, the economy, terrorism, the environment and other issues for the past decade or more. Results of a survey taken each year in January are provided in Table 8.2. These support the conclusions in Fig. 8.9. People do not view climate change as a 'top priority' and it falls near the bottom of

[10] *The Economist*, July 4, 2009 (pp. 24–25) with the full results available at (as viewed July 21, 2009): http://media.economist.com/media/pdf/Toplines20090701.pdf.

Table 8.2 U.S. Citizens' social, economic and environmental priorities, 2001–2010, percent considering each as a 'top priority'

Issue	2001	2002	2003	2004	2005	2006	2007	2008	2009	2010
1. Strengthening nation's economy	81	71	73	79	75	66	68	75	85	83
2. Improving the job situation	60	67	62	67	68	65	57	61	82	81
3. Defending US against terrorism	–	83	81	78	75	80	80	74	76	80
4. Securing social security	74	62	59	65	70	64	64	64	63	66
5. Improving education system	78	66	62	71	70	67	69	66	61	65
6. Securing Medicare	71	55	56	62	67	62	63	60	60	63
7. Reducing budget deficit	–	35	40	51	56	55	53	58	53	60
8. Reducing health care costs	–	–	–	–	–	–	68	69	59	57
9. Dealing with problems of poor	63	44	48	50	59	55	55	51	50	53
10. Strengthening the military	48	52	48	48	52	42	46	42	44	49
11. Dealing w US energy problem	–	42	40	46	47	58	57	59	60	49
12. Providing health insurance to uninsured	61	43	45	54	60	59	56	54	52	49
13. Reducing crime	76	53	47	53	53	62	62	54	46	49
14. Dealing with moral breakdown	51	45	39	45	41	47	47	43	45	45
15. Stricter rules for financial institutions	–	–	–	–	–	–	–	–	–	45
16. Protecting the environment	63	44	39	49	49	57	57	56	41	44
17. Reducing taxes	66	43	–	44	48	51	48	46	43	42
18. Illegal immigration	–	–	–	–	–	–	55	51	41	40
19. Reducing influence of lobbyists	–	–	–	–	–	–	35	39	36	36
20. Dealing with global trade	37	25	–	32	32	30	34	37	31	32
21. Dealing with global warming	–	–	–	–	–	–	38	35	30	28

Source: "Public's priorities for 2010: economy, jobs, terrorism – energy concerns fall, deficit concerns rise", January 25, 2010, the Pew Research Center For the People & the Press, a project of the Pew Research Center

a list of 21 issues to which survey respondents were asked to react. Since it was first introduced in the PEW survey in 2007, the number of individuals rating climate change as a 'top priority' has steadily fallen from 38 to 28% in 2010.

Finally, an early 2011 poll of Australian citizens conducted by *NEWSPOLL* and *THE AUSTRALIAN* newspaper to gauge attitudes towards climate change found that 78% of respondents believed climate change was currently taking place versus 16%

who did not (and 6% uncommitted).[11] Of those respondents indicating climate change was taking place, more than 92% attributed it partly or wholly to human activity; thus, 72% of those polled thought that human activities were responsible for global warming. However, only 39% of respondents indicated a willingness to pay more for energy even if it would slow climate change, and fully 60% were against the government's proposal to tax carbon.

Nations will agree to emission reductions in an effort to prevent global warming, but, as the above surveys show, they are unlikely to go much beyond what the Australian economist David Pannell refers to as 'tokenism.'[12] Countries will only agree to limit their emissions to the extent that is politically feasible; they will not agree to emission reductions that harm their economies or ones that they cannot circumvent, which is easy to do as there are no international governance structures in place to punish countries for not meeting targets, although environmentalists would like to see such structures in place.

Pannell gives a number of reasons why it is difficult to come to an international agreement to limit emissions to the extent required truly to mitigate global warming (leaving aside any controversy about its cause).[13]

- The nature of democratic politics in developed countries makes it hard to implement policies that cause some members of society to lose in the short term. Large numbers of losers in the long term are much easier to tolerate than small numbers in the short term.
- The main priority of developing countries will be on economic development and improvement of the welfare of their citizens, not on climate change.
- Mitigating climate change requires cooperation from all countries, but, because the atmosphere is a global public good, there is a strong incentive for one or more countries to 'free ride' and let other nations address the problem while they benefit.
- There is currently a lack of carbon-neutral energy technologies that are economically more attractive than high-carbon-emitting energy technologies when adopted at very large scales.

The 1997 Kyoto Protocol was an example of tokenism. Countries chose their own emission reduction targets (relative to 1990 emission levels) to be achieved during the 2008–2012 commitment period. Some countries achieved the targets, but for reasons unrelated to true efforts at reducing emissions. Both the financial crisis that led to stagnating or declining GDP in some countries and the mothballing of

---

[11] http://www.newspoll.com.au/image_uploads/101201_Climate_Change.pdf (viewed June 16, 2011).

[12] Pannell Discussions #160: http://cyllene.uwa.edu.au/~dpannell/pd/pd0160.htm (viewed 19 October 2009).

[13] See reference in previous note. The following is a paraphrase of Pannell's reasons for tokenism. Pannell also suggests that growing skepticism about global warming is an obstacle to an international agreement, which he attributes to over-arching scare mongering by advocates, the vociferousness of a minority of skeptical scientists, and social factors, including Christian and Islamic fundamentalism.

inefficient manufacturing facilities (as in eastern Germany after unification) and/or decrepit coal-fired generating facilities (United Kingdom) helped reduce emissions, but the general trend in developed countries was increased emissions over 1990. Further, as a result of economic expansion in developing countries, most notably China and India, global CO_2 emissions rose dramatically since 1990. However, this was also accompanied by a great improvement in the wellbeing of many poor people, significantly reducing the proportion of the world's population in absolute poverty (say, living on less than $1.25 per day).

The European Union has a 20–20–20 target: a 20% reduction in CO_2 emissions from 1990 levels by 2020, with 20% of energy to be produced from renewable sources.[14] At COP15 in Copenhagen in late 2009, the EU was prepared to impose a 30% reduction in CO_2 emissions by 2020, if there had been some sort of climate agreement. One reason is that, while the European Commission estimated in 2008 that the cost of meeting a 20% reduction target would be €70 billion, by 2010 this cost had fallen to €48 billion as a result of lower emissions caused by recession. The cost of meeting a 30% reduction target was estimated to be only €81 billion. If the reduction in CO_2 emissions is to be attributed to recession, neglected in the revised costs is the cost imposed by the recession. Further, the argument that a greener economy will lead to more jobs is also a fallacy, as was argued above.

All eyes are on the United States, however, because climate legislation in that country likely has more 'teeth' than that in any other country as a result of potential litigation by environmental groups. The courts would order the administration to take action to ensure that the law is enforced, which would mean that the government would either have to accept economic restrictions that could potentially harm the economy or pass legislation that undoes the previous legislation and thereby enables the administration from taking drastic action. The latter is generally something politicians wish to avoid. However, this also means it will be difficult to pass legislation to begin with.

The Kerry-Lieberman bill includes sanctions against countries that do not fall in line with their emissions reductions targets. Unfortunately, should legislation with a 'sanctions' clause come into effect, it could trigger a wave of protectionism that would cost the global economy vastly more than simple legislation to cut greenhouse gases.

The potential for politically-motivated mischief is great, especially where politicians feel pressured to act but recognize that the consequences of their actions fall upon a future government. The problem is succinctly described by Lomborg[15]:

> Japan's commitment in June [2009] to cut greenhouse gas levels 8 percent from its 1990 levels by 2020 was scoffed at for being far too little. Yet for Japan – which has led the world in improving energy efficiency – to have any hope of reaching its target, it needs to build nine new nuclear power plants and increase their use by one-third, construct more than 1 million new wind-turbines, install solar panels on nearly 3 million homes, double the

[14] Information in this paragraph is taken from *The Economist*, May 29, 2010, p. 53.

[15] Article in the *Washington Post*, September 28, 2009. At (as viewed October 12, 2009): www.washingtonpost.com/wp-dyn/content/article/2009/09/27/AR2009092701444_pf.html

> percentage of new homes that meet rigorous insulation standards, and increase sales of 'green' vehicles from 4 percent to 50 percent of its auto purchases. Japan's new prime minister was roundly lauded this month [September 2009] for promising a much stronger reduction, 25 percent, even though there is no obvious way to deliver on his promise. Expecting Japan, or any other nation, to achieve such far-fetched cuts is simply delusional. Imagine … that at the climate conference in Copenhagen in December [2009] every nation commits to reductions even larger than Japan's, designed to keep temperature increases under 2 degrees Celsius. The result will be a global price tag of $40 trillion in 2100, to avoid expected climate damage costing just $1.1 trillion, according to climate economist Richard Tol. … That phenomenal cost, calculated by all the main economic models, assumes that politicians across the globe will make the most effective, efficient choices. In the real world, where policies have many other objectives and legislation is easily filled with pork and payoffs, the deal easily gets worse.

One way to increase costs is to choose the wrong market instrument and governance structure for reducing emissions.

8.4 Discussion

William Nordhaus has not been afraid to make the case for a carbon tax. As noted in Chap. 7, he advocates a gradually rising carbon tax. In a September 2010 policy piece, Nordhaus (2010) argued that the "desirable features of any tax are that it raises revenues in a manner that has minimal distortionary effect on the economy and reinforces other objectives of national policy." According to Nordhaus, a carbon tax is particularly relevant at this point in history because, in the wake of the 2008–2009 financial crisis, it can be used to raise revenues to tackle the burgeoning U.S. debt, which is estimated to reach some 65–72% of U.S. GDP by 2015. A carbon tax has the potential to raise substantial revenue, is well understood, increases economic efficiency (as it tackles an undesirable externality), has potential health benefits (reducing emissions of CO_2 will also reduce emissions of other harmful pollutants, ceteris paribus), displaces regulatory inefficiencies (associated with attempts to subsidize and regulate greenhouse gas emissions), can be harmonized across countries, enables the U.S. to meet international CO_2-emission reduction targets, and captures the rents that would be lost to government under a grandfathered cap-and-trade scheme.

Some of the claims that Nordhaus makes in favor of a carbon tax are dubious (e.g., "substantial public health benefits"), but his main argument is that the tax can help reduce U.S. budget deficits while not hurting citizens any more, and probably much less, than any other policy. The ideal tax ramp and budget implications are provided in Table 8.3. The present value of the tax revenues over the period to 2030 is 15% (discounted at 5%) of 2010 GDP, or 35% if discounted over the period to 2050. Therefore, the carbon tax can be expected to make a significant contribution to reducing the U.S. budget deficit and debt.

Nordhaus also makes the case that the income redistributional effects of a carbon tax are minimal, or at least no worse than those associated with a value-added tax or payroll tax for social security purposes. The average household in the U.S. consumes 12,000 kilowatt hours (kWh) of electricity annually and pays an average of

Table 8.3 Ideal carbon tax ramp and budgetary implications for the U.S.

Year	Tax rate ($/t CO_2)	Revenues (2010 $\$\times10^9$)	Year	Tax rate ($/t CO_2)	Revenues (2010 $\$\times10^9$)
2005	0.00	0 (0.0%)	2025	63.00	282 (0.9%)
2015	25.00	123 (0.6%)	2030	89.80	386 (1.0%)
2020	39.70	184 (0.7%)	2035	128.10	528 (1.1%)

Notes: Adapted from Nordhaus (2010). Results assume inflation and real GDP growth of 2.5%. Revenues as a proportion of GDP are provided in parentheses

10¢ per kWh. If this power is generated solely by coal-fired plants, Nordhaus argues, the annual cost to a household would rise in 2015 from $1,200 to $1,500, or by 25% ($300). However, based on data in Table 8.1, this seems highly optimistic. From data in Table 8.1, a carbon tax of $25 per t$CO_2$ would increase the price that a household pays for electricity by 150%, or from $1,200 to $3,000 annually (assuming no reduction in use). The price of gasoline would rise by 15.1% adding nearly 14¢ to a gallon of gasoline and not the 7¢ indicated by Nordhaus.[16]

Although a carbon tax is probably the best instrument that governments have in their policy arsenal, it is unlikely, based on results reported in Fig. 8.9 (and Table 8.2), that citizens would willingly accept a carbon tax. Rather, they would view it as another attempt on the part of politicians to pay for wrongheaded policies related to the 2008–2009 financial crisis, and perhaps financing of the Iraq and Afghan wars, that led to the growing U.S. debt.

There is absolutely no way for the United States or Europe, or any other country, to reduce their emissions of carbon dioxide by 80% by 2050 and retain a standard of living even close to that which it has today. The same is true for 50% reductions in CO_2. Reductions in CO_2 on that scale are simply not achievable within a 40-year timeframe. Even reductions of as little as 25% will be difficult to achieve, and will be costly. They will require huge investments in nuclear power generation,[17] massive changes in transportation infrastructure, and impressive technical breakthroughs in everything from biofuels to battery technology. As argued in the next several chapters, it will be extremely costly to reduce CO_2 emissions and the potential for government (or policy) failure will be great, which will increase costs even further.

Yet, even if western countries are successful in reducing their emissions of greenhouse gases, the impact on climate change will be small. Growth in emissions by developing countries, especially China, India and Brazil, will easily and quickly exceed any reduction in emissions by rich countries (see Chap. 12). Fossil fuels are

[16] To be fair, the data in Table 8.1 are based on bottom-up types of calculations, while Nordhaus' estimates are derived from an integrated assessment model that takes into account some of the macroeconomic effects.

[17] It would seem that the March 11, 2011 earthquake off the coast of Japan and the subsequent tsunami that led to the failure of the Fukushima Dai-ichi nuclear power plant will make it more difficult to approve nuclear power installations in the future. Nuclear power is discussed further in Chap. 10.

abundant, ubiquitous and inexpensive relative to alternative energy sources; therefore, any country would be foolish to impair its economy by large-scale efforts to abandon them. As we show in the next chapters, many schemes to remove carbon dioxide from the atmosphere or reduce CO_2 emissions will yield a less-than-anticipated reduction in the carbon footprint while imposing higher social costs than proponents envisioned. Further, the efforts of any one country to tackle climate change are for naught, while developing countries are not about to jeopardize their development prospects for unproven benefits that might only accrue 100 years or more in the future.

It does not matter what rich countries do to reduce their emissions of carbon dioxide. Their efforts will have no impact on climate change, but they will have an adverse impact on their own citizens. Whether the climate change story is real or not, whether the climate model projections are accurate or not, fossil fuels will continue to be the major driver of economic growth and wealth into the foreseeable future. Hence, we now turn to efforts to change the CO_2 balance of the atmosphere.

References

Álvarez, G. C., Jara, R. M., Julián, J. R. R., & Bielsa, J. I. G. (2009, March). *Study of the effects on employment of public aid to renewable energy sources: Lessons from the Spanish renewables bubble*. Borrador draft. Retrieved July 9, 2010, from http//www.juandemariana.org/pdf/090327-employment-public-aid-renewable.pdf/

Beach, W. W., Kreutzer, D. W., Campbell, K. A., & Lieberman, B. (2009a, May 13). *The economic impact of Waxman-Markey* (Heritage Foundation WebMemo No. 2438, pp. 6). Viewed July 10, 2010, http://www.heritage.org/Research/EnergyandEnvironment/wm2438.cfm.

Beach, W. W., Kreutzer, D. W., Campbell, K. A., & Lieberman, B. (2009b, May 18). *Son of Waxman-Markey: More politics makes for a more costly bill* (Heritage Foundation WebMemo No. 2450). Viewed July 24, 2012, http://www.heritage.org/research/reports/2009/05/son-of-waxman-markey-more-politics-makes-for-a-more-costly-bill.

Bosetti, V., & Rose, S. K. (2011). Reducing carbon emissions from deforestation and forest degradation: Issues for policy design and implementation. *Environment and Development Economics, 16*(Special Issue 04), 357–360. doi:10.1017/S1355770X11000143.

Bovenberg, A. L., & Goulder, L. H. (1996). Optimal environmental taxation in the presence of other taxes: General-equilibrium analysis. *American Economic Review, 86*(4), 985–1000.

Congressional Budget Office. (2009a, June 5). Cost estimate: H.R. 2454 American clean energy and security act of 2009 (pp. 40). Viewed July 10, 2010. Dated June 5, 2009 http://www.cbo.gov/sites/default/files/cbofiles/ftpdocs/102xx/doc10262/hr2454.pdf.

Congressional Budget Office. (2009b, June 19). *The estimated costs to households from H.R. 2454 American clean energy and security act of 2009* (pp. 16). Viewed July 24, 2012. Retrieved from http://www.cbo.gov/ftpdocs/103xx/doc10327/06-19-CapAndTradeCosts.pdf.

Crutzen, P. J., Mosier, A. R., Smith, K. A., & Winiwarter, W. (2008). N2O release from agro-biofuel production negates global warming reduction by replacing fossil fuels. *Atmospheric Chemistry and Physics, 8*(2), 389–395.

Epstein, R. A. (1985). *Takings: Private property and the power of eminent domain*. Cambridge, MA: Harvard University Press.

Field, B., & Olewiler N. (2005). *Environmental Economics. Second Canadian Edition*. Toronto: McGraw-Hill.

Fukuyama, F. (1992). *The end of time and the last man*. New York: The Free Press.

Hart, O. D., Shleifer, A., & Vishny, R. W. (1997). The proper scope of government: Theory and an application to prisons. *Quarterly Journal of Economics, 112*(4), 1127–1161.

Helm, D. (2010). Government failure, rent-seeking and capture: The design of climate change policy. *Oxford Review of Economic Policy, 26*(2), 182–196.

IPCC WGI. (2007). *Climate change 2007. The physical science basis. Working Group I contribution to the fourth assessment report of the Intergovernmental Panel on Climate Change*. Cambridge: Cambridge University Press.

Klein, K., & LeRoy, D. G. (2007). *The biofuels frenzy: What's in it for Canadian agriculture?* Paper presented at the Conference of Alberta Institute of Agrologists, Banff. Viewed July 24, 2008, http://www.aic.ca/whatsnew_docs/Klein%20Final%20%234.pdf

La Porta, R., Lopez-de-Silanes, F., Shleifer, A., & Vishny, R. W. (1999). The quality of the government. *Journal of Law, Economics & Organization, 15*(1), 222–279.

Landes, D. S. (1998). *The wealth and poverty of nations*. New York: W.W. Norton & Company.

Lindzen, R. S., & Choi, Y.-S. (2009). On the determination of climate feedbacks from ERBE data. *Geophysical Research Letters, 36*, L16705. doi:10.1029/2009GL039628.

Lomborg, B. (Ed.). (2004). *Global crises, global solutions*. Cambridge: Cambridge University Press.

McKibbin, W., Wilcoxen, P., & Morris, A. (2009, June 8). *Consequences of cap and trade*. Retrieved July 24, 2009, from http://www.brookings.edu/~/media/Files/events/2009/0608_climate_change_economy/20090608_climate_change_economy.pdf

McKitrick, R. (2007). Why not tie carbon taxes to actual levels of warming? Both skeptics and alarmists should expect their wishes to be answered. *Financial Post Financial Post, June 12*. Viewed July 24, 2007, http://www.financialpost.com/story.html?id=d84e4100-44e4-4b96-940a-c7861a7e19ad&p=3.

McKitrick, R. (2011). A simple state-contingent pricing rule for complex intertemporal externalities. *Energy Economics, 33*(1), 111–120.

Morriss, A. P., Bogart, W. T., Dorchak, A., & Meiners, R. E. (2009). *Green jobs myth* (University of Illinois Law and Economics Research Paper Series No. LE09-001). Retrieved August 11, 2009, from http://ssrn.com/abstract=1358423

Nordhaus, W. D. (2010, September). Carbon taxes to move toward fiscal sustainability. *The Economists' Voice*. Viewed September 30, 2010. Retrieved from www.bepress.com/ev.

Ostrom, E. (2000). Social capital: A fad or a fundamental concept? In P. S. Dasgupta & I. Serageldin (Eds.), *Social capital: A multifaceted perspective* (pp. 172–214). Washington, DC: The World Bank.

Parry, I., Williams, R. C., III, & Goulder, L. H. (1999). When can carbon abatement policies increase welfare? The fundamental role of distorted factor markets. *Journal of Environmental Economics and Management, 37*(1), 52–84.

Peters-Stanley, M., Hamilton, K., Marcello, T., & Sjardin, M. (2011, June 2). Back to the future. State of the voluntary carbon markets 2011. *A report by Ecosystem Marketplace & Bloomberg New Energy Finance* (pp. 93). Washington, DC/New York: Ecosystem Marketplace and Bloomberg New Energy Finance.

Pizer, W. A. (1997, October). *Prices vs. quantities revisited: The case of climate change* (Discussion Paper 98-02, pp. 52). Washington, DC: Resources for the Future.

Schwartz, S. E. (2007). Heat capacity, time constant, and sensitivity of the Earth's climate system. *Journal of Geophysical Research, 112*, D24S05. doi:10.1029/2007JD008746.

Schwindt, R. (1992). *Report of the Commission of Inquiry into compensation for the taking of resource interests*. Victoria: Government of BC Printing Office.

Schwindt, R., & Globerman, S. (1996). Takings of private rights to natural resources: A policy analysis. *Canadian Public Policy – Analyse de Politiques, 22*(3), 205–224.

Searchinger, T. D., Heimlich, R., Houghton, R. A., Dong, F., Elobeid, A., Fabiosa, J., Tokgoz, S., Hayes, D., & Yu, T. (2008, February 29). Use of U.S. croplands for biofuels increases greenhouse gases through emissions from land-use change. *Science, 319*(5867), 1238–1240.

Searchinger, T. D., Hamburg, S. P., Melillo, J., Chameides, W., Havlik, P., Kammen, D. M., Likens, G. E., Lubowski, R. N., Obersteiner, M., Oppenheimer, M., Robertson, G. P.,

Schlesinger, W. H., & Tilman, G. D. (2009, October 23). Fixing a critical climate accounting error. *Science, 326*(5952), 527–528.

Shleifer, A., & Vishny, R. W. (1998). *The grabbing hand: Government pathologies and their cures*. Cambridge, MA: Harvard University Press.

Spencer, R. W. (2008). *Global warming as a natural response to cloud changes associated with the Pacific Decadal Oscillation (PDO)*. Retrieved August 13, 2009, from http://www.drroyspencer.com/research-articles/global-warming-as-a-natural-response/

Spencer, R. W., & Braswell, W. D. (2008). Potential biases in feedback diagnosis from observational data: A simple model demonstration. *Journal of Climate, 21*(21), 5624–5628. doi:10.1175/2008JCLI2253.1.

Spencer, R. W., Braswell, W. D., Christy, J. R., & Hnilo, J. (2007). Cloud and radiation budget changes associated with tropical intraseasonal oscillations. *Geophysical Research Letters, 34*, L15707. doi:10.1029/2007GL029698.

Tietenberg, T., Grubb, M., Michaelowa, A., Swift, B., & Zhang, Z. X. (2003). *International rules for greenhouse gas emissions trading: Defining the principles, modalities, rules and guidelines for verification, reporting and accountability* (United Nations Publication UNCTAD/GDS/GFSB/Misc.6, pp. 123). New York: United Nations.

U.S. Environmental Protection Agency. (2009, April 20). *EPA preliminary analysis of the Waxman-Markey discussion draft* (Office of Atmospheric Programs). Retrieved August 10, 2009, from http://www.epa.gov/climatechange/economics/pdfs/WM-Analysis.pdf

van Kooten, G. C. (2004). *Climate change economics: Why international accords fail*. Cheltenham/Northampton: Edward Elgar.

van Kooten, G. C. (2009). Biological carbon sequestration and carbon trading re-visited. *Climatic Change, 95*(3–4), 449–463.

van Kooten, G. C., & Arthur, L. M. (1997). Economic development with environmental security: Canadian-made strategies for co-existence. *American Journal of Agricultural Economics, 79*(5), 1508–1514.

van Kooten, G. C., & Scott, A. (1995, June). Constitutional crisis, the economics of environment and resource development in western Canada. *Canadian Public Policy – Analyse de Politiques, 21*(2), 233–249.

Weitzman, M. L. (1974). Prices vs quantities. *The Review of Economic Studies, 41*(128), 477–491.

Weitzman, M. L. (2002). Landing fees vs harvest quotas with uncertain fish stocks. *Journal of Environmental Economics and Management, 43*(2), 325–338.

Woerdman, E., Nentjes, A., & Couwenberg, O. (2009). Energy prices and emissions trading: Windfall profits from grandfathering? *European Journal of Law & Economics, 28*(2), 185–202.

Chapter 9
Avoiding Emissions Reduction: Terrestrial Carbon Sinks

On the one hand is the global scientific consensus, and on the other – given equal weight – are the crackpot theories of industry-financed deniers.

– Al Gore, Our Choice, p. 363.

Carbon flux in forest and agricultural ecosystems is primarily the result of land use, land-use change and forestry (LULUCF) activities (IPCC 2000). Under the Kyoto Protocol, such activities can lead to carbon offset credits and, supposedly, debits. Originally the European Union was opposed to the inclusion of sink credits because it was felt that, despite the potential of terrestrial carbon sinks to slow the rate of CO_2 buildup in the atmosphere, sinks were a distraction from the real need to reduce emissions of CO_2 from the burning of fossil fuels. Subsequent to the agreement struck at Kyoto in December, 1997 (recall the introductory paragraphs to Chap. 8), the United States argued that emissions reduction was too costly and that carbon sinks should be included as a means of meeting CO_2 emission reduction targets. A failure to include sinks at COP6 in The Hague in late 2000 was one of the factors that caused the U.S. to back out of the Kyoto agreement in early 2001. Later that same year, at COP 7 held at Marrakech in Morocco, carbon offsets from forest ecosystems were deemed eligible for meeting emission reduction targets and a more basic framework for including offset credits was struck (IPCC 2000).

Tree planting and activities that enhance tree growth clearly remove carbon from the atmosphere and store it in biomass, and thus should be eligible activities for creating carbon offset credits. However, since most countries have not embarked on large-scale afforestation and/or reforestation projects in the past decade, harvesting trees during the 5-year Kyoto commitment period (2008–2012) will cause them to have a debit on the afforestation-reforestation-deforestation (ARD) account. Therefore, the Marrakech Accord permitted countries, in the first commitment period only,[1]

[1] This is an interesting statement as the agreement reached at Marrakech implicitly assumes that the Kyoto process will be extended in the future, because, at the time, there existed no future commitment periods.

G.C. van Kooten, *Climate Change, Climate Science and Economics: Prospects for an Alternative Energy Future*, DOI 10.1007/978-94-007-4988-7_9,

to offset up to 9.0 megatons (Mt) of carbon, or 33 Mt CO_2, each year for 2008–2012 through (verified) forest management activities that enhance carbon uptake. If there is no ARD debit, then a country cannot claim the credit. In addition, some countries are able to claim carbon credits from business-as-usual forest management that need not be offset against ARD debits. Canada could claim 44 Mt CO_2 per year, the Russian Federation 121 Mt CO_2, Japan about 48 Mt CO_2, and other countries much lesser amounts.[2]

In addition to forest ecosystem sinks, agricultural activities that lead to enhanced soil organic carbon and/or more carbon stored in biomass could be used to claim offset credits. The activities include revegetation (establishment of vegetation that does not meet the definitions of afforestation and reforestation), cropland management (greater use of conservation tillage, more set asides) and grazing management (manipulation of the amount and type of vegetation and livestock produced). Activities that reduce deforestation and forest degradation are also being considered in addition to tree planting and forest management that enhances tree growth. This was briefly discussed in Chap. 8, but is considered further in Sect. 9.4 below.

A problem with the Kyoto agreement was the number of different instruments that countries could use to achieve their targets. Rich countries could reduce domestic CO_{2e} emissions or purchase allowances from other rich countries that reduced their emissions below the targeted level. They could also sequester carbon in domestic biological sinks, counting the CO_2 thus removed from the atmosphere towards their target. Alternatively, countries could purchase certified emission reduction credits (known as CERs) on the international market by investing in projects in developing countries through Kyoto's Clean Development Mechanism (CDM), or emission reduction units (ERUs) in economies in transition (former Soviet states) via Joint Implementation (JI); CERs are wholly credited to the rich country while ERUs are shared. The CDM (JI) enables rich countries to invest in environmental improvements (e.g., investments in new equipment that improves energy efficiency) developing (in-transition) countries, thereby receiving credit for the reduction in CO_2 and other greenhouse gas emissions. CERs and ERUs could also be earned for CO_2 removed from the atmosphere by a tree planting project.

In this chapter, we focus on biological sinks in greater detail, particularly forest sinks, because they shed light on the difficulties of implementing policies to reduce the CO_2 concentration in the atmosphere. In the next section, we examine the literature to determine what it might cost to sequester carbon in biological sinks, because we should have some idea of how sequestration activities compare with CO_2-emissions reduction on a cost basis. There is no sense investing in sequestration if the same climate benefits can be had at much lower cost. Then, in Sect. 9.2, we provide an example of how carbon uptake and release vary over time in a forest ecosystem if post-harvest carbon flux is also taken into account. Finally, in Sect. 9.3, we examine four major obstacles to the inclusion of biological sinks.

[2] Recall that the U.S. was no longer an active participant in the Kyoto process by the time of COP7, which was held at the end of 2001.

9.1 Costs of Creating CO_2 Emission Offsets

There is no doubt that tree planting and silvicultural investments that enhance tree growth remove CO_2 from the atmosphere and store the carbon in biomass. But are such activities competitive with emissions reduction? Are they financially attractive? To answer these questions requires information about the financial costs and carbon-uptake benefits of forest sector activities in various regions. If there are many studies that provide this information, they can be analyzed using meta-regression analysis.

A meta-regression analysis of the costs of sequestering carbon in forest sinks by van Kooten et al. (2009), and van Kooten and Sohngen (2007), suggests that, if emission offset credits trade at \$50 per tCO_2 (a rather high value), tree planting and other forest management activities are economically attractive. The meta-regression analysis is based on 68 studies with a total of 1,047 observations. A summary of the important results is provided in Table 9.1. If the opportunity cost of land is appropriately taken into account, forest management and forest conservation are not economically attractive activities. Assuming emission offset credits can be purchased for \$50 per tCO_2 or less, tree planting appears to be attractive only in the tropics, although there are clearly projects in other regions of the world where tree planting is attractive if trees are subsequently harvested to replace fossil fuels (reducing CO_2 emissions). Europe might be an exception because the opportunity cost of land is simply too high, although even there tree planting for bio-energy might be feasible on some marginal lands.

Van Kooten et al. (2004, 2009) considered many more studies than the 68 included in their final meta-regression analyses, but these failed to provide sufficient information to calculate the costs and benefits of carbon uptake. Examples regarding the inconsistency with which cost and carbon uptake data are reported can be found in a review of forestry projects conducted by the UN Food and Agriculture Organisation (FAO 2004). Indeed, even for the 68 studies included in the meta-regression analysis, many left out details concerning the time profile of carbon uptake, which is highly relevant to the assessment of a project. Importantly, studies also failed to take into account transaction costs. Overall, one cannot be confident that the values provided in Table 9.1 are not a gross underestimate of the actual costs of generating sink-based CO_2 offsets.

A similar story can be told about agricultural soil-carbon sinks. In order to increase soil organic carbon, farmers need to change their agronomic practices. In drier regions where tillage summer fallow is used to conserve soil moisture, this requires the use of chemical fallow or continuous cropping, or the return of cropped land to grassland. In other agricultural regions, a movement from conventional tillage to reduced tillage or no tillage (also known as zero tillage) can increase soil organic carbon. Reduced and zero tillage (ZT) increase soil carbon by increasing plant biomass and/or reducing rates of decay of organic matter, as does the replacement of tillage summer fallow by continuous cropping or chemical summer fallow. Are such practices worth pursuing, and can they result in significant reductions in carbon flux?

Table 9.1 Marginal costs of creating CO_2-offset credits through forestry, meta-regression analysis results ($/t$CO_2$)

Region and scenario	Study averages (n=68)	All observations (n=1,047)
Global	$28.85	$25.10
Planting	$0.26	−$4.93
Planting and opportunity cost of land	$29.80	$21.91
Planting, opportunity cost of land and fuel substitution	−$40.14	$19.88
Forest management	$88.47	$35.31
Forest management and opportunity cost of land	$118.01	$62.15
Forest management, opportunity cost of land and fuel substitution	$48.07	$60.12
Forest conservation	$158.28	$20.16
Forest conservation and opportunity cost of land	$187.82	$47.00
Europe	$173.26	$183.64
Planting and opportunity cost of land	$185.44	$180.14
Planting, opportunity cost of land and fuel substitution	$115.50	$178.11
Forest management and opportunity cost of land	$273.65	$220.38
Forest management, opportunity cost of land and fuel substitution	$203.71	$218.35
Tropics (CDM projects)	−$26.20	$4.04
Planting and opportunity cost of land	−$25.26	$0.85
Planting, opportunity cost of land and fuel substitution	−$95.20	−$1.18
Forest management and opportunity cost of land	$62.95	$41.09
Forest management, opportunity cost of land and fuel substitution	−$6.99	$39.06
Conservation	$103.22	−$0.90
Conservation and opportunity cost of land	$132.76	$25.94
Boreal region	$58.01	$8.77
Planting and opportunity cost of land	$70.19	$5.26
Planting, opportunity cost of land and fuel substitution	$0.25	$3.23
Forest management and opportunity cost of land	$158.40	$45.50
Forest management, opportunity cost of land and fuel substitution	$88.46	$43.47

Source: van Kooten et al. (2009) and van Kooten and Sohngen (2007)

Undoubtedly, there are soil erosion benefits from practicing reduced (conservation) tillage and zero tillage. In most cases, costs are also lower as a result of fewer field operations that offset higher chemical costs (although prices of herbicides are much lower than in the past), while planting of genetically modified crops that can withstand certain herbicides facilitated no tillage agriculture. As a result, the runoff of fertilizers and chemicals into surface waters has declined, while the extent of zero (and reduced) tillage area has increased significantly in the United States in the past several decades. In 1997 in the U.S., farmers employed conventional tillage on 36.5% of 294.7 million acres (119.3 million ha) planted to cropland; 26.2% was planted using reduced tillage and 15.6% using zero tillage, with other crop residue methods employed on the remaining land (Padgitt et al. 2000, p. 67). Not included were some 20 million acres of land left in tillage summer fallow in drier regions: 22% of all wheat planted in the U.S. in 1997 was part of a wheat-fallow rotation and, in some states, three-quarters of all wheat was part of a wheat-fallow rotation. Continuous cropping or chemical summer fallow ensure a higher level of soil organic carbon (SOC) compared to tillage summer fallow.

West and Marland (2002) use U.S. data on carbon uptake in soils, production of biomass, chemical and fuel use, machinery requirements and so on to compare conventional, reduced and zero tillage (ZT) in terms of their carbon flux. They provide a detailed carbon accounting for each practice, concluding that, due primarily to extra chemical use, reduced tillage does not differ significantly from conventional tillage in terms of carbon uptake benefits, but that ZT results in an average relative net carbon flux of –368 kg carbon per ha per year, denoted (C/ha/year). Of this amount, –337 kg C/ha/year is due to carbon sequestration in soil, –46 kg C/ha/year due to a reduction in machinery operations and +15 kg C /ha/year due to higher CO_2 emissions from an increase in the use of agricultural inputs. While annual savings in carbon emissions of 31 kg C/ha/year last indefinitely, accumulation of carbon in soil reaches equilibrium after 40 years. West and Marland suggest that the *rate* of uptake in soil is constant at 337 kg C/ha/year for the first 20 years and then declines linearly over the next 20 years. However, stored carbon can be released back into the atmosphere in as little as a year when conventional tillage is resumed (although it more likely takes several years).

West and Marland's estimates of carbon uptake by soils in the prairie region of Canada as a result of going from conventional to no tillage vary from 100 to 500 kg C/ha/year. Using these results and discount rates of 2 and 4%, van Kooten (2004) estimates that the net discounted carbon prevented from entering the atmosphere as a result of a shift to ZT from conventional tillage varies from about 4 tC/ha (14.5 tCO_2/ha) to at most 12.5 tC/ha (45.8 tCO_2/ha). Although seemingly high values, compared to forest plantations the amount of carbon that can potentially be prevented from entering the atmosphere by changing to zero tillage is small.

Research by Manley et al. (2005) comes to a more pessimistic conclusion even than West and Marland. They find that the costs on a per ton of carbon basis in going from conventional to zero tillage are enormous, and may even be infinite in some cases because there may be very little or no addition to SOC, particularly in North America's grain belt. Manley et al. conduct two meta-regression analyses to investigate the potential for the switch from conventional to zero tillage to create carbon offset credits that would be competitive with emission reductions. The first meta-analysis consisted of 51 studies and 374 separate observations comparing carbon accumulation under conventional tillage and ZT. A particularly important finding was that no-till cultivation may store no carbon at all if measurements are taken at sufficient depth. That is, the depth to which researchers measured SOC was important in determining whether there were carbon-sink gains from no-till agriculture. In some regions, including the Great Plains of North America, the carbon-uptake benefits of no tillage are non-existent. A possible explanation is that, under conventional tillage, crop residue is plowed under and carbon gets stored at the bottom of the

Table 9.2 Net costs of carbon sequestration, no till versus conventional tillage agriculture, North American regions, \$US2,010 per tCO_2

		Not corrected for emphemeral nature of carbon storage and at measured depth of soil		Corrected for emphemeral nature of carbon storage and at measured depth of soil	
Region	Crop	Shallow	Deep	Shallow	Deep
South	Wheat	\$3.40	\$4.26	\$68.04	\$85.29
	Other crop	\$0.66	\$0.66	\$13.12	\$13.26
Prairies	Wheat	\$127.18	∞	\$2,543.67	∞
	Other crop	\$49.83	\$70.25	\$996.55	\$1,404.94
Corn Belt	Wheat	\$48.03	\$62.98	\$960.50	\$1,259.52
	Other crop	\$28.42	\$29.21	\$568.35	\$584.11

Source: Adapted from Manley et al. (2005). Values in \$2,001 per tC were converted to \$2,010 per tCO_2 using the U.S. CPI (1.24) and conversion factor 12 tC–44 tCO_2

plow layer; with no-till, some carbon enters the upper layer of the soil pool, but as much CO_2 is lost from decaying residue as is lost from plowing in the case of conventional tillage.

In a second meta-regression analysis, Manley et al. examined 52 studies and 536 separate observations of the costs of switching from conventional tillage to zero tillage. Costs per ton of carbon dioxide uptake were determined by combining the two results, and are presented in Table 9.2. The viability of agricultural carbon sinks was found to vary by region and crop, with no-till representing a low-cost option in some regions (costs of less than \$10/tCO_2), but a high-cost option in others (costs greater than \$30/tCO_2). Nonetheless, in some limited circumstances no-till cultivation may yield a 'triple dividend' of carbon storage, increased returns and reduced soil erosion, but in most cases creating carbon offset credits in agricultural soils is not cost effective because reduced tillage practices store little or no carbon.

The studies included in the meta-regression analysis make no allowance for the ephemeral nature of the carbon sequestered in soil organic matter because of zero tillage agriculture. That is, soil scientists measured the difference between carbon levels in soil for zero versus conventional tillage, and the economic studies only looked at the extra costs (if any) imposed by zero tillage, so any carbon uptake was assumed to be permanent – its release at a later date is not penalized. As argued in Sect. 9.3 below, future release of carbon must be taken into account and, since the length of time that farmland will remain untilled is likely to be short, the original values need to be corrected to take this into account. In the two columns on the right in Table 9.2, the costs of carbon uptake are divided by 0.05 (see Table 9.6) to take into account the emphemeral nature of the agricultural carbon sinks. In that case, no-till agriculture is worthwhile undertaking only for some crops in the U.S. South, if the objective of reduced tillage is to remove CO_2 from the atmosphere.

Where continuous wheat, reduced (conservation) tillage and/or zero tillage are already in use, it is difficult to make the case that carbon offset credits are being created, because the so-called 'additionality' condition (discussed in Sect. 9.3) is violated. However, if landowners practicing conventional tillage can claim carbon offset credits by making a switch to reduced/conservation tillage to zero tillage (or to continuous cropping or to chemical as opposed to tillage fallow), it will be necessary to extend the claim to extant practitioners of reduced tillage, zero tillage and chemical summer fallow to prevent them from switching back to conventional practices to become eligible claimants in the future (see Lewandrowski et al. 2004, p. 11).

There is a further problem. The advantages of conservation and zero tillage are financial in the sense that there are fewer machinery operations. This cost saving offsets the cost of increased chemical use and the value of reduced crop yields (which may or may not be small). As more land is put into reduced or zero tillage and demand for energy crops increases, crop prices will rise. These factors will result in a greater loss in revenue from reduced crop yields, making conservation and zero tillage less attractive.

There are other ways to increase SOC besides reduced and zero tillage. These include increasing residue retention, regrowth of native vegetation, continuous cropping or reduced summer fallow, conversion from annual to perennial crops (e.g., including forages in a rotation), growing improved plant varieties, and so on. But these methods are often more costly to employ than simply switching from conventional to no till. In a comprehensive study of Australia's Carbon Farming Initiative, Kragt et al. (2011) find that the costs of increasing soil organic carbon in the central wheatbelt of Western Australia ranges from a low of \$40/tCO_2 to a high of over \$300/tCO_2 annually (with \$A = \$US1.06), depending on the pathway employed. Assuming a discount rate of 10%, this implies a cost of over \$400 per tCO_2 (and even more for lower discount rates)!

The conclusion from these analyses is that biological carbon sequestration activities are, for the most part, not competitive with CO_2-emission reduction efforts. An indication of comparable costs of reducing CO_2 emissions is provided by the prices found in markets where carbon offsets are traded, such as the European Trading System (see Fig. 8.5). As discussed in Sect. 9.3, there are other reasons to consider biological carbon sinks, but the case for including these activities in a policy arsenal for mitigating climate change does not appear to be a strong one. Before considering this further, we investigate carbon fluxes in a working forest where timber is harvested for products and/or bio-feedstock for generating electricity.

9.2 How Much Carbon Does a Biological Sink Sequester?

Suppose that we are interested in knowing the benefits of burning wood fiber for generating electricity. Our primary interest is related to the social costs and benefits, but we are also very much interested in CO_2 emissions reduction given the claim that bioenergy is carbon neutral. The neutrality argument only works when the timing of CO_2 removals from and emissions to the atmosphere do not matter. We argue otherwise: the timing of carbon flux is crucial. Therefore, it is imperative to weight CO_2 emissions and removals according to when they occur. This is the subject of the next subsection. Only then can we consider the role of carbon uptake in greater detail.

9.2.1 Discounting Physical Carbon

By discounting carbon, one acknowledges that it matters when CO_2 emissions or carbon uptake occurs – carbon sequestered today is more important and has greater potential benefits than that sequestered at some future time. Yet, the idea of discounting physical carbon is anathema to many who would discount only monetary values. However, the idea of weighting physical units accruing at different times is entrenched in the natural resource economics literature, going back to economists' definitions of conservation and depletion (Ciriacy-Wantrup 1968). One cannot obtain consistent estimates of the costs of carbon uptake unless both project costs and physical carbon are discounted, even if different rates of discount are employed for costs and carbon. To illustrate why, consider the following example.

Suppose a tree-planting project results in the reduction of CO_2-equivalent emissions of 2 t of carbon (tC) per year in perpetuity (e.g., biomass burning to produce energy previously produced using fossil fuels). In addition, the project has a permanent sink component that results in the storage of 5 tC/year for 10 years, after which time the sink component of the project reaches an equilibrium. How much carbon is stored? If all costs and uptake are put on an annual basis, we need to determine how much carbon is actually sequestered per year? Is it 2 tC or 7 tC/year? Clearly, 7 tC are sequestered for the first 10 years, but only 2 tC are sequestered annually after that time. Carbon sequestration, as stated on an annual basis, would either be that experienced in the first 10 years (7 tC/year) or in the infinite number of years to follow (2 tC/year). Suppose the discounted project costs amount to $4,000; these include the initial site preparation and planting costs plus any annual costs (maintenance, monitoring, etc.), appropriately discounted to the current period. If a 4% rate of discount is used, these costs can be annualized to $160 (=$4,000×0.04) – the amount that, if occurring each year in perpetuity, equals $4,000 in the current period. The costs of carbon uptake are then estimated to be $22.86/tC ($6.23/tCO_2) if it is assumed that 7 tC is sequestered annually and $80.00/tC ($21.82/tCO_2) if 2 tC is assumed to be sequestered each year. The former figure is used most frequently simply to make the project appear more desirable.

Suppose instead we intend to divide the $4,000 cost by the total undiscounted sum of carbon that the project sequesters. Since the amount of carbon sequestered is 7 tC/year for 10 years, followed by 2 tC/year in perpetuity, the total carbon absorbed is infinite, and the cost of carbon uptake would essentially be zero. To avoid an infinite sum of carbon uptake, an arbitrary planning horizon needs to be chosen. If the planning horizon is 30 years, 110 tC are sequestered and the average cost is calculated to be $36.36/tC ($9.91/tCO_2); if a 40-year planning horizon is chosen, 130 tC are removed from the atmosphere and the cost is $30.77/tC ($8.39/tCO_2). Thus, cost estimates are sensitive to the length of the planning horizon, which is not always made explicit in studies.

Consistent cost estimates that take into account all carbon sequestered plus the timing of uptake can only be achieved by discounting both costs and physical carbon. Suppose physical carbon is discounted at a lower rate (say, 2%) than that used to discount costs (4%). Then, over an infinite time horizon, the total discounted carbon saved via our hypothetical project amounts to 147.81 tC and the correct estimate of costs is \$27.06/tC (\$7.38/tCO_2). Reliance on annualized values is misleading in this case because costs and carbon are discounted at different rates. If carbon is annualized using a 2% rate, costs amount to \$54.12/tC [=\$160 ÷ (0.02 × 147.81 tC)], or \$14.77/tCO_2. If the same discount rate of 4% is employed for costs and carbon, the \$27.06/tC cost is the same regardless of whether costs and carbon are annualized.

The rate at which physical carbon should be discounted depends on what one assumes about the rate at which the damages caused by CO_2 emissions increase over time (Herzog et al. 2003; Richards 1997). If the damage function is linear so that marginal damages are constant – damages per unit of emissions remain the same as the concentration of atmospheric CO_2 increases – then the present value of reductions in the stock of atmospheric CO_2 declines at the social rate of discount. Hence, it is appropriate to discount future carbon uptake at the social rate of discount. "The more rapidly marginal damages increase, the less future carbon emissions reductions should be discounted" (Richards 1997, p. 291). Thus, use of a zero discount rate for physical carbon is tantamount to assuming that, as the concentration of atmospheric CO_2 increases, the damage per unit of CO_2 emissions increases at the same rate as the social rate of discount – an exponential damage function with damages growing at the same rate as the social rate of discount. A zero discount rate on physical carbon implies that there is no difference between removing a unit of carbon from the atmosphere today, tomorrow or at some future time; logically, then, it does not matter if the carbon is ever removed from the atmosphere. The point is that use of any rate of discount depends on what one assumes about the marginal damages from further CO_2 emissions or carbon removals.

The effect of discounting physical carbon is to increase the costs of creating carbon offset credits because discounting effectively results in 'less carbon' attributable to a project. Discounting financial outlays, on the other hand, reduces the cost of creating carbon offsets. Since most outlays occur early on in the life of a forest project, costs of creating carbon offsets are not as sensitive to the discount rate used for costs as to the discount rate used for carbon.

For many of the studies used in the meta-analyses of the costs of sequestering carbon (summarized in Sect. 9.1 above), the authors are unclear as to whether or not they did indeed discount carbon. In most but not all cases, discounting was implicit because carbon uptake was converted to monetary terms and then discounted. Failure to be explicit about the discounting of carbon can result in misleading cost estimates, thereby making it difficult for decision makers to choose appropriate policies for mitigating climate change.

In this section, we apply carbon discounting to determine the carbon fluxes attributable to forestry activities, including when these are combined with use of harvested timber for wood products or burning to generate electricity. Tracking carbon flux when timber is used for wood products, with carbon stored in such products, is particularly difficult for several reasons. First, the processes of creating wood products result in CO_2 emissions that could be substantial. Second, wood products substitute for steel, concrete and other materials in construction, and it is necessary to determine the extent of such substitution and the greenhouse gas emissions saved as a result. Of course, when timber prices go up as a result of increased demand for wood fiber as a bioenergy feedstock, or because increases in agricultural output prices lead to a reduction in forestland and timber output, other materials substitute for wood products and the added emissions of CO_{2e} must also be taken into account. This issue is addressed further in Sect. 9.3.

9.2.2 Carbon Flux in a Forest Ecosystem: Wood Products and Bioenergy

We begin by examining the carbon flux in forest ecosystems as a result of the activity of growing trees and then harvesting them for the purpose of burning in a power plant. We employ data for two relatively fast growing tree species (van Kooten and Folmer 2004, p. 367). For spruce, we use the following functional form,

$$v(t) = kt^a e^{-bt}, \tag{9.1}$$

where $v(t)$ is the volume of commercial timber (cubic meters per hectare, or m^3/ha) at time t, and parameter values are given by $k=0.25$, $a=2.00$ and $b=0.02$. For hybrid poplar, we use the Chapman-Richards functional form,

$$v(t) = \gamma(1 - e^{-\varepsilon t})^{\phi}, \tag{9.2}$$

with parameter values given by $\gamma = 300.00$, $\varepsilon=0.15$ and $\varphi=3.00$. The respective Faustmann or financial rotation ages are 23 years for spruce and 11 years for hybrid poplar, while maximum sustainable yield rotation ages are 50 and 13 years, respectively (van Kooten and Folmer 2004, p. 367).[3] Using a carbon-uptake model developed by van Kooten et al. (1999, 2000), we track carbon flux in the forest ecosystem.

For convenience, it is assumed that soil carbon remains unchanged over time, even though it will rise somewhat from the time trees are planted until they are harvested, then falling as a result of human disturbance due to harvest operations. Since both the commercial component of trees, known as the bole whose volume is actually measured by Eqs. (9.1) and (9.2), and the above-ground biomass are entirely removed at harvest by assumption, the only contribution to soil matter consists of dying leaves and fallen branches during the growing stage. Further, there is a point at which the soil carbon reaches a maximum and we assume that, since trees are planted on a previously forested site, this maximum has already been reached.[4]

---

[3] Tree growth could potentially be much higher than indicated in the text. For example, Clark Binkley (pers. comm., July 12, 2011) indicates that plantation forests in Chile used to produce biomass for energy can achieve growth rates of 100 m^3/ha/year, or uptake of more than 70 tCO_2/ha/year! The rotation age is only 3 years, so this is more like an agricultural crop.

[4] Asante et al. (2011) investigate soil carbon in forestry in greater detail. In their case, soil carbon declines over time, suggesting that the forest was in its original state so that soil carbon was above it long-term equilibrium for a managed forest.

The three factors that then determine the amount of carbon that is sequestered each year are the growth rate of the bole, the total amount of above-ground biomass relative to commercial biomass, and the specific gravity of the wood (i.e., the amount of carbon per unit of biomass). Total above-ground biomass (branches, leaves and bole) is 1.59 times the volume measured by Eqs. (9.1) and (9.2) (van Kooten et al. 1999, 2000). The amount of carbon in wood is approximately 190 kg/m^3 (Jessome 1977). The CO_2 sequestered in a given year through tree growth is given by:

$$\begin{aligned} CO_2 \text{ uptake} &= 1.59 \times \omega \frac{m^3}{ha} \times \frac{190\,kgC}{m^3} \times \frac{t}{1{,}000\,kg} \\ &\times \frac{44CO_2}{12C} = 1.1077\omega\, tCO_2\, ha^{-1}, \end{aligned} \tag{9.3}$$

where ω refers to the average annual growth in wood volume, or mean annual increment (mai), as determined by growth functions (9.1) or (9.2). Thus, the amount on the right-hand-side of (9.3) would be an annual amount of CO_2 sequestered in the forest ecosystem. Not included are the carbon fluxes at the time of harvest.

To calculate the CO_2 flux under conditions where trees are harvested and (1) used to produce long-lasting wood products or (2) burned to generate electricity, we employ the relationships found in Table 9.3.[5] Thus, for example, to determine the amount of CO_2 emissions saved when wood replaces coal in production of electricity, we first calculate the energy released when a hectare of wood is burned:

$$\begin{aligned} \text{Energy from burning wood} &= \frac{(n \times 1.59 \times \omega)m^3}{ha} \times \frac{1t\text{ wood}}{1.4m^3} \\ &\times \frac{15GJ}{t\text{ wood}} = 17.04 n\omega\text{ GJ}ha^{-1} \end{aligned} \tag{9.4}$$

where n is the number of years trees grow before they are harvested (or rotation age) and ω [= $v(n) \div n$] is the average growth each year, and remaining relations are found in Table 9.3. For spruce, $n=23$ years and $\omega=3.63$ m^3/year, with total above-ground biomass at harvest equal to 132.75 m^3/ha or 94.8 t wood/ha. For hybrid poplar, $n=11$ years and $\omega=14.38$ m^3/year, so total biomass at harvest equals

[5] All of the conversions in Table 9.3 are approximate and, in some cases, alternative values are used (as might be the case elsewhere in this book). For example, the energy released by burning various fuels will be different depending on the fuel and its quality (e.g., bituminous versus lignite coal), and on whether a low heating value (LHV) or high heating value (HHV) is employed. One place where energy and some other conversions are found is the website of the Oak Ridge National Laboratory (viewed March 21, 2010): http://bioenergy.ornl.gov/papers/misc/energy_conv.html. Data on conversions between bone dry biomass and volume of timber are found at (viewed March 21, 2010) http://www.globalwood.org/ tech/tech_wood_weights.htm, while remaining conversion data are provided in Niquidet et al. (2012).

Table 9.3 Some conversion factors and miscellaneous data employed in calculating benefits of replacing coal with wood in power generation

Wood conversions	*Energy conversions*
1 BDt = 2.65 m^3 (spruce)[a]	1 GJ (gigajoule) = 278 kWh (kilowatt hours)
1 BDt = 1.8 m^3 (poplar)[a]	1 kWh = 3.6 MJ (megajoule) = 3,413 Btu[b]
1 t wood = 1.4 m^3 solid wood	*Energy used in harvest*
	16 l diesel per bone dry ton (BDt)
Energy produced during burning	*Carbon content*
Wood: 15 GJ/t (20% moisture)	Coal: 746 kg/t
Coal: 27 GJ/t in power plants	Diesel/fuel oil: 2.77 kg C/3.79 l

Notes: [a] Wood residues of chips, shavings, sawdust, and, for spruce, bark
[b] Btu = British thermal unit
Sources: Energy/carbon conversions: http://bioenergy.ornl.gov/papers/misc/energy_conv.html
Wood conversions: http://www.globalwood.org/tech/tech_wood_weights.htm

Table 9.4 CO_2 emission savings from burning wood in a 100 MW capacity power plant, background information and results for two discount rates

Item	Spruce		Hybrid poplar	
Faustmann rotation age	23 years		11 years	
Mean annual increment (ω)	3.63 m^3/year		14.38 m^3/year	
Fiber available at harvest	132.75 m^3/ha		251.58 m^3/ha	
CO_2 emitted if harvested and burned	92.48 tCO_2/ha		175.27 tCO_2/ha	
Scenario analysis: Generating electricity in 100 MW capacity power plant				
Area required to support planting	153,000 ha		38,500 ha	
Discount rate for scenario (%)	*2*	*5*	*2*	*5*
Total discounted savings (Gt CO_2)	22.8	4.6	19.3	3.9
Savings over infinity per ha (tCO_2)	149.3	29.9	501.3	100.3
Annualized savings (tCO_2/ha)	3.0	1.5	10.0	5.0

251.58 m^3/ha or 179.7 t wood/ha (see Table 9.4). Thus, spruce produces 1,422.0 GJ of energy and releases 92.48 tCO_2/ha, while poplar produces 2695.5 GJ and releases 175.27 tCO_2/ha.

Turning to coal, we find that each metric ton (= 1,000 kg) of coal produces 27 GJ of energy while releasing 0.746 tC or 2.735 tCO_2. Suppose 1 ha of spruce is used to produce electricity in lieu of coal, and ignore the costs and CO_2 emissions associated with harvest and hauling of wood fiber as we ignore emissions associated with the mining and hauling of coal – essentially we assume these are offsetting. Then 52.7 t of coal are saved, amounting to a reduction in emissions of 144 tCO_2. If hybrid poplar is used as the bioenergy feedstock, comparable savings are 99.8 t coal and 273 tCO_2. If emissions of CO_2 from biomass burning are taken into account, total saving amount to about 51 tCO_2 for spruce and 98 tCO_2 for hybrid poplar. Clearly, all other things equal, hybrid poplar is preferred as a bioenergy feedstock to spruce.

The forgoing calculations are a very rough approximation because they do not take into account carbon discount rates and the time it takes to grow trees. To provide some indication of the impact these considerations have, consider a 100 MW capacity biomass-fired power plant. Assuming the plant runs at full capacity for the entire year, it would generate 2,628,000 megawatt hours (MWh) of electricity. From Table 9.3, the energy required to generate that much electricity would be 9,453,237 GJ which would require 630,250 t of wood annually, equivalent to harvesting 6650 ha of spruce or 3,500 ha of poplar per year. Thus, a spruce forest of at least 153,000 ha (1,530 km^2) or a poplar forest 38,500 ha (385 km^2), or some combination of the two, is required to keep the power plant in production. We use the term 'at least' to indicate that there is likely uncertainty related to the growth of trees, threat of wildfire, and so on, that should be taken into account.

If the power plant burned coal, it would require 350,000 t and release nearly 960,000 tCO_2 annually. By burning spruce, 873,000 tCO_2 are released annually, thereby saving 87,000 tCO_2; burning poplar releases 593,000 tCO_2, thereby saving 367,000 tCO_2/year. However, to obtain a true idea of the carbon saving over time, we need to discount carbon flux as to when it occurs. Each year 531,243 m^3 are added to the growing stock, amounting to an uptake of 370,000 tCO_2. To this one adds the saving from generating electricity from wood rather than coal, or 87,000 tCO_2, for a total annual saving of 457 Mt CO_2 (= 457,000 tCO_2). If the power was generated using wood from a hybrid poplar plantation, there would be an annual saving of 386 Mt CO_2. If the power plant were assumed to continue operating for an infinite time period (rebuilt as required, but ignoring the CO_2 emissions associated with such construction), the discounted reduction in CO_2 emissions or removals from the atmosphere is provided in Table 9.4. The annualized CO_2-emission reductions amount to 1.5–3.0 tCO_2/ha for spruce to 5–10 tCO_2/ha for hybrid poplar (Table 9.4).[6]

It is important to recognize that 385 km^2 of poplar plantation or 1,530 km^2 of spruce forest are required to provide sufficient biofeedstock for a 100 MW capacity power plant. Yet, annual emission reductions amount to a meager 1.5–10.1 tCO_2/ha if savings are taken over an infinite time horizon. Two sources of emissions have been ignored, however. First, when land is shifted out of agriculture into forestry, the carbon sequestered previously constitutes an opportunity cost that has not been taken into account. This is unlikely a problem in the case of spruce because the spruce forest would have been in existence in any event – and that is what we assumed. However, in the case of hybrid poplar, it tends to be grown as a plantation forest with short rotations. The land has generally been in native pasture that may have been storing a significant amount of carbon in soils; the release of that carbon would need to be taken into consideration. Further, it is likely that nitrogen fertilizer would be applied to hybrid poplar to provide the much higher mean annual increments compared to spruce (14.38 versus 3.63 m^3/year) and/or that irrigation is needed. Application of nitrogen fertilizers results in the release of N_2O into the atmosphere, which is a potent greenhouse gas (Crutzen et al. 2008), while irrigation requires electricity to pump water.

[6] The annualized values are obtained by multiplying the infinite amount of CO_2 saving per ha by the associated discount rate 2% and 5%.

Second, the area required to grow biofeedstock is enormous, and much larger than that of an open pit coal mine, for example. To harvest the trees and haul them to the power plant can result in a great deal of CO_2 emissions and, particularly, expense (see Niquidet et al., 2012).

We now consider what happens when we take harvest and hauling costs into account (but not other added costs or costs of hauling coal from the mine to the power plant).[7] Using data from Table 9.3, we can calculate the cumulative uptake of carbon in a forest ecosystem under the assumption that, when trees are harvested, all stored carbon is released back to the atmosphere. Clearly, without discounting physical removals from and emissions to the atmosphere, the total CO_2-sequestration benefit of forest activities is zero because as much CO_2 is emitted at harvest as is sequestered when the trees grow.

It is only when early removals of CO_2 from the atmosphere are preferred to later removals, or delaying CO_2 emissions is deemed beneficial, that forest activities are worthwhile considering. Therefore, we weight removals and emissions according to when they occur. The resulting discounted carbon flux for the simplest case of planting followed by later harvest (with no accounting for post-harvest fiber or CO_2 emissions associated with harvest and hauling activities) is illustrated in Fig. 9.1 for discount factors of 2 and 10%. The higher rate represents the discount rate one might use in assessing the financial feasibility of investments. Notice that, for the 10% discount rate, the discounted cumulative CO_2 uptake reaches equilibrium much quicker, but at a lower level, than with the 2% discount rate. Thus, choice of the rate used to discount physical carbon fluxes is important for determining the overall equivalent CO_2-emission reduction provided by forestry activities – the eventual carbon value of biological offset credits.

Now consider the case where harvested timber is used to produce wood products, such as lumber that is used to build houses. Again assume that there are no CO_2 emissions resulting from harvesting and hauling timber to mills. Further, assume that 50% of the carbon in the raw fiber ends up permanently stored in wood products. In that case, the total carbon sequestered will continue to increase in perpetuity if there is no discounting. With a discount factor for carbon of 2%, cumulative storage of carbon in perpetuity amounts to some 80 tCO_2/ha for spruce and 260 tCO_2/ha for hybrid poplar, as indicated in Fig. 9.2a. By raising the pickling factor to 0.8, cumulative carbon storage increases greatly to just over 100 tCO_2/ha for spruce and some 390 tCO_2/ha for hybrid poplar, as indicated in Fig. 9.2b.

In none of these examples do we account for CO_2 emissions resulting from the timber harvest operations. This is done in Table 9.5, which provides a summary of

[7] It makes sense to locate a power plant next to the coal mine (which is the case in Alberta). However, coal is sometimes shipped long distances. In some cases this is unavoidable because it is next to impossible to construct electrical transmission lines (e.g., Australian coal is shipped to other countries and used to generate electricity). In other cases, it might be preferable to generate electricity near the mine and ship it via high-voltage, direct current (HVDC) transmission lines (which experience least loss during transmission), thereby avoiding emissions from hauling coal long distance overland. This is the case in Ontario, where Alberta coal has been used to generate electricity.

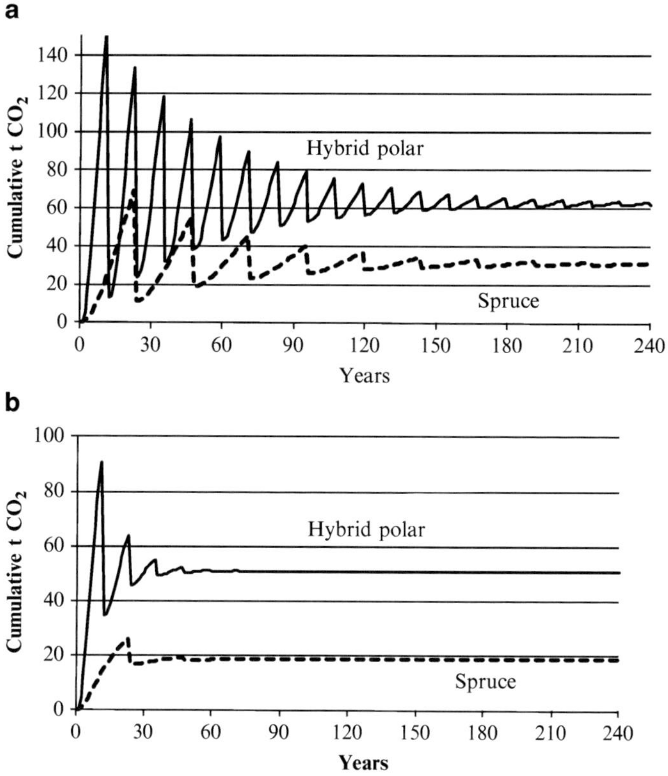

Fig. 9.1 (**a**) Cumulative carbon uptake in forest ecosystems, 2% discount factor, tCO_2/ha, (**b**) Cumulative carbon uptake in forest ecosystems, 10% discount factor, tCO_2/ha

cumulative and annual accumulations of carbon. Notice that, compared to sequestration and storage of carbon in products, harvest operations have only a slight impact on the overall carbon dioxide flux.

Finally, return to the case where post-harvest fiber is used to generate electricity. The results of this analysis are provided in Fig. 9.3 and the last two rows of Table 9.5.[8] These are at odds with the earlier analysis as they indicate cumulative carbon uptake that is only one-quarter (for spruce) to two-thirds (poplar) of that reported in Table 9.4. Part of the explanation for this difference is attributable to the inclusion of harvest-related CO_2 emissions and to the different method used to calculate the

[8] The only difference from the preceding analysis is that we calculate the energy from burning wood as follows: $\frac{(n \times 1.59 \times \omega)\text{m}^3}{\text{ha}} \times \frac{\text{Wood BDt}}{b\text{m}^3} \times \frac{20 \text{ GJ}}{\text{Wood BDt}} = \frac{31.8 n\omega}{b} \text{ GJ ha}^{-1}$, where b is the number of cubic meters of green wood required to make one bone dry ton, with $b = 2.65$ for spruce and $b = 1.80$ for poplar (see Table 9.3).

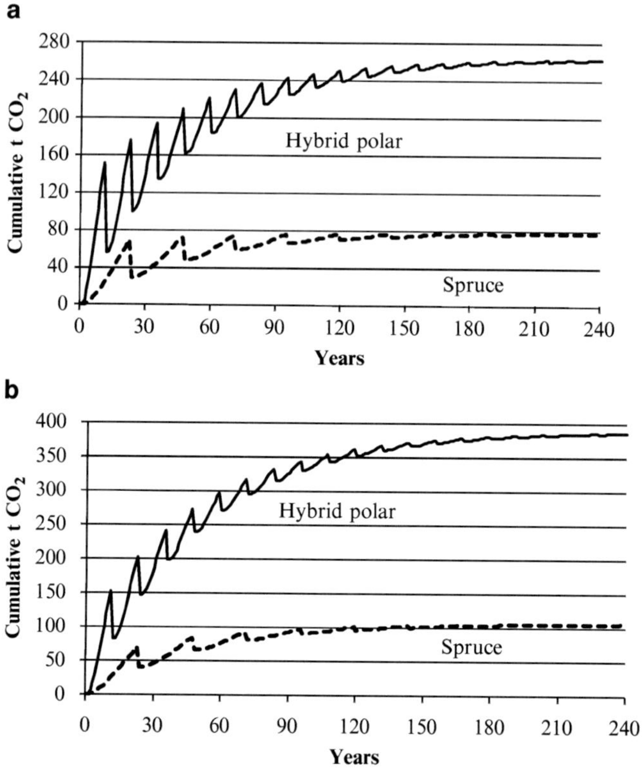

Fig. 9.2 (**a**) Cumulative carbon uptake in forest ecosystems with 2% discount factor, post-harvest use of fiber in wood products with pickling factor of 0.5, tCO_2/ha, (**b**) Cumulative carbon uptake in forest ecosystems with 2% discount factor, post-harvest use of fiber in wood products with pickling factor of 0.8, tCO_2/ha

Table 9.5 Cumulative and annualized CO_2 offsets created with spruce and hybrid poplar forests, carbon storage in forest products and biomass used to produce electricity in lieu of coal, 2% discount factor

	Spruce		Hybrid poplar	
Pickling factor	Cumulative	Annual	Cumulative	Annual
	(tCO_2/ha)			
No account of harvest-related emissions				
0.0	31.3	0.6	66.1	1.3
0.5	78.4	1.6	268.7	5.4
0.8	107.3	2.1	392.8	7.9
Including harvest-related emissions				
0.0	30.8	0.6	62.3	1.2
0.5	77.9	1.6	264.8	5.3
0.8	106.7	2.1	389.0	7.8
Biomass energy, including harvest-related emissions (no pickling)				
No fuel offset	30.8	0.6	62.3	1.2
Fuel offset	30.9	0.6	62.4	1.2

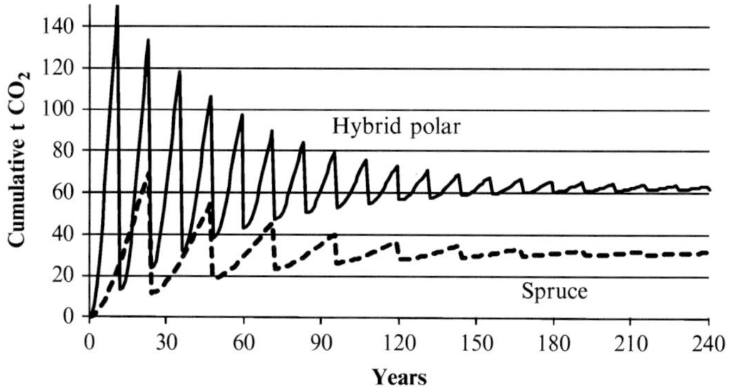

Fig. 9.3 Cumulative carbon uptake using a discount factor of 2%, with biomass used to generate electricity, and taking into account CO_2 released due to harvesting operations and emissions not released by burning coal, tCO_2/ha

energy substitutions. However, this would account for only a part of the difference. The other reason has to do with the fact that the earlier analysis was based on an entire forest continuously supplying wood fiber to a power plant as opposed to a single hectare of forest that grows over a period of 11 or 23 years and then is harvested, with post-harvest fiber used to generate electricity. Discounting then accounts for the difference. (In Chap. 10, we consider an example where wood biomass displaces both coal and natural gas in the generation of electricity.)

It is worth noting that it does not appear to make much difference whether post-harvest fiber is processed into wood products or used to generate electricity. What has not been taken into account in the current analyses, however, is a full accounting of wood products, both the greenhouse gas emissions released during their manufacture and the emissions saved because wood products replace concrete, steel and other building materials in construction. There is no easy way to account for these greenhouse gas fluxes as they depend on the forest ecosystem, the types of wood products created, the degree of substitution in construction and other uses, relative prices and so on. That is, the full accounting of greenhouse gas emissions is region and situation specific, and this is why no attempt was made to address this source of CO_2 flux here.[9]

What the foregoing examples illustrate is the great difficulty of properly accounting for all carbon fluxes associated with forest activities. This is a particular problem when carbon offsets generated by forestry activities are to be included in an emissions trading scheme, a topic discussed further in Sect. 9.3.

[9] The Consortium for Research on Renewable Industrial Materials (CORRIM) at the University of Washington in Seattle conducts research into wood products, their life-cycle emissions and the extent to which wood substitutes for other materials. Information can be found in two special issues of *Wood and Fiber Science* (v. 37, December 2005; v. 42, Supp. 1, 2010). An overview is provided by Lippke et al. (2010).

In an attempt to reduce reliance on fossil fuels, countries have increasingly turned to renewable sources of energy (see Chap. 10), including wood biomass. For example, western European countries – the EU-25 – have agreed on a binding target to achieve a 20% share of renewable energy sources in total energy consumption by 2020. In 2009, the share of renewables was approximately 7%. About two-thirds of the renewable energy is expected to come from biomass. If this is enforced, a Pöyry/McKinsey study forecasts an annual wood deficit for Europe of 200–260 million m^3 by 2020.[10] Canada is one of the world's largest producers and exporters of wood, but it only harvests about 200 million m^3/year. But the Europeans are not the only ones moving to greater reliance on wood for producing electricity.

The province of Ontario is looking to replace some or all of its coal-fired power with wood-pellet fired power, having provided significant feed-in tariffs for electricity produced from biomass. For biomass generators exceeding 10 MW capacity, the feed-in tariff is 13.0 ¢/kWh, while it is 13.8 ¢/kWh for smaller generators. Contracts are 20 years in length and subsidies (feed-in tariffs) are indexed to the Ontario Consumer Price Index. The tariff is also increased by a factor of 1.35 during peak hours (7:00 a.m. to 11:00 a.m. and 5:00 p.m. to 9:00 p.m.), but 0.90 for all off-peak hours. (Additional details of the feed-in tariff program are provided in Chap. 10, especially Table 10.3.) Already producers in British Columbia and Ontario are investing in wood pellet production for domestic use and export (see Stennes et al. 2010).[11]

Of course, the use of wood biomass to generate electricity is highly subsidized, but it has within it the seeds of its own demise: the price of wood products, including pellets, will increase offsetting the effectiveness of subsidies promoting energy production from wood. Further, higher prices will induce a shift in land use away from agricultural uses towards forestry. This shift in land use may be offset by biofuel subsidies that affect agriculture. Together wood biomass energy and agricultural biofuel programs will raise prices even more than such programs would do on their own. The results are higher feedstock prices, as well as higher wood product and agricultural output prices, and greater use of chemical inputs that offset the gains in reducing greenhouse gases that were the original intent of the programs. These issues are discussed further in the next section and later chapters.

9.3 Should Biological Carbon Offsets Be Included in the Climate Change Mitigation Arsenal?

It makes intuitive sense as a strategy for mitigating climate change to take account of carbon offsets generated by projects that promote tree growth or otherwise cause more carbon to be stored in biological ecosystems, including those that enhance

[10] The results were reported in presentations given in early 2010 by a financial analyst, Don Roberts, at Canada's CIBC bank.

[11] Wood pellets are easy to transport and can readily be used in lieu of coal in power plants; wood pellet production facilities are also simple to construct, and require relatively little capital investment.

soil organic carbon (IPCC 2000). The problem is that CO_2 removed from the atmosphere and stored in biological sinks can easily be released again to the atmosphere. Carbon sequestration is temporary and sometimes highly ephemeral. For example, one study found that all of the soil organic carbon stored as a result of 20 years of conservation tillage was released in a single year of conventional tillage (Lewandrowski et al. 2004). Likewise, tree plantations will release a substantial amount of their stored carbon once harvested, which could happen as soon as 5 years after first planting due to the use of fast-growing hybrid species. Yet, many scientists continue to be optimistic about terrestrial carbon sinks (IPCC 2000).

Of course, as we demonstrated in the previous section, forest activities that include post-harvest use of fiber for wood products or generation of electricity could reduce overall atmospheric CO_2. However, the vast areas of land required to bring about those benefits and the high relative costs of creating carbon offset credits (see Sect. 9.1), militate against the argument that forest activities in and of themselves can greatly contribute to climate mitigation.

Rather, the argument made in favor of including biological sinks hinges on the bridging benefits of such sinks – they would serve to remove CO_2 from the atmosphere during a transition period from an economy that is highly dependent on fossil fuels to one that is significantly less so.[12] The transition period would 'buy' time to enable society to develop the technologies required to reduce reliance on fossil fuels. Although the notion cannot be faulted, and the UNFCCC process commissioned a report promoting carbon sinks (IPCC 2000), four obstacles prevent implementation on a sufficiently large scale even though some projects will continue to be promoted. The obstacles are additionality, leakage, governance and duration. Each is considered in turn.

9.3.1 *Additionality*

It is difficult to determine whether an activity to reduce greenhouse gas emissions would not have been undertaken in the absence of policy to mitigate climate change – that it is truly additional. For example, the current UNFCCC rules permit payments to farmers who change the way they manage land so that it increases carbon uptake, such as when farmers move from conventional tillage to conservation or zero tillage agriculture. This increases soil organic matter and the carbon stored in soils. However, farmers have increasingly adopted conservation tillage practices because costs of chemicals to control weeds have fallen, genetically-modified crops are

[12] This argument has a counterpart in economics: Lenin and other communists argued that citizens were not yet capable of coping with or living in a purely socialist state, even though such a state was to their benefit; therefore, a transition period of dictatorship would be required (see Brown 2009). As argued below, the idea of a transition period during which sinks would sequester carbon is a solid one, but, in practice, the sink option is doomed by its drawbacks.

herbicide tolerant, fuel and certain machinery costs have risen (increasing tillage costs), and new cultivars reduce the impact of yield reductions associated with conservation tillage. If farmers adopt conservation tillage practices in the absence of specific payments for carbon uptake, they should not be provided with offset credits. Likewise, farmers who have planted shelterbelts should not be provided CO_2 subsidies unless it can be demonstrated that such shelterbelts are planted for the purpose of sequestering carbon and would not otherwise have been planted.

Determining whether large-scale tree planting projects are additional may be difficult. During the 1980s, Canada embarked on a major program to replant forestlands that had previously been harvested but had not regenerated 'valuable' species within a 15-year period. These lands were considered not sufficiently restocked, and substantial investments were made to clear weed species and establish more desirable ones. Had those trees been planted after 1990, the activity would have been eligible for carbon offset credits. Yet, this was clearly not an 'additional' activity.

The international community is on a slippery slope when sanctioning creation of carbon offsets from tree planting activities. Many such credits are nothing more than 'smoke and mirrors' that enable countries and firms to claim that they have reduced CO_2 emissions, but the claims are based on activities and investments that would have been undertaken in any event. That is, the country or firm has not made sufficient efforts to reduce actual CO_2 emissions.

There is no lack of schemes to generate carbon credits through forestry and agricultural activities. Even a cursory investigation finds there are many 'sellers' of such carbon offset credits. Several examples are cited here that illustrate the difficulty of judging whether forestry projects actually generate carbon offset credits for sale. In many cases, the projects cannot be considered as additional.

- "Greenfleet is an Approved Abatement Provider under the Australian Government's Greenhouse Friendly™ initiative … [whose] offset program has undergone independent scrutiny and meets appropriate standards. Greenfleet offsets greenhouse gas emissions by planting forests that soak up carbon dioxide from the atmosphere. Our forests are made up of a wide variety of Australian native trees that also help to reduce soil erosion and salinity, improve water quality and provide habitat for native animals."[13] How does it work? For $51 (tax deductible), Greenfleet will plant 17 native trees that will offset driving your car for 1 year; businesses can pay $A12.50 (US$13.60)/tCO_{2e} to offset their emissions, with such payments being tax deductible because Greenfleet is an approved abatement provider.[14] Greenfleet then uses funds to increase planting of native species in Australia (claiming to have planted six million since 1997), although no information is provided about the timing of carbon uptake and release, monitoring, et cetera. The Australian government's seal of approval is sufficient to ensure that Greenfleet's activities reduce atmospheric CO_2.

[13] See http://www.greenfleet.com.au/About_Greenfleet/index.aspx (viewed April 7, 2010).

[14] See http://www.greenfleet.com.au/Offset_emissions/index.aspx (viewed April 7, 2010).

- 'Trees for Life' is a conservation charity dedicated to the regeneration and restoration of the Caledonian Forest in the Highlands of Scotland. Interestingly, it claims that it is not possible to offset carbon emissions and become carbon neutral by planting trees.[15] It invites individuals and organizations to become 'carbon conscious' and uses the idea of a carbon footprint to solicit donations from individuals of £60 ($95), £140 ($220) and £280 ($440) depending on whether one's 'carbon footprint' is rated as light, medium or heavy (a guide is provided). Organizations are asked to contribute much more. For each £5 ($8) donation, 'Trees for Life' claims to plant one tree. No other details are available.
- The Haida-Gwaii Climate Forest Pilot Project off the northern coast of British Columbia intends to restore some 5,000–10,000 ha of degraded riparian habitat.[16] It hopes to fund the entire project by selling carbon credits, although alder that is "growing in an un-natural manner" would first need to be removed. The preferred mixed-conifer climax rainforest will eventually sequester 1928–2454 tCO_2/ha. Little in the way of cost data is provided and there is no indication about the timing of carbon uptake or potential future release, or loss of carbon from removing alder.

Given that the Haida Gwaii are committed to restoring ancient forests because they are part of their cultural heritage, and that 'Trees for Life' is committed to restoring the Caledonian Forest, the sale of carbon credits is part of a marketing technique to solicit funds for a project that may or may not have proceeded in any event. Much the same can be said about Greenfleet, but perhaps 'Trees for Life' is most honest in pointing out that there is a great deal of difficulty in offsetting CO_2 emissions via tree planting projects (see also below).

Some of the now many biological sink projects available to would-be purchasers of emission offsets provide clear carbon uptake benefits, but others are more dubious in nature. In some cases, projects are promoted because it happens to be convenient at the time. For example, the Little Red River Cree Nation in northern Alberta sought tradable carbon permits for delaying timber harvests, a delay caused by low prices associated with a decline in lumber demand; the request was subsequently turned down by the Canadian government. In other situations, such as that of a community group in Powell River, British Columbia, sale of carbon credits is necessary to help fund activities to prevent the harvest of coastal rainforest.[17] The latter constitutes a forest conservation activity that may well generate real carbon-uptake services, although one might want to consider whether the site could not generate even more carbon-uptake services if it were harvested and replanted (as required by law) – in which case it would not be eligible for carbon offset credits in any event. Likewise, tree planting projects that would proceed at a slower pace without carbon payments, such as those mentioned above, might well generate legitimate carbon offset credits.

---

[15] See http://www.treesforlife.org.uk/tfl.global_warming.html (viewed April 7, 2010).

[16] See http://www.haidaclimate.com/ (viewed April 7, 2010; originally viewed September 7, 2008).

[17] In both cases, the author was originally approached via telephone to help argue the case.

The problem here is that, in principle, a country should get credit only for sequestration above and beyond what occurs in the absence of carbon-uptake incentives, which has been described as the ‘additionality’ condition. If it can be demonstrated that a forest would be harvested and converted to another use in the absence of a specific policy to prevent this from happening, the additionality condition is met. Carbon sequestered as a result of incremental forest management activities (e.g., juvenile spacing, commercial thinning, fire control, fertilization) would be eligible for carbon credits, but only if the activities would not otherwise have been undertaken (say, to provide higher returns or maintain market share). Similarly, afforestation projects are additional if they provide environmental benefits (e.g., regulation of water flow and quality, wildlife habitat) not captured by the landowner and would not be undertaken in the absence of economic incentives, such as subsidy payments or an ability to sell carbon offset credits. If governments have a law that requires replanting following harvest, then additionality is ensured if no carbon offset credits can be generated from such forests.

When do LULUCF projects meet the additionality requirement and how do you tell? The United Nations Framework Convention on Climate Change (UNFCCC) does provide some guidelines to deal with additionality and other problems related to the certification of emission reductions via forestry activities (e.g., UNFCCC 2006), but problems remain. Plus, not only is it difficult to determine whether a carbon sequestration project is additional, but many other aspects of a carbon sequestration project are unknown and perhaps unknowable. Even when projects are declared legitimate by a certifying authority, information about the amount and timing of carbon uptake, release due to harvests or unexpected denudation by wildfire, pests or disease, and other aspects of the project is often lacking. For many CDM-initiated forestry activities that seek to create carbon credits, for example, projects fail to identify all of the carbon sequestration costs, the future path of carbon uptake and harvests, the risks of forest denudation, and so forth. Yet, in some cases projects are simply ‘picked up’ by companies seeking to improve their corporate image (which is what ‘Trees for Life’ is hoping for). Recall from Chap. 8 that many, often dubious, voluntary credits are sold to companies wishing to be ‘carbon neutral, and are sometimes even sold in legitimate markets. To some extent, the UN climate process has facilitated the sale of a variety of carbon offset credits by avoiding some of the difficult issues, as discussed after our consideration of the duration issue.

9.3.2 Leakage

Payments that promote direct changes in land uses for the purpose of carbon sequestration often result in indirect changes in land use that release CO_2, something known as a ‘leakage’. Examples of leakages occur at the micro and macro levels. At the micro-level, a landowner who is paid to plant trees might compensate for the loss in agricultural output by cutting trees at another location. At a macro-scale, tree planting causes agricultural output to decline, raising prices and causing landowners to expand cultivation onto marginal lands currently in permanent pasture or forest, thereby releasing CO_2. In the context of CDM forestry projects, the United Nations defines a 'leakage' as "the increase in greenhouse gas emissions by sources which occurs outside the boundary of an afforestation or reforestation project activity ... which is measurable and attributable to the afforestation or reforestation project activity" (UNFCCC 2006). Leakage estimates for forestry projects are exceedingly wide (5–93%), suggesting that project developers need to carefully consider leakages when designing carbon sequestration projects (Murray et al. 2004; Sohngen and Brown 2006; Wear and Murray 2004).[18] Wear and Murray (2004), for example, examine the effect of U.S. policy to address spotted owl habitat in the U.S. Pacific Northwest. This was done by setting aside (taking out of the harvest land base) certain old-growth forests on federal National Forests. The researchers found that, because of induced harvests elsewhere, the leakage resulting from the avoided CO_2 emissions from not harvesting the old-growth trees that had been set aside is 43.3% if only increased harvests on private forestlands in the western U.S. are taken into account. This increases to 57.7% if additional harvests elsewhere in the U.S. are accounted for, and to 84.4% if increased harvests in Canada are also included. Although the authors did not consider imports of lumber and other wood products from areas outside North America, these may also have increased. Overall, the leakage likely amounted to more than 85% of the CO_2 emissions avoided by not harvesting old-growth forest in the Pacific Northwest to protect spotted owl.

Leakages are often ignored when individual projects to create terrestrial offset credits are evaluated, but failure to include a 25% leakage factor, for example, underestimates costs by one-third (Boyland 2006). Nonetheless, leakages are generally ignored in bottom-up (or engineering cost) analyses.[19] Since top-down models take into account changes in prices and, thereby, indirect effects on land use, one would expect estimates of carbon uptake costs from bottom-up (technology) models to be lower than those from top-down models. As indicated below, the UN's climate process does require that projects take into account leakage; whether that is done in practice is another matter.

9.3.3 Transaction Costs and Governance: Carbon Sinks with Cap and Trade

Transaction costs refer to the costs of measuring, monitoring, enforcing and negotiating trades, while governance structures are the means by which trades are made. Both are affected by the institutional framework that exists in a country or, in the case of

[18] Leakage estimates for conservation tillage are substantially less than this (Pattanayak et al. 2005).

[19] Van Kooten and Folmer (2004) and van Kooten et al. (2004, 2009) could find no evidence that bottom-up studies had accounted for leakages. Hence, costs of carbon-uptake reported in Sect. 9.1 needed to be raised by at least one third.

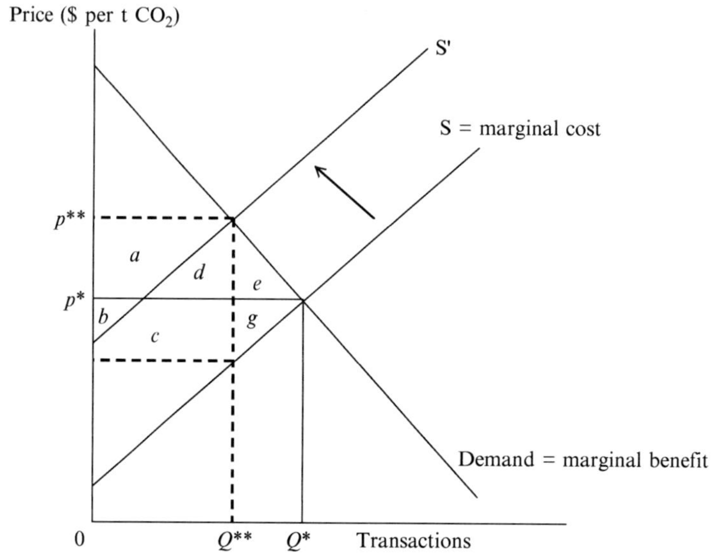

Fig. 9.4 Effect of transaction costs on trade in CO_2 offsets

international trading, by the nature of agreements between independent jurisdictions. Included in the institutional framework are such things as social capital, rule of law (independence of the judiciary) and freedom to engage in trade, while international agreements rely primarily on trust and the ability of one or more parties to an agreement to make credible threats should other parties not comply. The ability to impose credible threats on Kyoto-ratifying countries that fail to meet their obligations is pretty well non-existent – the offending state is required to reduce emissions even further than it would otherwise in a future, unspecified commitment period. It is at the individual country level, or at the level of a bloc such as the European Union, that one is most likely to encounter credible offset trading schemes that involve biological sinks. For example, the EU's emission trading system (ETS) permits the use of offset credits from LULUCF projects, including projects in developing countries that are certified under the CDM.

The effect of transaction costs can be illustrated with the aid of Fig. 9.4. For simplicity, assume we are only interested in the sales of temporary or biological sink CO_2-offset credits so that S refers to the supply and D to the demand for sink offset credits. In the absence of transaction costs, market equilibrium occurs where Q^* offsets are sold. As shown by Bovenberg (2002), transaction costs occur on both sides of the market, with respect to purchasers of offset credits and suppliers. For the current purposes, we consider only the supply side.

In Fig. 9.4, transaction costs cause the supply curve to shift upwards to S′ with equilibrium now equal to Q^{**}. Importantly, the price of biological sink credits has risen, implying that large industrial firms that need to offset emissions will shift

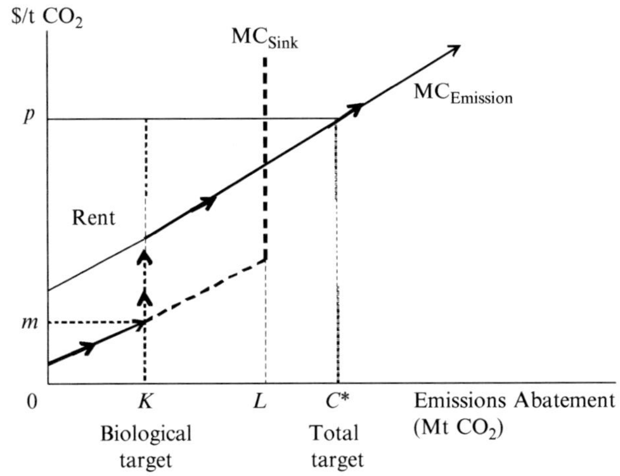

Fig. 9.5 Caps on terrestrial and total carbon: The arbitrage gap

purchases toward emission reduction offsets rather than sink offset credits. Total transaction costs amount to area $a+b+c+d$, but there is also a cost equal to area $e+g$ that constitutes the loss to society because some transactions are crowded out – emitters wishing to purchase offset credits must seek a more expensive means to meet emission reduction targets.

Are transaction costs a significant obstacle to biological CO_2-offsets? Relevant research reported by Slangen et al. (2008, pp. 204–205) indicates that these amount to one-quarter or more of the costs of providing nature services. Since transaction costs were ignored in the meta-regression results reported in Table 9.1, it is clear that transaction costs make biological sinks a lot less attractive, with forestry projects in the tropics (essentially via the CDM) and perhaps some fuel substitution projects in the boreal region left as the only options capable of competing with emissions reduction.

Biological CO_2 offsets pose a problem for emission trading schemes as it is difficult to compare and smoothly trade off temporary sink offsets against emissions reduction. In Fig. 9.5, the total amount by which CO_2 emissions must be abated (C^*) represents a country's internationally agreed upon target. Given the marginal cost of abatement function for emissions reduction as indicated ($MC_{Emission}$), the price of permits would equal p if only emission allowances were considered. Now, the availability of biological sink activities could lead a country to exceed its emissions reduction target by permitting sink offsets to substitute for emissions reduction. In the limit, if there are sufficient domestic and offshore carbon offsets available, a country could potentially maintain or exceed current levels of emissions and comply with emissions reduction targets. This is even more the case for a single large emitter. Therefore, it may be necessary to place a cap or target on the use of biological-sink generated CO_2 offsets (set at K in Fig. 9.5), as well as a target on emissions reduction that equals the difference between the overall targeted emissions reduction and the CO_2 offsets permitted through biological activities (=*C**–*K*).

Suppose that the emissions reduction target is set at *C**–*K*. Then, given the marginal cost of abatement function for emissions ($MC_{Emission}$), and with biological sequestration less costly than emissions reduction, the terrestrial option is always chosen over emissions reduction as long as the sink target is not reached. Arrows in Fig. 9.5 are used to indicate the direction along which costs should increase as a country mitigates its CO_2 emissions – this denotes the overall marginal cost of abatement function.

If, on the other hand, the sequestration option is more costly everywhere than emissions reduction, as suggested by the results in Table 9.1 (and particularly if transaction costs are included), then the biological sink option will not be chosen unless there are political reasons for doing so. For example, forest activities may be subsidized or CO_2 fluxes are not all fully taken into account (as discussed further below). In that case, setting a target for emissions reduction that is less than *C** will result in society not achieving the desired or targeted reduction in atmospheric CO_2.

Setting a target for biological carbon sequestration could also lead to potential political maneuvering. The existence of a gap at point *K* (the vertical section in the abatement function) implies that, unless there is competitive bidding to supply sink-generated offset credits, a scarcity rent is created and there is room for higher cost sequestration projects to push out lower cost ones as a result of rent seeking behavior.

Finally, a domestic carbon-trading scheme will also need to specify limits on the use of other Kyoto-type instruments over and above domestic biological sink offsets. Indeed, using the same reasoning as above, limits will also be needed on the credits available through the CDM and JI. If carbon trading is international in scope, caps on various mechanisms for creating credits will be required, at both the country level and supra-national levels. Further, given the difficulty of comparing tree planting and forest conservation programs (van Kooten 2009a, b; van Kooten and Sohngen 2007), it is also wise to set global and domestic caps on the amount of CO_2-offset credits that can be earned by preventing deforestation. Otherwise, every woodlot owner will want to receive offset credits whenever harvests are postponed for whatever reason (see Sect. 9.4 below).

9.3.4 Duration: Comparing Carbon Credits Across Temporary Projects

Consider a comparison between two climate change mitigation options, neither of which results in permanent removal of CO_2 from the atmosphere. Suppose that the more permanent of the two, say a policy that leads to a lower current rate of CO_2 emissions, leads to an increase in CO_2 emissions *N* years from now; the more ephemeral project generates temporary offset credits through sequestration of CO_2 in a forest ecosystem, but releases the CO_2 in n years. (The comparison could just as well be between two carbon sequestration projects of different durations.) What then is the value of a forest-sink offset credit relative to an emissions reduction credit? Suppose that a unit of CO_2 not in the atmosphere is currently worth $\$q$, but that the shadow price rises at an annual rate $\gamma < r$, where r is the discount rate. Then the value of emissions reduction is:

$$P = \sum_{t=1}^{N} \frac{(1+\gamma)^t q}{(1+r)^t} = \frac{1+\gamma}{r-\gamma} q \left[1 - \left(\frac{1+\gamma}{1+r}\right)^N\right], \tag{9.5}$$

while a sink offset would be worth some proportion α of the emissions reduction:

$$\alpha P = \sum_{t=1}^{n} \frac{(1+\gamma)^t q}{(1+r)^t} = \frac{1+\gamma}{r-\gamma} q \left[1 - \left(\frac{1+\gamma}{1+r}\right)^n\right]. \tag{9.6}$$

Upon taking the ratio of (9.6) to (9.5) and simplifying, we obtain the value of 'temporary' relative to 'permanent' storage:

$$\alpha = \frac{1 - \left(\frac{1+\gamma}{1+r}\right)^n}{1 - \left(\frac{1+\gamma}{1+r}\right)^N}, \tag{9.7}$$

which depends on the discount rate (r), the time it takes a ton of CO_2 stored in a forest ecosystem to return to the atmosphere (n), and the time it takes a ton of CO_2 not emitted today to increase emissions at a future date (N). Notice that the value does not depend on the price of carbon (q). As shown in Table 9.6, the proportional value of a sink credit to an emissions reduction credit (α) varies depending on the relationship between n and N, the discount rate, and the growth rate (γ) in damages from rising atmospheric CO_2.

The forgoing results have important policy implications that relate to the duration problem. It is clear that sink offset credits cannot be traded one-for-one for emissions reduction credits, even if the latter are not considered permanent; nor can credits from different sink projects be traded one-for-one without some adjustment for duration (say using Table 9.6). The conversion rate will depend on the length of time that each project keeps CO_2 out of the atmosphere, and, crucially, on the discount rate. For example, if a sequestration project can ensure that carbon remains sequestered for 10 years, it is worth only 0.11 of an emission reduction that ensures no future increase in emissions for 200 years if the discount rate (r) is 2% and the growth rate of damages (γ) is 1% (Table 9.6).

When the damages from atmospheric concentrations of CO_2 (shadow carbon prices) rise over time, the value of temporary sequestration will fall relative to permanent emissions reduction. However, while the demand for both temporary and permanent offsets is expected to increase as the price of a permanent emissions

Table 9.6 Value of ephemeral relative to permanent carbon credits, various scenarios

	$N=100$ years discount rate			$N=200$ years discount rate			$N=500$ years discount rate		
n to *N* ratio	2%	5%	10%	2%	5%	10%	2%	5%	10%
Growth rate of shadow price of carbon, $\gamma=0$									
0.01	0.023	0.048	0.091	0.040	0.093	0.174	0.094	0.216	0.379
0.05	0.109	0.218	0.379	0.183	0.386	0.614	0.390	0.705	0.908
0.10	0.208	0.389	0.615	0.333	0.623	0.851	0.629	0.913	0.991
0.15	0.298	0.523	0.761	0.457	0.769	0.943	0.774	0.974	0.999
0.20	0.379	0.628	0.851	0.558	0.858	0.978	0.862	0.992	1.000
0.25	0.453	0.710	0.908	0.641	0.913	0.991	0.916	0.998	1.000
0.30	0.520	0.775	0.943	0.709	0.947	0.997	0.949	0.999	1.000
Growth rate of shadow price of carbon, $\gamma=0.01$									
0.01	0.016	0.039	0.082	0.023	0.075	0.157	0.048	0.177	0.347
0.05	0.077	0.180	0.347	0.109	0.322	0.574	0.220	0.621	0.882
0.10	0.150	0.329	0.574	0.208	0.540	0.819	0.392	0.857	0.986
0.15	0.219	0.451	0.722	0.297	0.688	0.923	0.526	0.946	0.998
0.20	0.285	0.551	0.819	0.378	0.789	0.967	0.631	0.979	1.000
0.25	0.348	0.634	0.882	0.452	0.857	0.986	0.713	0.992	1.000
0.30	0.408	0.703	0.923	0.519	0.903	0.994	0.778	0.997	1.000

Growth rate of shadow price of carbon, $\gamma = 0.02$									
0.01	n.a.	0.030	0.073	n.a.	0.056	0.140	n.a.	0.135	0.314
0.05	n.a.	0.143	0.315	n.a.	0.252	0.530	n.a.	0.516	0.849
0.10	n.a.	0.266	0.530	n.a.	0.441	0.779	n.a.	0.765	0.977
0.15	n.a.	0.373	0.678	n.a.	0.583	0.896	n.a.	0.886	0.997
0.20	n.a.	0.466	0.780	n.a.	0.688	0.951	n.a.	0.945	0.999
0.25	n.a.	0.546	0.849	n.a.	0.768	0.977	n.a.	0.973	1.000
0.30	n.a.	0.615	0.897	n.a.	0.827	0.989	n.a.	0.987	1.000
Growth rate of shadow price of carbon, $\gamma = 0.04$									
0.01	n.a.	0.015	0.055	n.a.	0.022	0.106	n.a.	0.047	0.245
0.05	n.a.	0.076	0.245	n.a.	0.107	0.429	n.a.	0.215	0.754
0.10	n.a.	0.148	0.431	n.a.	0.204	0.674	n.a.	0.383	0.939
0.15	n.a.	0.217	0.571	n.a.	0.293	0.814	n.a.	0.516	0.985
0.20	n.a.	0.283	0.677	n.a.	0.373	0.894	n.a.	0.621	0.996
0.25	n.a.	0.345	0.757	n.a.	0.446	0.939	n.a.	0.704	0.999
0.30	n.a.	0.405	0.817	n.a.	0.512	0.965	n.a.	0.768	1.000

Notes: n.a. indicates not applicable as calculation cannot be made
Source: van Kooten (2009a)

reduction credit rises, the supply of temporary credits from biological activities might rise or fall as γ increases – it depends on the effect that an increase in the rate of damages has on the price of permanent offsets, and the costs of inputs into biological sequestration (if any), and how these in turn relate to the effect of relative price or duration on supply.

To judge sink projects in the absence of market data requires that a policy analyst interested in cost-benefit analysis make arbitrary judgments about the discount rate, the rate of increase in damages, and the conversion rate between different biological sink projects to account for differing durations. These are over and above assumptions and uncertainty related to vegetation growth rates, uptake of carbon in soils, wildfire, disease, pests and so forth, the majority of which are not explicitly spelled out in most analyses of terrestrial sink projects.

We do not know the rate at which economic damages increase as more anthropogenic emissions of CO_2 enter the atmosphere. If the rate of increase in damages equals or exceeds the discount rate, then CO_2 offset credits from sink activities are only worth n/N of an emissions-reduction credit. This is equivalent to assuming a zero discount rate for physical carbon. But this implies that temporary offsets from biological sink activities are overvalued because, as $N \rightarrow \infty$, the relative value of a temporary offset credit falls to zero. It is reasonable to assume that $N \rightarrow \infty$ if an emissions-reduction policy results in behavioral changes that cause permanent reductions in CO_2 emissions (e.g., car manufacturers stop producing SUVs as people demand smaller vehicles).

Given the difficulty of determining not only the discount rate and the growth rate in damages, but also the uncertainty surrounding n and N, it will simply not be possible for the authority to determine a conversion factor between activities leading to carbon credits of differing duration. Perhaps one can rely on the market to determine conversion rates, but even the market will have difficulty resolving all uncertainty, and can only do so if the authority sets rules for trading off temporary and permanent credits. These are necessarily arbitrary and, given high transaction costs associated with the creation (measurement, monitoring and trading) of biological offsets, sink credits are likely to sell more cheaply than warranted. Emitters will substitute cheap sink credits for more expensive emissions reduction credits, which reduces their incentive to invest in technologies to increase efficiency. As a consequence, the 'bridging' feature of biological sink activities actually serves to reduce incentives to conduct needed research and development (R&D) – it increases the length of the bridging interval required.

Finally, a country that uses carbon sequestration credits to achieve some proportion of its CO_2 emissions-reduction target during Kyoto's first commitment period has avoided emissions reductions. If it is to remain committed to long-term climate mitigation, however, the country must increase its emissions-reduction target in the next commitment period. It must meet that target plus the shortfall from the previous period – it still needs to reduce the emissions that were covered by forestry activities. Further, the country is technically liable for ensuring that the stored carbon remains there, which will be difficult given the non-permanence of forest sinks. For example, suppose a country relies on forest sinks for one-third of a 6% reduction in emissions, or 2% of emission reduction is achieved using forestry activities. Suppose the country then commits to a further 7% reduction in its emissions for the second commitment period. It must then reduce emissions in the second commitment period by an incredible 11%. Why? In the first period the country has only reduced emissions by 4%, but must reduce emissions by 9% during its second period commitment to meet the 13% overall target by the end of the second commitment period. But, because a forest sink will release its carbon to the atmosphere, the country must also cover that loss, which amounts to a further 2% reduction in emissions. The temporal shifting in the emissions-reduction burden caused by reliance on carbon sinks therefore results in an onerous obligation for future generations, one they may not be willing to accept.

9.3.5 Certifying Offset Credits from Forestry Activities

A major component of the Kyoto process is the ability that countries have to purchase emission offsets. A variety of different policy instruments have been developed by the UNFCCC process for this purpose.

- An 'emission reduction unit' (ERU) equals a reduction of 1 metric ton of carbon dioxide equivalent (tCO_{2e}) from the atmosphere, either by reduced emissions or removal via a sink activity. ERUs are earned in countries in transition through JI.
- A 'removal unit' (RMU) refers to the removal of 1 tCO_2 from the atmosphere via sequestration.
- An 'assigned amount unit' (AAU) refers to CO_2 emission credits that result when a Kyoto-ratifying country reduces its emissions by more than its Kyoto target. AAUs are available for sale to countries that are unable to achieve their targets and can use them to meet their obligations. It is primarily Russia and the countries of eastern Europe that are able to sell AAUs but other countries have been reluctant to do so because such purchases are viewed simply as an undesirable income transfer.
- A 'certified emission reduction' (CER) simply refers to a carbon offset has been certified by a country's Designated National Authority (DNA). AAUs are created in developed countries, while a CER is generated in developing countries under the CDM mechanism.

Despite this categorization, biological sink activities pose a difficult problem because of issues concerning additionality, leakage, duration and governance.

Recall that biological activities were permitted as a result of the 2001 COP meeting in Marrakesh, which essentially ratified the findings of the IPCC (2000) report on land use, land-use change and forestry (LULUCF). At the 2003 COP in Milan, countries came up with two new instruments to address the unique problems that biological activities posed, thereby helping to facilitate the use of biological activities undertaken in developing countries through the CDM. The two instruments were the temporary certified emission reduction (tCER) and long-term certified emission reduction (lCER).

A large industrial emitter or country can reduce emissions by one tCO_2, say, or, in lieu of actually reducing emissions, several other options open to it. First, it can simply 'purchase' offsets that are purely of domestic origin – perhaps from firms that have met their target (if the country employs a cap-and-trade scheme), paying a tax (if a country uses a carbon tax), or investing in a domestic tree planting or biofuels project that is agreed to by the national authority. Alternatively, the emitter can purchase an ERU, AAU or CER that offsets the CO_2 emission. The price of such an offset can range from several dollars per tCO_2 to perhaps \$30/tCO_2 or more. Say the firm can purchase emission offsets for \$30/tCO_2. The firm can also rent or lease a less permanent offset. Thus, a tCER can be rented each year – an annual payment is made rather than a one-time payment. If a permanent emissions reduction is purchased for \$30/tCO_2 and the discount rate is 5% (and markets function perfectly), then a temporary credit should be available for purchase for no more than \$1.50; if the cost is higher, the emitter will choose the more permanent option.

In the context of forestry, the tCER operates like an annual rental of a permanent offset or CER. The lCER, on the other hand, appears to operate as something in between the annual rental and the permanent reduction. Both instruments were created in response to the duration problem and, secondly, to reduce governance issues (particularly transaction costs). Consider a supplier of forest-based carbon offset credits under the CDM. The supplier must initiate a valid reforestation or afforestation project, although conservation projects are now also being considered (see Sect. 9.4 below).

Consider Fig. 9.6 where a forest owner plants trees to create carbon offset credits. According to rules initially developed in 2003 at the COP held in Milan, the forest landowner must first ensure that the afforestation or reforestation project is eligible. She then chooses the initial time when tCERs are offered for sale. In the figure, the landowner chooses T_1 as the first time to enroll tCERs for sale. At that time, the number of eligible tCERs for sale is given by $tCER_1$, which is equal to the total amount of carbon sequestered from time 0 to T_1 as a result of the tree planting decision (with sequestration determined against a baseline non-project use of the land). Thus $tCER_1$ emission reductions can be sold each year for 5 years, despite the fact that the site will continue sequestering carbon (at least for a number of years) beyond T_1 as indicted in Fig. 9.6.

After 5 years, the amount of carbon available on the site is re-evaluated, with the landowner eligible once again to sell whatever carbon is available on the site at time $T_1+5=T_2$. In this case, the eligible amount is given by $tCER_2$ (> $tCER_1$), which can then be sold in each year of the next 5-year period. Notice that the forest yields varying carbon offset credits, depending on tree growth and even harvests. Thus, 10 years after the initial sale of carbon offset credits, and T_1+10 (=T_3) years after initial tree planting, the tCERs available for sale has fallen dramatically to $tCER_3$ as a result of an intervening harvest. The sequestered carbon subsequently lost to the atmosphere as a result of harvests is completely ignored, except for the fact that the landowner's ability to sell tCERs has fallen.

It is also important to recognize that year 0 might be 1990, the year that the Kyoto process recognizes as the base line. Thus, any tree planting that occurred on

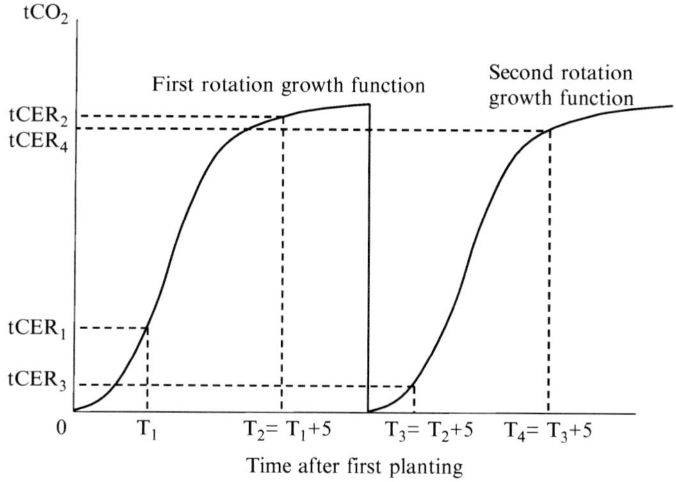

Fig. 9.6 Defining tCERs from forestry activities

or after that date, and meets the afforestation or reforestation eligibility criterion, could generate tCERs, and the amount could now be substantial as 30 years of growth might have occurred.

Of course, the landowner could also sell more permanent lCERs. These do not equal the amount of carbon on the stand but, rather, the change in carbon over the project life. In the context of Fig. 9.6, an lCER might equal $tCER_2 - tCER_1$, or for a different time period, $tCER_4 - tCER_3$. The landowner can sell an lCER that can be used by a buyer to set against today's emissions. However, the lCER remains a temporary fix. The purchaser is responsible at the end of the period (at time T_2 if the lCER represents the carbon sequestered between T_1 and T_2) for covering the lCER by buying a permanent emissions reduction credit or further temporary carbon offset credits.

Suppose a sequestration project involves planting spruce. From Table 9.4, each year a stand of spruce will grow an average of 3.63 m^3/ha, sequestering $\approx$4 tCO_2/ha if all above ground biomass is taken into account. The emitter can purchase tCERs amounting to 4 tCO_2/ha each year to offset 4 tCO_2 of emissions. Alternatively, the emitter can purchase an lCER of 20 tCO_2 (representing accumulated carbon over a 5-year commitment period) that can be used to offset today's emissions of an equivalent amount. In the latter case, however, the landowner or purchaser would in principle be responsible for ensuring that the carbon stays sequestered during the commitment period, and that in future commitment periods an equal amount of carbon is stored or that a permanent emission offset of 20 tCO_2 is purchased at the end of 5 years.

The price of the lCER would likely equal that of any other emission reduction (say \$30/tCO_2) unless the purchaser is legally responsible for the potential release of the carbon at the end of the period. In that case, the price would be discounted compared to permanent emissions reduction credits. The price of a tCER would similarly need to be discounted. In addition to the discount rate, the landowner's and purchaser's risks and evaluation of the probability that they will be held to task in a future commitment period will determine the relative prices between tCERs and lCERs, and between these and emissions reduction credits. A guide as to what relative prices should be is found in Table 9.6, but, given UNFCCC rules governing LULUCF projects, other governance issues, transaction costs and so on, there is no guarantee that the ratios provided there will be a guide for determining potential tradeoffs related to duration.

Finally, it should be noted that the UNFCCC rules governing LULUCF activities under the CDM, which also affects how domestic LULUCF projects in developed countries are treated, are meant to facilitate the integration of biological sinks into a climate change mitigation strategy as required by the 2001 COP agreements made at Marrakesh. Under UNFCCC guidelines, a designated operational entity (DOE) validates that the requirements for eligibility under the CDM are met. "The DOE shall, based on its verification report, certify in writing that, since its start, the afforestation or reforestation project activity under the CDM has achieved the net anthropogenic greenhouse gas removals by sinks" (UNFCCC 2006). But this does not necessarily mean that biological sink activities play a meaningful role in reducing atmospheric CO_2.

The problems of additionality, leakage and duration remain unaddressed, although UNFCCC rulemaking has addressed governance and transaction costs (even though the potential for corruption of DOEs remains). Despite all the efforts to facilitate LULUFC activities under the CDM, the first forestry project was approved only in November 2006, but it was not until January 2009 that the second project was approved. There then followed 11 additional approvals (with another two near completion of the approval process) by the end of January 2010.[20] Clearly, there remains some dissatisfaction with the role of biological sinks in reducing atmospheric CO_2 levels.[21]

9.4 Further Discussion

Terrestrial ecosystem activities that are used to generate CO_2 offset credits are a distraction from the actual job of mitigating climate change. While there is no question that carbon can be stored in biological sinks, and that care should be taken to foster such sinks and ensure that carbon is not unwontedly and needlessly released

---

[20] See http://cdm.unfccc.int/Projects/projsearch.html (viewed April 7, 2010). A number of agricultural projects have also been approved under the CDM but these deal primarily with livestock wastes (e.g., reduction of methane emissions) and use of wastes and residuals for generating electricity or biofuels. Land use and land-use activities were absent.

[21] Research reported in van Kooten and Sohngen (2007), van Kooten et al. (2009), and van Kooten (2009a, b) questions the validity of claims made by project proponents that forestry activities actually sequester the amounts of carbon claimed.

(e.g., via deforestation), the primary focus of climate change mitigation should be on policies that reduce greenhouse gas emissions. There are several reasons for this. First, measurement, monitoring and verification of sink activities is particularly difficult, resulting in high transaction costs that need to be added to the price at which temporary credits will trade. Transaction costs are sufficiently large so that most sink projects are simply not economically viable, at least based on estimates provided in Tables 9.1 and 9.2 that do not include transaction costs. This problem is compounded further if stored carbon must be accounted for in perpetuity.

If transaction costs related to terrestrial carbon sinks are the only concern, a straightforward way of reducing them is to employ contracts between the authority (government, trading exchange) and a landowner. The contract specifies the change in land use that the landowner will implement and an accompanying schedule of carbon flux – a schedule of annual CO_2 uptake for each year a land use is in place and the amount released when the land use changes, either to the previous use or some other. It also specifies the length of time that the land is to be kept in its new use and the penalties if the contracted-for use changes (including denudation due to wildfire, pests and disease). Further, there will be a requirement that the landowner pay a penalty or purchase emissions reduction offsets at the end of the contract period. Transaction costs are minimized because only land uses need to be monitored, not the CO_2 flux or anything else. There is only the cost of writing a contract. Under these circumstances, contracts can be traded in carbon markets, although it is likely that few landowners would undertake to purchase such contracts as the costs of providing the required services might be too high (see Tables 9.1 and 9.2). At international negotiations, countries could set separate targets for emissions reduction and biological sink activities, again using contracts with landowners to minimize transaction costs and facilitating exchange of contracts in a separate market.

Second, while it makes some sense to encourage carbon sinks because they offer a bridge to enable development of technologies with lower fuel emissions (e.g., more efficient vehicles), their existence as a sort of escape valve that enables countries and large emitters to delay or even avoid emissions reduction results in reduced incentives to invest in new technologies.

Third, rent seeking by opportunistic sellers of carbon credits, and even by environmental groups, highlights another important problem: terrestrial sinks remove CO_2 from the atmosphere at different rates and store it for varying lengths of time, with both removal rates and storage times embodying significant uncertainty. This facilitates the marketing of dubious sink offset credits. While this duration problem can readily be solved (e.g., taxing emissions and subsidizing removals at the time they occur), given the high transaction costs of including sink activities and the reluctance of countries to make sinks work, the only conclusion is that great care must be taken, and appropriate institutions put in place, before terrestrial ecosystem sink activities can be included in a carbon trading system.

Fourth, concerns about tropical deforestation have more recently led many commentators to commend the use of forest conservation in developing countries as an additional tool for addressing global warming, because deforestation accounts for more than one-quarter of all anthropogenic emissions. As already discussed in Chap. 8, activities that Reduce Emissions from Deforestation and forest Degradation (REDD) are seriously being considered as a method for earning certified emission reduction (CER) credits under the CDM. Some analysts are positive about its prospects (Wibe and Gong 2010), others are more cautious (Angelsen 2010). The problem with conservation or avoidance of deforestation is easy to illustrate. Anyone who has forestland or native grassland that stores substantial amounts of carbon can threaten to release the carbon by cutting trees unless they receive a payment for the carbon stored in it. Unless eligible forests are identified beforehand, which will result in vigorous rent seeking that increases transaction costs, the process is subject to abuse, unable to address issues related to additionality or leakage. It is much better to focus only on emissions to and removals from the atmosphere, and even then there remain problems.

Consider REDD in a way that is analogous to emissions trading. Under cap and trade, credits can only be earned by a country or an emitter if emissions are below a target. Without the target, emissions avoidance is nothing more than avoidance of debits; true credits can only be earned by removing CO_2 from the atmosphere. While it may be possible to mitigate CO_2 emissions by delaying (perhaps indefinitely) deforestation, there can be no credit for doing so unless there is some target level of deforestation so that, just as in the case of emissions avoidance, one gets credits by being below the target. Otherwise, the only benefit accrues from the avoidance of debits related to the carbon dioxide released when forests are cut.

It now appears that the definition of REDD credits will be extended to include sustainable management of forests, forest conservation and the enhancement of forest carbon stocks, collectively known as REDD+, thereby linking the UN's climate agreement with its Convention on Biological Diversity (Caparrós and Jacquemont 2003; Secretariat of the Convention on Biological Diversity 2009). Increasingly, climate negotiators appear willing to accept REDD+ activities as potential emissions offset credits to the extent that these activities also enhance biodiversity. The idea is that REDD+ generates co-benefits of forest conservation, including biodiversity preservation and other ecosystem services deliveries. Since deforestation and biodiversity are a greater problem in developing countries, and because industrial nations are also interested in providing indirect development aid through the CDM, only REDD+ projects in developing countries merit attention, although these still need to be approved under the CDM. However, as discussed in Chap. 8, voluntary emission reduction credits can be earned through REDD+ activities, including REDD+ activities in developed countries. Indeed, some voluntary carbon offset credits may well be appearing in legitimate markets.

By considering REDD+ activities in lieu of emissions reduction, climate negotiators (and by implication the countries they represent) have implicitly if not explicitly indicated that 'climate change' is not really about global warming. Rather, because biodiversity can be traded off against emissions, the issue of climate change is about the role that humans have on the environment. Climate change is simply a euphemism, or more palatable means of presenting the case, for minimizing the human footprint upon the globe. In this view, climate change is an ideology or environmental religion (Nelson 2010) that regards human activities as a grave threat to the Earth's ecosystems, and that these activities have to be severely curtailed. Climate change politics is certainly not about global warming.

Finally, consider the two problems that might be of greatest concern: (i) the ephemeral nature of biological sinks makes it difficult to compare biological activities with emissions reduction, while (ii) keeping track of CO_2 uptake and release in biological sinks requires measurement and monitoring, which are imprecise and expensive. An appropriate way to deal with the first issue – the problem of duration – is to count removals of CO_2 from the atmosphere and emissions reduction on the same footing. A debit occurs whenever an anthropogenic activity releases CO_2 into the atmosphere, regardless of the source. A credit is earned by removing CO_2 from the atmosphere and storing it in a terrestrial sink. The credit is the mirror image of an emissions reduction – one removes CO_2 from the atmosphere, the other avoids putting it there to begin with. Thus, if a forest is harvested, any carbon not stored in products but released to the atmosphere is debited (in the same way as emissions from fossil fuels). Likewise, any carbon released by decay of wood products, or any soil carbon released to the atmosphere, is counted as a debit at the time of release. If harvested fiber is burned in lieu of fossil fuels, a debit is also incurred but it is offset by the credit earned when growing biomass removes CO_2 from the atmosphere: The main benefit from biomass energy production is the reduction in CO_2 emissions from fossil fuel burning. An appropriate way to proceed, then, is for a country to tax debits and subsidize credits (as proposed by van Kooten et al. 1995).

References

Angelsen, A. (2010). The 3 REDD 'I's. *Journal of Forest Economics, 16*(4), 253–256.

Asante, P., Armstrong, G. W., & Adamowicz, W. L. (2011). Carbon sequestration and the optimal forest harvest decision: A dynamic programming approach considering biomass and dead organic matter. *Journal of Forest Economics, 17*(1), 3–17.

Bovenberg, A. L. (2002). Norms, values and technological change. *De Economist, 150*(5), 521–553.

Boyland, M. (2006). The economics of using forests to increase carbon storage. *Canadian Journal of Forest Research, 36*(9), 2223–2234.

Brown, A. (2009). *The rise and fall of communism*. Toronto: Doubleday Canada.

Caparrós, A., & Jacquemont, F. (2003). Conflicts between biodiversity and carbon sequestration programs: Economic and legal implications. *Ecological Economics, 46*(1), 143–157.

Ciriacy-Wantrup, S. V. (1968). *Resource conservation: Economics and policies* (3rd ed.). Berkeley: University of California, Agricultural Experiment Station.

Crutzen, P. J., Mosier, A. R., Smith, K. A., & Winiwarter, W. (2008). N2O release from agro-biofuel production negates global warming reduction by replacing fossil fuels. *Atmospheric Chemistry and Physics, 8*(2), 389–395.

FAO. (2004). *A review of carbon sequestration projects* (AGL Miscellaneous Paper AGL/MISC/37/2004, p. 83). Rome: Land and Plant Service, Land and Water Development Division, Food and Agriculture Organization of the United Nations.

Herzog, H., Caldeira, K., & Reilly, J. (2003). An issue of permanence: Assessing the effectiveness of temporary carbon storage. *Climatic Change, 59*(3), 293–310.

IPCC. (2000). *Land use, land-use change and forestry*. New York: Cambridge University Press.

IPCC. (2001). *The Marrakesh accords and the Marrakesh declaration*. Marrakesh, MA: COP7, IPCC. Viewed July 12, 2012. Retrieved from http://unfccc.int/cop7/documents/accords_draft.pdf.

Jessome, A. P. (1977). *Strength and related properties of wood grown in Canada* (Forestry Technical Report 21). Ottawa, ON: Eastern Forest Products Laboratory.

Kragt, M. E., Pannell, D. J., & Robertson, M. (2011). *Easy winnings? The economics of carbon sequestration in agricultural soils* (Working Paper 1111. 21 pp.). Crawley: School of Agricultural and Resource Economics, University of Western Australia.

Lewandrowski, J., Peters, M., Jones, C., House, R., Sperow, M., Eve, M., & Paustian, K. (2004, April). *Economics of sequestering carbon in the U.S. agricultural sector* (Technical Bulletin TB-1909, 61 pp.). Washington, DC: Economic Research Service, USDA.

Lippke, B., Wilson, J., Meil, J., & Taylor, A. (2010, March). Characterizing the importance of carbon stored in wood products. *Wood & Fiber Science Supplement, 1*(42), 5–14.

Manley, J., van Kooten, G. C., Moeltner, K., & Johnson, D. W. (2005). Creating carbon offsets in agriculture through zero tillage: A meta-analysis of costs and carbon benefits. *Climatic Change, 68*(1–2), 41–65.

Murray, B. C., McCarl, B. A., & Lee, H.-C. (2004). Estimating leakage from carbon sequestration programs. *Land Economics, 80*(1), 109–124.

Nelson, R. H. (2010). *The new holy wars: Economic religion versus environmental religion in contemporary America.* University Park: Pennsylvania State University Press.

Niquidet, K., Stennes, B., & van Kooten, G.C. (2012). Bio-energy from mountain pine beetle timber and forest residuals: The economics story, *Canadian Journal of Agricultural Economics, 60*(2), 195–210.

Padgitt, M., Newton, D., Penn, R., & Sandretto, C. (2000, September). *Production practices for major crops in U.S. agriculture, 1990–97* (Statistical Bulletin No. 969, pp. 114). Washington, DC: Resource Economics Division, Economic Research Service, U.S. Department of Agriculture.

Pattanayak, S., McCarl, B., Sommer, A., Murray, B., Bondelid, T., Gillig, D., & DeAngelo, B. (2005, August). Water quality co-effects of greenhouse gas mitigation in U.S. agriculture. *Climatic Change, 71*(3), 341–372.

Richards, K. R. (1997). The time value of carbon in bottom-up studies. *Critical Reviews in Environmental Science and Technology, 27*(Supp. 1), 279–292.

Secretariat of the Convention on Biological Diversity. (2009). *Connecting biodiversity and climate change mitigation and adaptation: Report of the second ad hoc technical expert group on biodiversity and climate change* (CBD Technical Series No. 41, p. 126). Montreal, QC: Secretariat of the Convention on Biological Diversity.

Slangen, L. H. G., Loucks, L. A., & Slangen, A. H. L. (2008). *Institutional economics and economic organization theory: An integrated approach.* Wageningen: Wageningen Academic Publishers.

Sohngen, B., & Brown, S. (2006). The cost and quantity of carbon sequestration by extending the forest rotation age. Draft. *AED Economics*, Ohio State University. Retrieved April 13, 2010, from http://aede.osu.edu/people/sohngen.1/forests/Cost_of_Aging_v2.pdf

Stennes, B., Niquidet, K., & van Kooten, G. C. (2010). Implications of expanding bioenergy production from wood in British Columbia: An application of a regional wood fiber allocation model. *Forest Science, 56*(4), 366–378.

UNFCCC. (2006). United Nations Framework Convention on Climate Change report of the conference of the parties serving as the meeting of the parties to the Kyoto Protocol on its first session, held at Montreal from November 28 to December 10, 2005. Addendum. Part two: Action taken by the conference of the parties serving as the meeting of the parties to the Kyoto. Viewed July 25, 2012, http://cdm.unfccc.int/Reference/COPMOP/08a01.pdf.

van Kooten, G. C. (2004). *Climate change economics. Why international accords fail.* Cheltenham/Northampton: Edward Elgar.

van Kooten, G. C. (2009a). Biological carbon sequestration and carbon trading re-visited. *Climatic Change, 95*(3–4), 449–463.

van Kooten, G. C. (2009b). Biological carbon sinks: Transaction costs and governance. *The Forestry Chronicle, 85*(3), 372–376.

van Kooten, G. C., & Folmer, H. (2004). *Land and forest economics.* Cheltenham: Edward Elgar.

van Kooten, G. C., & Sohngen, B. (2007). Economics of forest carbon sinks: A review. *International Review of Environmental & Resource Economics, 1*(3), 237–269.

van Kooten, G. C., Binkley, C. S., & Delcourt, G. (1995). Effect of carbon taxes and subsidies on optimal forest rotation age and supply of carbon services. *American Journal of Agricultural Economics, 77*(2), 365–374.

van Kooten, G. C., Stennes, B., Krcmar-Nozic, E., & van Gorkom, R. (1999). Economics of fossil fuel substitution and wood product sinks when trees are planted to sequester carbon on agricultural lands in western Canada. *Canadian Journal of Forest Research, 29*(11), 1669–1678.

van Kooten, G. C., Stennes, B., Krcmar-Nozic, E., & van Gorkom, R. (2000). Economics of afforestation for carbon sequestration in western Canada. *The Forestry Chronicle, 76*(1), 165–172.

van Kooten, G. C., Eagle, A. J., Manley, J., & Smolak, T. (2004). How costly are carbon offsets? A meta-analysis of carbon forest sinks. *Environmental Science & Policy, 7*(4), 239–251.

van Kooten, G. C., Laaksonen-Craig, S., & Wang, Y. (2009). A meta-regression analysis of forest carbon offset costs. *Canadian Journal of Forest Research, 39*(11), 2153–2167.

Wear, D. N., & Murray, B. C. (2004). Federal timber restrictions, interregional spillovers, and the impact on U.S. softwood markets. *Journal of Environmental Economics and Management, 47*(2), 307–330.

West, T. O., & Marland, G. (2002). A synthesis of carbon sequestration, carbon emissions, and net carbon flux in agriculture: Comparing tillage practices in the United States. *Agriculture, Ecosystems, and Environment, 91*(1–3), 217–232.

Wibe, S., & Gong, P. (2010). Can forests save the world? *Journal of Forest Economics, 16*(3), 177–178.

Chapter 10
Economic Growth, Energy and Climate Change

The evidence was the leaked e-mails of the University of East Anglia's Climate Research Unit (CRU), which are now subject to several official investigations, forcing the head of the CRU to step aside. The e-mails tell a lurid tale of unbecoming, unwarranted, organized and fierce hostility to skeptical climatic researchers, as well as data tampering, anti-scientific secrecy, manipulations of scientific journals, and distortions of peer review that make George Orwell look like a prophet. This could be dismissed as an isolated case if the CRU were some marginal backwater. But what was produced there was central to the scientific case, such as it was, mounted by the United Nations' Intergovernmental Panel on Climate Change.

– Christopher Essex, Toronto Sun, February 22, 2010.

James Hansen, Al Gore and other proponents of catastrophic anthropogenic global warming have called for drastic action to reduce greenhouse gas emissions. In an effort to do so, as already noted in Chap. 8, the leaders of the G8 countries agreed in 2009 to limit the increase in global average temperature to no more than 2 °C. They would do this by reducing their own greenhouse gas emissions by 80% or more, and global emissions by 50%, by 2050. To achieve these targets, it is necessary to radically transform the fundamental driver of global economies – the energy system. The main obstacle is the abundance and ubiquity of fossil fuels, which can be expected to power the industrialized nations and the economies of aspiring industrial economies into the foreseeable future. Realistically, global fossil fuel use will continue to grow and remain the primary energy source for much of the next century (Bryce 2010; Duderstadt et al. 2009; International Energy Agency 2009; Smil 2003).

The extent to which this prognosis will change depends on factors that are impossible to predict in advance. These include primarily the willingness of countries to spend vast sums on programs to reduce reliance on fossil fuels – to forgo cheap fossil fuel energy that emits CO_2 for much more expensive non-carbon

G.C. van Kooten, *Climate Change, Climate Science and Economics: Prospects for an Alternative Energy Future*, DOI 10.1007/978-94-007-4988-7_10,

energy sources, such as wind, solar, hydro, wave and tidal power, and, of course, nuclear power. They depend on the ability of governments to convince their citizens to accept large increases in energy prices and thereby reduced standards of living. They depend on high prices of fossil fuels relative to other energy options, and on very iffy and uncertain technological breakthroughs – a wicked uncertainty that the tools of economics cannot handle very well. Economists cannot predict technical advances, nor can others, because they depend on the minds and resourcefulness of citizens, and on educational, cultural and governance institutions.

President Obama announced on various occasions that the United States would embark on two new research programs that would enable America to retain its technological advantage over other countries – a program that would eventually put a man on Mars and a program to de-carbonize the U.S. economy, especially the electricity sector.[1] The President is counting on spinoff benefits of the kind that have characterized the U.S. industrial-military complex for the past 50 years and perhaps longer if research related to World War II is taken into account. Government funded military and space research under the Defense Advanced Research Projects Agency (DARPA),[2] originally created in 1958 as the Advanced Research Projects Agency (ARPA) in response to the Russian launch of Sputnik, led to technologies – the internet, micro chips, modern food processing and fast-food technologies, spandex, cell phones, et cetera – that are now common place.[3]

The impetus to rid the economy of fossil fuels might indeed change the playing field against fossil fuels. As we saw in Chap. 7, it is the type of research envisioned by Martin Weitzman, who favors a 'put-a-man-on-the-moon' type of R&D program for finding a technological solution that will enable humans to control the climate, and it is R&D that the Copenhagen Consensus expert panel prefers over other options for addressing climate change (Lomborg 2010). In this chapter, we address questions related to these efforts. What are the global challenges facing the energy sector in converting global (not just U.S.) economies from a fossil fuel basis to a non-fossil fuel basis? What are the prospects and the potential costs? Will the new technologies and energy sources reduce the anthropogenic component of global warming?

---

[1] Regarding Mars, Obama is quoted by the BBC (April 15, 2010) as saying: "By the mid-2030s, I believe we can send humans to orbit Mars and return them safely to Earth. And a landing on Mars will follow. And I expect to be around to see it" (http://news.bbc.co.uk/2/hi/8623691.stm, viewed April 21, 2010). Regarding Obama's desire to make the U.S. a leader in tackling climate change, thereby creating new technologies and jobs, see "Energy and Environment," White House, posted April 11, 2010 (http://www.whitehouse.gov/issues/energy-and-environment, viewed April 21, 2010).

[2] See http://www.darpa.mil/. "DARPA defines its mission as preventing technological surprise for the United States and to create technological surprise for adversaries" (DARPA: developing the wild, the wacky and wicked cool for 50 years, by M. Cooney at http://www.networkworld.com/community/node/24814, viewed April 20, 2010).

[3] Nowak (2010) provides an entertaining but compelling argument that technological advances are the result of large-scale, government-sponsored research efforts (often related to the military), followed by adaptations by private-sector companies to make them marketable. These companies often participated in the original research and receive their primary benefits from developing applications for non-military, final consumers.

10.1 Energy and the Economy

While good governance (low corruption, effective rule of law, etc.) is crucial to economic growth, economic development cannot occur without expanding energy use – rich countries are rich because they used and continue to use large amounts of energy to create wealth and provide citizens a high standard of material wellbeing (Smil 2003). By 2030, global energy use is expected to increase by nearly 50% over what is was in 2005; this will require the equivalent of one new 1,000 megawatt (MW) power generating plant coming on stream every day for the next 20 years just to satisfy growth in electricity demand (Duderstadt et al. 2009, p. 9). Likewise, the International Energy Agency (2010b) projects that, unless governments implement major policies to reduce energy use and carbon dioxide emissions, energy consumption will increase by 40% between 2007 and 2030, with three-quarters of this growth coming from fossil fuels. The lower 40% versus 50% projection is the result of taking into account the impact of the 2008 financial crisis and subsequent recession in North America and Europe.

The majority of growth in energy use will come from developing countries, especially China and India that together account for about one-third of the world's population. In 2010, Chinese emissions of greenhouse gases surpassed those of the U.S., although per capita emissions remain glaringly lower. Attempts by rich countries to reign in economic growth in developing countries for the purpose of mitigating climate change will be strongly resisted, although rich country subsidies for clean energy and investments in renewable energy will be welcomed by poorer nations. Energy policies that lower rates of economic growth in developing countries will simply perpetuate the misery of millions of people who live in poverty.

While clean and renewable energy sources can contribute to the energy needs of developing nations, economic growth will depend primarily on traditional sources of energy, such as coal, oil and natural gas, because they are relatively cheap and ubiquitous, and are a great improvement over heating with wood biomass, agricultural wastes, dung, et cetera, especially from a health standpoint. In Sects. 10.2 and 10.3, we consider in more detail the alternatives to fossil fuels and increased emissions of CO_2. In this section, we provide an overview of global energy use and trade.

10.1.1 Global Energy Markets

Fossil fuels are the most important source of energy in the world. This is clear when we look at energy used in the global generation of electricity (Fig. 10.1) and global consumption of energy (Fig. 10.2). Approximately two-thirds of electricity is produced from fossil fuels, while the remainder comes primarily from hydro and nuclear sources. Geothermal, biomass, solar, wind and other sources contribute a meager 2.8% of the energy required to produce electricity. World consumption of energy from various sources has increased steadily, with renewable sources of energy only gaining traction since the mid 1990s. This is evident from Fig. 10.3.

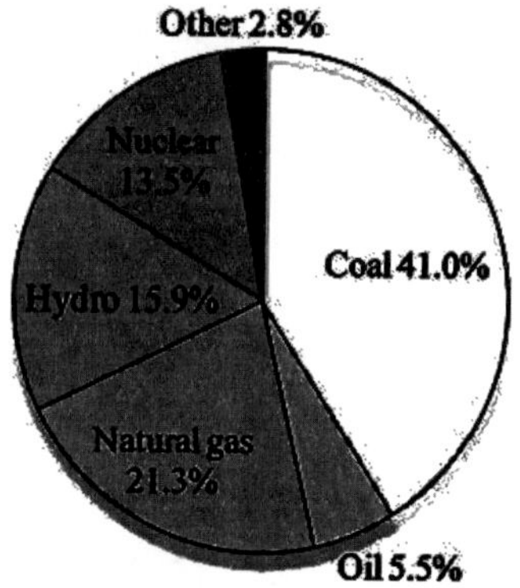

Fig. 10.1 Global electricity production by energy source, 2008, total = 20,181 TWh (Source: Key World Energy Statistics 2010 © OECD/International Energy Agency 2011, p. 24)

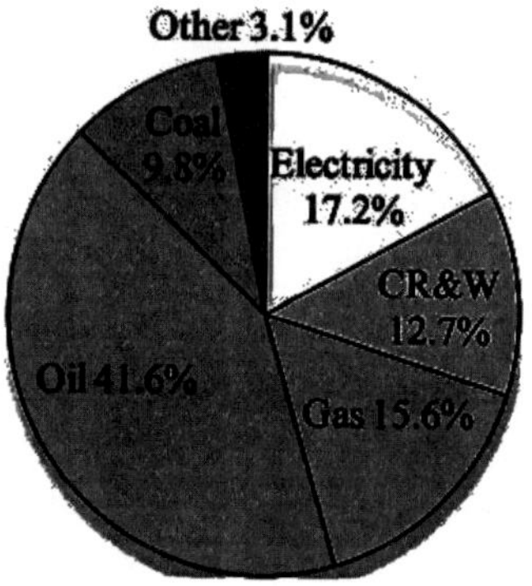

Fig. 10.2 Global energy consumption by source, 2008, percent, total = 8,428 Mtoe (Source: Key World Energy Statistics 2010 © OECD/International Energy Agency 2011, p. 28)

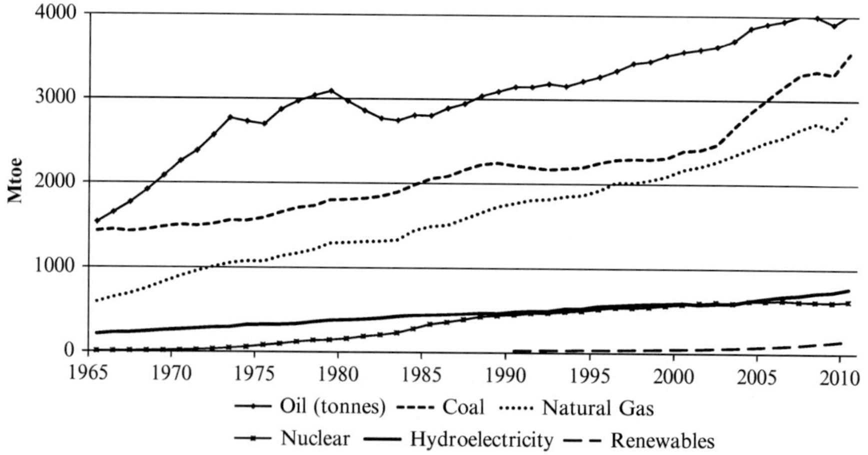

Fig. 10.3 Global energy consumption by source, 1965–2010, Mtoe (Data are from the BP Statistical Review of World Energy as found at (viewed December 2, 2011): http://www.bp.com/statisticalreview)

Table 10.1 Ten largest electricity producers, total and by fossil fuel energy source, 2008 (TWh)

Total		Coal/peat		Gas		Oil	
Country	TWh	Country	TWh	Country	TWh	Country	TWh
U.S.	4,344	China	2,733	U.S.	911	Japan	139
China	3,457	U.S.	2,133	Russia	495	Saudi Arabia	116
Japan	1,075	India	569	Japan	283	U.S.	58
Russia	1,038	Germany	291	UK	177	Mexico	49
India	830	Japan	288	Iran	173	Indonesia	43
Canada	651	S. Africa	241	Italy	173	Iraq	36
Germany	631	Australia	198	Mexico	131	Kuwait	36
France	570	Russia	197	Spain	122	Iran	36
Brazil	463	Korea	192	Thailand	102	India	34
Korea	444	Poland	143	Turkey	99	Pakistan	32
ROW	6,678	ROW	1,278	ROW	1,635	ROW	532
Total	20,181	Total	8,263	Total	4,301	Total	1,111

Source: Adapted from Key World Energy Statistics 2010 © OECD/International Energy Agency (2011), pp. 25 and 27
Notes: ROW refers to Rest of World

To obtain some notion regarding which countries generate the most electricity and the importance of coal in the global electricity generating mix, consider Table 10.1. More than 20,000 terawatt hours (TWh), or 20 petawatt hours (PWh),[4] of electricity were generated in 2008, the latest year for which statistics are available from the International Energy Agency (2011). This represents an increase of 2.1% over the previous year. Although the proportion of energy accounted for by all sources fell, except for hydroelectricity and 'other', there was an absolute increase in power generation from each source.

Notice that the U.S. and China are the largest producers of electricity and also the largest producers of coal-fired power. The U.S. is also the largest producer of electricity from natural gas and third largest user of oil (mainly from diesel) for electricity. Other large industrial nations generate large amounts of electricity, with many relying on coal (Fig. 10.1). Canada is the sixth largest producer, but much of it comes from hydro sources and a significant amount (≈ 25 TWh annually) is exported to the U.S. Clearly, rich countries are rich because they consume large amounts of energy, especially electricity.

Oil dominates total global consumption of energy, primarily because it is used for transportation and, to a much lesser degree, generation of electricity – mainly in diesel generators, although there are some generation facilities that rely on oil. With the exception of Japan, the countries that generate the most electricity from oil also tend to be major oil producers. The major producers, exporters and importers of crude oil are indicated in Table 10.2, as are the amounts involved.

[4] A watt (W) equals 1 joule (J) per second. A kilowatt (kW) equals 1,000 W; megawatt (MW) = 10^6 W; gigawatt (GW) = 10^9 W; terawatt (TW) = 10^{12} W; petawatt (PW) = 10^{15} W. Kilo is abbreviated with k and equals 10^3; Mega (M, 10^6); Giga (G, 10^9); Tera (T, 10^{12}).

Table 10.2 Ten largest global producers, exporters and importers of crude oil (Mt)[a]

Production		Net exports		Net imports	
Country	Mt	Country	Mt	Country	Mt
Russia	494	Saudi Arabia	355	U.S.	564
Saudi Arabia	452	Russia	241	Japan	199
U.S.	320	Iran	120	China	175
Iran	206	UAE	108	India	128
China	194	Nigeria	102	Korea	116
Canada	152	Angola	92	Germany	105
Mexico	146	Norway	90	Italy	88
Venezuela	126	Kuwait	89	France	83
Kuwait	124	Iraq	88	Spain	61
UAE	120	Venezuela	74	Netherlands	57
ROW	1,509	ROW	593	ROW	514
Total	3,843	Total	1,952	Total	2,090

Source: Adapted from Key World Energy Statistics 2010 © OECD/International Energy Agency (2011), p. 11
Notes: [a]Production statistics for 2009; exports and imports for 2008

Although Canada is not indicated as a major exporter, because the data on exports are for 2008, it is expected to move up the table in the future because of large oil sands development. Notice that both the United States and China are major oil producers, but they are also major importers because of the size of their economies.

Together fossil fuels (coal, oil and natural gas) account for 78.7% of total global energy consumption once account is taken of fossil fuels used in the generation of electricity. If combustibles, renewables and waste (CR&W)[5] are included, then 91.3% of all energy used globally comes from sources that emit CO_2. Of the remainder, 5.1% comes from hydro and nuclear sources, leaving less than 4% from solar, geothermal, wind, and tidal sources. Clearly, reducing reliance on fossil fuels in a big way presents a tremendous challenge.

Because fossil fuels are readily available, policies to replace them will likely require a combination of large subsidies (e.g., to producers of alternative fuels), regulations forcing firms and individuals to rely more on non-fossil fuel sources, publicly-funded R&D, and taxes or cap-and-trade schemes that drive up fossil fuel prices to the point where it makes economic sense for consumers to switch to alternative energy sources. However, there are limits to the amounts governments will pay to subsidize development of non-carbon sources of energy and to citizens' willingness to accept huge increases in the price of energy when cheaper fossil fuel alternatives are available. As argued in Chap. 7, it is morally objectionable to raise energy costs when poor people already pay too much for energy (Prins et al. 2010).

[5]This includes wastes from sawmilling and pulping, wood burned in stoves of subsistence farmers in developing countries, and wastes used for space heating and cooking.

One argument used to justify public spending on alternative energy is that the globe will run out of fossil fuels and that we need to prepare for that eventuality. For example, there are predictions that the world's oil production will soon attain 'Hubbert's peak' and begin to decline (Deffeyes 2003). Hubbert's peak is predicated on the notion that prices and technology remain unchanged. But Hubbert's peak will shift outwards with improvements in technology and higher prices. Indeed, from an economic standpoint, the idea that we will run out of oil (or gas or coal) is simply nonsense. We will never run out of oil, gas or coal. As these resources become increasingly scarcer, supply and demand intersect at increasingly higher prices to ensure that the market clears – there is always enough of the resource to meet demand. The higher prices will, in turn, signal scarcity and thereby induce technological innovations that will increase supply, reduce demand and/or lead to new sources of energy. This is evident from recent advances that have greatly expanded exploitable reserves of oil and natural gas.

Recent increases in the supply of oil have come from the Alberta oil sands and deep-water drilling.[6] Indeed, the planet is endowed with plentiful sources of oil and natural gas, so much so that we do not have an energy problem, but, rather, an energy strategy problem see Saleri (2011).

New natural gas drilling technologies were developed in North America beginning in the mid 1990s. These included horizontal drilling and fracturing of rock formations that enable gas to be extracted from shale formations in particular. This has resulted in massive upgrades in recoverable reserves and a surfeit of gas. Shale is globally ubiquitous and the drilling methods developed in North American can easily be repeated elsewhere. Global reserves of unconventional gas are estimated at about 1×10^{12} (trillion) cubic meters, or about five times as large as proven recoverable conventional reserves.[7] In terms of reducing CO_2 output, these developments position natural gas as the most likely alternative to coal for generating electricity (see Table 8.1).

At the same time, there have been advances in transportation and other technologies that reduce the amounts of energy to produce the same levels of economic services. Vehicles can travel farther on the same amount of fuel, new public transportation infrastructure has been built to reduce demand for fuel, and hybrid and electric vehicles are being brought to market. For example, automobiles in the United States

[6] Deep-water drilling will continue despite the massive oil spill resulting from the British Petroleum disaster in the Gulf of Mexico in 2010. If drilling is prevented in the U.S., this does not mean it will not be pursued by other countries. For example, Exxon has been prevented from exploiting deep water oil deposits in the Gulf and reserves off the northern coast of Alaska, but in late summer 2011 it reached an agreement with Russia to lend its technology to the exploitation of oil deposits off the northern shores of Siberia. American firms are involved in deep-sea drilling off the coast of Brazil. In Alberta, environmental concerns related to the oil sands development are increasingly addressed by new investments in technology and methods for restoring the environment. As the price of oil rises, willingness to incur costs to prevent environmental damage increase.

[7] See (viewed July 15, 2010): http://www.dawn.com/wps/wcm/connect/dawn-content-library/dawn/the-newspaper/letters-to-the-editor/breakthrough-in-gas-technology-240.

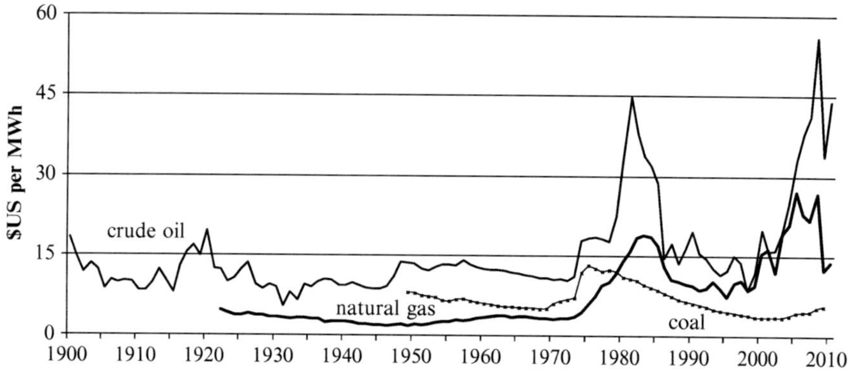

Fig. 10.4 U.S. oil, natural gas and coal prices, \$2,010/MWh, 1900–2010

require an average of 10 liters to drive 100 km, with those in Germany only slightly lower. Automobiles now coming onto the French market have a fuel economy of 5 liters per 100 km, despite relying on internal combustion engines, while economy might get down to 3 liters per 100 km as a result of better engines, lighter vehicles, et cetera (Gerondeau 2010, pp. 100–106). At the same time, costs of space heating have fallen as buildings have become 'greener.'

Costs of producing electricity from alternative wind and solar sources have fallen dramatically as well, while new geothermal, tidal, wave and other renewable energy technologies are in various stages of development. Advances in nuclear power generation technology and experience also continue, particularly with regards to performance and safety (Ansolabehere et al. 2003; Deutch et al. 2009). However, most of the renewable portfolio standards (RPS) programs implemented by many countries to address concerns about climate change "tend to exclude two important low-carbon technologies, nuclear and coal with CO_2 sequestration, confusing the objective of reducing carbon emissions with encouraging renewable energy in electricity generation" (Deutch et al. 2009, p. 9). That is, renewable energy is considered to reduce CO_2 emissions (even though biomass burning releases CO_2), while nuclear and so called 'clean coal' (coal with carbon capture and storage) are not.

What has driven these developments? First and foremost, market signals have played an important role. Consider Fig. 10.4.[8] In real terms, oil prices reached an initial high in 1981, but peaked again at an even higher level in 2008, before falling to slightly below the earlier peak by the end of the period. Natural gas prices peaked in 1983, and again in 2005 and 2008, before plunging as a result of recession and new developments in drilling technology. While oil and gas prices are historically above their levels in the period before the first 'oil crisis' in 1973, which was brought

[8] In Fig. 10.4, prices are from the U.S. Energy Information Administration (EIA), various tables viewed September 16, 2011. Nominal prices are deflated using the U.S. CPI and measures are converted to a per MWh basis from barrels (oil), cubic feet (natural gas) and short tons (coal) using conversion factors also available on the EIA website.

on by the exercise of monopoly power on the part of the Organization of Petroleum Exporting Countries (OPEC), they have exhibited more erratic movement since then (Fig. 10.4). Coal prices, on the other hand, peaked in 1975 and have since declined gradually to about one-third their historic high.

More recently, environmental concerns and political factors have prevented the expansion of drilling activities, while economic growth in developing countries, particularly China, has expanded demand. Together these factors led to higher real prices of oil. The same was true for natural gas, although natural gas prices are now falling as a result of the discovery of vast new reserves of unconventional gas. Anticipation of continued higher oil prices in the future has spurred technological changes, greater conservation and a switch to alternative fuels, including natural gas. The other incentive has been government policies, particularly subsidies. In the final analysis, however, the low price of coal relative to other energy sources remains a powerful incentive for countries to install coal-fired generating capacity, which is exactly what developing countries such as India and China are doing.

10.1.2 Renewable Energy Policy

Various countries are hoping to wean their economies off fossil fuels and thereby reduce CO_2 emissions. These countries have established renewable energy targets (renewable portfolio standards) and are in the process of implementing policies to meet targets – subsidizing the production of electricity from renewable sources and production of biofuels for transportation, and/or mandating levels of renewable energy so that costs are borne by consumers rather than the treasury. For example, a jurisdiction can require renewable standards for gasoline and diesel fuel, which will ensure that 20 or 40% (or some other proportion) of the fuel sold at the pump consists of biofuels. Electrical system operators may be required to purchase some minimum proportion of their power from renewable generating sources, or a country may mandate that a minimum proportion of electrical generating capacity come from renewable sources.

10.1.2.1 Scrambling to Reduce CO_2 Emissions: The Renewable Target Game

Many jurisdictions have now passed laws requiring that renewable targets be met. The European Union has set a target requiring 20% of total energy be derived from renewable energy sources by 2020, although only some 7% of energy was derived from renewable sources in 2009. To meet these targets, many countries will rely primarily on wind and energy from biomass. As noted in Chap. 9, an EU wood deficit of 200–260 million m^3 is forecast for 2020 as a result. An ECE/FAO report estimates that there will be a global wood deficit of 320–450 million m^3 annually simply to satisfy planned demand for energy plus a growing wood-based industry.[9] The deficit is almost double Canada's annual harvest of about 200 million m^3/year. Global wood fiber prices will certainly increase, resulting in potentially detrimental changes in land use.

The EU is also targeting vehicular use of renewables. By 2020, 10% of the fuel used for transportation must be a biofuel. As discussed in the next section, this target creates externalities that militate against such plans.

As an EU member, the United Kingdom's climate change mitigation plan also requires an increase in the share of renewable energy to 20% by 2020 (although 15% was originally targeted) from approximately 1% in 2006. The target requires that 35% of electricity generated in the UK come from renewable sources by 2020, compared to about 5% in 2007. Germany, on the other hand, has more ambitious climate goals than other EU members – a 40% reduction in greenhouse gas emissions from 1990 levels by 2020 (double the EU target). In addition, it aims to have 30% of its electricity generated from renewable sources by 2020, compared with 15.6% in 2009.[10] The latter target will be difficult to attain given that an earlier government had determined to cease nuclear power generation by 2022, although it accounted for 22.6% of consumption in 2009. Environmentalists and the near meltdown of Japan's Fukushima Dai-ichi nuclear power plant in the wake of the March 2011 earthquake and tsunami will make it difficult to extend this deadline.

The United States has yet to pass comprehensive climate change legislation (as noted in Chap. 8), but its farm legislation requires the production of 36 billion gallons of renewable fuels by 2022, including 21 billion gallons of 'advanced' (non-corn starch) biofuels. Some 50 Mt of wood is to be converted to fuel by 2012, with a targeted 70–100 Mt by 2020; the Biomass Crop Assistance Program (announced June 8, 2009) will provide subsidy of $45/t. This has the potential to result in an annual subsidy of $4.5 billion by 2020.

The Kerry-Lieberman-Graham bill promoted by the Obama administration in early 2010 sought to cut greenhouse gas emissions by 17% from 2005 levels by 2020 and by more than 80% from 2005 levels by 2050.[11] One of the bill's sponsors, Senator Lindsey Graham, a Republican who collaborated with the Democrats in crafting the energy bill, withdrew his support because of polls showing that 71% of Americans opposed higher gasoline prices to combat climate change (see also Chap. 8). Americans also view the cap-and-trade scheme proposed in this legislation as the equivalent of a carbon tax. The Renewable Electricity Standard bill sponsored by Senator Jeff Bingham requires electricity producers to generate 15% of their power from wind and other renewable sources by 2021.

Even China hopes to produce 10% of all its energy needs from renewables by 2010, with a target of 15% by 2020. Most of this will come from farm biomass and forest plantations. However, it will be a logistical challenge to transport

---

[9] Results reported by Don Roberts, CIBC, in presentations given in early 2010.

[10] See *The Economist*, September 4, 2010, pp. 53–54.

[11] Information based on an editorial in *The Washington Times*, April 27, 2010, entitled "Meltdown of the climate-change bill." Senator Graham subsequently dropped his sponsorship of the bill.

150,000–200,000 tons of bulky straw each year from thousands of 0.15 ha farms to fuel a large number of 25 MW capacity power plants. The target of planting 13.3 million ha of forests for bio-feedstock will be accomplished with help from rich countries through the Clean Development Mechanism. In effect, these efforts could be counted twice – they enable China to meet its renewable energy targets, while making it possible for developed countries that purchase CDM offset credits to achieve their targets as well (at least until changes are made to the system of crediting offsets).

Other countries have their own targets. Like the U.S., Canada is in the process of increasing biofuel production, but it also has a target to eliminate all coal-fired power generation by 2020. Both targets will be extremely difficult to meet, requiring large subsidies that will see electricity prices rise, greater reliance on natural gas, and the expansion of nuclear generating capacity, which is increasingly unlikely in the short term. Consider the case of Ontario as an example of the direction policy has taken in efforts to increase generation of electricity from renewable energy sources.

10.1.2.2 Feed-in Tariffs: The Case of Ontario

Because electricity grids have their own peculiar dynamics (discussed in Chap. 11), feed-in tariffs have been used in many countries to increase the use of renewable energy sources in the generation of electricity. One of the most ambitious attempts to affect power generation from renewable sources was launched by the Ontario government when it passed the *Green Energy and Green Economy Act* on May 14, 2009. Its feed-in tariff (FIT) schedule is provided in Table 10.3. The important thing to note is that the feed-in tariffs are indexed to inflation, with the exception of solar power (mainly because the subsidy is high to begin with and prices of solar panels are expected to fall dramatically in the future).

At the end of 2010, installed coal-fired generating capacity amounted to nearly 4,500 megawatts (MW), compared with 11,400 MW of nuclear capacity and nearly 9,000 MW available from combined-cycle gas turbines (CCGT). Thus, Ontario's installed base-load capacity amounts to 24,900 MW (Fox 2011). In addition, the system has 7,900 MW of hydroelectric and 500 MW of open-cycle gas turbine (OCGT) capacity, which can be used to meet peak load (although actual hydro generating capacity will vary depending on the amount of water in the reservoirs at any given time). Finally, Ontario has 1,046 MW of wind capacity, with another 765 MW scheduled to come on stream by 2013.

To provide some notion of the potential subsidies under Ontario's FIT program, it is necessary to have information about wholesale electricity rates and generating sources. Electricity demand in July is about equal to that in January because summer days are hot and humid and electricity is used for cooling. During July 2011, hourly demand averaged 17,874 MW with demand peaking at 25,450 MW.[12]

[12] Information in this paragraph is from (viewed September 16, 2011): http://www.ieso.ca/imoweb/siteShared/monthly_update.asp?sid=ic.

Table 10.3 Ontario power authority's feed-in tariff (FIT) program for renewable energy projects, base date: September 30, 2009

Renewable type	Size (capacity of generating plant)	Contract price (¢/kWh)[a]
Biomass[b]	≤10 MW	13.8
	>10 MW	13.0
Landfill gas[b]	≤10 MW	11.1
	>10 MW	10.3
Biogas[b]		
On-farm	≤100 kW	19.5
On-farm	>100 kW, ≤250 kW	18.5
Biogas	≤500kW	16.0
Biogas	>500 kW, ≤10 MW	14.7
Biogas	>10 MW	12.2
Wind		
On-shore	Any size	13.5
Off-shore	Any size	19.0
Solar		
Roof/ground	≤10 kW	80.2
Roof top	>10 kW, ≤250 kW	71.3
Roof top	>250 kW, ≤ 500 kW	63.5
Roof top	>500 kW	53.9
Ground mount	>10 kW, ≤10 MW	44.3
Water power[b]	≤10 MW	13.1
	>10 MW, ≤50 MW	12.2

Source: http://fit.powerauthority.on.ca/Storage/99/10863_FIT_Pricing_Schedule_for_website.pdf (viewed April 21, 2010) and http://fit.powerauthority.on.ca/sites/default/files/FIT Price Schedule_June 3 2011.pdf (viewed November 30, 2011)
Notes: [a]Generally a 20-year contract with 2–3-year lead time; for hydro, 40-year contracts. Indexed by the Ontario CPI, except solar
[b]Performance factor: 1.35 peak, 0.90 off peak

Total demand for the month of July amounted to 13,298 gigawatt hours (GWh), half of which was satisfied using nuclear assets, while coal accounted for 8% of generation, hydropower for 20%, wind resources for 17%, and the remainder was supplied by natural gas or imported. Weighted hourly wholesale price averaged 3.71¢/kWh; the weighted average price was 4.31¢/kWh during peak times (8:00AM to 8:00PM) and 3.14¢/kWh during off-peak times (8:00PM to 8:00AM). During the first 7 months of 2011, 83,536 GWh of power was generated and the weighted average hourly wholesale price electricity was 3.71¢/kWh.

Assume that an annual demand of 145,000 GWh needs to be met; baseload represents approximately 96,000 GWh, or nearly two-thirds of total demand. Currently, nuclear generating assets supply about half of the total load, with coal accounting for perhaps 10% as the province seeks to reduce reliance on coal. The remaining supply comes from variable wind and hydroelectric and natural gas sources, with the latter two helping to meet baseload demand and contribute to reserves (see Chap. 11).

Table 10.4 Current and projected installed generating capacity for Ontario, and associated CO_2 emission factors

Generator	Installed capacity (MW) Current	Projected[b]	Capacity factor[a]	Emissions (tCO_2/MWh)
Nuclear	11,400	9,900	0.80	0.01–0.06
Biomass	0	3,000	0.75	0.23
Coal	4,500	0	0.75	0.86–1.13
Wind	1,046	3,000	0.18	0–0.02
Hydro	7,900	7,900	0.40	0–0.03
CCGT	9,000	9,000	0.70	0.42–0.58
OCGT	500	500	0.25	0.50–0.75

Source: Emissions are derived from Fox (2011) and van Kooten (2010). Other data based on author calculations

Notes: [a]Capacity factor is the annual output from the electrical generating source divided by installed capacity multiplied by the number of hours in a year. Assumptions based on approximate supply from indicated source

[b]Author projections for base case scenario

The Ontario government will decommission all 4,500 MW of coal-fired capacity by 2014, and 1,500 MW of nuclear capacity by 2018 (Fox 2011). Some of the coal-fired power plants will be converted to burn biomass. For the current analysis, as a base-case scenario, assume that 3,000 MW of the coal capacity is replaced by biomass and that the remainder plus the decommissioned nuclear capacity is replaced with wind. Thus, 3,000 MW of wind capacity will be added to the approximately 1,000 MW already in place. Because wind is intermittent, we further assume that its capacity factor is 18%; although capacity factors for wind sites exceed 30% in cases, the overall capacity factor falls as increasingly marginal wind sites are exploited (see also Table 11.1). In the sensitivity analysis, we also consider a capacity factor of 25%. Whether the capacity factor is 18 or 25%, annual wind generated power amounts to no more than 7 GWh and, given that wind power must be delivered when it is produced, this leaves more than 138,000 GWh to be met by other generating assets. A summary of current and base-case projected available capacity by generating source and CO_2 emissions by energy source are provided in Table 10.4.

The costs of the FIT program to Ontario consist of a subsidy component and the costs of removing extant generating assets and replacing these with more expensive assets for producing electricity. If the amount of the subsidy is added to the latter cost, there is some double counting because FIT payments compensate producers who generate power from a more costly source. The FIT payment constitutes an income transfer from society to power producers – it is only a cost to the public treasury (taxpayers) or electricity users if the cost of the subsidy is passed along to ratepayers; however, the costs of removing coal-fired and nuclear assets, converting old or installing new generating plants, and integrating electricity from renewable sources into the electricity grid represent the true cost to society. (The difference

Table 10.5 Construction and operating costs of various generating assets, $2,010

		Construction costs ($/kWh)			Variable costs ($/MWh)	
Asset	Yrs to build	Overnight[a]	Fixed O&M	Decommission cost as % of overnight	O&M	Fuel
Nuclear	7	3,000	110	0.42	2.04–4.03	6.2
Biomass	2	640	30	0.24	4.25–9.05	90.0
Coal	4	1,780	30	0.24	4.25–9.05	20.0
Wind	3	1,300	28	0.10	0–0.20	0.0
Hydro	4	2,100	40	1.50	5.00–15.00	2.0
CCGT	3	970	18	0.10	3.43–6.45	37.0
OCGT	2	700	16	0.10	14.70	41.0

Source: Author calculations based on Fox (2011), van Kooten (2010) and Chap. 9
Notes: [a]Overnight costs are the total costs of labor, materials, etc. required to build the facility immediately or overnight. Hence, they need to be adjusted for the construction time. A rough approximation might be to divide the overnight cost by the time required to build the plant and then discount the stream of costs to the present. For wind, an alternative value is an overnight cost of $2,440/kWh, which is used for sensitivity purposes because it is the price Ontario would pay for wind installations as the Ontario FIT legislation requires a certain proportion of manufacturing to occur in the province. For biomass, the overnight cost refers to the cost of converting coal-fired generation to biomass

between income transfers and true costs was discussed in Chap. 6.) The background data needed to calculate the potential size of the subsidy payment and the true costs to society are provided in Table 10.5.

A simple model coded in Excel is used to calculate the annual costs to the electric operating system of supplying 145,000 GWh of power. The amount of wind power supplied to the grid is first subtracted, because wind power is non-dispatchable and must be accepted by the grid when it is produced. Then annual expected generation from nuclear assets is subtracted. With the current generating mix, it is assumed that remaining baseload is met from coal-fired plants; when biomass is introduced, the biomass facility is assumed to contribute all the power it can generate. Then the contribution from hydroelectric and OCGT resources is subtracted. Finally, any left-over load is supplied from CCGT assets; these assets are also assumed to constitute the primary system reserves.

Once the contribution of each asset is determined, system costs are calculated as are total emissions of CO_2 using data from Tables 10.4 and 10.5. The subsidy is then calculated as the difference between the feed-in tariff and the average wholesale price for the year, assumed to be 4¢/kWh (slightly higher than indicated above). Subsidies amount to $95/MWh for on-shore wind, $115/MWh for off-shore wind, and $90/MWh for biomass.

The size of the annual subsidy, the annual system (social) costs and the costs of reducing CO_2 emissions are calculated using the with-without principal of cost-benefit analysis (Chap. 6). The costs/payments with the FIT policy in place are compared to those without the policy. The results for selected scenarios are provided in Table 10.6. Also provided in the table are the changes in the electricity

Table 10.6 Economic and subsidy costs of Ontario's Feed-In Tariff Program, costs of reducing CO_2 emissions, and change in electricity from affected generation sources, various scenarios[a]

Item	Base case	Higher discount rate[b]	More biomass[c]	More wind[d]
Economic cost ($ millions)				
Operating	$258.9–636.0	$461.7–1,090.9	$184.3–541.9	$335.8–732.5
Subsidy	$2,122.5	$2,122.5	$2,956.5	$1,296.8
Emissions reduction (Mt CO_2)				
	7.90–12.14	7.90–12.14	9.05–14.96	6.78–9.36
Cost of reducing emissions ($/t$CO_2$)				
Operating	$21.33–80.55	$38.04–138.15	$12.32–59.90	$35.89–108.08
Subsidy	$174.89–268.79	$174.89–268.79	$197.66–326.81	$138.60–191.35
Change in power generation (GWh)[e]				
Biomass	19,710	19,710	32,850	6,570
Wind	3,081	3,081	–	6,235
CCGT	4,190	4,190	–5,869	14,176

Source: Author's calculations

Notes: [a]A range of values is provided for emissions per MWh (Table 10.4) and O&M costs (Table 10.5); values in the table reflect the low and high values associated with these ranges

[b]10% discount rate compared to 5% in base case scenario

[c]Installed biomass generating capacity increased to 5,000 MW compared to 3,000 MW in base case scenario; wind capacity set at current level of 1,046 MW

[d]Installed wind capacity increased to 5,000 MW compared to 3,000 MW in base case scenario; biomass generating capacity set at 1,000 MW

[e]Change from 2010 asset mix; output from other generating assets does not change over scenarios

produced by the generating assets affected by the feed-in tariffs. Since 1,500 MW of nuclear and 4,500 MW of coal-fired generating capacity are assumed to be removed, in all scenarios annual nuclear electricity output is 10,512 GWh and coal output 16,469 GWh below 2010 levels. Annual electricity from hydro and OCGT assets remains unchanged for reasons discussed above.

Three scenarios are not provided in Table 10.6. The first involves a reduction in the construction costs of wind turbines from $2,440/kW of installed capacity to $1,300/kW. In that case, costs of reducing CO_2 emissions fall slightly to $17.76–75.06/t$CO_2$, compared to $21.33–80.55 in the base case. Second, if we assume that 90% of wind energy is produced from on-shore turbines, rather than two-thirds as in the base case, system costs remain unchanged but the subsidy per tCO_2 drops slightly – by between $3.22 and $4.93, or by less than 2%. The final scenario not provided in the table relates to the higher capacity factor for wind turbines. This situation is discussed in more detail below.

Results provided in Table 10.6 suggest that subsidies could amount to $1.3–$3.0 billion annually, which will put a severe strain on the provincial treasury. The actual economic cost to society is an order of magnitude smaller, however; it amounts to some $184 to $733 million (assuming a 5% discount rate), still not an insignificant sum. The amount of the subsidy and the economic costs of removing nuclear and coal-fired generating assets and replacing them with wind and biomass facilities depend on the extent to which these renewables substitute for fossil fuel generation. A scenario with greater wind generation and less biomass results in a much lower subsidy but higher cost to society than one where more biomass and less wind generating capacity is installed. The reason why the economic cost to society is higher while the subsidy is lower is discussed in Chap. 11.

We also calculate carbon fluxes and the costs of reducing CO_2 emissions. In essence, by substituting fossil fuel energy with renewable sources in the generation of electricity, the Ontario government will pay a subsidy ranging from some \$174/tCO_2 to more than \$300/tCO_2. However, the actual cost to society of reducing emissions tends to be much lower, ranging from \$12.32/tCO_2 to \$108.08/tCO_2 (assuming a discount rate of 5%). Greater reliance on biomass generation compared to wind leads to lower costs of reducing CO_2 emissions, although it requires a greater subsidy payment. Nonetheless, the economic costs of reducing CO_2 emissions by replacing coal plants with biomass generating facilities might be competitive with other means of reducing carbon dioxide emissions, including purchase of offset credits on the European carbon market (see Fig. 8.5). This assumes, however, that an adequate wood supply exists and that wood fiber prices do not increase as a result of expanding biomass generating capacity.

Because of wind's intermittency, the capacity factor of wind farms will vary from 1 min to the next and from 1 year to the next. In the current model, the total electricity derived from natural gas and wind during the year will remain constant at 28,229 GWh. Thus, as the capacity factor of wind increases, for example, the generation of electricity from natural gas decreases; the subsidy paid by the government also increases, but the reduction in the actual cost of reducing CO_2 emissions is small because emissions from natural gas generation are smaller than if wind replaced coal. Thus, the average economic cost of emissions reduction declines \$54.83/tCO_2 if the capacity factor is 0.10 to \$50.05/tCO_2 if it is 0.20 and to \$46.04/tCO_2 if the capacity factor is 0.30.

Four additional points are worth mentioning. First, the analysis in Table 10.6 is crude as it excludes costs associated with intermittent or erratic sources of power generation, some of which are quite high (see below). Second, when FIT program costs and other costs (e.g., increased grid management costs due to intermittency, costs of building transmission lines) are taken into account, there likely exist cheaper ways to reduce CO_2 emissions, including purchase of certified emission reduction credits on carbon markets. Third, Ontario is not the only jurisdiction to employ feed-in tariffs. Germany has also implemented lucrative feed-in tariffs to support renewable power generation, with costs borne by consumers. Feed-in tariffs are guaranteed for 20 years, the levies vary from 21¢/kWh for offshore wind turbines to 46¢/kWh for roof-mounted solar panels. German subsidies to wind, solar and hydro generation amounted to \$7.3 billion in 2009 and will rise to \$11.3 billion by the end of 2010.[13] However, Germany's FITs for solar PV have fallen dramatically in 2011

[13] See http://www.upi.com/Science_News/Resource-Wars/2010/10/05/Solar-boom-drives-up-German-power-price/UPI-74351286299555/ (viewed 11 October 2010).

as a result of opposition to higher rates by electricity customers and the difficulties experienced by energy providers incorporating erratic solar power into their networks, as almost half the world's solar panels are mounted on German roofs and not linked directly to the grid.[14]

Finally, Ontario's FIT program was supposed to result in huge investments in green energy and 50,000 new jobs by the end of 2012. One company that took up the province's offer was Samsung, which agreed to build four wind and solar plants. By early 2011, Samsung had invested $100-million, which included construction of a local plant to produce wind towers, since the FIT program requires that 60% of equipment be sourced from within the province. However, the buy-local provision has been challenged at the WTO by Japan, the U.S. and Europe. If the challenge stands, Ontario's ability to achieve its employment targets will fail. By mid 2011 (about 2 years into the program), the FIT subsidies resulted in 15 projects with an installed capacity of 1,050 MW, below the rate at which Alberta is installing wind capacity but without subsidies.[15]

10.1.3 Transportation and the Hydrogen Economy

Before leaving this section, it is worthwhile mentioning the 'hydrogen economy'. Until about 2007, everyone raged about how society would reduce the world's reliance on fossil fuels by moving to hydrogen as quickly as possible (Scott 2007). Hydrogen was touted as the best alternative fuel to gasoline because it could be made from water. However, this requires a large amount of electricity, which, it was argued would come from renewable and nuclear sources of energy. Meanwhile, in practice, any hydrogen that was produced relied on methane (CH_4) as the source for hydrogen; it became increasingly evident that producing hydrogen from water (H_2O) was difficult and expensive, as cheap renewable and nuclear power generation were turning out to be more difficult to bring on stream than originally envisioned (as argued in the next section and Chap. 11).

The advantage of the hydrogen economy is that electricity generation, transportation and space heating would be integrated in a way that obviates the need to distinguish renewable energy sources according to the purpose to which they can be ascribed. This is illustrated in Fig. 10.5, which provides an indication of how fuel sources might be integrated in the hydrogen economy of the future, where hydrogen is the energy currency rather than electricity.

The main disadvantage of the hydrogen future is that hydrogen is a currency, not an energy source. It relies mightily on electricity, but the electricity needs to be

[14] See http://www.earthtimes.org/articles/news/363164,producers-agree-subsidy-cuts.html (viewed March 15, 2011).

[15] Source (accessed 14 June 2011): http://www.theglobeandmail.com/report-on-business/mcguintys-explosion-of-green-energy-would-you-believe-implosion/article2057633/

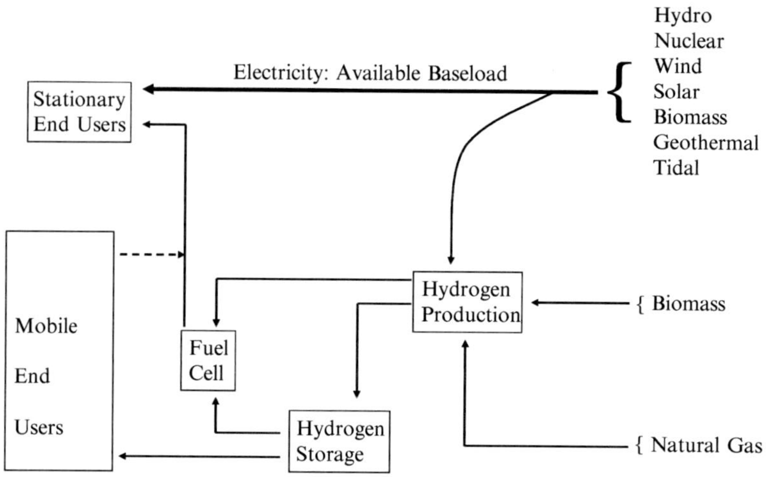

Fig. 10.5 The hydrogen economy

generated from an energy source such as coal, natural gas, diesel, water, uranium, wind, waves, tides, the sun, biomass or some other energy form. Reliance on fossil fuels does not simply disappear – other sources of energy still need to be developed. Further, it now appears that hybrid and all-electric vehicles have overtaken hydrogen fuel-cell vehicles in the imagination of the public and policymakers. In essence, the hydrogen economy is still some considerable distance in the future, perhaps 50 or more years, if ever.

What then are the options being considered by various jurisdictions for reducing carbon dioxide emissions in the generation of electricity? These range from continued reliance on fossil fuels, but then in ways that reduce emissions, to greater reliance on nuclear and a variety of renewable energy alternatives. In the remainder of this chapter, we consider first options related to coal, natural gas and nuclear energy, and then renewable energy sources. Then, in Chap. 11, we focus exclusively on electricity markets and the potential for renewable energy, most specifically power generated from the wind, to penetrate and compete as an alternative energy source.

10.2 Fossil Fuel and Nuclear Options for Reducing CO_2 Emissions

Fossil fuels are cheap and ubiquitous, and it is unlikely that such cheap and abundant resources can be denied their role in the generation of electricity; it simply makes no economic sense to leave valuable resources in the ground, and it is likely that someone will ultimately exploit the associated rents (Gerondeau 2010).

Even when efforts are made to reduce reliance on fossil fuels, options for their continued exploitation remain. The same is true of nuclear power. In this section, we examine in more detail the options for 'clean' coal, natural gas and nuclear power for generating electricity.

10.2.1 Clean Coal

Carbon capture and storage (CCS) is associated with so-called 'clean coal.' CCS involves removing CO_2 from the flue gas and pumping it into an underground reservoir. As of 2007, there were four industrial CCS projects in operation. Two projects are located off the Norwegian coast, on the Norwegian shelf or Utsira formation in the North Sea. Natural gas from the Sleipner gas field contains nearly 10% CO_2 and, to avoid paying carbon taxes, Norway's Statoil pumps the waste CO_2 into a deep underground saline aquifer. Since 1996, it has pumped annually about one million metric tons of carbon dioxide (1 Mt CO_2) into the aquifer. A similar project at the Snøhvit gas field in the Barents Sea stores 700,000 tCO_2 per year.

The largest CCS project is found at Weyburn in southeastern Saskatchewan, Canada. Since 2000, the Weyburn-Midale CO_2 Project has injected some 1.5 Mt of CO_2 underground for enhanced oil recovery, with the CO_2 coming from the Dakota Gasification Company plant in Beulah, North Dakota.[16] The North Dakota company had produced methane gas from coal for 30 years while the oil field was discovered in 1954 and thus had also been in operation for quite some time.

A fourth project at In Salah in Algeria is much like the two Norwegian projects. CO_2 is removed from natural gas and re-injected underground, thereby preventing 1.2 Mt CO_2 from entering the atmosphere.

Many other CCS projects are now under consideration or under construction. For example, in Saskatchewan the electrical system operator, SaskPower, is providing $1.4 billion in subsidies to convert one of its coal-fired generators at the Boundary Dam Power Station to capture CO_2 and pump it underground to enhance oil recovery near Estevan. SaskPower hopes to generate 115–120 MW of base-load electricity from clean coal, thereby avoiding the need to shut down its facility. Although only a demonstration project that received the go ahead in early 2010, it is believed that upwards of 10 Mt CO_2 can be stored underground. Given that Canada hopes to eliminate coal-fired power plants, CCS projects related to coal are likely to constitute a stop-gap measure, especially in Saskatchewan which had invested heavily in coal generated power in recent decades.

[16] A graduate student associated with the Institute for Integrated Energy Systems at the University of Victoria told the author that, after working with other engineers on measuring the success of CO_2 storage, it appeared they could not track the eventual destination of CO_2, except for that which actually enhanced oil recovery. There was no guarantee in other words that CO_2 did not leak out of the underground formation at some unknown location.

The province of Alberta has announced it would provide $2 billion for carbon capture and storage projects. CCS is required to offset emissions related to oil sands development. Germany, Australia, China and the United States are also looking into clean coal, while Norway, the Netherlands and possibly British Columbia are looking into CCS as they develop natural gas fields that contain high proportions of CO_2.

A pilot 30-MW CSS facility in operation in Germany since 2008 trucks captured CO_2 150 km every day to a storage facility; this activity itself releases CO_2. The company operating the power plant, Vattenfall, had to re-think its plans to construct a large-scale facility because residents feared release of CO_2 from the nearby underground storage facility that had been identified.

While CCS might well be technically feasible on a large scale, it certainly will not be economically feasible. There are two crucial obstacles. First, removing CO_2 from the flue gas and then compressing, storing, transporting and finally pumping the carbon dioxide into a permanent underground storage facility is extremely costly. For a coal-fired power plant, output would have to increase by 28% just to cover the energy required to remove the CO_2, although some of this can be done in off-peak hours when it is difficult to ramp down power output. Since not all regions have readily available places to store CO_2, it will be necessary to build a large pipeline transmission infrastructure and/or pipeline infrastructure plus storage and ship loading and offloading facilities.

Suppose that the objective is to capture and store just 10% of the world's CO_2 emissions, or about 3 Gt CO_2. Bryce (2010, pp. 162–165) estimates that, if CO_2 is compressed at 1,000 pounds per square inch (psi), or 68 atmosphere (atm),[17] it would amount to an oil equivalent volume of 81.8 million barrels per day. If all of this CO_2 were to be moved by ship, it would require filling 41 very large crude carriers (each holding about two million barrels) each and every day. Of course, much of the CO_2 would simply be transported by pipeline to a suitable underground location, but clearly not all. Even if only a quarter had to be shipped, this would require loading ten supertankers per day. Clearly, carbon capture and storage is a very expensive, and probably unrealistic, proposition.

But it is the second issue that is the real obstacle to large-scale CCS. There is always a risk that captured CO_2 is released, which could potentially lead to large loss of life, as when an underwater landslide in 1986 naturally 'burped' a large mass of CO_2 from Lake Nyos in Cameroon, forming a low-lying cloud that killed over 1,700 people before it dispersed. Unless carbon dioxide storage occurs in remote regions, which increases its costs, people would need to be compensated to have a storage facility nearby. Research pertaining to the transportation and storage of nuclear wastes indicates that this could be an enormous cost (see Riddel and Shaw 2003).

[17] 1 atm = 14.696 psi = 101,325 Pascal (Pa), where 1 Pa = 1 kg/m/s^2 = 1 kg/m^2. Note that CO_2 reaches a supercritical stage (where it becomes liquid) at about 70 Pa (measured at 31 °C), but to get it there would take a great deal of energy.

In essence, the only real options appear to be those of conservation (e.g., via smart grids), greater reliance on natural gas and/or nuclear power, or development of alternative renewable sources of energy.

10.2.2 Natural Gas

During the 1990s, a Texas oil and gas well driller, George Mitchell, experimented with various techniques to cause gas to flow from shale deposits. In 1997, he and his crews found that, if water under extreme pressure was injected into wells along with sand and certain chemicals, this caused the gas to flow.[18] This process is known as hydraulic 'fracturing' or simply 'fracking.' Then, in 2003, they discovered horizontal drilling. Thereby, they could drill down 1 km or more and then turn the drills sideways, and drill horizontally (laterally) for several km. At various locations along the lateral (about every 120 m), the rock formation could be 'fractured' by injecting water and sand. The water would force openings in the rock, which were then occupied by grains of sand that, along with the chemicals, facilitated the flow of natural gas.

As a result of horizontal drilling and hydraulic fracking that opened up the pores to allow gas to flow, the Texas' Barnett shale vaulted into the top ten of the globe's natural gas fields. Its recoverable reserves of unconventional or shale gas are estimated to be about 44 trillion cubic feet (1.25 trillion m^3), or an energy equivalent of eight billion barrels of oil. This compares with the six-billion barrel, East Texas oil field discovered in 1931, which was the largest oil field in the world at that time.

Further, recoverable reserves of unconventional gas in the United States are now estimated at 861.7 trillion cubic feet, or 24.4 trillion m^3 (*The Economist*, August 6, 2011, pp. 51–53). Already the Marcellus shale that underlies parts of Pennsylvania and New York is considered to be the largest in the U.S.; in 2009, natural gas produced solely in Pennsylvania was worth more than $3.5 billion. In 2009, its exploitation generated some $400 million in state and local tax revenues, and created 48,000 new jobs. See Reed (2011) and Special Report on Gas in The Economist, July 14, 2012.There has simply been an overwhelmingly large (more than fourfold) increase in recoverable gas reserves in the United States since 1989. Further, unconventional gas can be found elsewhere in the world as the technological advance resulting from lateral drilling methods and fracking formations can be adopted in other locations. The ratio of technically recoverable shale-gas resources to reserves of conventional gas is 6.1 for Canada, 3.2 for the U.S., 3.6 for Australia, 12.0 for China and over 54 for Argentina (*The Economist*, August 6, 2011). Global reserves of unconventional gas are estimated to be more than five times as large as 2009 conventional reserves.

[18] Chemicals constitute about 1% of the volume of water. There remains some concern that chemicals could enter the water supply, but this is unlikely because wells are significantly deeper than the porous layers from which water may be taken.

Table 10.7 A comparison of the potential release of greenhouse gases from various fossil fuels

Item	Chemical structure
Natural gas	
75% methane	CH_4
15% ethanol	C_2H_6
10% other hydrocarbons	
Hydrocarbons	
Propane	C_3H_8
Butane	C_4H_{10}
Octane	C_8H_{18}
Benzene	C_6H_6
Hexane	C_6H_{14}
Naphthalene	$C_{10}H_8$
Bituminous coal	
Carbon (C)	75–90%
Hydrogen (H)	4.5–5.5%
Nitrogen (N)	1.0–1.5%
Sulfur (S)	1–2%
Oxygen (O)	5–20%
Ash	2–10%
Moisture	1–10%
Coal[a]	C_nH_m (n>m, n large, m small)
Glucose	$C_6H_{12}O_6$
Gasoline (average)	C_8H_{18} Range: C_6H_{14} to $C_{12}H_{26}$
Diesel	$C_{16}H_{34}$

Source: Author's own construction from internet sources
Notes: [a]Macromolecules consisting of clusters of aromatic coal linked by bridges of sulfur, oxygen or other element(s)

Given the tremendous increase in global natural gas reserves that the new technology has brought about, many countries will pursue a strategy of substituting highly energy-efficient natural gas for coal in the production of electricity. Natural gas power plants can be simply and quickly built; the up-front construction costs of gas plants is half or less than that of coal plants, and much lower that of nuclear, solar, wind or other power generating facilities (Table 10.5; NEA and IEA 2005). Hence, it is not surprising that countries are opting for natural gas, although in some cases the decision to build natural gas power plants is the result of political indecision concerning the extension of old or construction of new nuclear power plants – the only types of generating assets that can be built quickly enough are gas plants.

Finally, as shown in Table 10.7, natural gas is generally composed of methane (CH_4), ethanol (C_2H_6) and other hydrocarbons. Consequently, compared to coal, it releases much less CO_2 into the atmosphere. However, some unconventional gas contains higher quantities CO_2 that get released as part of the production process;

but a greater problem for climate change is the increased leakage of methane that occurs with fracking compared to conventional production. Indeed, studies suggest that, unless leakage rates for new natural gas sources can be kept below 2%, substituting gas for coal will have little impact on projected global warming because methane is a more potent greenhouse gas than carbon dioxide – the warming effect associated with the leakage of methane offsets the reduction in CO_2 emissions from substituting gas for coal (Howarth et al. 2011; Wigley 2011). Methane leaked from fracking operations has also found its way into groundwater (which is not a large problem as far as drinking water is concerned), although the bigger fear that some of the chemicals in the fracking fluids would contaminate the groundwater has not materialized (Osborn et al. 2011).

In essence, natural gas may well be the primary fuel that powers the world economy in the twenty-first century, but it is not the magic solution to the climate change conundrum – adequate energy and reduced emissions of greenhouse gases – that some suggest it might be (e.g., Bryce 2010). The verdict on that score remains. Nonetheless, as argued in Sect. 10.4, natural gas might be a bridge to a nuclear future. At the very least, unconventional natural gas can, in the near to intermediate future, provide greater energy security for the U.S. and European Union, and perhaps China and India, as it can reduce their reliance on energy from a limited number of less politically stable states.

10.2.3 Nuclear Power

Together the United States and France produce some 47% of global nuclear energy output, and account for 45% of installed capacity (Table 10.8). More than three-quarters of France's domestic consumption of electricity comes from its nuclear power plants and it exports nuclear power to other countries.

It is difficult for a country to expand reliance on nuclear energy much beyond that experienced by France because nuclear plants are baseload, so peaking gas plants or hydro facilities are needed to address short periods of high demand. France avoids some of its need for peaking capacity by selling nuclear power to other European countries, especially ones such as the Netherlands that are looking to reduce their CO_2 emissions and are closing coal and/or gas plants.

The current top nuclear power producing countries and those of the future are found in Table 10.8. The rest of the world accounts for only 10% of global nuclear power production and only 9% of generating capacity, although 14% of current construction and nearly one-quarter of planned and proposed future expansion of capacity is accounted for by countries not listed in Table 10.8.

It is interesting that rich democratic countries are the least likely to expand their nuclear generating capacity in response to climate change, and are even less likely to do so after the nuclear incident at Fukushima in Japan in March 2011. In contrast, countries like China, India, Russia and others are much less likely to scale down their nuclear expansion plans. These countries simply need electricity for economic

Table 10.8 Nuclear power production and capacity, top 14 producers (alphabetically ordered)[a]

Country	Energy produced 2009 (TWh)	Active 2011 #	Active 2011 Capacity (MWe)	Under construction #	Under construction Capacity (MWe)	Planned and proposed #	Planned and proposed Capacity (MWe)	% of domestic use[b]
Canada	85.3	18	12,679	2	1,500	6	7,100	14.4
China	65.7	13	10,234	27	29,790	160	165,830	2.0
France	391.7	58	63,130	1	1,720	2	2,820	77.1
Germany	127.7	17	20,339	0		0	0	23.5
India	14.8	20	4,385	5	3,900	58	64,700	n.a.
Japan	263.1	55	47,348	2	2,756	15	20,532	24.0
Russia	152.8	32	23,084	10	8,960	44	44,000	15.7
S. Africa	11.6	2	1,800	0		6	9,600	n.a.
S. Korea	141.1	21	18,716	5	5,800	6	8,400	34.0
Spain	50.6	8	7,448	0		0	0	n.a.
Sweden	50.0	10	9,399	0		0	0	42.6
Ukraine	77.9	15	13,168	0		22	26,700	46.7
UK	62.9	19	10,962	0		13	18,680	16.1
U.S.	798.7	104	101,229	1	1,218	32	45,662	19.3
Global	**2,560.1**	**443**	**377,791**	**62**	**64,374**	**484**	**547,762**	**13.8**

Notes: [a]Source: *The Economist* (March 26, 2011, pp. 79–81), www.economist.com/reactors2011 (viewed March 30, 2011), except final column

[b]Data for 2008 except UK data are for 2007; n.a. indicates not available from source. Source: Key World Energy Statistics 2009 © OECD/International Energy Agency (2010a), p. 17, and Key World Energy Statistics 2010 © OECD/International Energy Agency (2011), p. 17

growth, and see nuclear energy as a means of reducing reliance on foreign supplies of coal and natural gas and less as a strategy for de-carbonizing their economies, although that reason is used to defend nuclear expansion in global forums.

Generation of electricity from nuclear energy is likely to remain confined to a small group of countries that includes emerging economies such as China, India, Russia and Brazil. Yet, nuclear power is a sensible and realistic and, perhaps, the only option for achieving CO_2 emission-reduction targets of 50% or more. For a country, such as Canada, 70% of electricity demand is already met from hydro and nuclear sources; because it is difficult to expand hydro capacity and, given the obstacles posed by biomass energy, Canada might wish to expand its nuclear capacity in order to mitigate climate change.

How realistic is the nuclear option? Despite its promise, there are severe challenges facing expansion of nuclear energy. Nuclear wastes, the potential risk of enriched nuclear material being used by terrorists, high construction costs, cost over-runs, and general opposition to nuclear power plants by citizens, and especially environmental groups, militate against nuclear power. Storage of wastes in central facilities such as Nevada's Yucca Mountain makes sense as the amount involved is relatively quite small (no more than the volume of a large room), while the status quo of storing wastes on site is likely riskier.

Given that far less than 5% of the available energy in nuclear fuel is used to generate power, enriching the spent uranium fuel can extend the usefulness of the fuel and, eventually, reduce its radioactive half life and the amount of wastes. Because enrichment leads to bomb grade material, the U.S. government in particular has sought to prevent further refinement or recycling of spent fuel, preferring instead to store the more radioactive material. Sweden has been much more open to the upgrading and recycling of fuel.

Although nuclear wastes and fear of nuclear weapons proliferation cast doubt upon the future of nuclear energy, the most likely obstacle to its greater use might be fear of a catastrophic meltdown. This is particularly the case after the near partial meltdown at three nuclear reactors at Japan's Fukushima Dai-ichi nuclear power plant in March 2011. However, it is necessary to put this incident in perspective. There have been two previous major nuclear plant incidences: Three Mile Island in Pennsylvania in the U.S. (March 1979) and Chernobyl in the Ukraine (April 1986). At Three Mile Island, the nuclear plant experienced a partial meltdown, some radiation was released, and nearby residents were evacuated, but no one was killed. Chernobly was the result of an unauthorized technical experiment; 31 people died in the aftermath of an explosion and several hundred thousand were exposed to high levels of radiation. In Japan, a major earthquake and subsequent tsunami overwhelmed the safeguards at the plant. Between 1950 and 2000, there were 16 other nuclear accidents: One occurred on a Soviet nuclear submarine and killed eight (July 1961). Three technicians died at an experimental reactor in Idaho (January 1961) and several hundred may have died of radiation sickness in a Soviet incident during the winter of 1957–1958, but little is known about this incidence.[19] Compared to deaths from coal mining and drilling for oil and gas, nuclear power remains remarkably safe.

Despite these obstacles, some countries will necessarily choose to expand reliance on nuclear energy to meet greenhouse gas emission targets and concerns about energy security. As of 2009, 44 nuclear power plants were under construction globally – 11 in China, eight in Russia, six in India, five in Korea, two in each of the Ukraine, Bulgaria, Taiwan and Japan, and one in each of Argentina, Finland, France, Iran, Pakistan and the United States (Deutch et al. 2009). Estimates provided by Deutch et al. (2009) indicate that the life-cycle costs of producing nuclear energy are 8.4¢/kWh, compared with 6.2¢/kWh for coal and 6.5¢/kWh for gas, although the latter costs would rise to 8.3¢/kWh and 7.4¢/kWh, respectively, if a charge of $25 per tCO_2 emissions were imposed.[20] Further, if the added risks of capital used in building nuclear reactors were eliminated, so that the carrying costs of capital investments were the same as those of coal and gas plants, nuclear energy would cost 6.6¢/kWh rather than 8.4¢/kWh.

[19] See http://www.atomicarchive.com/Reports/Japan/Accidents.shtml (viewed March 15, 2011).

[20] These costs are significantly higher than those reported in the earlier MIT study (Ansolabehere et al. 2003), but are probably higher than they would be today given that construction costs have declined since the financial crisis. This needs to be taken into account in the following discussion as well.

Table 10.9 Lifetime generation costs by generating type ($/MWh)[a]

Generating type[b]	Midpoint	Low	High
Wind onshore	68.08	36.39	168.71
Wind offshore	78.54	59.09	144.38
Solar thermal	193.64	193.64	315.20
Solar PV	192.21	141.10	2,195.39
Run of river/small hydro	108.28	46.45	283.02
Large-scale hydro	53.12	53.12	99.33
Nuclear	30.71	24.34	80.26
Coal (lignite)	39.35	34.40	75.35
Coal (high quality)	31.90	30.30	80.85
Coal (integrated coal gas)	44.73	31.94	69.15
Gas (CCGT)	54.62	44.69	73.24
Gas (open)	54.64	54.64	57.33
CHP (using CCGT)	55.12	33.11	94.65
CHP (using coal)	39.09	29.25	54.87
CHP (using other fuel)	40.01	34.40	116.42
Waste incineration	11.39	–4.68	61.19
Biomass	48.74	43.64	117.59

Source: Author calculations from various disparate sources

Notes: [a]The costs include capital, operating and maintenance, and fuel costs over the lifetime of a power plant, discounted to the present and 'levelized' over the expected output of the generating source over its lifetime. Values are in 2008 US dollars. The midpoint value is based on a 5% discount rate, as is the low value (except in the case of high quality coal); the high value is derived using a 10% discount rate

[b]Open-cycle gas turbines lose exhaust heat but can respond quickly to changes in demand; closed-cycle gas turbines (CCGT) recycle exhaust heat, which makes them suitable as base-load plants but makes it more difficult for them to ramp up and down. Combined heat and power (CHP) occurs when exhaust heat from space heating is used to generate power; such power is usually available at night and in colder climates

It is difficult to compare costs of producing electricity from renewable sources with those from traditional sources. Using data from a survey conducted by the International Energy Agency (IEA) (2005), it is possible to provide some comparison of costs on a per megawatt hour (MWh) basis. Estimates are provided in Table 10.9.[21] These indicate that electricity generated from renewable energy sources is significantly higher than that from traditional sources. Waste incineration is only the lowest cost means of generating electricity if there is a payment to dispose of municipal and industrial waste (which explains the negative value in the table, indicating a benefit). Further, the contribution of wastes to total electricity generation will be small, which is also true of combined heat and power (CHP). Coal and nuclear energy are the lowest cost realistic options. Gas is more expensive because of high fuel costs, but gas plants are cheap to build and are needed for fast response to shifts in load.

[21] These can be compared with other estimates, such as those in Table 10.5.

Proponents of renewable energy generation argue that the costs in Table 10.9 do not reflect externality costs, in particular the costs associated with CO_2 emissions (and other pollutants) from fossil fuel plants and the health and safety risks associated with nuclear power. Assuming that coal emits 0.5–0.6 tCO_2/MWh of electricity – an emission level that is dropping as more efficient plants come on line – it would take a carbon tax well above what CO_2 emissions have been trading for under the Europe's Emission Trading System (or the Chicago Climate Exchange, CCX, prior to its demise) before even wind energy is competitive with coal. But there remains another problem: With the exception of biomass and large-scale hydro, only nuclear and closed-cycle gas turbine (CCGT) plants can replace coal because, without storage, intermittent sources of power cannot serve baseload needs (van Kooten 2010).

10.3 Renewable Alternatives to Fossil Fuels[22]

In the electricity sector, renewable sources of energy include large-scale hydro, small-scale run-of-river hydro, geothermal, wind, tidal, solar, wave, municipal solid wastes, farm wastes and biomass. In addition, biomass is increasingly considered for the production of biofuels (ethanol and biodiesel) for transportation. Some of these sources are severely constrained.

Consider first biomass and biofuels. While there has been a great deal of emphasis on the use of terrestrial carbon sink credits for offsetting fossil fuel emissions of CO_2, the costs of sequestering carbon in agricultural and forest ecosystems are generally quite a bit higher than emission-reduction options. There are some fundamental problems with the use of terrestrial sinks that make them a very dubious means of mitigating climate change; these include their ephemeral nature, high monitoring and transaction costs in establishing CO_2 baselines and fluxes, and the potential for corruption. These issues were discussed in detail in Chap. 9.

In this section, we want to expand the discussion of renewable energy sources to focus on their prospects. Although biofeedstocks and bioenergy were discussed in the previous chapter, we return to this theme mainly with regards to biofuels. We then consider the prospects of other renewable sources of energy, including hydropower, and intermittent resources, such as wind, wave, tidal and solar power. (Nuclear power was discussed in the previous section.) Given the scope of our discussion, we provide only a broad brush analysis of the challenges society faces in turning a fossil fuel based economy into one that is much less so. However, we leave to the next chapter a more detailed discussion concerning the economics of electricity and the integration of renewable energy sources into electricity grids.

---

[22] Unless otherwise indicated, much of the material for this section comes from graduate student research, seminars and discussions at the University of Victoria's Institute for Integrated Energy Systems (http://www.iesvic.uvic.ca/).

10.3.1 Biomass and Biofuels

Current policies to mitigate climate change have focused on the potential of using biomass to generate electricity or as a liquid fuel instead of gasoline or diesel.

10.3.1.1 Biomass for Generating Electricity

Increasing electrical power production from forest biomass, sawmill residue, and 'black liquor' from pulp mills is constrained by high transportation costs and competition for residual fiber that makes forest biomass an expensive source of energy. Consider the example of British Columbia, which is a major forest products exporting jurisdiction.

Because of the extent of mountain pine beetle damage to forests in the interior of British Columbia, many commentators felt that an obvious use of beetle-killed trees would be power generation. Early studies that examined the costs of producing electricity from dead trees concluded that this could be done with little in the way of government subsidies. These analyses were based on average past costs of harvesting and hauling timber from the harvesting site to sawmills. However, when one takes into account the rising costs of hauling timber as more remote timber damaged sites need to be harvested, marginal costs rise rapidly with truck cycle times (the time required to travel to and from the harvesting site) of 9 h or more (Niquidet et al. 2012). An electrical generating facility turns out to be only a marginally attractive option for reducing CO_2 emissions when feedstock costs are low; but, as feedstock costs alone rise from an equivalent of 4–8.5¢/kWh, biomass power is no longer an economically viable option.

Producing char from biomass through a process known as pyrolysis (a form of incineration that chemically decomposes organic matter by heat but without oxygen) suffers from similar problems, although high transportation costs might be mitigated somewhat by producing char on site. Nonetheless, the amount of char available for generating electricity will be negligible in comparison to what is needed and there are concerns that the process produces hazardous wastes.

Perhaps the best option for generating electricity from wood biomass is wood pellets. Wood pellet production plants are relatively inexpensive to construct and can, in some instances, be moved quite easily to new locations (although they are not mobile enough to be located at the harvesting site). Wood pellets can be used directly in coal-fired power plants with little or no adjustments to the burners – pellets can be pulverized much like coal and pellets are preferred over wood chips (which are used for pulp). Wood pellet stoves are also popular for space heating in residential homes.

Because of their flexibility, relatively low production costs, and government programs and subsidies, demand for pellets rose sharply, particularly in northern Europe as a result of subsidies beginning in 2005. As a result, British Columbia's wood pellet production capacity had risen to about one million tons by 2010. But, as noted earlier, as demand for pellets, char and other energy uses of wood biomass increase, prices will rise making them less attractive as an alternative form of energy.

Using a regional fiber allocation and transportation (mathematical programming) model, Stennes et al. (2010) demonstrate a major drawback of timber feedstocks. As one of the largest lumber producing and exporting jurisdictions in the world, British Columbia's forest resources are enormous and one would think that these resources would form a logical foundation for a thriving bioenergy sector. Lumber is far and away the most lucrative product that is produced in the province. Chips from sawmilling operations form the mainstay of the province's pulp industry. Other sawmill residues (bark, sawdust, etc.) are already allocated by mills to on-site space heating and power production, with some excess chips and residues used to make wood pellets, oriented strand board and other products. Competition for sawmill residuals occurs between pulp mills and other wood product manufacturers as well as heating and electricity. There is some leeway to increase available wood waste by hauling roadside and other waste from harvest operations to electricity generation and other facilities that might be able to use them. The important point to note is that any residuals and other wood waste are available at a reasonable cost only as a result of timber harvests for sawmilling purposes (Bogle and van Kooten 2010; Niquidet et al. 2012).

When account is taken of the demand and supply of wood fiber for all its different purposes, and when costs of transporting various types of fiber from one location in the province to another, there is little wiggle room. Indeed, the government might wish to implement policies, such as direct construction subsidies or feed-in tariffs, to increase power generation or wood pellet production from a wood biomass feedstock, but this will only lead to increased demand for fiber. This causes prices of wood residuals and wood 'waste' to increase, driving out existing users such as pulp mills, or the bioenergy producers themselves, depending on their ability to compete (Stennes et al. 2010). For example, pulp prices were under $500/t several years ago, but reached $1,000/t in 2010. Pulp producers can out bid energy producers for wood fiber at high pulp prices but have a harder time competing at lower prices, especially if bioenergy producers are subsidized.

What is often neglected in discussions of biofuels and biomass-fired power generation is the fact that biomass and biofuels are not carbon neutral as is often claimed. The combustion of biofuels and biomass releases carbon dioxide, indeed more than what is released from fossils fuels to generate an equivalent amount of energy. It is only when crops and trees grow that carbon dioxide is removed from the atmosphere, and this can take quite a long time in the case of trees. Further, CO_2 and other greenhouse gases are emitted in the harvest and hauling of biomass, and their conversion to fuel or power.

When wood biomass is burned in lieu of coal, say, more pollutants are released than with coal. In addition, more CO_2 is released in gathering biomass across a large landscape than is the case with coal as coal deposits are concentrated near a particular location. Thus, there is an increase in the release of carbon dioxide, not a reduction. The reduction comes only as trees grow, which could take as much as 80 years. To mitigate the length of the growing season, fast-growing tree species, such as hybrid poplar, can be grown, or alternative plants such as switchgrass can be used as a biomass fuel. While this tilts the greenhouse gas emissions more favorably towards biomass burning, nitrogen fertilizer is often required to spur growth, and NO_x is a more potent greenhouse gas than CO_2.

Finally, land is the most important factor in the production of biofuels. Increased demand for energy crops reduces cultivated area devoted to food production as land is diverted into energy crops (Searchinger et al. 2008), thereby increasing the carbon footprint. Overall, then, the process of generating electricity from biomass is hardly carbon neutral.

10.3.1.2 Biofuels

Wood biomass could also be converted to a biofuel, perhaps even on site. Pyrolysis can be used on-site to produce a bio-oil that requires further refining, but there remain concerns about hazardous wastes associated with pyrolysis. Alternatively, wood biomass can be hauled to a central location and used to produce cellulosic ethanol, but, in some jurisdictions, this requires transporting fiber over long distances and again results in competition for fiber that drives up fiber prices. A better option for cellulosic ethanol is to grow switchgrass, willow or hybrid poplar on large plantations. Where corn can be grown for ethanol, or grains for biodiesel, these are preferred to biomass plantations for cellulosic ethanol because fuel yields are higher and costs are lower.

Biofuels are also not neutral with respect to greenhouse gas emissions; CO_2 is released whenever biofuels are burned, and often more CO_2 is released to generate the same amount of energy compared with fossil fuels (as discussed in Chap. 9). The biomass needs to be harvested, transported and processed, which contributes to CO_2 emissions. Only the growth of plants and trees removes CO_2 from the atmosphere, and such growth takes time – a lot in some regions – or inputs of chemical fertilizers (whose production, transport and application also release greenhouse gases). While ethanol can be burned in place of gasoline, its energy content is only about two-thirds that of gasoline. Further, compared to fossil fuels, the growth and processing of energy crops requires enormous amounts of water, some of which comes from non-renewable aquifers (Bryce 2008, pp. 183, 191).

Nobel laureate Paul Crutzen focused only on the climate effects from nitrogen (N) fertilization associated with the production of energy crops for biofuels. He and his colleagues showed that, "depending on N content, the current use of several agricultural crops for energy production, at current total nitrogen use efficiencies, can lead to N_2O emissions large enough to cause climate warming instead of cooling by 'saved fossil CO_2'" (Crutzen et al. 2008, p. 393). This is illustrated in Table 10.10. Given current nitrogen-use efficiencies in agriculture, the increased nitrogen emissions from the fertilizer used to grow energy crops offset the reduction in CO_2 emissions from the gasoline that the biofuel replaces. If ethanol came from sugar cane, the contribution of the biofuel to global warming was between 0.5 and

Table 10.10 Net climate warming relative to fossil fuel CO_2 savings

Crop	Biofuel	Nitrogen use efficiency 0.4	0.6	50% of N harvested for biofuels production replaces crops that need N fertilizer
Rapeseed (canola)	Bio-diesel	1.0–1.7	0.7–1.2	0.5–0.9
Maize (corn)	Bio-ethanol	0.9–1.5	0.6–1.0	0.4–0.7
Sugar cane	Bio-ethanol	0.5–0.9	0.4–0.6	0.3–0.4

Source: Derived from data provided in Crutzen et al. (2008)
Notes: Climate warming occurs if values exceed 1.0. Current nitrogen use efficiency is around 0.4

0.9, where a value above 1.0 indicates increased release of greenhouse gases (greater warming rather than cooling); if ethanol came from corn, the warming factor was 0.9–1.5; but, if the biofuel came from canola, it resulted in no benefit as the greenhouse gases released exceeded those associated with the fuel that was replaced (factor of 1.0–1.7). Only if the nitrogen use efficiency could be increased from about 0.4 to 0.6 might maize-ethanol or canola-biodiesel be climate neutral or beneficial.

Further, land is the most important factor in the production of biofuels. Increased demand for energy crops reduces cultivated area devoted to food production as land is diverted into energy crops (Searchinger et al. 2008). It also increases the carbon footprint. For example, biodiesel produced from rapeseed (also known as canola) in Europe has an indirect carbon footprint of 150.3 kg/GJ, compared with 100.3 kg of carbon per GJ for ethanol produced from European-grown sugar beet. In contrast, conventional diesel or gasoline releases only 85 kg of carbon per GJ (including CO_2 released during refining). This compares with carbon footprints of 82.3 and 73.6 kg/GJ for imports of ethanol from Latin American sugar cane and from Southeast Asian palm oil, respectively (see Harrison 2010). The energy densities of various energy sources in terms of land use are provided in Table 10.11. The energy density of corn ethanol in terms of land use is given an index of 1.0. We then compare the energy density of other renewable and fossil fuel energy sources in terms of this definition of energy density. Compared to corn ethanol, other renewable energy sources are preferred: biomass power plants, wind and solar energy are much more efficient in terms of land use. However, none compare to fossil fuels (where wells are used to remove the energy source from underground), but nuclear power turns out to be most energy efficient in terms of land use requirements.

Finally, food prices have risen dramatically in recent years, partly as a result of the diversion of grains to biofuels. In 2004, 2% of world grain production was used to produce biofuels, while virtually no vegetable oils (e.g., corn, canola) were diverted to biofuels. As a result of government policies, by 2010 6.5% of world grain production and 8% of vegetable oils went to produce biofuels, with governments hoping to triple this amount by 2020 (see Searchinger 2011). Clearly, this will have an impact on food prices, especially harming those who spend about half of their incomes on food.

Table 10.11 Energy densities: comparison of the physical area required to produce energy from selected sources

Energy source	Energy density	Index
Corn ethanol	0.05 W/m^2	1.0
Biomass-fuelled power plant	0.4 W/m^2	8.1
Wind turbines	1.2 W/m^2	24.6
Oil stripper well[a] producing two barrels per day	5.5 W/m^2	115.4
Solar PV	6.7 W/m^2	138.5
Oil stripper well[a] producing ten barrels per day	27.0 W/m^2	577.0
Gas stripper well[a] producing 60,000 ft^3/day	28.0 W/m^2	590.4
Average U.S. natural gas well, 115,000 ft^3/day	287.5 W/m^2	1,105.8
Nuclear power plant[b]	56.0 W/m^2	1,153.8

Source: Derived from data in Bryce (2010, pp. 91–93)
Notes: [a]A stripper well is one that has passed its peak production (or never was a large producer) but continues to pump oil or gas. Stripper wells are defined by their maximum output – ten barrels per day for oil wells and 60,000 ft^3/day for gas wells
[b]Based on a 4,860 ha location in Texas, although the power plant occupies only a very small area within the property

The objective of biofuel policies is to reduce global warming and its negative impacts on developing countries in particular. Ironically, by raising food prices, biofuel policies actually lead to increased deaths among the globe's poorest people. Indeed, Goklany (2011) estimates that the increase in biofuel production between 2004 and 2010 increased the number of people in absolute poverty (below $1.25 per day) from 798 million (without biofuel policies) to 830–834 million, or by 32–36 million. He then estimated that, for 2010, biofuel policies led to 192,000 additional deaths, or 52,000 deaths if it was assumed that the biofuel policies were able to roll back global warming to 1990 levels.[23]

From a policy perspective, biological methods are not an efficient means of addressing climate change, although promising research into various biological organisms that make this process more efficient is ongoing. These may very well come to fruition, but it could be several decades before such options are commercially viable. However, energy from biological organisms does not appear to be a major component of governments' policy arsenals for combating climate change. Landfill gas generated from solid waste is another potential source for generating electricity, but, even if it is employed on a large scale, its contribution to the globe's electricity needs would be extremely small. The same holds for the incineration of municipal wastes.

[23] It should be noted that the mortality figures are crude estimates, with the difference between the two estimates attributed to high estimated rates of mortality assumed to be associated with global warming (but see Chap. 7).

Table 10.12 Hydro electric power production and capacity, 2007

Country	Production (TWh)	Capacity (GW)[a]	% of domestic consumption
China	585	149	16.9
Canada	383	73	58.7
Brazil	370	77	79.8
United States	282	100	6.5
Russia	167	47	16.0
Norway	141	29	98.5
India	114	36	13.8
Venezuela	87	n.a.	72.8
Japan	83	47	7.7
Sweden	69	n.a.	46.1
Rest of World	1,007	366	13.6
World	**3,288**	**924**	**16.2**

Source: Key World Energy Statistics 2010 © OECD/International Energy Agency (2011), p. 19
Notes: [a]Data for 2007; n.a. not available

10.3.2 Water Power or Hydraulics

A number of countries have developed their hydraulic resources to build large-scale hydropower facilities. With the so-called 'three gorges' dam (affecting the Upper Mekong, Yangtze and Salween Rivers), China now has the greatest hydro capacity in the world (Table 10.12). In 2007, hydro production only accounted for 14.8% of China's consumption of electricity. This is much less than the proportions accounted for by hydro in Norway (98%), Brazil (84%), Venezuela (72%) and Canada (57%). Both India and Russia relied on hydropower to a greater extent than China, and Russia despite its relatively abundant fossil fuel resources.

Large-scale hydro remains one of the best options for generating 'clean' electricity, but its main drawbacks relate to inadequate runoff for power generation (especially in regions where water is scarce, intermittent and/or unreliable) and negative environmental externalities (changes in the aquatic ecosystem, impediments to fish migration, land inundation by reservoirs, etc.). Environmentalists oppose large-scale hydro development, particularly in developing countries because of the ecological damage it causes, while even small-scale, run-of-river projects have been opposed in rich countries on environmental grounds. Because of strong environmental opposition against hydropower developments, hydropower's future contribution to increases in overall generating capacity will inevitably remain limited in scope. Expansion of water power is not expected to be a large contributor to the mitigation of climate change.

Although unlikely to contribute much in the way of additional clean power, existing large-scale hydro and strategic expansions of reservoir storage capacity (which raises generating capacity) might serve an important purpose when combined with intermittent sources of energy, namely, wind, tidal and solar sources. For example, wind-generated power is often available at night, when base-load power plants are able to supply all demand. Wind energy would then need to be curtailed (wasted) or, where possible (and it may not always be possible), base-load plants would need to reduce output, causing them to operate inefficiently. If a base-load plant is coal fired, inefficient operation implies that CO_2 emissions are not reduced one-for-one as wind replaces coal. In some cases, the tradeoff is so poor that CO_2 emissions are hardly reduced whatsoever. This problem can be overcome if adequate transmission capacity exists so that the excess wind-generated power could be stored behind hydro dams by displacing electricity demand met by hydropower. This is the case in northern Europe, where excess wind power generated at night in Denmark is exported to Norway, with hydropower imported from Norway during peak daytime hours.

Similar relationships are found elsewhere. In Canada, for example, the provinces of Quebec and British Columbia rely almost exclusively on hydropower, while the respective neighboring provinces of Ontario and Alberta generate significant base-load power from coal (or nuclear in Ontario's case). Ontario and Alberta are both expanding their installed wind capacity. During nighttime, off-peak hours, excess wind and/or base-load power from Ontario (Alberta) is sold to Quebec (British Columbia), with hydropower sold back during peak periods. Given that the rents from these transactions have accrued to the provinces with hydro assets, Ontario and Alberta have been less than keen to upgrade the transmission interties, preferring to look at other possible solutions to the storage problem.

In all three cases, there are net economic and climate benefits from the development of higher capacity transmission interties; or, in the case of northern Europe, simply more interties between jurisdictions where wind power is generated (northern Germany, other parts of Denmark) and those with hydro resources (Norway and Sweden). The main obstacle is the lack of incentives for the wind-generating region to 'dump' power into the region with storage, as the latter captures all the rents from such an exchange. This is a game theory problem: If institutions can be developed that facilitate the sharing of both the economic rents and the climate benefits (emission reduction credits), the jurisdictions have the incentive to better integrate the operations of their electricity grids (including construction or upgrading of transmission interties) so that overall CO_2 emissions are minimized. We return to the issues of hydraulic storage, transmission and power generation in Chap. 11.

10.3.3 Geothermal

Deep in the earth, the temperatures are much higher than on the surface. In these places, the magna of volcanoes forms. In some places, heat escapes from underground through vents or geysers and can be captured to generate electricity or used for space heating. The country that relies most on such geothermal energy is Iceland. Proposals to drill deep into the earth and capture heat for power generation suggest that this is a viable source of energy from an engineering standpoint. However, economic considerations will prevent the use of geothermal energy on a sufficiently large scale to make a dent in the globe's energy supply in the foreseeable future.

10.3.4 Generating Electricity from Intermittent Energy Sources

There exists a number of promising renewable energy sources that could at some time in the future make a significant contribution to global electrical energy needs. However, the likelihood that these will have a major impact in the short or medium term (5–50 years) is small. It is evident from Figs. 10.1 and 10.2 that non-conventional sources of energy constitute only about 4% of global consumption. Raising that to 20% or more constitutes an enormous challenge, especially in a world where energy demand is rapidly increasing as a result of economic development in countries such as India and China. Simply expanding the use of renewable energy and then incorporating renewable energy sources into energy systems will prove difficult, not least because an expansion in the use of renewables will lead to increases in their prices (as we noted with regard to wood biomass).

Among alternative energy sources, tidal and wave energy are promising, especially considering the potential energy that might be harnessed. Tidal energy is considered particularly desirable because of its regularity and predictability. While some tidal barrage systems are in place and experiments are underway with tidal turbines (which function much like wind turbines), huge technological and cost obstacles still need to be overcome. This is even more the case for wave energy conversion systems, which simultaneously suffer from unpredictability and intermittency. For both wave and tidal systems, costs of transmission lines can be prohibitive.

Solar energy is another promising energy source. The energy or irradiance from the sun averages some 1.366 kW/m^2, or 174 PW for the entire globe, but it is difficult to convert to usable energy. Other than through plant photosynthesis, there are two ways to harness this solar energy: (1) solar photovoltaic (PV) converts the sun's energy directly into electricity, while (2) solar heaters warm water (swimming pools, water tanks, etc.). Solar heaters convert up to 60% of the sun's energy into heat, while PV cells convert only 12–15% of the energy into electricity, although PV laboratory prototypes are reaching 30% efficiency. One problem with solar electricity is its prohibitive capital costs, which amount to some $13,000–$15,000/kilowatt (kW) of installed capacity (International Energy Agency (IEA) 2005); costs have fallen significantly in the past few years, but they remain prohibitive compared to other renewable energy sources. In addition, solar power is intermittent (e.g., output is greatly reduced on cloudy days), unavailable at night, and, in high latitudes, less available in winter when demand is high than in summer (due to shorter days).

Nonetheless, for remote locations that receive plenty of sunshine and are not connected to an electrical grid, the costs of constructing transmission lines to bring in outside power might make solar PV and solar heaters a viable option.

Given the current drawbacks of many other renewable sources of energy, wind energy appears to be the renewable alternative of choice when it comes to the generation of electricity. Given that other renewable sources of energy are not as prevalent as wind in generating electricity, we use wind as our primary example in Chap. 11 where we examine in more detail the integration of renewable energy into electricity markets.

10.4 Discussion

There are three views regarding the energy future and how societies can best reduce reliance on fossil fuels and thereby emissions of CO_2: (1) energy conservation; (2) optimism regarding the future of technological change and the potential for renewable energy; and (3) the gradual shift to a nuclear economy where nuclear energy is not the sole but certainly the primary energy source. Of course, there are a myriad of other possible paths that can be followed over the next century as it is next to impossible to predict what will happen. Further, elements of all three positions will come into play, and governments should not bank on any one to the exclusion of others. Subsidies should not be provided to construct infrastructure that favors one over the other, but funds should be available for research and development. The objective of such an R&D program should be focused more on finding alternatives to fossil fuels rather than options for removing CO_2 from current emissions through carbon capture and storage, whether in terrestrial biological sinks or underground.

10.4.1 Energy Conservation: Rebound and Backfire

There is certainly room for energy conservation. People in North America live in large houses that not only provide each individual with a much larger living area than was enjoyed by previous generations, but ceilings also tend to be higher than in the past to give a feeling of space. This implies that the volume of air that needs to be heated in winter or cooled in summer is much greater than before. Automobiles also tend to be much larger than necessary, as witness the large sport utility and off-road vehicles, the vast majority of which are never used for the purpose for which they were designed. It is little wonder that one popular writer on energy matters, Peter Tertzakian (2009), thinks we are living beyond our means – that there is a lot of energy fat that can be cut.

In many ways, Tertzakian is correct. Western society could reduce its energy consumption by relying on smaller vehicles and smaller houses, and adopting energy innovations such as fluorescent light bulbs. It could enable the 'smart grid,' which, as discussed in the next chapter, would potentially allow the electricity provider to control some of your appliances, including the thermostat setting in your house. This latter technology would be resisted by many because it crosses the line regarding personal freedom – a citizen should be free to pay more for electricity to heat or cool their home to a level that he or she considers comfortable without coercion. Tertzakian even argues that energy saving could be realized if people simply change their mindset. While all of this is true, evidence regarding voluntary efforts to reduce energy use are likely to fail.[24] However, the savings that can be realized are small as these sorts of individual actions are tiny compared to overall emissions.

To get people to change their behavior requires economic incentives such as carbon taxes, 'over-size' taxes on extra-large, gas-guzzling vehicles and/or homes above a certain area or volume, and so on. Unfortunately for the conservation proponents, citizens will oppose legislation that leads to higher energy costs, as we saw in Chap. 8. But there is another problem with reliance on energy conservation.

Earlier we had discussed the term leakage in reference to the offsetting increase in carbon dioxide emissions resulting from land use, land-use change and forestry projects. For example, suppose that a large industrial emitter purchases certified emission offset credits by participating in a large tree planting project that is truly additional. Suppose further that the project removes land from agriculture. In response to reduced agricultural output, forestland in another location will be converted to agriculture and/or farmers will increase their use of nitrogen fertilizer to compensate for lost production. In the longer run, other forest landowners will harvest trees earlier than otherwise in anticipation of lower prices as additional logs are eventually marketed from the afforestation project. These offsetting activities increase greenhouse gas emissions – N_2O from using more fertilizer and CO_2 from harvesting trees. As noted, leakages could be as high as 90% of the greenhouse gas savings from the original project.

Economists have likewise begun to estimate the unanticipated effects of energy efficiency or energy conservation projects. Energy efficiency projects reduce the price of the goods and services derived from energy, and economic actors respond in a variety of ways to these changing prices, thereby increasing energy use. The unexpected effects are as follows:

1. Energy efficiency upgrades themselves require energy to produce and install, offsetting some of the energy savings.
2. There is a direct substitution effect when the prices of goods and services produced from energy fall – there is an increase in energy consumption simply

[24] An example was the Government of Canada's effort in 2002–2004 to meet part of its commitment to reduce CO_2 emissions through voluntary action. The government assumed each citizen would reduce emissions by 1 t ('one-ton challenge' advertising campaign), thereby reducing emissions by 30 Mt annually (or almost one-quarter of the total needed to achieve the Kyoto target). The program failed miserably because citizens had no incentive to reduce emissions and often had no idea how personally they could achieve the objective.

because it is now cheaper. That is, the person who replaces an automobile with one that guzzles less gasoline will tend to drive more, while someone who insulates their home may subsequently turn the thermostat higher because the overall cost of heating her home has fallen. This increases energy demand. Since approximately two-thirds of all energy in an economy is used to produce goods and services, producers will substitute now cheaper energy services for other goods and services (like materials, labor or capital).

3. There is also an income effect: It says that people will spend the energy savings (increased income) on more goods and services, whose provision requires expenditure of energy
4. Any remaining savings in energy costs will be re-spent throughout the economy, increasing demand for goods and services that require energy as an input – an indirect re-spending effect.
5. Finally, at a macro-economic level, energy efficiency improves the productivity of the economy (more output from a given level of energy inputs). Because productivity is a key driver of economic growth, such growth in turn drives up energy demand.

These unanticipated effects result in a rebound, or an increase in carbon dioxide emissions that offsets the initial reduction in emissions brought about by the energy efficiency measure. Estimates suggest that this rebound averages some 60%; that is, only 40% of a project's anticipated emissions reduction is actually realized (Jenkins et al. 2011).[25] Clearly, the size of the rebound effect varies from one measure to another, and may even exceed the original reduction in CO_2 emissions, in which case the term 'backfire' is used. For example, studies of energy efficiency projects that impact electric utilities find that there is a significant backfire effect – in the long term, for every tCO_2 saved through energy efficiency improvements, 1.2 extra tons of CO_2 are eventually emitted (Jenkins et al. 2011). Studies using computable general equilibrium models report economy-wide rebound effects generally exceeding 50% in developed countries, with some even finding backfire. Rebound and backfire are even greater in developing countries (primarily as a result of the fifth factor identified above).

It is not just energy efficiency measures that give rise to unanticipated rebound effects. In Chap. 11, we find that wind, solar and other intermittent sources of energy result in offsetting increases in CO_2 emissions as base-load and other power plants already on the grid operate at reduced capacity, burning more fossil fuels per unit of output. With intermittent energy sources, reserve requirements need to be higher for the same overall level of installed capacity, with these additional generating facilities usually burning fossil fuels. The construction of wind turbines, solar panels or solar heating systems, reserve capacity and added transmission lines increases

[25] Jenkins et al. (2011) provide a literature review and summary of numerous published and unpublished studies of rebound and backfire effects. They find that empirical research into this phenomenon is only now emerging, and that, "as efforts to analyze rebound effects expand in scope and complexity, … the scale of rebound observed generally becomes larger and larger" (p. 25).

emissions of greenhouse gases, as does the harvest, collection and transportation of biomass for use in biomass generators. To the extent that subsidies for renewable energy lead to premature mothballing of existing (mainly coal-fired) generating capacity, the greenhouse gas emissions with decommissioning plants must be considered to offset gains from renewable generating capacity. Likewise, in Sect. 10.3, we saw that there was a large rebound effect associated with the production of biofuels.

The economy-wide reduction in greenhouse gas emissions is not easy to compute even for a straightforward carbon tax or cap-and-trade scheme. The tax or cost of purchasing emission permits leads to higher energy prices and a shift away from fossil fuels and, thereby, a reduction in CO_2 emissions. However, the tax revenue or rent accruing to holders of emission permits is subsequently spent on goods and services, which in turn are produced using energy inputs. It is true that there is a substitution away from goods and services that are energy-intensive towards ones that are not, but the extent to which this occurs is tempered by the spending of tax revenue and/or rents from tradable permits. The size of the rebound is likely greater than 30% for a domestic economy. However, to the extent that other jurisdictions, especially developing countries, do not impose similar taxes or cap and trade, prices for fossil fuels decline relative to what they would be in the absence of such climate mitigation policies. This encourages them to invest more in fossil fuel intensive energy technologies, thus offsetting the emission reduction benefits of the tax or quota policies.

10.4.2 Renewable Energy

We have already shown that, with the exception of large-scale hydropower, renewable sources of energy currently contribute little to the global energy supply. This is not about to change in the near future, despite massive investments in wind energy and biofuels, and ongoing research into solar, tidal, wave, geothermal and other energy sources as well as storage options that facilitate use of intermittent sources of power such as wind (as discussed in Chap. 11). Yet, there remains some optimism. For example, Gerondeau (2010) argues that large swaths of land in various regions around the globe are currently unproductive and can, with some investment, be converted to food and energy crops.

An example is provided by a story in *The Economist* (August 28, 2010, pp. 58–60) on Brazilian agriculture. Large areas of land are currently capable of supporting little more than a few cattle, despite the availability of sufficient precipitation and/or water for irrigation. In Brazil, it is estimated that 300–400 million hectares are in such a state. However, in the *cerrado* or savannah region in the state of Piaui, and well outside the Amazon basin, land has been made productive through the adoption of a systemic approach to agriculture – a combination of investment in the application of lime to reduce the land's acidity and research to create a tropical variety of soybean. In addition to soybeans, the region also grows cotton and maize (corn).

When prices are good, landowners in the region produce soybeans for export; when prices are disadvantageous, the region's crops go to make biofuels.

The argument is that the green revolution has not yet run its course. Not only do research and development lead to greater productivity of land, as in Brazil, but genetically modified organisms (GMOs) have the potential to increase the yields of food and energy crops. (Indeed, many agricultural experts do not think it will be possible to feed a future global population of nine billion or more people without relying on GMOs.) Likewise, genetically modified trees grow faster and R&D might succeed in converting cellulosic fibers to biofuels at lower cost. Even algae could be modified to provide greater yields and, with R&D, used to produce biofuels. And yet, when land is used to grow energy crops, it causes the prices of food and commercial timber products to increase, thus distorting land uses in ways that undermine the CO_2-emissions reduction benefits associated with bioenergy. Further, land distortions impose untold costs on the environment, including loss of wildlife habitat, while promoting growth of algae for food or energy reduces the oxygen available for fish and other aquatic species.

The main problem with energy crops is that biological sources of energy are not sufficiently 'dense' or concentrated – they require too much land (or water) to produce energy, while collecting energy from a large, dispersed area costs energy. Efforts by peasants to gather wood, livestock dung and crop residues for heating purposes are an illustrative example of the problem. While technology certainly can improve upon this relation, the peasant would benefit most from access to affordable electricity or fossil fuels. Overall, bioenergy is a reasonable alternative to fossil fuels for transportation and generation of electricity, but its usefulness on a large scale is limited by its unavoidably low energy density.

Wind and solar power are worthwhile renewable alternatives that also suffer from low energy density, while wave and tidal are costly even if their costs can be greatly reduced, because the construction of transmission lines is a mitigating factor. Given that wind is currently popular, and its costs appear to be within a range that enables them to compete with other sources of electricity generation, it is considered further in Chap. 11.

10.4.3 Natural Gas to Nuclear (N2N)

A final option for reducing CO_2 emissions while still providing enough energy to support economic growth and development is to facilitate the movement toward a nuclear future. Nuclear energy appears to be the only viable alternative to fossil fuels that will enable societies to meet their greenhouse gas emission targets. Nuclear technology is well established and, despite perceptions (noted in Sect. 10.2), a safe and reliable option for generating electricity. Because less than 5% of the energy available in uranium is employed in a nuclear power plant, recycling (refining and upgrading) the nuclear material after a 'first-pass' can lead to a second and even third use of the material to generate power. After a sufficient number of passes, the remaining waste is less radioactive and safer. It might even be possible to store the final waste material in the same mines from which the uranium was originally removed. The problem is that, during the upgrading stages, plutonium is produced, which could be used in nuclear bombs. By keeping meticulous track of the nuclear material, this problem can be overcome as the amount of fuel is significantly smaller than that required in coal-fired plants, for example.

Given that it will take time to build nuclear power plants, establish adequate international safeguards on nuclear materials and conduct research on nuclear technology, it is necessary to bridge from current reliance on coal and oil to nuclear power. Among others, Bryce (2010) suggests that natural gas serves as the most likely bridge. Natural gas releases less CO_2 per unit of heat generated than other fossil fuels and biomass. The technology for using it for transportation already exists, while on-board storage is continually improving. Combined-cycle gas turbines can replace coal- and oil-fired power plants, which are a major source of CO_2 emissions. Further, as already noted, horizontal drilling and hydraulic fracturing have made available for exploitation vast reserves of shale gas, so natural gas scarcity is not a problem in the foreseeable future. However, as discussed in Sect. 10.2, extraction of shale gas might lead to leakages of methane that, if too great, could mitigate the beneficial effects of reduced carbon dioxide emissions. Gas is ubiquitous and not confined to the Middle East, for example, and can easily be transported in pipelines and in liquefied form.

It is clear that the greenhouse emission reduction targets proposed by most developed countries are simply not achievable without nuclear energy, which is why many other scientists favor this option (see Scott 2007). It is also why the prominent environmentalist responsible for the Gaia Hypothesis, James Lovelock, initially came out in support of nuclear energy,[26] but subsequently backed away from this option (and any renewable solution to global warming) arguing instead that the human population needs to be drastically curtailed (Lovelock 2009). However, any attempt to increase reliance on nuclear energy and other non-carbon sources of energy or increase conservation of energy will require huge investments in R&D. Yet, in the United States, for example, energy output is $1.27 trillion annually, but R&D spending on energy is only $3.8 billion, of which the U.S. government supplies $1.4 billion. Government spending on energy R&D is only one-fifth of what it was in the 1970s and 1980s, and well below the $20–30 billion annually recommended by the Brookings Institute (Duderstadt et al. 2009).

In the end, the greatest obstacle to the carbon-neutral economy of the future may not be the lack of policy initiatives. There are many of those. Rather, it may be the failure of governments and their citizens to examine the problem realistically.

---

[26] As viewed August 20, 2009: http://www.ecolo.org/media/articles/articles.in.english/love-indep-24-05-04.htm. There is a network of environmentalists favoring nuclear energy; see http://www.ecolo.org/ (viewed August 20, 2009).

References

Ansolabehere, S., Deutch, J., Driscoll, M., Gray, P. E., Holdren, J. P., Joskow, P. L., Lester, R. L., Moniz, E. J., Todreas, N. E., Hottle, N., Jones, C., & Parent, E. (2003). *The future of nuclear power: An interdisciplinary MIT study*. Cambridge, MA: Massachusetts Institute of Technology.

Bogle, T., & van Kooten, G. C. (2010). *What makes mountain pine beetle a tricky pest? Optimal harvest when facing beetle attack in a mixed species forest* (REPA Working Paper 2011-07). Victoria: Resource Economics & Policy Analysis Research Group, University of Victoria. Retrieved from http://web.uvic.ca/~repa

Bryce, R. (2008). *Gusher of lies: The dangerous delusions of 'energy independence'*. New York: Public Affairs.

Bryce, R. (2010). *Power hungry: The myths of "green" energy and the real fuels of the future*. New York: Public Affairs.

Crutzen, P. J., Mosier, A. R., Smith, K. A., & Winiwarter, W. (2008). N_2O release from agro-biofuel production negates global warming reduction by replacing fossil fuels. *Atmospheric Chemistry and Physics, 8*(2), 389–395.

Deffeyes, K. S. (2003). *Hubbert's peak: The impending world old shortage*. Princeton, NJ: Princeton University Press.

Deutch, J. M., Forsberg, C. W., Kadak, A. C., Kazimi, M. S., Moniz, E. J., Parsons, J. E., Du, Y., & Pierpoint, L. (2009). *Update of MIT 2003 future of nuclear power: An interdisciplinary MIT study*. Cambridge, MA: Massachusetts Institute of Technology. Retrieved from http://web.mit.edu/nuclearpower/

Duderstadt, J., Was, G., McGrath, R., Muro, M., Corradini, M., Katehi, L., Shangraw, R., & Sarzynski, A. (2009). *Energy discovery-innovation institutes: A step toward America's energy sustainability*. Retrieved from http://www.brookings.edu/~/media/Files/rc/reports/2009/0209_energy_innovation_muro/0209_energy_innovation_muro_full.pdf

Fox, C. (2011). *The effects of CO2 abatement policies on power system expansion*. Unpublished MA thesis, Department of Mechanical Engineering. University of Victoria. Victoria, BC. Retrieved from http://www.iesvic.uvic.ca/

Gerondeau, C. (2010). *Climate: The great delusion – A study of the climatic, economic and political unrealities*. London: Stacey International.

Goklany, I. M. (2011). Could biofuel policies increase death and disease in developing countries? *Journal of American Physicians and Surgeons, 16*(1), 9–13.

Harrison, P. (2010, April 21). *Once-hidden EU report reveals damage from biodiesel*. Retrieved from http://www.reuters.com/assets/print?aid=USTRE63K2CB20100421

Howarth, R. W., Santoro, R., & Ingraffea, A. (2011). Methane and the greenhouse-gas footprint of natural gas from shale formations. *Climatic Change, 106*(4), 679–690.

International Energy Agency. (2005). *Projected costs of generating electricity*. Paris: Nuclear Energy Agency, OECD/IEA. 2005 update.

International Energy Agency. (2009). *World energy outlook 2008*. Paris: OECD/IEA.

International Energy Agency. (2010a). *Key world energy statistics 2009*. Paris: OECD/IEA.

International Energy Agency. (2010b). *World energy outlook 2009: Executive summary*. Paris: OECD/IEA.

International Energy Agency. (2011). *Key world energy statistics 2010*. Paris: OECD/IEA.

Jenkins, J., Nordhaus, T., & Shellenberger, M. (2011, February). *Energy emergence: Rebound and backfire as emergent phenomena* (p. 59). Oakland: Breakthrough Institute.

Lomborg, B. (Ed.). (2010). *Smart solutions to climate change: Comparing costs and benefits*. Cambridge: Cambridge University Press.

Lovelock, J. R. (2009). *The vanishing face of Gaia*. Toronto: Penguin.

NEA & IEA. (2005). *Projected costs of generating electricity*. Paris: Nuclear Energy Agency, International Energy Agency, OECD. 2005 update.

Niquidet, K., Stennes, B., & van Kooten, G. C. (2012). Bio-energy from Mountain Pine Beetle timber and forest residuals: The economics story. *Canadian Journal of Agricultural Economics, 60*(2), 195–210

Nowak, P. (2010). *Sex, bombs and burgers: How war, porn and fast food created technology as we know it*. Toronto: Viking Canada/Penguin Group.

Osborn, S. G., Vengosh, A., Warner, N. R., & Jackson, R. B. (2011). Methane contamination of drinking water accompanying gas-well drilling and hydraulic fracturing. *Proceedings of the National Academy of Sciences, 108*(20), 8172–8176.

Prins, G., Galiana, I., Green, C., Grundmann, R., Hulme, M., Korhola, A., Laird, F., Nordhaus, T., Pielke, R. A., Jr., Rayner, S., Sarewitz, D., Shellenberger, M., Stehr, N., & Tezuka, H. (2010, May). *The Hartwell paper: A new direction for climate policy after the crash of 2009* (p. 42). London: Institute for Science, Innovation & Society, University of Oxford. LSE Mackinder Programme, London School of Economics and Political Science.

Reed, T. (2011, April 11). *Washington Examiner*. Viewed April 20, 2011, from http://washington-examiner.com/opinion/op-eds/2011/04/new-york-and-america-can-profit-fracking-marcellus-shale

Riddel, M., & Shaw, W. D. (2003). Option wealth and bequest values: The value of protecting future generations from the health risks of nuclear waste storage. *Land Economics, 79*(4), 537–548.

Saleri, N. G. (2011, March 9). Our man-made energy crisis. *Wall Street Journal*. Viewed March 17, 2011, from http://online.wsj.com/article/SB10001424052748703386704576186622682563228.html?mod=djemEdito

Scott, D. S. (2007). *Smelling land: The hydrogen defense against climate catastrophe*. Ottawa: Hydrogen Association and Natural Resources Canada.

Searchinger, T. D. (2011, February 11). *How biofuels contribute to the food crisis*. Retrieved from www.washingtonpost.com/wp-dyn/content/article/2011/02/10/AR2011021006323.html

Searchinger, T. D., Heimlich, R., Houghton, R. A., Dong, F., Elobeid, A., Fabiosa, J., Tokgoz, S., Hayes, D., & Yu, T. (2008, February 29). Use of U.S. croplands for biofuels increases greenhouse gases through emissions from land-use change. *Science, 319*(5867), 1238–1240.

Smil, V. (2003). *Energy at the crossroads: Global perspectives and uncertainties*. Cambridge, MA: MIT Press.

Stennes, B., Niquidet, K., & van Kooten, G. C. (2010). Implications of expanding bioenergy production from wood in British Columbia: An application of a regional wood fiber allocation model. *Forest Science, 56*(4), 366–378.

Tertzakian, P. (2009). *The end of energy obesity: Breaking today's energy addition for a prosperous and secure tomorrow*. New York: Wiley (with K. Hollihan).

van Kooten, G. C. (2010). Wind power: The economic impact of intermittency. *Letters in Spatial & Resource Sciences, 3*(1), 1–17.

Wigley, T. M. L. (2011). Coal to gas: The influence of methane leakage. *Climatic Change, 108*(3), 601–608. doi:10.1007/s10584-011-0217-3.

Chapter 11
Electricity Markets and Wind Energy

Science is the belief in the ignorance of experts.

– Richard Feynman

Electricity generation relies primarily on fossil fuels as its energy source, which implies that it is a major source of CO_2 emissions. The share of power generation in global CO_2 emissions increased from 36% (8.8 $GtCO_2$) in 1990 to 41% (11.0 $GtCO_2$) in 2005 (Malla 2009). In the United States, the electric generating sector accounted for over 40% of total domestic CO_2 emissions in 2009, representing a steady increase from 32% in 1980, which is unsurprising as two-thirds of U.S. electricity generation is derived from fossil fuels.[1] As noted in Chap. 10, there are differences in the generating mixes of the electricity grids in various jurisdictions and even within countries, with some mixes much more fossil-fuel intensive than others. Across countries, for example, Canada's electricity sector is only 24% fossil-fuel based, whereas Australia and China produce 93 and 82% of their electricity from fossil fuels, respectively.

Because electricity is such an important commodity and because of its important role in emitting CO_2, it is well worth considering ways of reducing fossil fuel electricity generation. The benefits of producing power from renewable energy sources will vary depending on the makeup of the electricity grid. If hydroelectricity is the primary generating source, investments in generating facilities that employ non-hydro renewable sources of energy might reduce CO_2 emissions very little, or not at all. For example, if in-stream values (e.g., fish habitat) are important, water that would otherwise run through a turbine and generate electricity may need to be spilled when hydropower is replaced by power from another renewable source. If biomass is used to generate electricity, for example, the benefits of CO_2 emission

[1] U.S. Energy Information Administration 2010. *Annual Energy Review 2009*. Carbon Dioxide Emissions from Energy Consumption by Sector (Table 12.2). At http://www.eia.doe.gov/aer/txt/ptb1202.html (viewed October 18, 2011).

G.C. van Kooten, *Climate Change, Climate Science and Economics: Prospects for an Alternative Energy Future*, DOI 10.1007/978-94-007-4988-7_11,

reduction are small (as shown in Chap. 10), so much so that, if it replaces hydropower, a wind source or even natural gas, the costs of emissions reduction may be so high on a per ton of CO_2 basis that alternative means of addressing climate change concerns are preferred on economic efficiency grounds.

In this chapter, we want to investigate electricity markets and the potential for renewable energy to make a contribution to a global reduction of greenhouse gas emissions. We focus on wind energy for reasons mentioned in Chap. 10, namely, that wind energy is the fastest growing and likely the most cost-effective renewable option outside of hydropower. Here wind is used to represent the potential of other renewable energy sources because, with the exception of biomass and large-scale hydro, renewable energy sources are best characterized as an intermittent or variable source of electricity.

We begin in the next section by describing how electricity markets function. We examine supply and demand for electricity, the role of reserves, and what it means to implement a 'smart grid.' Then, beginning in Sect. 11.2, we turn our attention to wind power. We describe the particular nuances of wind power in Sect. 11.2, and then illustrate the problems of integrating wind into existing electricity grids in Sect. 11.3. One purpose is to examine the challenges that wind power poses and provide some indication of the potential costs of integrating wind into power grids, and particularly the costs of reducing CO_2 emissions. The chapter ends with some concluding observations and policy implications.

11.1 Electricity Markets

Electricity is an unusual commodity in that production and consumption occur simultaneously and at every instant in time. That is, unlike a normal market where there is a mechanism that enables consumers and producers to 'discover' the market clearing price over a period of time, the market for electricity must clear continuously and instantaneously. Nonetheless, supply and demand for electricity remain the essential means for describing the underlying process that enables the electricity grid to function.

11.1.1 Investment in Generating Capacity and the Load Duration Curve

A convenient method of presenting the annual demand for electricity in a region is via the load duration curve. The annual load duration curve is constructed by arranging the hourly loads for a particular year from highest to lowest. The minimum load during the year is referred to as the base load because it is the minimum power that must be delivered to the grid at every moment throughout the year. Unlike the base load, which is invariant throughout the year, the peak load represents the maximum

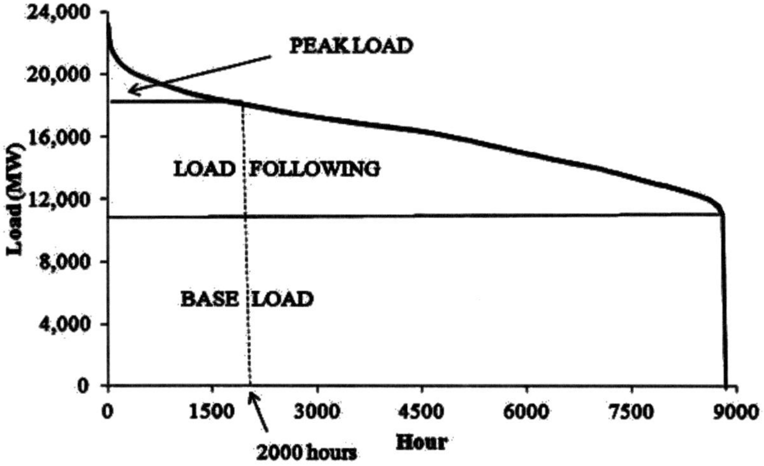

Fig. 11.1 Annual load duration curve illustrating concepts of base and peak load

power that must be delivered to the grid at some point throughout the year. While peak demand varies daily reaching its highest levels in the late afternoon or early evening, or on work-day mornings, there is one hour during the year when electricity demand reaches an apex. Depending on the system, this will occur during a particularly cold hour during the daily peak period when power is required for heating, or during a particularly warm period when electricity is needed for cooling.

An example of an annual load duration curve is provided in Fig. 11.1 for the Ontario power grid for 2008. (As this was a leap year, the number of hours on the horizontal axis is 8,784 rather than 8,760.) A peak load of 23,168 MW occurred on June 9, 2008 at 3PM, indicating that power was required for cooling and not heating. Ontario's base load during 2008 was 11,042 MW, accounting for 96,993 GWh (68%) of Ontario's total annual demand of 142,272 GWh. Using the definition of peak load in Fig. 11.1, peak demand amounted to 2,442 GWh, or only 1.7% of total annual demand; 'load following' represented 42,837 GWh (30% of annual demand) and is usually met by generation assets that have some ability to adjust output, although not quite as rapidly as required to meet peak demand. As discussed in the next subsection, load following assets might consist of spinning reserves, hydro assets, fast-responding natural gas plants, et cetera, or even base load facilities at certain times. Since the load duration curve can be used to analyze investments in generating facilities, it is discussed further below.

Compared to Ontario, Alberta's industrial base is more focused on resource extraction, especially oil and gas. Further, Albertans heat their homes primarily with natural gas while heat and humidity are less of a problem during the summer – less electricity is required for heating and cooling than in Ontario. As a result, one expects the difference between peak load and base load to be smaller in Alberta than Ontario, and this is the case. Based on 2008 load data for the two electricity systems, one finds the ratio between peak load demand and base load demand to be 2.1 for

Table 11.1 Proportion of installed generating capacity by type, Alberta, Ontario and British Columbia, 2010

	Province		
Type	Alberta (%)	Ontario (%)	British Columbia (%)
Nuclear	–	33	–
Coal	46	13	–
Gas	39	28	9
Hydro	7	23	88
Biomass	2	–	1
Wind	6	3	2

Ontario and 1.5 for Alberta. This implies that Ontario will need to rely on 'peaking' plants – fast-responding gas or diesel plants and hydroelectric facilities – to a greater extent than Alberta. That this is the case is indicated in Table 11.1.

As of the end of 2010, total installed capacity in Alberta amounted to 13,520 MW, while that in Ontario amounted to 34,400 MW; in 2008, peak loads for Alberta and Ontario were 10,347 and 23,168 MW, respectively. While the ratio of peak load to installed capacity is much higher for Alberta than Ontario (0.765 versus 0.673), Ontario is in the process of phasing out its coal plants and reducing its nuclear capacity by more than 10%, replacing these with biomass and wind generating assets (see Chap. 10).

Total installed generating capacity by type is also provided in Table 11.1 for British Columbia, which is characterized by it large dependence on hydroelectric sources. Hydropower is used to meet base load, load following and even peak demand, although a small amount of natural gas generation is sometimes needed to meet peak load.

The load duration curve can be used, along with information about the overnight construction cost, to guide a system operator regarding the technology mix to use – the asset mix that leads to the lowest long-term costs. This is illustrated with the aid of Fig. 11.2. Using information about the overnight construction cost and operating costs from Table 10.5, it is possible to build 'screening curves' as shown in the upper half of the figure. For a given technology, whether nuclear, coal, diesel, hydro, gas, et cetera, the screening curve indicates how the cost of generating electricity varies with the capacity factor. The capacity factor for a particular generation technology is simply the amount of electricity that the technology generates in a given period divided by what it is capable of generating in that period. For a given technology, the fixed (overnight construction) cost, appropriately adjusted to reflect lifetime expected operation of a plant (and denoted C_j, j=C, F and P for coal, load following and peaking plants), determines the intersection of the screening curve on the vertical axis, while operating costs determine the slope. As discussed by Stoft (2002, pp. 33–45), where the screening curves intersect determines the choice of technology – the prices of construction material, labor, fuel and so on determine the choice of technologies that an operating area will employ to meet demand.

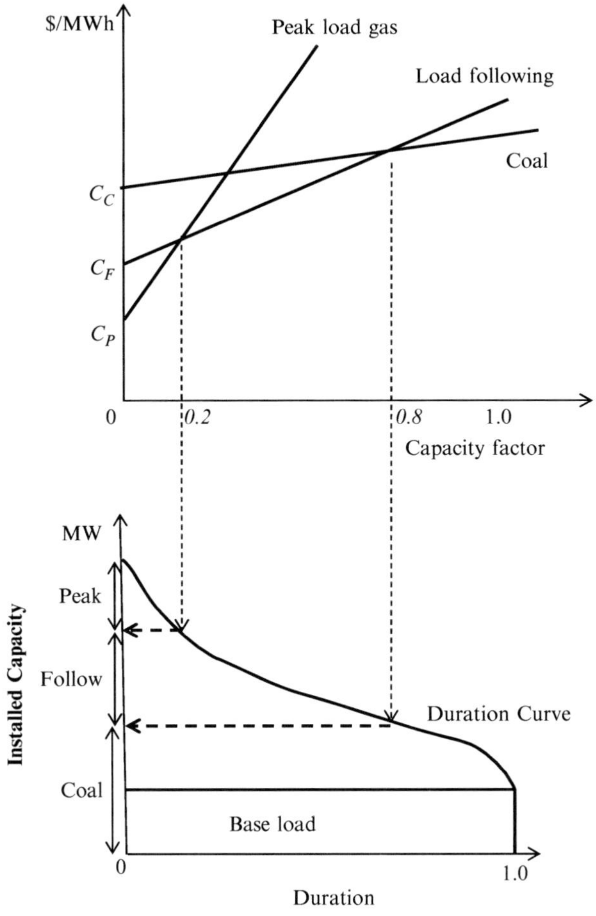

Fig. 11.2 Screening curves, load duration and determination of generation mix

Based on the screening curves provided in the upper half of Fig. 11.2, all loads with duration greater than 80% should be met using base-load coal-fired power plants. For the illustration in Fig. 11.2, coal should be used to satisfy all base load demand, plus some of the load following demand. That is, the least-cost method of generating electricity is to rely on coal-fired (or other base-load) power plants, if these can be operated nearly continuously, so that they attain an average capacity of 80% or more. In the example, the system operator would invest in coal plants that would then satisfy base load demand plus some proportion of load following demand. However, if loads include a significant component of intermediate variability (requiring load following), the generating assets required to satisfy this demand variability need to be designed so they can operate efficiently between 20 and 80% capacity over the long run. A variety of such assets may be needed, perhaps including some coal-fired generators (as indicated in the figure); investments in load following capacity of different types are chosen based on their abilities to address particular ramping speeds. But only gas and diesel generating assets (peaking plants), and hydroelectric facilities, can deal with rapid changes in load; peaking plants need to ramp rapidly up and down, but are only required for a very short time during the year (less than 20% of the time). Peak plants trade off low construction costs against high operating costs.

The actual capacity of each technology that should be installed depends on the load duration curve. In the lower half of Fig. 11.2, the load duration curve (similar to that in Fig. 11.1) is normalized by dividing through by the total number of hours in the year (8,760). For the capacity factors associated with the intersection of the screening curves, it is possible to derive the 'desired' installed capacity for each technology as indicated in Fig. 11.2. In this example, to keep overall system costs at their lowest in the long term, the authority would install the greatest amount of coal capacity, with the least amount of peaking capacity.

Although a useful guide for policy makers and researchers, there are three problems with this approach. First, because investments in generating capacity are long term, the system operator or investor must have some notion of what the load duration curve might look like in the future. Second, if deregulation of wholesale prices also leads to real-time pricing at the retail level, then the load duration curve can no longer be considered fixed. Finally, when wind, solar and other renewable sources of energy are used to generate electricity, with the result that their supply is intermittent and non-dispatchable (must run), the load duration curve also shifts about. The latter two concerns are discussed in the next subsections.

11.1.2 Demand Management

Consumers have rarely been asked to respond to changes in wholesale prices of electricity; in most jurisdictions they face the same price throughout the day and year. The retail price is generally set by a regulator, with price changes occurring only when the regulator permits the system operator to do so. Prices are regulated because production, transmission and delivery of electricity are inherently monopolistic activities, at least historically. The generation of electricity and its delivery to the customer were considered to be the function of a single firm – a monopolistic activity that then had to be provided publicly or regulated, or both. However, many jurisdictions have recently separated generation, transmission and delivery to varying degrees, and the need for regulation or public ownership has started to disappear.

The first step in the process of deregulation is to separate ownership of power generation from transmission and delivery, thereby creating a wholesale market for electricity. An independent (private or public) electrical system operator (ESO) allocates power generation across various generating assets, and arranges its transmission and delivery to customers. While the wholesale price might fluctuate widely as power generating companies compete to sell electricity, the retail price is still set by the regulator. In a fully deregulated system, one would expect the retail price to fluctuate hourly with the wholesale price, with the difference between the two reflecting the cost of transmission and delivery. The difference between wholesale and retail prices should remain fixed unless competition for use of transmission lines or congestion cause transmission costs to change; thus, an increasing cost of using transmission lines might add to the retail price at peak periods.

If retail prices are fixed, the demand function is essentially a vertical line at any given time – load does not respond to changes in wholesale prices. One way to affect consumer demand is to employ a tiered system whereby rates rise (or fall) with increased (decreased) overall usage over a specified period. Rather than redistribute some load from peak to off-peak hours, a tiered system of prices can reduce or increase demand, depending on circumstances and the prices of alternative energy sources. For example, an increase in demand can occur if a large customer is well below the use that would take it to the next higher-price tier. Suppose the consumer heats water using natural gas and currently does not reach the next price level in its use of electricity. If gas prices are sufficiently high, it will pay for the consumer to heat water using electricity rather than natural gas (assuming the flexibility to do so). Electricity will be used to heat water to the point where the power usage encounters the higher-price threshold of the next tier.

With the exception of tiered pricing that relies on measures of cumulative power use, 'smart' controls are required to implement any other pricing scheme. Smart controls send signals to the system operator via the internet, say. Unless smart controls are built onto transmission lines, or smart meters are installed at consumption points, the operator of the transmission system is unaware of when and where power outages occur or even theft of power. Such events must be communicated to the operator. Without smart controls that receive price signals and adjust electrical use accordingly, consumers are simply unable to respond to time-of-use price signals.

When it comes to affecting demand, it is important to distinguish between efforts to shift load from peak periods to off peak periods and attempts to influence demand (conserve electricity). In the absence of wide-spread installation of smart meters, the system operator can provide incentives only to the largest industrial and commercial customers. By installing smart meters on the premises of large customers, pricing can be used to get them to switch demand from peak times to off-peak times, for example. In practice, however, the purpose of such incentives is primarily to shift load as opposed to reducing consumption.

If smart controls are more widely available so that residential users and small businesses have smart meters, time-of-use pricing can be implemented. However, it would be too expensive in terms of time and effort for the vast majority of consumers to respond to anything but simple pricing schemes, such as those that distinguish between daytime and nighttime use; a simple pricing scheme such as this is about all that small customers can handle, and even then it requires the installation of smart meters at each customer's location to identify when electricity is consumed. But daytime-nighttime pricing schemes do little more than shift the load from day to night. Nonetheless, any scheme that 'shaves' (reduces) the peak load, even if it only shifts demand to off-peak times, can result in substantial cost saving as less overall and reserve generating capacities are required. (Reserves are discussed in the next subsection.)

Load 'shedding' is a different proposition: An ESO will need to shed load in an emergency when the system demand exceeds generation. This can be done via built-in incentives or, more often, contracts between the operator and large customers. Again, the purpose here is not to conserve energy or influence demand as much as it is to reduce system management costs.

Demand for electricity can only truly be affected by implementing time-of-use or real time pricing at the retail level. As noted, however, the vast majority of customers cannot respond to time of use prices. In order for demand to respond to constantly varying price signals, it is likely necessary to implement a 'smart grid' – something beyond just smart meters. Smart grids require smart appliances, including smart thermostats, which can respond to signals from off site. There is much discussion about smart grids, but there are some obstacles to its implementation. The computer chips in smart meters and smart appliances send and receive signals, usually in conjunction with the internet. These enable the system operator (or anyone else with access) to monitor electricity use, signal appliances to go on or off, and adjust thermostats – control devices from a distance. For example, appliances such as dishwashers, washing machines, clothes dryers and heaters could be turned off or on depending on the price of electricity. At times of excessive load or when a generator fails, the system operator could curtail consumers' use of electricity or signal certain appliances to shut down.

While not all electronic devices have smart technology embedded in them, and installing smart devices could be expensive, perhaps the greatest obstacle to smart grids might be concerns about privacy. One option in this case might be to allow consumers to opt out of the smart grid, but at a cost (e.g., higher overall average electricity rates). Alternatively, customers might provide only partial off-site control of appliances (e.g., a thermostat could not be set below a particular reading from off site, operation of a clothes dryer could only be delayed for a pre-set number of hours, etc.).

It is fair to conclude, at this point, that prices likely vary little at the retail level and, further, that the demand for electricity is likely to remain relatively inelastic even if real-time pricing was implemented. The reason is that electrical equipment provides health-care and life-saving benefits, reduces menial household chores (which enables women to participate in the labor force), and contributes to an improvement in our overall wellbeing. Based on cross-section and time series analyses, the short-run elasticity of demand is often assumed to be about –0.3 (U.S. Energy Information Administration 2010, p. 26), while it is between –1.5 and –0.5 in the long run.[2] This implies that a 1% increase in the price of electricity only results in a 0.3% reduction in demand in the short run, and a reduction of 0.5–1.5% in the long run.

---

[2] Estimates of both the short- and long-run price elasticities of demand for electricity vary widely. In a meta-regression analysis of studies of U.S. residential demand for electricity, Espey and Espey (2004) concluded that the best estimate of short-run and long-run elasticities were –0.28 and –0.81. For example, a co-integration study found long-run price elasticity to be –0.5 (Silk and Joutz 1997). However, a more recent Swiss study found long-run price elasticity of demand to range from –1.27 to over –2.0, with demand more elastic during peak than off-peak periods (Filippini 2010).

11.1.3 Electricity Supply and the Wholesale Market

To examine the supply side, assume an electricity system that is deregulated at the wholesale level. The system operator requires owners of generating facilities to commit to produce electricity at a given hour 1 day (24 h) ahead of actual delivery. Each generator will offer to produce a certain amount of electricity at a particular price, knowing that the final price received is the market-clearing price for that hour. In essence, a power plant will offer units of electricity at a single price (or variety of prices if costs of producing electricity differ across units) to be produced and delivered on a specified hour the next day. This is known as day ahead, unit commitment. Of course, as the hour approaches for which an owner of a generating facility has committed to deliver power more information about the status of its own and other suppliers' generators and the evolution of prices becomes known. Therefore, generators are able to make changes to their offers up to 2 h before delivery. The extent of permitted changes is increasingly constrained by penalties as the hour nears.

What do offers to supply electricity look like? Base load nuclear and coal-fired power plants, and for some grids base-load hydropower dams, will bid in lowest. Indeed, for base-load facilities that cannot readily change their power output, or can do so only at high cost, the optimal strategy is to provide very low (even zero) price bids to ensure that they can deliver power to the grid. Open-cycle gas turbine (OCGT) peaking plants will want to bid in at their true marginal cost of production, which is determined primarily by the price they have to pay for fuel and, thus, whether gas is purchased under contract or in the spot market. The facilities that provide the highest bids are those that wish to export electricity; by setting price high, their output is unlikely to be chosen by the system operator and can thus be exported. (Importers will want to set their prices low to guarantee that the imported power will be chosen.) In between the extreme prices are found a variety of generating facilities, such as biomass plants, combined-cycle gas turbine (CCGT) plants, and sub-units of extant plants that are at different levels of readiness, maintenance, et cetera.[3] Once the ESO has all of the information regarding the amounts of electricity that the various components of the generating system are willing to supply and their associated prices, a *market (economic) merit order* is developed to allocate power across generators. An example is illustrated in Fig. 11.3.

In Fig. 11.3, the supply curve is given by the market merit order. Base-load nuclear and coal facilities bid in at the lowest price, followed by CCGT and other generating facilities as indicated. The market clearing price is determined by the

[3] Unlike open-cycle gas turbine (OCGT) plants, combined-cycle gas turbine (CCGT) plants can be base load. CCGT plants capture heat that would escape out of the stack in an open-cycle system to generate additional electricity. While CCGT plants can be built to ramp more quickly, there is always a tradeoff that adds to cost. Even coal-fired generators can be built to better track changes in output from variable generating sources, but again at increased cost in terms of reduced efficiency and greater wear and tear of equipment. Likewise, biomass fueled plants are often base load, although their capacity tends to be much smaller than that of coal, nuclear and CCGT plants. Finally, load might be above base load for long periods during winter or summer.

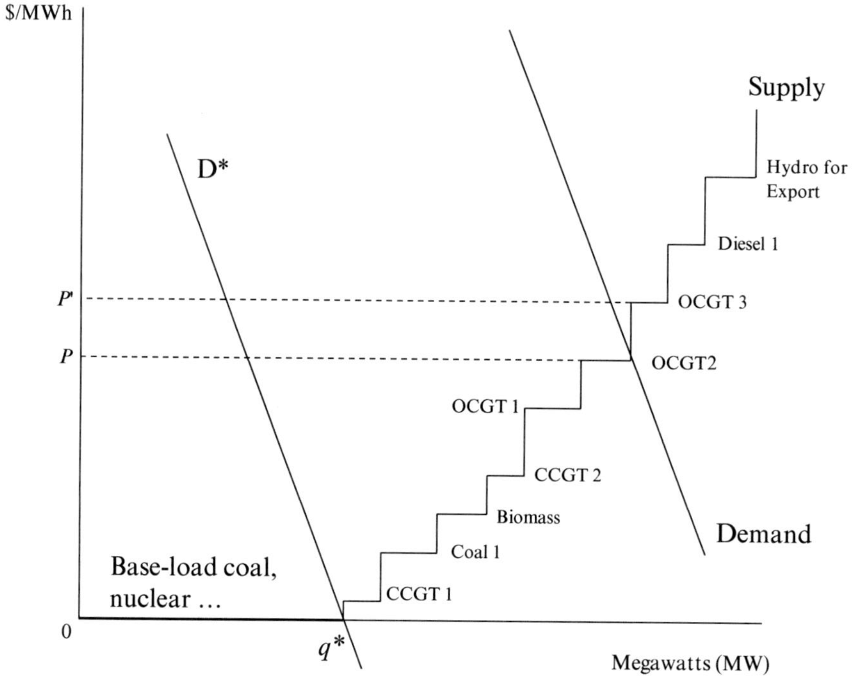

Fig. 11.3 Market merit order

location of the demand curve at that hour. Assuming the demand curve on the right, the market price *P* is given by the marginal open-cycle natural gas plant (OCGT 2). All generators get paid the average price *P* for the period in question. Base-load facilities bid in at zero price to avoid incurring the high costs of curtailing output, but knowing they will receive *P*. If the demand curve is D* then the wholesale price is zero and only base-load facilities generate electricity. Assuming that investments in base-load capacity were determined by the minimum load through a year, demand would never be less than D*, with q^* representing the system's minimum load. To reiterate, base-load plants would bid in at a zero price despite potentially earning no revenue; this is to avoid high costs of ramping production or, worse, dumping power in an emergency-like situation (i.e., instantaneously reducing pressure in the boiler).

If the transmission infrastructure somehow impedes OCGT 2 (or some other plant) from delivering power, then OCGT 3 determines the market clearing price, which becomes P′. This higher average system price distorts incentives. As a result, some systems have gone to location-specific pricing, with the prices that generators receive established at a local or regional center within the ESO's operating area rather than averaged over the entire operating area. Knowing this, the bidding strategy likely changes, both in the market for power delivered to the grid and in the market for ancillary services (discussed below). One advantage of location-specific pricing is that it leads to differences in prices, which, in turn, provide

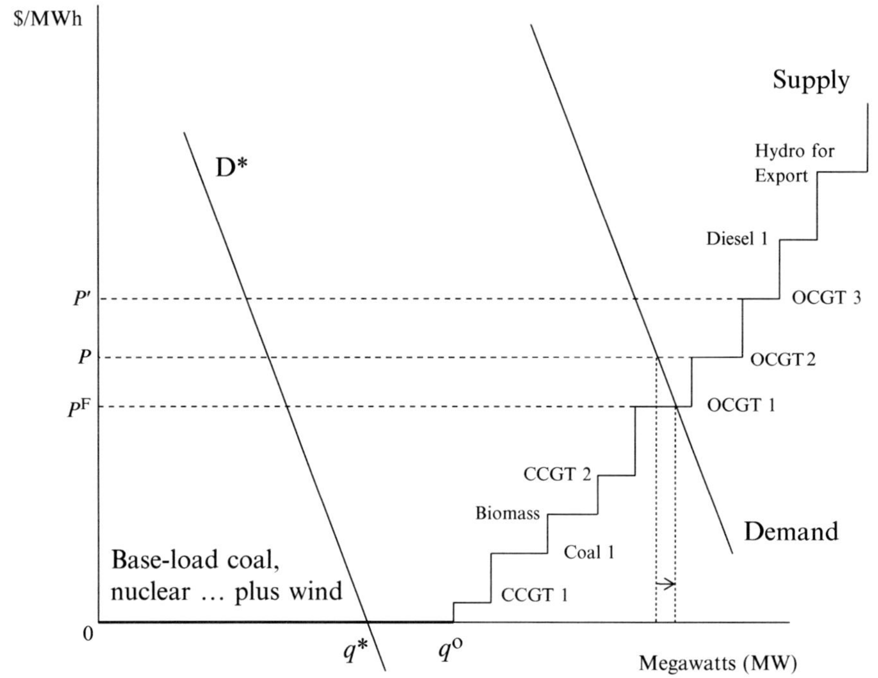

Fig. 11.4 Market merit order with an intermittent renewable

incentives for arbitrage through the upgrading or construction of transmission lines connecting regions.

Now suppose a feed-in tariff for biomass-generated power increases biomass generating capacity. In terms of Fig. 11.3, biomass would drop in the merit order because of the subsidy and more would be available. This could result in moving CCGT 2 or even CCGT 1 and CCGT 2 'higher up' in the merit order – essentially the bid prices would remain the same but biomass will be able to bid in at a lower price. All other generators would be chosen later in the merit order, with OCGT 1 or even 'Coal 1' becoming the marginal power plant if sufficient additional biomass capacity becomes available. The price of electricity would fall, ceteris paribus. If biomass generation becomes base load, it will be necessary to displace some nuclear and/or coal base-load capacity. This might be desirable except that, as noted in Chap. 10, there may be constraints on wood fiber availability.

The picture changes completely when wind, solar, run-of-river or other variable generating capacity is introduced into the electricity system as a result of subisidy or a feed-in tariff. The situation can be illustrated with the aid of Fig. 11.4. The only difference between Figs. 11.3 and 11.4 is the addition of q^*q^0 amount of power from variable sources (hereafter referred to as wind). This shifts the supply curve in Fig. 11.3 to the right by amount q^*q^0. Now, with the original demand curve (right most one in Fig. 11.4), it is no longer OCGT 2 that is the marginal producer of electricity; rather, it is the plant with a lower marginal cost, OCGT 1. The market clearing price of electricity for that hour falls from P to P^F. The feed-in tariff lowers the price of electricity, which will induce consumers to purchase more of it (as indicated by the arrow).

What does one do with the wind energy q^*q^0 if the demand in a given hour is D*? Clearly, either the wind must be curtailed (wasted) or base-load output reduced. Base-load hydropower can easily be reduced, as discussed below, so consider only a system with base-load thermal generating capacity. If q^*q^0 could be reliably produced in every period, so it can be considered part base load production, then some coal or nuclear base-load capacity becomes redundant and can be eliminated – an ideal outcome.[4] However, wind generated power is not reliable and thus cannot replace thermal base-load capacity, except at some cost.

Suppose base load capacity is reduced by the amount q^*q^0. Then, whenever wind power is less than q^*q^0, this is the same as shifting the supply curve in Fig. 11.3 to the left, which would raise the market price for every hour that wind is less than q^*q^0, while lowering price if wind output exceeds q^*q^0. Thus, the effect of a feed-in tariff for wind (or solar, wave, tidal, etc.) is to increase price volatility if thermal base-load capacity is driven from the system; if thermal base-load capacity is not driven from the system, electricity prices will generally be lower, but base load plants will need to ramp up or down if wind energy is non-dispatchable (i.e., considered to be must run), which will increase their operating costs (van Kooten 2010). Alternatively, if wind is considered dispatchable, wind will need to be curtailed or wasted (as illustrated in Sect. 11.2).

The situation is somewhat different in a system with significant hydroelectric capacity, because hydropower can provide base-load power and serve the peak load and reserve markets. The presence of significant hydro capacity enables a system to absorb wind power that might overwhelm the ability of a system with a high thermal capacity in the generating mix to absorb it, or raise system costs by too much in doing so. That is, the existence of hydro reservoirs enables a system to store wind energy that would be wasted in systems lacking hydro generating capacity in the mix. However, there must be times when this stored wind energy is required to meet load, perhaps at peak load times. The problem is that stored energy might not always be used in the future as a result of in-stream (aquatic life) or other (flood control) requirements. In that case, wind generated power simply replaces another renewable source of power, with the consequence that renewable energy is wasted.

11.1.4 Reserve Markets

Because demand and supply of electricity must balance at all times, there is one further aspect to electricity markets and that is the need for operating reserves – there is a market for ancillary services. Ancillary services are not homogeneous, and

[4] Of course, with concern about climate change, the optimal solution would be to reduce coal-fired capacity.

even how they are defined and handled may differ across jurisdictions. Regulatory (fast-response) services are needed to address second-by-second, minute-by-minute fluctuations in demand and supply so that grid reliability is maintained – that the grid delivers 120 V at 60 Hz (in North America) or 240 V at 50 Hz (Europe). Regulating reserves operate over a time frame of several seconds to as much as 10–15 min. Such short-term fluctuations are generally met by the on-line generators themselves, as base-load plants are able to vary their outputs slightly over some range. In addition, some open-cycle gas and diesel generators are operating below capacity or on standby. By adjusting the fuel going to the turbine (in essence applying more or less pressure to the 'gas pedal'), these units can readily adjust output. For example, generator OCGT 1 in Fig. 11.4 is not operating at full capacity and can easily adjust supply (e.g., by even more than the amount indicated by the arrow). Some generators will simply be idling in standby mode, not delivering electricity to the grid; these are referred to as 'spinning reserve,' as distinguished from units that are operating below capacity.

Storage devices, such as batteries and flywheels, might also be used in a regulatory capacity as might hydropower. Automated generation control (AGC), which is also known as regulation, is used to manage small fluctuations in the supply-load balance.

Load following reserves are those that are required to follow shifts in load on time frames that usually do not exceed 10 min, and have much in common with regulatory reserves. The difference is that load following requirements are generally anticipated (as discussed in conjunction with Fig. 11.2), while regulating reserves are designed primarily to deal with unanticipated changes in load and supply. However, the distinction is often opaque.

Finally, contingency reserves are required to meet a situation where power from any given generator is suddenly lost for whatever reason. Contingency reserves are designed to handle emergencies – the contingency that a power plant goes 'off line' and is unable to provide the electricity that it had committed. For example, the Western Electricity Coordinating Council (WECC, www.wecc.biz) requires that contingency reserves be sufficient to cover the most severe potential loss (loss of the largest generating unit) plus some proportion of the total production from hydro and thermal sources. The market for contingency reserves is indicated in Fig. 11.5.

Suppose that the merit-order demand for contingency reserves in a given hour is denoted D^C, which is determined by the conditions set out by WECC. The various units bid their reserves much like they do in the establishment of the merit order in Fig. 11.3. In the ancillary market, the OCGT and diesel peakers will now want to bid in at a low price because they are the ones that can get off the mark the quickest. Likewise, the bid price of the hydro contingent reserves will be low, perhaps even zero (in which case they only obtain price P in Fig. 11.3 for any power sold), but the prices bid by peakers OCGT 3 and Diesel 1 will also be low because they know that, when there is a demand for ancillary services, they will receive at least the price determined by the marginal generator (OCGT 2 in Fig. 11.3) plus their own bid in the ancillary market. Base-load plants, on the other hand, will bid in very high, if at all, because they can only ramp up output at great expense. The actual bid price will depend on the strategy of the owners of the various units, which would depend on

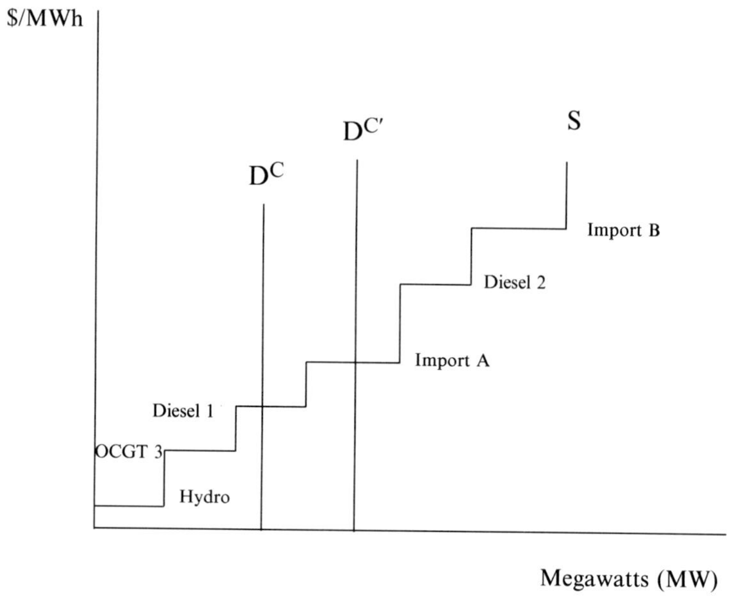

Fig. 11.5 Market merit order for ancillary contingent reserves

the anticipated state of the units at the time. A market for ancillary contingent reserves is established in a fashion similar to that of the real-time market, except that units receive the market-determined capacity payment for their reserve position plus the market clearing price for any electricity they are asked to dispatch. Note that the payment in Fig. 11.3 is per MWh while it is per MW of capacity in Fig. 11.4.

Hydroelectricity is a particularly good provider of ancillary services, although it can also provide base load power. Hydropower can bid in as low-cost provider in the generating services market or as a low-cost provider of ancillary services. It can play either role, although the makeup of the hydroelectric facilities in the system will determine the role it actually plays. For example, in British Columbia, large hydro dams make it ideal for base-load power, with an open-cycle gas facility providing power in the rare instances when load cannot be met from hydro plus imports (see Table 11.1). In the Alberta, on the other hand, there is only a limited ability to store water, with reservoirs tending to be small relative to the needs of the grid. Hence, hydropower is used almost solely for meeting peak load demand.

In the previous discussion of intermittent renewable sources of energy, it was assumed that wind output was predictable. However, when wind enters the system, there is a real risk that output from this source falls (or rises) dramatically and unexpectedly during the course of an hour. This means that the system operator must not only meet the requirements of the WECC, but also have additional contingency reserves that address the variability in wind. This is seen by the shift of demand for contingent reserves from D^C to $D^{C'}$ in Fig. 11.5. By installing wind capacity, contingency reserves could increase by upwards of 10% and regulating reserves even more so, increasing the amount the ESO has to pay for reserves.

11.2 Economics of Integrating Wind Energy into Electricity Grids

Installed global wind generating capacity has expanded rapidly over the past three decades. At the end of June 2011, it reached 215 GW (Fig. 11.6), of which only 1.6% was installed off shore. In rank order, China, the U.S., Germany, Spain and India accounted for 73.9% of global wind power capacity (WWEA 2011). With the exception of China and India, and a few other countries, very little electricity is produced from wind in developing countries, and especially in the least developed countries, although wind is used in many poor countries to drive mechanical devices such as water pumps. Once China and India are excluded, developing countries only accounted for 1.6% (3,108 MW) of global wind generating capacity at the end of 2010. Because of their rapid economic growth and accompanying need for electricity, China and India have taken particular advantage of Kyoto's Clean Development Mechanism (CDM), which allows rich countries to purchase carbon offsets in developing countries. Yet, according to the China Electricity Council, less than 70% of the wind capacity installed at the end of 2010 was delivering power to the grid because of lack of connectivity (WWEA 2011).

Over the period 1990 to mid 2011, growth in wind generating capacity averaged about 25% per annum, and more than 28% since 2000. It is not surprising, therefore, that the growth in capacity is forecast to continue at above 20% until 2015, although this will depend on government subsidies to wind power production. Despite these very high rates of growth over the past several decades, the current role of wind power in meeting global electricity demand is almost negligible as it accounts for much less than 2% of the global electricity supply (Figs. 10.1 and 10.2). What then are the prospects for wind energy? And what are the obstacles?

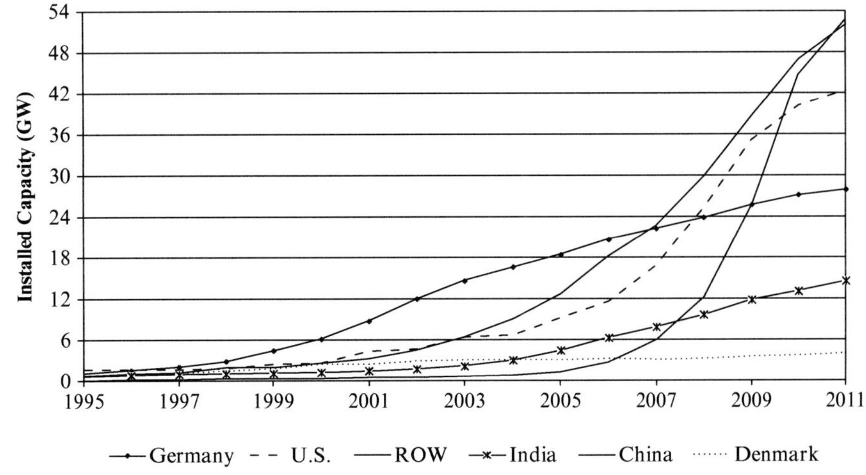

Fig. 11.6 Expansion of global wind generating capacity, 1995 to June 2011

Some quick answers to these questions are as follows. First, it is unlikely that, even under the most optimistic estimates, wind will account for more than 5% of total global electricity production (van Kooten and Timilsina 2009). Second, wind energy requires storage, is unreliable, costly to install, harmful to some wildlife (e.g., birds), noisy, visually unattractive, and, above all, destabilizing of existing electricity grids. Wind turbines only produce about one-fifth to one-quarter of their rated output because of vagaries in wind, while attempts to reduce intermittency by scattering wind farms across a large geographic area and integrating wind power into a 'super grid' have not overcome the grid instability that occurs when wind penetration reaches about 30%.[5]

In summary, the economics of wind-generated energy restrict its potential, essentially deflating the euphoria that is often brought to this renewable energy source. This is not to deny the role that wind energy can play. For example, van Kooten and Wong (2010), and others, have demonstrated that there are potentially huge savings to be had from investing in wind turbines under certain circumstances (discussed further below). But, in order to understand the limitations of wind energy, one needs to recall the discussion in Sect. 11.1 about how the electricity grid functions, and then consider the challenges that this poses for wind power. To determine this, we now investigate in more detail the integration of wind power into electricity grids.

11.2.1 Integration of Wind Power into Electricity Grids

Intermittency is the greatest obstacle to the seamless integration of wind generated power into electricity grids. When there is no wind, no power is generated; the wind comes and goes, and does not always blow with the same intensity – it is a whimsical source of power. Wind power enters an electricity grid whenever there is adequate wind; unless provision exists to curtail wind generation, any electricity generated by wind turbines is 'must run' – it must be delivered to the grid; it is said to be non-dispatchable. Because of intermittency, the supply of wind power fluctuates to a degree not experienced with traditional generating facilities. In addition, there is often a mismatch between periods of high demand and high wind output. As a result, wind disrupts load. This is seen in Fig. 11.7 where actual wind output as a percent of Alberta's load is plotted at 10-min intervals for January and July, 2010.[6] Installed wind capacity at the end of 2010 (803 MW) is just under 8% of peak load (10,277 MW). In Fig. 11.7, we see that wind output will rapidly increase from zero to more than 6% of load at various times in winter and in summer.

---

[5] Most of these results are based on various modeling exercises (e.g., Lund 2005; Maddaloni et al. 2008a, b; Prescott and van Kooten 2009; van Kooten 2010; White 2004). For example, despite generating 25% of its electricity from wind, Denmark's wind energy accounts for only a 4% reduction in CO_2 emissions, something also found to be the case in Germany (Inhaber 2011).

[6] Available at http://www.aeso.ca/gridoperations/20544.html (viewed February 13, 2012).

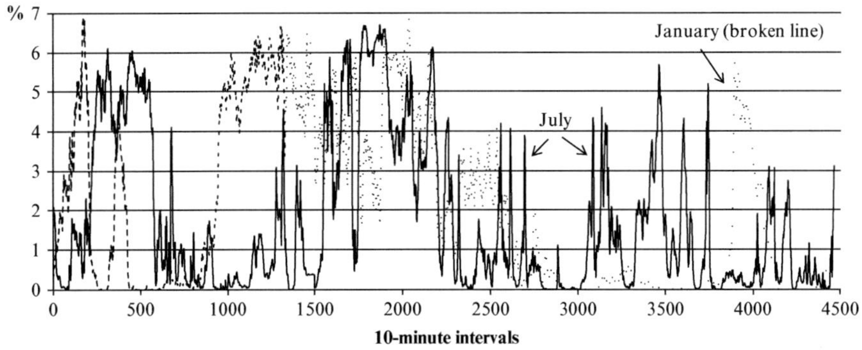

Fig. 11.7 Wind output as a percent of load, Alberta, 2010, 10-min intervals

Although wind energy producers are able to provide forecasts of the likely amount of wind power they can deliver to the grid at a given hour the next day, it is clear from the previous figure that the variance of such forecasts will be large. As discussed in the previous section, producers of wind energy will bid their expected power into the merit-order at the lowest price, although they can change the expected quantity up to 2 h before delivery. Nonetheless, there is no guarantee that the amount of power bid into the system (even 1 h in advance) can actually be delivered, whether it will exceed the bid amount or be below it. As an incentive, some European systems impose a penalty on wind producers if they exceed the stated amount or come in below that amount.

Other system operators simply force wind to be dispatchable, requiring wind farm operators to curtail wind output when there is too much, rather than requiring other generators to curtail their power. When wind power is below the expected amount, the system operator must call upon reserves and may charge the wind supplier for this added cost.

Unless wind power can be stored in reservoirs behind large hydro dams, wind requires fast-responding, OCGT (peak gas) plants as backup. However, since wind energy will first displace electricity produced by fast-responding gas (see below), it makes investments in peak load capacity less attractive as peak plants will be called upon less often when wind is in the system than when there is no wind-generated capacity. Even adding a more stable renewable source, such as tidal power, does little to address the problem of intermittency (Monahan and van Kooten 2010).

As noted in our discussion concerning Fig. 11.4, the location of the supply function and the eventual market clearing price in each hour becomes uncertain as more wind is bid into the market. This uncertainty has a cost. The direct costs of wind power include those associated with the construction of wind turbines, including the cost of purchasing or renting land, the upgrading and construction of transmission lines, and the environmental costs related to bird kills and impact on human health, which can be significant and are often ignored (Bryce 2010, pp. 85, 121–124). The indirect costs associated with intermittency are, most notably, (1) the costs of additional system reserves to cover intermittency, and (2) the extra costs associated with balancing or managing generating assets when power from one (or more) generation sources fluctuates. Both types of indirect costs depend crucially on the existing mix of generating assets. Further, it is important to recognize that, compared to other generators, wind power facilities tend to operate well below capacity.

11.2.2 Capacity Factors

Consider first the so-called 'capacity factor.' If 1 MW of wind generating capacity is installed, the potential power that can be generated annually is given by the number of hours in a year multiplied by the generating capacity. For a turbine with a rated capacity of 1 MW, regardless of the energy source, the potential power output is 8,760 MWh. For coal and nuclear plants, the power actually generated will equal somewhere between 85% and as much as 95% of potential. This is the capacity factor, which should not be confused with the efficiency of a generator. Because wind is highly variable, the average capacity factor of a wind farm is usually well below 30%, generally near 20%. Thus, rather than generating 8,760 MWh of electricity, an average of some 1,750 MWh gets generated with the actual amount varying greatly from one year to the next. Of course, capacity factors at some wind locations exceed 30% and on occasion even 40%, but that is the exception rather than the rule.

To illustrate the types of capacity factors one might encounter, consider the data in Table 11.2. The table provides estimates of capacity factors for wind-generated power in selected countries and the global average for 2005, as well as capacity factors for areas in Alberta and British Columbia east of the Rocky Mountains, and high-wind areas elsewhere in BC west of the Rocky Mountains. Although the country-level information in the table is based only on a single year, the results are illustrative nonetheless. They clearly demonstrate that capacity factors can often be quite low, and are usually lower than expected, even for good wind site locations (Bryce 2010, pp. 96–97). In 2005, the highest capacity factor recorded was 38.1% at a site in Morocco. Capacity factors at sites in France and Portugal averaged less than 15%, and the global average was only 19.6%.

The region east of the Rocky Mountains in western Canada is considered to have some of the world's best wind-generating potential because of prevailing winds off the mountains. Capacity factors have been calculated for northeastern British Columbia from 10 years of wind speed measurements and for southern Alberta using actual output from wind farms.[7] Capacity factors range from 16.8 to 36.6% for the region (Table 11.2). The northeastern British Columbia and southern Alberta sites are directly east and near the Rocky Mountains, but they are about 1,000 km apart. Despite this, there are times when very little wind power will enter the grid, and many hours where there is none (van Kooten 2010). There is also good wind

[7] Data can be found at http://web.uvic.ca/~kooten/documents/LSRS2009WindData.xls.

Table 11.2 Capacity factors for selected countries and Western Canada wind sites

Site	Capacity (MW)	Production (GWh)	Capacity factor (%)
Selected countries and global average (2004/2005)[a]			
U.S.	5,740	17,003.2	28.8
UK on-shore	1,651	3,574.1	27.2
UK off-shore	304	648.2	28.7
Denmark	3,128	6,613.8	24.1
Spain	11,615	22,197.8	21.8
Portugal	1,022	1,769.5	19.8
The Netherlands	1,219	2,067.4	19.3
Germany	20,622	30,502.3	16.9
India	4,430	6,167.0	15.9
Italy	1,718	2,347.7	15.6
Poland	153	192.7	14.6
France	757	954.8	14.5
GLOBAL	59,051	101,256.8	19.6
Sites in Western Canada, East of the Rocky Mountains			
Alberta (2010)	722	9,311.0	27.3
Peace River (BC)[b]	130	347.7	31.1
Sites in British Columbia, Canada			
Southern Interior (BC)[b]	135	277.6	23.7
North Coast (BC)[b]	335	886.2	30.3
Vancouver Island (BC)[b]	100	240.5	27.5

Notes: [a]Source: Global data are from http://lightbucket.wordpress.com/2008/03/13/the-capacity-factor-of-wind-power/ (viewed October 25, 2011)

[b]BC refers to British Columbia. Regional data for BC are based on theoretical power production using wind speed records for the period 1998–2008 (DNV Global Energy Concepts 2009).

potential along Canada's west coast. Theoretical models that use 10 years of wind data indicate that average capacity factors on the North Coast and on Vancouver Island rival those at sites east of the Rocky Mountains.

11.2.3 Reserve Requirements

Next consider reserve requirements. As noted in Sect. 11.1, greater system balancing (regulating) reserves are required with wind than would normally be the case if an equivalent amount of thermal or hydro capacity were installed. This is true even after one adjusts for the lower capacity factors associated with wind. The reliability of power from wind farms is lower than that of thermal or hydro sources because of the high variability associated with wind power, and this variability must be compensated for by greater system reserves.

How large must the additional reserves be? According to Gross et al. (2006, 2007) and assuming no correlation between demand and the supply of electricity from wind turbines, additional reserve requirements might be relatively small. For Britain they find that the standard deviations of wind fluctuations amount to 1.4% of installed wind capacity for a 30-min time horizon and 9.3% of installed capacity over a 4-h time period. For the shorter time horizon, regulating or fast-response reserves are affected, while contingency or standing reserves are affected in the case of the longer time horizon. Suppose required reserves equal $\pm 3(\sigma_s^2+\sigma_d^2+\sigma_w^2)^{1/2}$, where σ_s, σ_d and σ_w are the standard deviations of supply, demand and wind fluctuations, respectively. For the UK, wind intermittency requires a 15.8% increase in regulating reserves and an 8.1% increase in contingency reserves, resulting in an overall increase in total generating capacity of 3%.

For Alberta, reserves would need to increase from 2,581 to 2,631 MW, or by only 50 MW because wind accounts for only 6% of installed capacity. However, because wind can meet nearly 7% of load at one time and none at all 30 min later, this is equivalent to the loss of a generator with a capacity of 760 MW. Operating procedures require a system operator to be able to have adequate reserves in place to meet this contingency. Given that the largest single asset in the Alberta system is a 450 MW unit, it will be necessary to increase contingent reserves by 310 MW, or nearly 70%![8]

While the increases in reserve requirements associated with wind power are not onerous, they are also not insignificant. Yet, they may be an underestimate because a correlation between wind output and load cannot be ruled out entirely. There is evidence to indicate that there is a strong negative correlation between wind output and load – wind production of electricity is consistently greater at night when demand is low than during the day (see Pitt et al. 2005).

11.2.4 Modeling the Management of an Electricity Grid

In addition to the need for greater system reserves, there is a second cost associated with the need to retain system balance – the added cost of managing the grid. How the grid is managed depends on the policy implemented by the authority. If the grid operator is required to take any wind power that is offered (wind is 'must run' or non-dispatchable), extant generators may need to operate at partial capacity, although they must be ready to dispatch power to the grid in the event of a decline in wind availability. Peak-load generating assets are better able to follow changes in wind by ramping up and down than are other assets. If assets are unable to match the ups and downs in wind power availability, there will be excess power in the system

[8] A consulting report, entitled "Alberta 10 year generation outlook" (AMEC Americas Ltd., Calgary, AB, October 2006), assumes system reserves rise from 10% of installed capacity to 15.5% when wind is present.

that must be sold to another operator, usually at low cost. With non-dispatchable wind power entering a grid, there is an economic cost because other generators in the system operate more often below their optimal efficiency ratings or instantaneous capacity factors. In addition, wind variability causes peak-load diesel and OCGT plants to stop and start more frequently, which increases operating and maintenance (O&M) costs.

A mathematical programming model of an electricity grid can be used to address these issues. Programming models assume that load and wind power availability are known beforehand (referred to as 'rational expectations'), and account for the need to balance output from existing generators on the grid (Prescott and van Kooten 2009; Prescott et al. 2007). Costs of new transmission lines from wind farms to an established grid are ignored for convenience, although construction costs could be sufficiently high to militate against investments in wind generation. Also, such grid management models do not take into account additional investment in reserves or backup generation. Further, the mathematical optimization models used in this chapter are linear, with constant marginal generation costs and simple capacity limits and ramping constraints. Linear models are often sufficiently robust and useful when the intention is primarily to investigate the effects of government policies. It turns out that the main conclusions from linear models with rational expectations are reinforced if nonlinearities and uncertainty are added (see Maddaloni et al. 2008a, b; Weber 2005).

It is difficult to maintain system reliability when conventional generation is replaced by non-dispatchable wind power (ESB National Grid 2004; Liik et al. 2003; Lund 2005). To illustrate this problem and, at the same time, provide estimates of the costs of reducing CO_2 emissions, we examine integration of wind into three grids with different generating mixes – those of Alberta, Ontario and British Columbia (Table 11.1). We could think of the three generating mixes in Table 11.1 as 'high fossil fuel' (Alberta), 'typical' (Ontario) and 'high hydro' (British Columbia). The high hydro mix contains nearly 90% hydroelectric generation with the other 10% allocated between natural gas and renewable biomass and wind. Typical is made up of 33% nuclear generation, 13% pulverized coal and 28% gas, with the remainder consisting of hydro (23%) and wind (3%). Finally, the high fossil fuel mix is 45% coal fired and 40% gas fired, with some hydro, biomass and wind but no nuclear power generation. Finally, given the minor role of biomass, we combine it with coal generation for convenience.

In addition to examining the effects of different generating mixes on the costs of reducing carbon dioxide emissions, we also consider how a carbon tax might affect the addition and removal of generating capacity. In particular, we examine how a rising carbon tax in Alberta's high fossil fuel mix facilitates removal of coal-fired capacity and the addition of wind turbines. As part of the analysis, we also investigate the potential role of nuclear power vis-à-vis wind. The model employed in this case is similar to that used to investigate the impact of increasing wind power in various generating mixes.

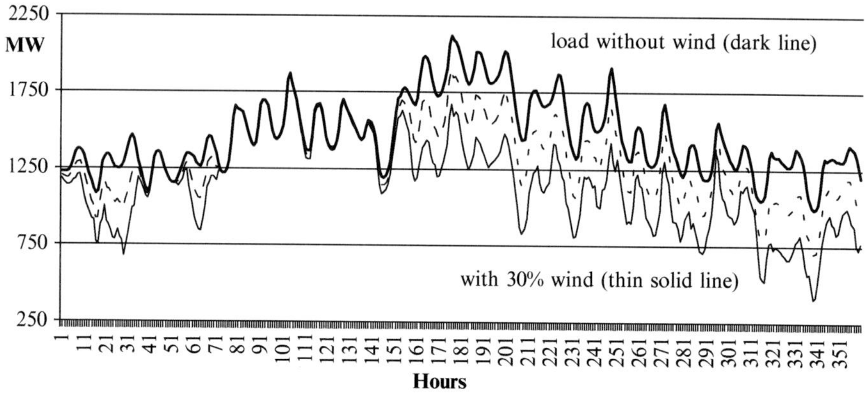

Fig. 11.8 Load facing traditional generators, January 1–15, 2010

11.2.5 Integrating Wind Energy into Various Generating Mixes

The costs and benefits of introducing wind power into an electricity grid depend on the generating mix of the system. To address this issue and provide some notion of the costs of reducing CO_2 emissions, we employ the mathematical programming model described in the Appendix to this chapter to investigate grids with generating mixes approximating those in Table 11.1. We take into account fuel and operating and maintenance (O&M) costs, costs of adding new wind capacity, and life-cycle CO_2 emissions. Information on costs and emissions is provided in Table 10.5. Linearity permits optimization over a full year or 8,760 h. Reserve requirements are ignored.

We use hourly load data for 2010 from the Electric Reliability Council of Texas (ERCOT) system, and wind data from sites in western Canada. The ERCOT load data are standardized to a peak load of 2,500 MW (multiplying load data by 2,500 MW and dividing by ERCOT peak load of 65,782 MW). Hourly wind power output consists of aggregated Alberta 10-min wind data for 2010 (Fig. 11.7), normalized to the output of a single 2.3 MW turbine.[9] Net load equals demand minus wind output. Because wind is more variable than load, net load becomes increasingly variable as wind penetration increases, where penetration is defined as the ratio of installed wind capacity to peak load. This is seen in Fig. 11.8, where the non-wind load and net loads with 15 and 30% wind penetrations are plotted for the first 15 days in January. For the entire year, the coefficient of variation of the load (defined as the standard deviation divided by the mean) is 0.24, while it is 0.27 for wind penetration of 15% and 0.32 for 30% penetration. Thus, the higher the extent of wind penetration, the greater is the volatility of the remaining load.

[9] Data found at http://www.aeso.ca/gridoperations/20544.html (viewed February 15, 2012).

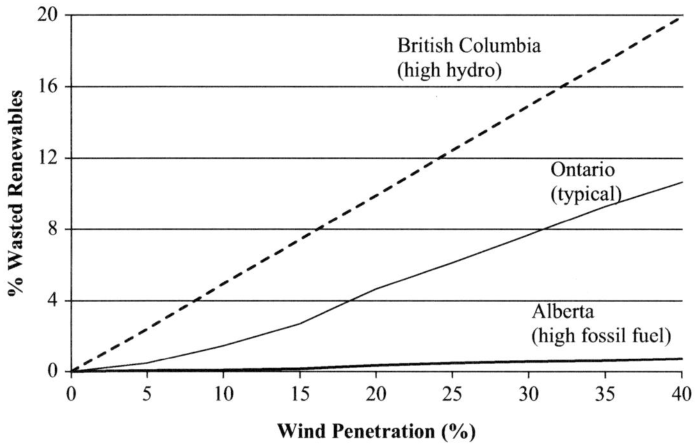

Fig. 11.9 Displacement of renewable energy as a function of wind penetration

We would like to use our model to answer some policy questions. The central question of concern is the following: What is the expected cost of reducing CO_2 emissions by building and operating wind turbines to generate electricity? What are the impacts of wind turbines on existing generating facilities? What if any are the limits to substituting fossil fuel generated electricity with wind power? Some of these questions, but particularly the question regarding the impact of a carbon tax on addition and removal of capacity is addressed in the next subsection.

The simulation results can best be seen through the lens of wasted renewables, or the amount of hydroelectricity that is replaced by wind energy – in essence, water is spilled over the dam without generating electricity. This is indicated in Fig. 11.9 for the three generation mixes. Notice that, for the high-fossil fuel system with little hydro capacity, more fossil-fuel generation is replaced by wind energy. Consequently, although the average cost of adding wind capacity is quite high, the marginal cost remains negative (i.e., there is a benefit to further investment in wind capacity) until wind penetration reaches 33.5%; beyond 35% penetration, the marginal cost of adding wind becomes exorbitant and it is cheaper to purchase carbon offsets. For the more typical Ontario mix, the tipping point occurs at less than 25% penetration, while for the high-hydro mix that characterizes British Columbia, the tipping point is already reached at 5% penetration.

Despite perfect foresight regarding wind availability, generators cannot adjust their output quickly enough to prevent unnecessary generation, unless there is sufficient hydro generating capacity. Hydroelectric units can be adjusted on extremely short notice. Because of its hydro assets, for example, British Columbia is better able to track load from 1 h to the next than other systems. As a result, there are no additional carbon dioxide emissions due to inefficient thermal generation. However, unless an export market for electricity is available (and that depends on transmission lines) or the system's hydro capacity is unable to meet load, there is

Table 11.3 Optimal installed capacity for various levels of the carbon tax, Alberta Grid, 2010[a]

Tax ($/t$CO_2$)	# of wind turbines	Coal (MW)	Natural gas (MW)	Peak gas (MW)	Biomass (MW)
$0	350	6,240	3,800	1,500	310
$25	350	6,240	3,566	0	0
$50	1,840	6,240	3,290	0	0
$75	6,561	4,587	4,323	506	0
$100	7,551	4,007	4,722	549	0
$150	8,585	2,409	5,477	799	345
$200	9,317	1,839	5,799	1,044	345

Source: Author calculations
Notes: [a]Based on wind profile found in Figs. 11.7 and 11.8. There is no limit on the number of turbines that can be installed, starting installed capacities approximate those of Alberta in 2010; Alberta's 2008 load increased by 4.5% is used to represent 2010 load.

little benefit to adding wind capacity to the system as wind simply replaces renewable hydro. In a system that relies primarily on fossil fuels, but also has some reservoir capacity, the benefits of such capacity are much greater. This is the case for the Alberta system, where wind takes advantage of reservoir storage, despite some replacement of renewable hydro. Therefore, we next investigate the Alberta system in greater detail.

11.2.6 Adding and Removing Generating Assets: Nuclear vs Wind

Now consider Alberta's fossil-based generating system. What would be the optimal generating mix in the longer term if the costs of carbon dioxide emissions on the global commons are taken into account? To examine this issue, we modify the model in the Appendix to this chapter so that investors are encouraged to remove generating assets that emit CO_2 while adding low-emission ones. The instrument used in the model is a tax on CO_2 that is set at various levels to determine the extent to which the current mix is suboptimal.

In our model, actual Alberta load data are used in addition to information on existing generation. Therefore, we include an exogenously given run-of-river hydro power series, but develop a model of a reservoir-dam facility with a capacity of 250 MW that provides endogenously determined hydroelectric output. The first row of Table 11.3 provides the installed capacities of the other generators in the current mix. We permit investment in biomass because coal-fired plants could potentially be converted to use biomass as an energy source with lower CO_2 emissions. However, wind is the most suitable alternative to thermal generation because of its negligible CO_2 emissions. The model chooses the number of wind turbines to install (or remove). Finally, cost and emissions data are found in Table 10.5.

Table 11.4 Electricity output by generating assets for various levels of the carbon tax, Alberta Grid, 2010 (GWh)[a]

CO_2 tax	Wind	Coal	Natural gas	Peak gas	Hydro	Biomass
$0	1,911	62,344	6,008	0	1,745	0
$25	1,911	54,568	13,791	0	1,738	0
$50	10,045	53,027	7,315	0	1,662	0
$75	35,816	10,424	24,917	402	1,539	0
$100	41,221	6,359	24,548	586	1,472	0
$150	46,866	1,679	24,415	1,071	1,352	670
$200	50,862	650	23,345	1,045	1,257	716

Source: Author calculations
Notes:
[a]See note on Table 11.3. Output based on available generating assets. Must run electricity adds 886 GWh/year

Table 11.5 Wasted energy from wind penetration into the Alberta Electricity Grid by Source, 2010 (GWh)[a]

Tax ($/t$CO_2$)	Renewable[b]	Thermal[c]	Total
$0	0	0	0
$25	7	0	7
$50	83	41	124
$75	206	1,090	1,296
$100	273	2,178	2,451
$150	393	4,045	4,438
$200	488	5,867	6,355

Notes:
[a]Generation in excessive of base generation of 72,894 GWh
[b]Hydroelectricity that was not generated although available
[c]Thermal generation that was generated but not used

The simulation results are provided in Tables 11.3 and 11.4. When the carbon tax is $25/t of CO_2 (or when the tax is $0/t$CO_2$), the model removes the more expensive biomass, peak gas and some CCGT assets from the mix, adding no other assets since the load can be fully met without them. The reason why Alberta includes these 'extra' gas and biomass assets has to do with reserve requirements, insufficient transmission capacity between north and south (gas generation is mainly in the south), and the use of biomass from sawmilling operations, none of which is modeled here. As the carbon tax rises, greater numbers of wind turbines are installed, coal-fired capacity and generation are significantly reduced, but peak and CCGT capacities are increased. At carbon taxes of $150 and 200, biomass even joins the generating mix. Notice that, at the highest tax, almost 70% of the 72,894 GWh of electricity demand is met by wind, while coal accounts for only 650 GWh of generation (Table 11.4); but an intolerable number of wind turbines would be required.

A more interesting result pertains to the extent of wasted energy as wind penetration increases. As indicated in Table 11.5, upwards of 8.7% of the energy needed by Alberta consumers might be wasted because it replaces hydro output (7.7% of the

Table 11.6 Optimal installed capacity for various levels of the carbon tax when nuclear construction is permitted, Alberta Electricity System, 2010[a]

	Installed capacity (MW)			Generation (GWh)[b]		
Tax ($/t$CO_2$)	Nuclear	Coal	Natural gas	Nuclear	Coal	Natural gas
$0	0	6,240	3,800	0	62,344	6,008
$25	1,202	6,240	2,364	10,489	53,460	4,402
$50	7,305	1,950	897	63,214	4,295	836
$75	7,700	0	2,106	65,567	0	2,778
$100	7,844	0	1,962	66,240	0	2,104
$150	8,040	0	1,766	66,993	0	1,352
$200	8,159	0	1,647	67,353	0	992

Source: Author calculations
Notes:
[a]See notes on Table 11.3.
[b]Output based on available generating assets. Wind and 'must run' electricity add 2,797 GWh/year

wasted renewable energy) or because it results in inefficiencies in the operation of thermal power plants (92.3%).

Finally, consider the option of bringing nuclear power into the generation mix. As indicated in Table 11.6, nuclear capacity and output replaces both coal and gas generating assets. Some (peak and CCGT) natural gas remains in the system for load following and peak demand purposes, but substantially less than what would be required in the case of wind (compare Tables 11.4 and 11.6). Although not shown, all hydroelectricity and wind generated power are utilized, but there is no investment in new wind turbines. These findings are identical to those of Fox (2011), who used screening curves and the Ontario grid and generating mix. It is clear, therefore, that nuclear power is preferred to wind power and that the reason has to do with the intermittency problem and low capacity factors associated with wind energy, both of which serve to increase the costs of wind above those of nuclear power.

11.3 Concluding Observations

The U.S. Department of Energy (2008) indicates that wind power could reduce CO_2 emissions at a cost of $5.70/t$CO_2$. Most studies find quite the opposite, however. For example, a German study by Rosen et al. (2007) found costs of reducing CO_2 emissions rise from €87.70/tCO_2 to €125.71/tCO_2 and then to €171.47/tCO_2 as wind power production increases from 12.0 TWh (6 GW installed capacity in 2000) to 34.9 TWh (17.3 GW 2005) and 50.4 TWh (22.4 GW 2010) corresponding to respective wind penetrations of about 8, 23 and 29%. Our results were somewhat more favorable towards wind, but optimistically one could not expect costs to be competitive with alternative means of reducing carbon dioxide emissions at wind penetrations of more than about 30%; indeed, in most cases, costs would be considered exorbitant at much lower levels of penetration.

Several factors must be aligned before wind energy can reduce system-wide CO_2 emissions at reasonable cost. These relate to the load and wind profiles, and crucially the existing generating mix into which wind power is to be integrated. Operating constraints for coal- and gas-fired base load generation lead to overproduction of electricity during certain periods, because units cannot ramp up and down quickly enough when wind energy is available. This results in less emission reductions than anticipated. Wind integration into a system that has high nuclear and/or hydroelectric generating capacity might also see fewer CO_2 benefits than anticipated as wind displaces non-CO_2 emitting sources, despite the ability of some hydro facilities to fluctuate as quickly as wind. Hydro storage is an advantage, but not always. The research indicates that a high degree of wind penetrability is feasible (negative to low costs of reducing CO_2 emissions) for flexible grids that have sufficient hydro for storage and relatively fast-responding gas plants that track changes in load minus non-dispatchable wind, while keeping base-load nuclear and coal power plants operating efficiently (with only minor changes in output).

Rather than allowing extant generators to vary their output, thus increasing system costs, an alternative policy is to make wind power dispatchable by requiring wind operators to reduce output (by 'feathering' wind turbines or simply stopping blades from rotating) whenever the grid operator is unable to absorb the extra electricity. In this case, output from base-load plants is effectively given precedence over wind generated power because such plants cannot be ramped up and down, the ramping costs are too great, and/or excess power cannot be stored or sold. In many instances, wind variability can only be handled by selling electricity to other jurisdictions or forcing wind plants to reduce output if necessary. In Alberta, for example, further expansion of wind farms was initially permitted only after developers agreed to control power output so that wind power was no longer 'must run.' This policy makes investments in wind farms must less attractive and is usually unacceptable to environmental groups (as wind energy might be wasted).

Another possibility is to permit wind farms only if they come with adequate storage, which generally means they need to be connected to large-scale hydro facilities that have adequate reservoir capacity, or are bundled with a peaker plant. With respect to the latter, the output of a wind facility would be reliable because any shortfall in wind output would be covered by natural gas. The only drawback is that wind variability tends to increase the costs of a peak gas plant because more frequent stops and starts are required.

Placement of several or many wind farms across a sufficiently large geographic area is also a possibility that has been promoted for mitigating wind's intermittency. To overcome variability, it is argued, wind farms can be located across as large a geographic area as possible, with their combined output integrated into a large grid. By establishing wind farms across the entire country, onshore and offshore, the United Kingdom hopes to minimize the problems associated with intermittency. Further, by connecting all countries of Europe and placing wind farms throughout the continent as well as in Britain and Ireland, the hope is to increase the ability to employ wind generated power. But, as demonstrated by Oswald et al. (2008), large weather systems can influence the British Isles and the European continent simultaneously. Oswald and his colleagues demonstrated that at 18:00 hours on February 2, 2006, electricity demand in the United Kingdom peaked, but wind power was zero (indeed wind farms added to the load at that time). At the same time, wind power output in Germany, Spain and Ireland was also extremely low – 4.3, 2.2 and 10.6% of capacities, respectively. Something similar occurs with respect to wind farms located some 1,000 km apart in the Great Plains of Canada near the Rocky Mountains (van Kooten 2010). Thus, even a super grid with many wind farms scattered over a large landscape cannot avoid the problems associated with intermittency, including the need to manage delivery of power from various non-wind power generators.

The best strategy for integrating intermittent wind and other renewable resources into electricity grids is to provide incentives that cause the intermittent resources to take into account the costs they impose upon the grid. Some European jurisdictions already penalize wind power providers if they deliver more or less than an agreed upon amount of electricity to the grid – they incur a penalty for variability. This might cause producers to waste renewable energy if they exceed the limit, or pay a fee if they are under it. However, it also provides strong incentives to find ways to store electricity, or invest in reserve capacity.

It is also possible that special ancillary markets develop to mitigate intermittency. This provides the same incentives as a penalty regime. Payments for backup services provide service providers with incentives to store electricity and/or ensure sufficient backup services are available at lowest cost.

Finally, upon examining the potential of wind energy to meet global society's energy needs, Wang and Prinn (2010) conclude that, if 10% of global energy is to come from wind turbines by 2100, it would require some 13 million turbines that occupy an area on the order of a continent. Wind turbines themselves would cause surface warming exceeding 1°C over land installations, and alter climate (clouds and precipitation) well beyond the regions where turbines are located – reducing convective precipitation in the Northern Hemisphere and enhancing convective precipitation in the Southern Hemisphere. Wind turbines on such a massive scale would also lead to undesired environmental impacts and increase energy costs because of the need for backup generation, onsite energy storage and very costly long-distance power transmission lines. Not surprisingly, therefore, we found that nuclear power might even be preferred to wind energy if the objective is solely to reduce carbon dioxide emissions.

Then what about wind? While a clean source of energy, wind power must be able to compete in the market place. It must be able to compete in the production of electricity without subsidies of any form. But other generating sources must also compete without subsidies – the playing field must be level and the role of government is to ensure that this is indeed the case. The government should not be in the business of trying to pick winners. Under these circumstances and because of problems with intermittency and possible externality effects (bird kills, adverse impacts on human health), the future role of wind power might be limited. As with any good thing, there comes a point where more may not be in the best interests of society – where the marginal social benefit from installing more wind capacity equals the marginal social cost. A buoyant and optimistic wind sector is of the opinion that that point is still far in the future. This might be true, but it may also be the case that the bubble is about to burst. Only time will tell.

Appendix

A grid management model can be represented mathematically by the following linear programming problem:

$$\underset{Q_{t,i}}{\text{Minimize}}\ TC = \underset{Q_{t,i}}{\text{Minimize}} \sum_{t=1}^{24\times d} \left[\sum_{i} (OM_i + b_i) Q_{ti} \right] + \sum_{i} (a_i - d_i)\Delta C_i, \quad (11.1)$$

where TC is total cost (\$); i refers to the generation source (*viz.*, natural gas, coal, nuclear, wind, hydro); d is the number of days (365 for a 1-year model); t is the number of hours (8,760); Q_{ti} is the amount of electricity produced by generator i in hour t (MW); OM_i is operating and maintenance cost of generator i (\$/MWh); and b_i is the variable fuel cost of producing electricity using generator i (\$/MWh), which is assumed constant for all levels of output. In addition, we define D_t to be the demand or load that has to be met in hour t (MW); C_i is the capacity of generating source i (MW); and T_i is the amount of time it takes to ramp up production from plant i. Finally, the last term in Eq. (11.1) permits the addition or removal of generating assets, where a_i and d_i refer to the annualized cost of adding or decommissioning assets (\$/MW), and ΔC_i is the capacity added or removed.

Objective function (11.1) is optimized subject to the following constraints:

$$\text{Demand met in every hour:}\quad \sum_i Q_{t,i} + \sum_r Q_{t,r} \geq D_t, \forall t = 1,\ldots,24\times d \quad (11.2)$$

$$\text{Ramping - up constraint:}\quad Q_{t,i} - Q_{(t-1),i} \geq \frac{C_i}{T_i}, \forall i \quad (11.3)$$

$$\text{Ramping - down constraint:}\quad Q_{t,i} - Q_{(t-1),i} \geq -\frac{C_i}{T_i}, \forall i \quad (11.4)$$

$$\text{Capacity constraints:}\quad Q_{t,i} \geq C_i, \forall i \quad (11.5)$$

$$\text{Non - negativity:}\quad Q_{t,i} \geq 0 \quad (11.6)$$

Two further constraints are not shown, but they require that thermal nuclear and coal-fired power plants are kept running at 50% or more of their capacity to avoid shutting down base-load plants.

A carbon tax can be included in the objective function by adding the following term to Eq. (11.1): $\tau \sum_{t=1}^{24\times d} \left[\sum_k \varphi_k Q_{t,k} \right]$, where τ is a carbon tax ($/t$CO_2$) and ϕ_k is the CO_2 required to produce a MWh of electricity from generation source k.

The linear programming model can be made somewhat more realistic by including nonlinear elements in the objective function and/or additional constraints that provide more detail regarding generator operations or transmission links. Research by Maddaloni et al. (2008a) indicates that this improves the results slightly, but reduces the number of periods for which the model will solve. In principle, the outcomes from a linear programming model are sufficient to guide policymakers. Further, linear programming models can readily be modified to address risk and uncertainty, and nonlinear functions can be approximated with linear functions as needed (see Louck et al. 1981).

References

Bryce, R. (2010). *Power hungry: The myths of "green" energy and the real fuels of the future*. New York: Public Affairs.

DNV Global Energy Concepts. (2009). *BC Hydro wind data study* (Report #CSRP0009-A, pp. 52). Seattle, WA: DNV Global Energy Concepts Inc.

ESB National Grid. (2004, February). *Impact of wind power generation in Ireland on the operation of conventional plant and the economic implications* (42 pp.). Dublin, Ireland. Retrieved July 15, 2010, from http://www.eirgrid.com/media/2004 wind impact report (for updated 2007 report).

Espey, J. A., & Espey, M. (2004). Turning on the lights: A meta-analysis of residential electricity demand elasticities. *Journal of Agricultural and Applied Economics, 36*(1), 65–81.

Filippini, M. (2010, July). *Short and long-run time-of-use price elasticities in Swiss residential electricity demand* (CEPE Working Paper No. 76, 21 pp.). Zurich: Center for Energy Policy and Economics, Swiss Federal Institutes of Technology. Retrieved from www.cepe.ethz.ch

Fox, C. (2011). *The effects of CO2 abatement policies on power system expansion*. Unpublished MA thesis. Victoria: Department of Mechanical Engineering. University of Victoria. Retrieved January 13, 2012, from http://www.iesvic.uvic.ca/

Gross, R., Heptonstall, P., Anderson, D., Green, T., Leach, M., & Skea, J. (2006, March). *The costs and impacts of intermittency: An assessment of the evidence on the costs and impacts of intermittent generation on the British electricity network* (pp. 96). London: UK Energy Research Centre.

Gross, R., Heptonstall, P., Leach, M., Anderson, D., Green, T., & Skea, J. (2007). Renewables and the grid: Understanding intermittency. *ICE Proceedings, Energy, 160*(1), 31–41.

Inhaber, H. (2011). Why wind power does not deliver the expected emissions reductions. *Renewable & Sustainable Energy Reviews, 15*(6), 2557–2562.

Liik, O., Oidram, R., & Keel, M. (2003). *Estimation of real emissions reduction caused by wind generators*. Paper presented at the IIASA, International Energy Workshop, Laxenburg, AT.

Louck, D. P., Stedinger, J. R., & Haith, D. A. (1981). *Water resource systems planning and analysis*. Englewood Cliffs: Prentice Hall.

Lund, H. (2005). Large-scale integration of wind power into different energy systems. *Energy, 30*(13), 2402–2412.

Maddaloni, J. D., Rowe, A. M., & van Kooten, G. C. (2008a). Network constrained wind integration on Vancouver Island. *Energy Policy, 36*(2), 591–602.

Maddaloni, J. D., Rowe, A. M., & van Kooten, G. C. (2008b). Wind integration into various generation mixtures. *Renewable Energy, 34*(3), 807–814.

Malla, S. (2009). CO_2 emissions from electricity generation in seven Asia-Pacific and North American countries: A decomposition analysis. *Energy Policy, 37*(1), 1–9.

Monahan, K., & van Kooten, G. C. (2010). The economics of tidal stream and wind power: An application to generating mixes in Canada. *Environmental Economics, 1*(1), 90–99.

Oswald, J., Raine, M., & Ashraf-Ball, H. (2008). Will British weather provide reliable electricity? *Energy Policy, 36*(8), 3202–3215.

Pitt, L., van Kooten, G. C., Love, M., & Djilali, N. (2005). *Utility-scale wind power: Impacts of increased penetration* (Paper No. IGEC-097). Proceedings of the international green energy conference. Waterloo: University of Waterloo.

Prescott, R., & van Kooten, G. C. (2009). Economic costs of managing of an electricity grid with increasing wind power penetration. *Climate Policy, 9*(2), 155–168.

Prescott, R., van Kooten, G. C., & Zhu, H. (2007). The potential for wind energy meeting electricity needs on Vancouver Island. *Energy and Environment, 18*(6), 723–746.

Rosen, J., Tietze-Stöckinger, I., & Rentz, O. (2007). Model-based analysis of effects from large-scale wind power production. *Energy, 32*(4), 575–583.

Silk, J. I., & Joutz, F. L. (1997). Short and long-run elasticities in US residential electricity demand: A co-integration approach. *Energy Economics, 19*(4), 493–513.

Stoft, S. (2002). *Power system economics: Designing markets for electricity*. Piscataway: IEEE Press.

U.S. Energy Information Administration. (2010). *Assumptions to the annual energy outlook 2010.* Retrieved September 16, 2010, from http://www.eia.doe.gov/oiaf/aeo/assumption/pdf/0554%282010%29.pdf

van Kooten, G. C. (2010). Wind power: The economic impact of intermittency. *Letters in Spatial & Resource Sciences, 3*(1), 1–17.

van Kooten, G. C., & Timilsina, G. R. (2009, March). *Wind power development: Economics and policies* (Policy Research Working Paper 4868). Washington, DC: The World Bank, Development Research Group, Environment and Energy Team.

van Kooten, G. C., & Wong, L. (2010). Economics of wind power when national grids are unreliable. *Energy Policy, 38*(4), 1991–1998.

Wang, C., & Prinn, R. G. (2010). Potential climatic impacts and reliability of very large-scale wind farms. *Atmospheric Chemistry and Physics, 10*(4), 2053–2061.

Weber, C. (2005). *Uncertainty in the electric power industry: Methods and models for decision support*. New York: Springer.

White, D. J. (2004). Danish wind: Too good to be true? *The Utilities Journal, 7*(July), 37–39.

WWEA. (2011, April). *World wind energy report 2010* (pp. 23). Bonn: World Wind Energy Association.

Chapter 12
Climate Change Policy Encounters the Real World

All solutions to environmental collapse offered by politicians require tough government controls on individuals.

– James Wanliss, physicist

Climate policy is all about distributing income from poor people in rich countries to rich people in poor countries.

– Fred Singer, climate scientist and global warming sceptic

Economists have made all kinds of contributions to the literature on climate change, including contributions to climate science itself. This is one of the conclusions that can be drawn from the preceding chapters. Econometricians with particular expertise in statistics have investigated claims that current temperatures are the warmest in the past two millennia. Overall, this claim has been found wanting. Indeed, along with a growing number of researchers, econometricians have determined that the underlying tree-ring, ice-core, lake-bed sediment and other proxy data for such a claim have been erroneously interpreted and that the so-called 'hockey stick' (1,200 or more years of unchanging average global temperatures followed by a rapid rise in temperature since the mid to late 1800s) does not characterize the historical record. The Medieval Warm Period and the Little Ice Age were real events that affected all regions of the Earth at the same time. Econometricians have also found that, even if the validity of the proxy record is taken as correct (that there is a type of hockey stick phenomenon), there is no statistically significant difference between current temperatures and historical ones. That is, econometricians have concluded that current warming falls within historical temperature variations and cannot be regarded as unusual.

Economists have also examined the observational record – temperature reconstructions based on data from weather stations. These temperature reconstructions have been made for two reasons. First, temperature data are available over a large landscape, with better coverage in some areas (Europe, United States) than in others. There are vast spatial gaps in the record, and the numbers of reporting weather

G.C. van Kooten, *Climate Change, Climate Science and Economics: Prospects for an Alternative Energy Future*, DOI 10.1007/978-94-007-4988-7_12,

stations have changed over time. Therefore, it has been necessary to aggregate weather data across the landscape so that continuous time series of weather data, principally temperature and precipitation, are available for each grid point on Earth. In Chap. 2, we demonstrated the difficulties associated with doing this.

Second, reconstructions need to remove the effects of socioeconomic or non-climate factors; that is, they need to adjust for the heat island effect. For example, temperature readings taken in the same place over a long period of time will be influenced by changes in the immediate environment – construction of buildings, pavement, air conditioning units, exhaust from fossil fuel burning, and other economic activities will cause temperatures to rise independent of any climate effect. Hence, land temperature records constructed by the Climate Research Unit (CRU) at East Anglia University, by NOAA and NASA, and by the Berkeley Earth Surface Temperature project should no longer be affected by non-climate factors as the reconstructions have specifically corrected for such influences. However, as discussed in Chap. 3, statistical analysis indicates that perhaps half of the warming found in the record can still be attributed to socioeconomic and non-climate factors.

Economists have also been involved in climate modeling. The IPCC's special report on emission scenarios begins with assumptions concerning the future evolution of population, technological change, economic growth and the convergence of per capita incomes between rich and poor countries. Various economic models employ these assumptions to derive potential paths of human greenhouse gas emissions. As explained in Chap. 4, these emission paths are the basis for climate models that then project the future course of temperatures and precipitation (including at the regional level). While simple climate models only examine the Earth's energy balance, more complicated models couple atmospheric circulation models with ocean circulation models, and even interactions with the terrestrial ecosystem and the economy. In Chap. 4, we also discussed model complexity, pointing out that a more complex model may well be able to duplicate a past climate (although this often requires fine tuning of model parameters), but this level of detail makes such models less able to predict the future climate. Computational economists have questioned not only the ability of models to find solutions and the (in)stability of such solutions, but also the validity of predictions based on any model that is not grounded in observation.

There is mounting evidence that climate models do not predict very well. In some cases, statistical models based on actual observational data predict much better than climate models. And sometimes the best models for predicting the future course of temperatures are simple energy balance models. In Chap. 5, we investigated alternative explanations of climate change, ones that were not necessarily based on anthropogenic emissions of CO_2 and other greenhouse gases. We found that some of these provide a much better explanation of the observational record including, in particular, explanations of data from satellites. We also examined theories that still need to be verified. The point is simply this: There exist very good explanations of climate change outside of the IPCC story that it is driven solely by human emissions of greenhouse gases. It would be a travesty to ignore such alternatives because most economic policies, especially ones that seek to reduce reliance on fossil fuels, will harm the poorest members of global society the most. If there remains a good chance that global warming is due to natural causes, the precautionary principle suggests that fossil fuels be made readily available to these members of society at lowest possible cost.

Economists would argue that they have a comparative advantage over other professions in the evaluation of economic policy. This advantage, if it exists at all, stems from the fact that economists have a well developed theory for measuring costs and benefits (Chap. 6). This enables them to identify the gainers and losers from global warming, and, importantly, the gainers and losers of policies to mitigate climate change. It also enables them to determine the extent of losses from climate change (as Nordhaus has done) and which policies are most efficient (result in the least cost to society) in addressing climate change. Yet, the complexity of the climate change issue has resulted in debates about the appropriate policy strategy to take.

The general consensus among economists is that a carbon tax is preferred to emissions trading. The problems with emissions trading are several: If rights to emit carbon dioxide are grandfathered, the large emitters stand to gain a windfall. Grandfathering of emission rights may be difficult to avoid, however, if an emissions trading (cap-and-trade) scheme is to be politically acceptable. Yet, given the nature of uncertainty, a quota is less efficient than a tax (as discussed in Chap. 8). It may be necessary to adjust quota on an ad hoc basis to prevent social costs from becoming unacceptable, while a carbon tax can simply start low and slowly increase over time as the emissions response is revealed. As evidence from the European Union's Emission Trading System indicates, ad hoc adjustments to quota can be troubling, with a quota system far more susceptible to political manipulation than a tax.

In terms of the rate at which a tax or quota should be increased over time, the consensus is a slow policy ramp to avoid getting locked into a particular technology. This might be even more the case if it turns out that climate change is not going to be as rapid as the climate models indicate, or that the global warming story or theory is not quite correct as told – that the feedback in climate models needed to obtain rapid increases in global average temperatures is not correct. However, a few economists, most notably Nicholas Stern, have argued that it is imperative to implement policies immediately to prevent more CO_2 from entering the atmosphere. The problem is that his view is simply unrealistic – it runs contrary to what is happening in the real world.

There are other options to government action. As indicated in Chap. 7, a panel of economists found that, for addressing climate change, research and development, adaptation, and technology transfer are preferred to carbon taxes (and to subsidies, cap-and-trade and regulation as well). The economists are knowledgeable about the grabbing hand model of government, and the potential adverse economic impacts of government tax breaks, subsidies and renewable standards associated with climate change legislation. Some of the most common adverse effects of climate legislation are as follows:

- Wealth on the order of hundreds of millions of dollars annually is transferred from taxpayers and electricity customers to the owners of wind farms and other renewable facilities and to their financial partners.

- Capital investment gets misdirected to the construction of energy facilities, especially high cost wind farms and solar facilities (and wave, tidal and other renewable projects) that produce only small amounts of electricity, which is intermittent and unreliable. Such power has lower value to society because of its unreliability and because it tends to be produced at night, and not in colder months or other periods of high electricity demand when electricity has higher real value.
- Biofuel policies and biofuel subsidies raise the price of food, harm the poorest in society and do very little to mitigate carbon dioxide emissions; indeed, biofuel programs do more harm to the environment than good.
- Other resources including human talent are diverted. Those in the private sector with resources to invest can obtain larger returns with less risk by investing in projects that benefit from generous government tax breaks and subsidies, thereby forgoing the opportunity to invest in potentially productive and innovative endeavors in the private sector where risks are higher and returns not guaranteed.

As demonstrated in Chaps. 10 and 11, there are no real alternatives to fossil fuel burning, at least not in the near to intermediate future. Unless politicians are willing to stop economic growth and, indeed, de-industrialize the global economy and make everyone less well off, something that is not likely to gain support among citizens, global warming is inevitable – assuming that the climate models are correct.

We conclude that there is a lot of rhetoric associated with climate change and greenhouse gas emission reduction targets (Chaps. 7 and 8). While some reduction in CO_2 might be attainable, the targets being proposed in the post-Kyoto world are simply not rooted in reality. The reality is that the rich countries have only had limited success at achieving the much lower Kyoto targets because of a financial crisis and recession. Unless energy production is drastically curtailed or there is a huge immediate investment in nuclear energy, or both, targets cannot possibly be met. In the meantime, subsidies and legislation under consideration will lock several generations into energy systems that are detrimental to their interests and harmful to the least well off (and to the environment).

The reality is that, if access to cheap energy is curtailed, economic development in places such as Africa and India will be set back; however, if access to cheap energy is curtailed only in rich countries, developing countries will benefit as the prices they face fall, but then the benefits of reducing CO_2 emissions would be greatly offset. People are not willing to pay the high price needed to reduce greenhouse gas emissions to the degree advocated by climate scientists and environmentalists, which is why democratically-elected politicians have tended to postpone the pain until after the next cycle of elections or even farther in the future.

The reality is that, contrary to economic wisdom, rich countries have been spending beyond their means, with large annual deficits and debts upwards of 100% or more of their gross domestic product. Deficit spending occurred during good times as well as bad, rather than running counter to the business cycle as recommended by Keynesian orthodoxy. With large debts, countries have little option but to pursue economic growth that results in greater energy use, with increases in energy consumption met primarily by lower cost fossil fuels as opposed to renewable energy sources. Suppose a country's debt is 100% of its GDP and that the interest rate it faces is 3%. Simply to keep up with interest payments, the country will need to grow by 3% annually; to reduce the debt, it will have to grow even faster.

Everything hinges on addressing some fundamental questions. Is climate change really happening? If so, does it matter? Is concern about climate change actually about something else – about human impact on the environment – and not about climate per se? We conclude this book by providing some answers to these questions.

12.1 Malthus Revisited: Crying Wolf?

Economic policy and climate change are not really about what temperatures and precipitation might do in the future. The subject of climate change is not really about measuring temperatures and whether these are higher today than they ever have been in the past two millenia. Nor is it about climate modeling, polar bears, sea level rise, hurricanes and what not. Climate change is about the impact that humans have on the environment, whether human are a blight on nature or a steward acting to protect and enhance the capacity of nature to provide goods and services that improve the circumstances of life on Earth. It is about whether nature is meant to serve humankind or to somehow be worshipped for its own intrinsic value. The notion that human activities are leading to catastrophic global warming and that policies can be implemented to arrest and even reverse climate change leads to an inevitable clash of ideologies.

An indication of the view that climate change has gone beyond science and economic policy into the realm of ideology comes from comments made by the co-chair of IPCC Working Group III ('Mitigation of Climate Change'), Ottmar Edenhofer, during an interview.[1] He was quoted as saying:

> Basically it's a big mistake to discuss climate policy separately from the major themes of globalization. The climate summit in Cancun at the end of [November 2010] is not a climate conference, but one of the largest economic conferences since the Second World War. … [O]ne must say clearly that we redistribute de facto the world's wealth by climate policy. … One has to free oneself from the illusion that international climate policy is environmental policy. This has almost nothing to do with environmental policy anymore. … [T]here is always the risk that individual rationality leads to collective stupidity. Therefore, one cannot solve the climate problem alone, but it has to be linked to other problems. There must be penalties and incentives: global CO_2 tariffs and technology transfer.

[1] See http://thegwpf.org/ipcc-news/1877-ipcc-official-climate-policy-is-redistributing-the-worlds-wealth.html (viewed November 25, 2010). Edenhofer is the chief economist of the Potsdam Institute for Climate Impact Research and Professor of the Economics of Climate Change at the Berlin Institute of Technology. The interview was in German but was translated.

The ideology is clear: a global system of government is required to implement climate policy and redistribute wealth, presumably to the poor, although the history of governments helping the poor serves more as a warning against such a course of action rather than an endorsement (e.g., see Brown 2009).

If one takes the view that humans are a blot on the landscape of Earth, the only real solution is to get rid of people – to reduce the global population drastically by choosing nature over humans. This view goes back at least 1,800 years to Tertullian, who wrote that humans "weigh upon the world; its resource hardly suffice to support us. As our needs grow larger, so do our protests, that already nature does not sustain us. In truth, plague, famine, wars and earthquakes must be regarded as a blessing to civilization, since they prune away the luxurian growth of the human race" (Beisner 1997, p. 97). This is something that Paul Ehrlich (Ehrlich and Ehrlich 1972, 1990, 1991) could have written in our own day. It is the view that James Lovelock takes with respect to climate change; he takes this view after first favoring the nuclear option but then concluding that only a reduction of the global population by some two-thirds would stop humanity's destruction of nature. It is opposed by solid academic research conducted by Julian Simon (1996) and Bjørn Lomborg (2001, 2007), and many others.

Nothing is new about this debate. In the modern era it has been ongoing since the time of Thomas Malthus in the eighteenth century. Malthus postulated that population growth proceeded geometrically, while growth in agricultural output (natural resources) was arithmetic. As a consequence, human procreation would ensure that people would remain in perpetual poverty – always on the verge of starvation. This view was given a scientific aura in the 1950s and 1960s when Jay Forrester pioneered computer simulation modeling using what became known as 'Systems Dynamics.'[2] In an early application of this computer modeling approach, Forrester (1970) investigated the interactions among human activities and resource availability. In a subsequent Club of Rome sponsored book by Donella Meadows and her colleagues (Meadows et al. 1992, 1972), computer modeling was used to demonstrate the inevitability of systems collapse – mass starvation because population growth had outstripped food production, and collapse of political systems as natural resource scarcity increased strife among nations and between people of the same nation (viz., class conflict). Of course, different assumptions about the model parameters led to opposite outcomes, but these were deemed as less likely.

Economic growth and increasing prosperity, despite large increases in population that resulted from increased lifespans and reduced deaths at birth, proved the models wrong. While the models predicted economic collapse by the early twenty-first century, a collapse brought about by resource scarcity and environmental degradation, the data indicate that resources have not become scarcer over time, but more abundant, that corruption and failed economic systems are culpable in making people's lives miserable, and that environmental degradation is not the root cause of

[2] A history of this approach is available at (as viewed October 21, 2010): http://www.systemdynamics.org/DL-IntroSysDyn/origin.htm.

human misery. Bad governance, lack of trust, low levels of social capital, failure to enforce the 'rule of law,' and other political and economic ills destroy people's lives and make them poor; poverty cannot be attributed to environmental degradation. Nor can environmental degradation be attributed to wealth; indeed, the overwhelming evidence indicates that the quality of the environment has improved as people became wealthier.

Even population growth is forecast to slow and then fall, with earlier predictions that the global population would increase to ten billion or more people by mid century now dramatically reduced. In many rich countries, there is concern that population decline will lead to an inability to provide for those reaching retirement age (because many pension plans, especially public sector plans, are 'pay-as-you-go' schemes). Despite the evidence, doomsayers such as Ehrlich, Meadows and Al Gore continue to predict environmental collapse due to an increasing population.

Climate modeling has replaced the now discredited systems dynamics approach to the interaction between humans and the environment. Climate modeling exudes much greater confidence that scientists have the relationship between human activities and their impact on the environment correct. Whether this is true or not, one cannot help but have the feeling that we have been here before. The threat of resource scarcity and environmental doom now masquerades as catastrophic anthropogenic climate change. Again, excessive numbers of people are the primary cause, although the options are less stark than before. Rather than simply getting rid of people, there is the possibility that we can solve the problem by reducing our reliance on fossil fuels, that we can find a technological fix, that intermediary solutions (such as carbon sinks) exist, and that coordinated government action can lead us into some promised future utopia. Even the threat to individual freedom and the environment posed by greater government involvement, as most evident in a Soviet style system (Brown 2009), can be mitigated through the use of carbon markets, or so we are led to believe.

As demonstrated throughout this book, it is not entirely clear that past temperatures have been lower than current ones – that current (high) temperatures are not simply a part of humanity's historical experience. It is not clear that projections of future temperature from climate models are realistic, given that climate models tend to predict higher temperatures than models based on observational (empirical) evidence. Nor is it clear that the costs of climate change exceed the benefits – humans might prefer a warmer climate if given the choice.

The Earth has been warmer and wetter than at present for some 80% or more of the time, and the Earth's atmospheric concentration of CO_2 has also been much higher in the past than now. The decrease in atmospheric CO_2 was the result of long-term, biota-assisted sequestration into carbonate rocks and altered rocks. Some of this carbon dioxide is now being released back into the atmosphere as a result of fossil fuel burning, an activity that has brought about tremendous advances in standards of living across the globe. Even so, the projected future levels of CO_2 in the atmosphere are nowhere near as high as in former times, when the oceans were no more acidic than now, there was no runaway greenhouse effect, and the rates of change in temperature, sea level and ice were not that much different than today.

Paleoclimatic evidence suggests that major ice ages began when atmospheric CO_2 was higher than now, with increases in CO_2 following temperature rises during deglaciation. In the recent instrumental record, temperature decreases (1880–1910, 1940–1976, 1998–present) run counter to the increase in atmospheric CO_2 and there is no evidence to suggest that greater atmospheric CO_2 causes temperatures to rise. Indeed, there are many factors that affect the Earth's climate, not just greenhouse gases emitted as a result of human activities. Solar activity, the intensity of cosmic rays striking Earth, changes in the Earth's orbit and orientation (e.g., Milankovitch cycles), oceanic events (PDO, ENSO and others), lunar tides and other factors affect the Earth's climate in ways that remain poorly understood. Climate models throw no new light on climate processes because they attribute climate change solely to anthropogenic emissions of carbon dioxide and other greenhouse gases. This is both naïve and irresponsible.

Despite predictions from climate models, it is reality that will have the final say. This aspect is often neglected in the rush to do something. Unless one looks at things through the lens of the real world, and real possibilities, one might impose a greater burden on citizens and the environment than need be the case. In the rush to action, one could bring about political strife that leads to the adverse consequences that have been attributed to anthropogenic global warming, but that are brought about in its absence.

Given that the culprit behind global warming is considered to be the burning of fossil fuels, let us consider them in more detail. Should we reduce fossil fuel use? Can we even do so? Until the science and the economic issues are settled, it seems convoluted to be concerned about one side of the global warming debate but not the other – about the potential damages from global warming and not the benefits.

12.2 A Dilemma for the United Nations

Western countries are increasingly engaged in one of the greatest economic policy experiments ever conducted during peacetime. In the midst of a major recession, the United States, Canada, Europe, Japan and Australia, to one degree or another, are implementing climate policies in an attempt to change the very foundation upon which economic prosperity has been built. They are using the legislative, regulatory and spending powers of the state in an effort to make their economies carbon neutral and nuclear free. The risks associated with these policies are several:

1. It may be impossible to use the powers of the state to bring about the desired end – an economy that greatly reduces reliance on fossil fuels while eschewing nuclear energy. The nuclear option is not considered a viable alternative to fossil fuels for many reasons, but the main one is environmental opposition.
2. The cost to the economy may be too great: it is possible that the economy spirals downward and everyone is made worse off. Not only do per capita incomes decline but health, life expectancy and other measures of wellbeing decline as well, with the least well off in society suffering the most.

3. Given that predictions of climate change are based solely on computer models, and not observational evidence, it is possible that projections of global warming are wrong.
4. Even if projections from climate models are correct, the world might actually be a better place because it is generally warmer. Evidence from the Medieval Warm Period (MWP) and the Little Ice Age (LIA), as provided in the writings of Brian Fagan, Emmanuel Le Roy Ladurie, Jared Diamond and many others, indicates that there were more crop failures, disease and general misery during the LIA than during the MWP.

In February 2010, a group of climate economists and policy experts met at Hartwell House, Buckinghamshire, England, under the auspices of Oxford University and the London School of Economics to reexamine global climate policy (Prins et al. 2010). The background to the meeting was the failure of countries to agree to limit global emissions of CO_2 at the 15th Conference of the Parties to the UN Framework Convention on Climate Change (UNFCCC) at Copenhagen in late 2009. The economists recognized that fossil fuels are both too cheap and too expensive. They are too cheap because they impose a global externality by way of CO_2 emissions that lead to climate change, but they are also too expensive because many poor people lack access to sufficient energy to enable them to escape poverty.

The Economist (September 25, 2010, p. 117) reported that, in 2009, 1.44 billion people lacked access to electricity and all but three million of these lived outside the developed countries. Worse yet, some 2.7 billion still cook their food on inefficient stoves that use dung, crop residues and fuel wood. It is estimated that perhaps two million people die prematurely each year because of health problems associated with biomass-burning stoves (p. 72). Collection of biomass for burning occupies much time (mainly of women and children), robs cropland of important nutrients that can only partly be replaced by artificial fertilizers from offsite, and causes deforestation. It also results in emissions of black carbon that enhance global warming and increase the amount of melting from glaciers.

One-quarter to one-third of the world's population needs to be provided with electricity and high-density energy, such as can currently only be found in fossil fuels, so that they can live decent lives and have some hope that their children will lead a better life than they. It would be immoral to deny the poor the ability to develop by curtailing their access to cheap energy, all in pursuit of an environmental objective that only interests one billion rich people. Ottmar Edenhofer's idea of transferring massive amounts of wealth from rich to poor countries through climate policy is simply untenable as past experience with development aid has demonstrated all too well – Africa remains a continent left behind despite having been the recipient of massive transfers of income (Moyo 2009).

12.2.1 The Dilemma

We now have a *huge dilemma*: We can pursue the rich world's environmental climate objective only by denying developing countries the cheap energy needed for economic development. It is important to recognize that there are sufficient fossil fuels and they can be made available cheaply enough to drive the economic development of the least developed nations. The problem is not lack of resources; it is the obstacles that both rich and poor countries put in the way of exploration, development, transportation and distribution of energy. Rich countries block exploitation of all sorts of natural resources on the grounds of their potential adverse environmental impacts, while poor governance, corruption and failure of rule of law hinder all aspects of the energy supply chain, resulting in huge waste.[3] Sources of energy are plentiful enough to drive economic development, and they can be made available at low cost to developing countries. The problems are a lack of will to do so and the fact that the energy resources are in the form of hydrocarbons.

The same issue of *The Economist* (September 25, 2010) also had a lead article pertaining to the UN's Millennium Development Goals (MDGs) that, among other targets, aim to halve the number of people living below $1.25 per day by 2015 (United Nations 2011a). That target and other MDG targets seem to be within reach because of economic growth in China. Despite this, nearly one billion people continue to live in abject poverty. Interestingly, the UN's MDGs do not talk about economic development, but economic growth is the only way to meet the MDG's targets. Economic development cannot occur without energy – vast amounts of which are required when we consider that one-quarter to one-third of the world's population lacks access to electricity. Unfortunately, high-quality, high-density energy is only available from fossil fuels.

The dilemma is of course that the rich countries have agreed, through the United Nations, to provide aid to poor countries so that their standards of living will converge to those of the developed world. But, by signing onto the UN's Framework Convention on Climate Change process, rich countries have also agreed to decarbonize the global economy. These objectives are incompatible. China and India recognize this all too well, which is why they refuse to allow rich countries to seduce them into limiting their greenhouse gas emissions. The incompatibility between these goals led to the debacle at the climate-change summit in Copenhagen in late 2009, the failure to reach agreement at a subsequent conference in Cancun, Mexico (2010), and the mixed success in Durban, South Africa (December 2011).[4]

---

[3] One example of such waste is the flaring (burning) of natural gas from oil wells. It is done because it is too expensive to market, largely due to lack of transportation infrastructure. The practice was halted in the U.S. in 1947, but it continues in petroleum producing countries of the Middle East, North Africa, Russia and the petro states around the Caspian Sea. Each year some 150 billion cubic meters of natural gas is flared globally, equivalent to about 30% of European consumption (Bryce 2010, pp. 226–227, 297).

[4] At Durban nations agreed to continue the Kyoto process but extend it to developing countries, which now account for 58% of global CO_2 emissions. However, whether the agreement is legally binding is (purposely) vague, emissions reduction targets are to be negotiated by 2015, the agreement is not to come into effect until 2020, and there is to be a Green Climate Fund that, beginning in 2020, will provide poor countries with $100 billion annually to help them reduce emissions and adapt to climate change (*The Economist*, December 17, 2011, p. 138). Much remains to be negotiated, but it would appear that, if climate scientists are correct, global warming will continue for some time to come.

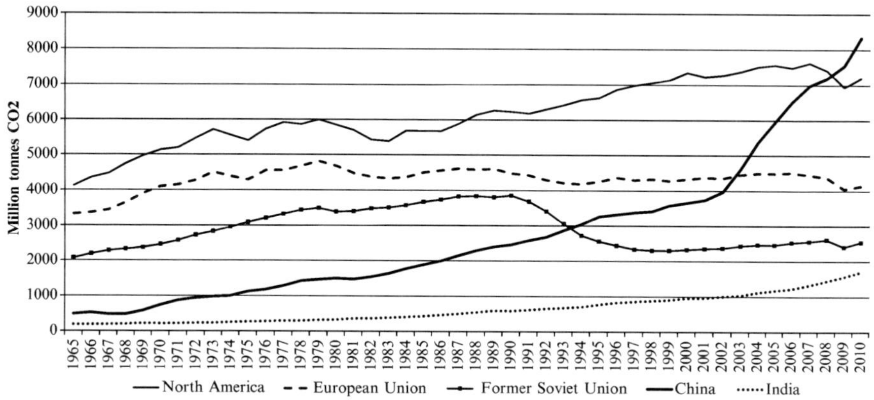

Fig. 12.1 CO_2 emissions from fossil fuel burning, selected regions, 1965–2010

12.2.2 Heads in the Sand: The Ostrich Effect

What has been the response of the developed countries to the aforementioned dilemma? Surprisingly, rather than focus efforts on helping poor countries access sources of energy that would enable the economic growth required for these economies to adapt to the negative effects of climate change, rich countries are acting as if there is no dilemma whatsoever. They are ramping up efforts to de-carbonize their own economies while continuing to threaten and cajole developing countries into doing the same. The developing countries have simply rejected such efforts, continuing to expand their energy consumption and CO_2 emissions as fast as they can. China is in the forefront, with India coming on and others likely to follow in the not-too-distant future.

Consider the evidence. A graph of carbon dioxide emissions from fossil fuel burning for selected regions or countries is provided in Fig. 12.1.[5] By 2010, U.S. CO_2 emissions were 12.9% higher than its 1990 baseline emissions under the Kyoto Protocol; Germany's emissions fell by 19.6% and those of the UK and the Russian Federation by 12.0% and one-third, respectively, while Japanese emissions rose by 13.0%. However, it is important to remember that German, UK and Russian emissions fell because of extenuating circumstances related to German reunification (and the closing of many inefficient manufacturing and power generating facilities in the East), the closure of inefficient coal mines and coal-fired power plants in the UK, and the break-up of the Soviet empire and its subsequent economic demise. EU emissions as a whole fell by only 7.6%, with emissions rising in each of these

[5] Data for Figures 12.1 and 12.2 are from the BP Statistical Review of World Energy June 2011 as found at (viewed December 2, 2011): http://www.bp.com/statisticalreview. Corroborating and more detailed historical data are available from T. Boden, G. Marland and B. Andres at the Carbon Dioxide Information Analysis Center, Oak Ridge National Laboratory in Tennessee as found at: http://cdiac.ornl.gov/trends/emis/meth_reg.html.

regions or countries in the most recent year. With the exception of the ex-Soviet States and a few western European countries, most notably Germany and the UK, countries have generally failed to achieve their Kyoto targets.

Global carbon dioxide emissions increased by 46.6% between 1990 and 2010. Clearly, CO_2 emissions increased throughout the developing world as evident from emissions data for China and India. However, the emissions data in Fig. 12.1 fail to take into account the shift in manufacturing from countries such as the UK and Japan to developing countries such as Indonesia and Malaysia. That is, for many rich countries, emissions embodied in imports have risen to such an extent that, if appropriately accounted for, these countries' overall emissions would be significantly higher today than they were in 1990.

Coal, oil and natural gas consumption for selected regions are graphed for the period 1990–2010 in Fig. 12.2. Coal is primarily used for generating electricity and making steel, while gas is used for generating electricity and for space heating. The main use of oil is transportation, although oil is also important for space heating, electricity and the manufacture of a host of products that include plastics. Global consumption of fossil fuels has risen tremendously during the past decade; coal consumption rose by 60.1%, oil by 23.4% and natural gas by 61.5%.

The pattern of coal consumption is shown in Fig. 12.2a. Consumption by the U.S. and Japan (not shown) has remained relatively flat between 1990 and 2010, while it declined in the EU. The two principal reasons for the decline in EU consumption relate to closure of inefficient manufacturing due to German unification and closure of inefficient coal collieries in the UK. Again, some coal-consuming activities, especially steel making, have shifted abroad. Coal consumption by the 'rest of the world' declined in the early 1990s because of the demise of the Soviet empire, but it has risen significantly since then. Indian consumption of coal, for example, rose slowly over the period and should overtake that of the U.S. within the next several years. However, the important thing to note in Fig. 12.2a is Chinese coal consumption, which has increased some threefold since 2000.

A similar picture emerges from Fig. 12.2b and c. While oil use in the U.S. and EU has remained relatively stable, or fallen slightly, since 1990, global consumption has increased as a result of increased use elsewhere. Meanwhile, consumption of natural gas since 1990 has increased in both the U.S. and EU, by 25.7 and 50.7%, respectively, but that in other countries increased substantially more. Indeed, the countries that agreed to reduce domestic emissions of greenhouse gases at Kyoto in 1997 no longer make much impact on changes in fossil fuel use. The world has changed, with Brazil, China, India and other developing countries slowly moving out of poverty, much as expected under the UN's Millennium Development Goals (MDGs).

Chinese emissions of CO_2 from other sources are growing as well. China is adding 5,000 km of freeways every year, and Beijing now has an incredible nine ring roads. Clearly, the country is gearing up for greater use of automobiles. Indian emissions are also set to take off as that country develops. During the past two decades, Chinese CO_2 emissions from coal grew by an average 6.2%, while those of India grew by an average 5.7%, although baseline emissions are much lower in India than China.

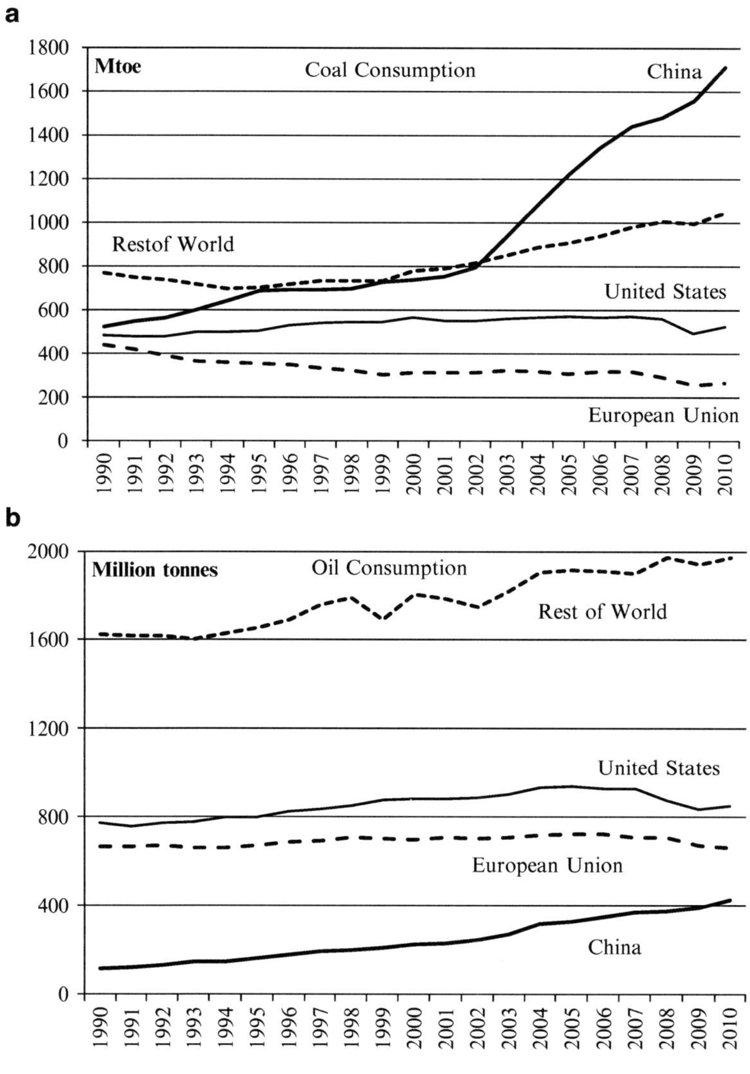

Fig. 12.2 Coal, oil and natural gas consumption, selected regions, 1990–2010

Installed electrical generating capacity in most countries has remained relatively unchanged over the period 1990–2008, with the exception of the United States and China (Fig. 12.3). U.S. capacity has increased by some 276 GW (or 38%), while that of China increased by a whopping 659 GW (578%) and India by 102 GW (237%). Notice that Canada's generating capacity is about the same as that of Germany; Canada is a larger exporter of hydropower to the U.S., while Germany imports power from other European states. Despite the increases in Chinese CO_2 emissions, the country is the world's leader in generating electricity from renewable sources,

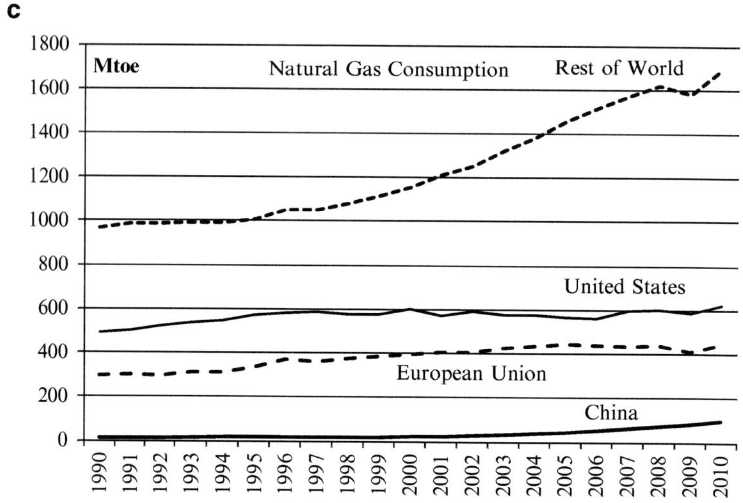

Fig. 12.2 (continued)

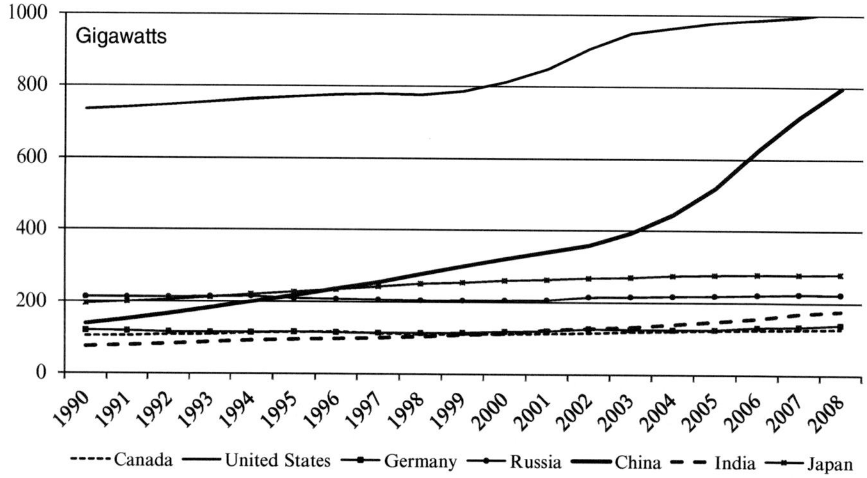

Fig. 12.3 Installed generating capacity, selected countries, 1990–2008

having overtaken the United States in 2004. China is also looking to nuclear power, with 27 nuclear power plants currently under construction and 160 planned or proposed (Table 10.8), each with a capacity of about 1,000 MW. This has made China a power house in the production of wind turbines and an exporter of nuclear technology, sometimes to unsavory states. Not surprisingly, China now appears amenable to continuing the Kyoto process, because it has been a major beneficiary under Kyoto. It sells carbon offsets to rich countries, exports solar panels and wind turbine parts (see below), benefits from access to rich-country energy technologies, and uses the carbon card to justify the expansion and promotion of nuclear technology.

One thing is abundantly clear from the graphs. No matter what rich western countries are doing about CO_2 emissions, global emissions of CO_2 will continue to rise inexorably. Nothing the Americans do, nothing the Europeans do, nothing the Japanese do can prevent global warming. Consider this: In just over 2 years, the *increase* in Chinese emissions of CO_2 from coal generation alone has exceeded the emissions of greenhouse gases, measured in CO_2 equivalence, of the entire Canadian economy. China is currently adding 1,000 MW of installed coal-fired generating capacity every week, and Chinese consumption of coal in 2010 exceeded the total consumption of Germany, Russia, India, Japan and the United States combined! Despite this, China's generating capacity lags that of the United States by more than 20%, although total generation of electricity lags that of the U.S. by less than 5%, because the U.S. imports electricity from its neighbor, Canada, while China has no such option.

The response of rich nations has been to stick to the ill-advised UNFCCC Kyoto process as the roadmap to follow, and attempt to impose it upon the rest of the globe. In September 2010, U.S. Senators again introduced a bill establishing a Renewable Energy Standard (RES). It would require 3% of electricity to be generated from renewable sources by 2012 and 15% by 2021, and similar to the generous feed-in tariffs provided by the province of Ontario, it would provide huge subsidies to wind and solar companies. The costs to the Ontario treasury of its feed-in tariff program were estimated in Chap. 10 at well over $1.5 billion per year, although budgetary pressures will cause politicians to pass costs onto electricity consumers in the form of large rate hikes or simply not enter into agreements with proponents wishing to benefit from Ontario's largesse. In terms of climate change, the Ontario program reduces emissions at a cost of hundreds of dollars per ton of CO_2 (when CO_2 emission credits are trading for about \$25/tCO_2 in Europe), but does absolutely nothing to forestall global warming because of what is happening in China, India and elsewhere. The same can be expected of the U.S. program and similar programs in Europe, where targets require countries to achieve 20% renewable energy in the production of electricity by 2020.

Despite the fact that none of these programs, even collectively, can impact climate change, why do governments continue to pursue them? One reason is the mistaken notion that large subsidies will lead to greater employment and the development of a renewable energy sector that is a global leader. Every country believes it will be the global leader in the development of wind turbines and/or solar panels. However, research indicates that public funds directed at the renewable energy sector actually reduce employment by crowding out private sector investment or public infrastructural investments elsewhere in the economy (e.g., investments in transportation infrastructure that reduce costs of moving goods and people) (Álvarez et al. 2009; Morriss et al. 2009). Indeed, it appears that the main winner from efforts by countries to expand wind and solar output are the Chinese. China currently controls the supply of rare earth minerals which are used to make solar panels and parts of wind turbines, among other things (*The Economist*, October 2, 2010, p. 64).

China restricts exports of these minerals because it desires to export the manufactured products in which they are found (Humphries 2010). China gains from rich-country subsidies to solar and wind producers, while its economic growth is the main factor driving up greenhouse gas emissions. But who can deny China its economic growth given that individuals are lifted out of poverty in a major way.

The other reason for pursuing the Kyoto roadmap comes from environmental groups and the media, which together have convinced politicians to do something about reducing greenhouse gas emissions and the so-called carbon footprint. Doing something, anything, is not always wise. Economists have long known that governments cannot pick winners and, worse, government subsidies can lock-in technologies that become a hindrance to more efficient energy use rather than a solution. An example is biofuels.

Production of energy crops raises land and food prices and leads to deforestation (Searchinger et al. 2008). Ethanol from corn and biodiesel from canola actually increase rather than decrease greenhouse gas emissions (Crutzen et al. 2008); they raise food prices causing greater numbers of poor to suffer. Biofuels are not competitive with petroleum even when oil prices are at $150 per barrel, which is why subsidies (or mandates) are needed. When subsidies end, the production facilities and distribution network remain in place, and constitute a sunk cost. Even though the original investors may go bankrupt, production of biofuels continues. The same is true of wind farms and solar facilities. Alternatively, when subsidies end, facilities are abandoned and the public is left the task and expense of cleaning them up. Meanwhile, subsidies to solar, wind or some other renewable option tilt the playing field against new energy sources or technologies that are better at reducing greenhouse gas emissions and improving energy efficiency.

12.3 Concluding Discussion

Governments have been culpable in failing to consider options for addressing global warming other than reductions in emissions of greenhouse gases. A government often sets unrealistic emissions reduction targets, knowing full well that it will be up to a future government to deal with the consequences. For example, Canada's annual CO_2 emissions currently amount to approximately 730 Mt, more than 20% above what they were in 1990. Canada's Kyoto target is 571 Mt CO_2 by the end of 2012; clearly, there is no way that Canada will achieve its commitment and, if you read the literature at the time that Canada agreed to reduce its emissions by 6% from the 1990 level, you recognize that policymakers were well aware that the country could not achieve this target (van Kooten 2004, pp. 107–120). Indeed, the previous government knew full well it would never meet the target, even when it ratified the Kyoto Agreement. Canada simply wanted to be seen as a leader on the international stage.

Other countries are just as guilty. In Chap. 9, we noted that many emission targets proposed by U.S. senators and representatives in the Congress, and by the G8 and the European Union, are simply not based in reality. Not only are the required technologies inadequate, but extant options such as wind and solar power are viewed through rose-colored glasses. As demonstrated in Chap. 11, simply installing renewable generating capacity and assuming it replaces an equivalent coal-fired capacity does not bring about an equivalent reduction in fossil-fuel CO_2 emissions. A claim that a country has increased its reliance on renewable energy from 20 to 45% needs to be examined in detail.[6] If a country installed significant wind generating capacity in recent years, it may well be that renewable sources of energy account for 45% of total generating capacity, but delivered power will be substantially less.

In the real world, renewable energy sources cannot easily substitute for fossil fuels, and countries are discovering that the costs of greater reliance on renewable energy are much higher than anticipated. Nor can one introduce nuclear generating capacity quickly enough to meet government mandated targets, even if all objections to the use of nuclear power are set aside. And, finally, none of the targets envisioned by rich countries will do anything to prevent global warming. For example, China's growth in emissions of greenhouse gases will more than offset any reductions achieved by rich countries. But it may also be the case that anthropogenic emissions of carbon dioxide and other greenhouse gases only tell part of the story of climate change, and that emissions reduction will not serve to lower global temperatures to the extent envisioned by climate modelers.

Now, any reasonable person would take the view that Canada, the U.S., Europe and other rich countries should not spend untold millions of dollars to reduce their CO_2 emissions. As shown in the preceding chapters, much of the money is simply wasted – but not all. It is likely beneficial to spend some money on research and development into alternative energy sources, and to implement a forward looking energy policy, one that leads to greater output per unit of energy input. However, we need to be realistic. Otherwise, we violate the precautionary principle, but in a way that is totally unexpected!

Although the current focus of mitigation policy is emissions reduction, there are other ways to address global warming, if the concern is truly to avoid higher perhaps even catastrophic temperatures, although there are few climate scientists who would seriously suggest that a catastrophe is imminent. For example, engineered solutions to the climate problem that do not involve reducing carbon dioxide and other greenhouse gas emissions are increasingly considered to be viable. So too is adaptation, an option that has been sorely neglected. Climate scenarios that predict high average global temperatures in the future are based on the assumption that per capita incomes in poor countries will converge to those of rich countries, that the developed countries

[6] In an extremely optimistic report that might be considered more propaganda than science, the United Nations (2011b. pp. xiii, 54) points to Portugal as an example of a country quickly shifting from fossil fuels to clean sources of energy. Portugal increased its share of renewable power from 17 to 45% in just 5 years (2005–2010), but it did so by relying more on hydropower (more hydroelectric capacity was brought on line, although construction may have exceeded 5 years) and likely wind power. In both cases, installed capacity should not be confused with actual generation. Insufficient water and lack of wind reduce the ability of renewable sources to generate power, something that is not a problem for coal, gas or nuclear power plants.

will be substantially richer than now, and that there will be little phasing out of fossil fuels. Few regions in the world would remain so poor that they could not possibly adapt to a warmer world, while higher per capita incomes will reduce the carbon intensity of economies.

Perhaps the main conclusion to be drawn from the analyses and discussions in this book is that adaptation to climate change is the most sensible policy approach to take. "The green pressure groups and politicians who have driven the debate on climate change have often been loth to see attention paid to adaptation, on the ground that the more people thought about it, the less motivated they would be to push ahead with emissions reduction" (*The Economist* November 29, 2010, p. 86). This is not to suggest that mitigation should be ignored completely; there is some room for mitigation, although many economists would not consider large-scale expenditures to mitigate climate change worthwhile.

Mitigation is likely to be too costly from a social perspective, with most of the costs borne by the poor. And mitigation is unlikely to be effective in preventing global warming – the concentration of carbon dioxide in the atmospheric keeps increasing, human emissions show no sign of slowing, and international negotiations to halt climate change are floundering. Therefore, climate engineering and adaptation are preferred to mitigation for dealing with uncertainty, particularly since the science is less than adequate for making firm statements about what is likely to happen to the Earth's climate in the future. But then 'muddling through' (Lindblom 1959) is not such a bad approach to policy in such circumstances.

References

Álvarez, G. C., Jara, R. M., Julián, J. R. R., & Bielsa, J. I. G. (2009, March). *Study of the effects on employment of public aid to renewable energy sources*. Lessons from the Spanish renewables bubble. Borrador draft.. Retrieved July 9, 2010, from http//www.juandemariana.org/pdf/090327-employment-public-aid-renewable.pdf/

Beisner, E. C. (1997). *Where garden meets wilderness*. Grand Rapids: William B. Eerdmans Publishing.

Brown, A. (2009). *The rise and fall of communism*. Toronto: Doubleday Canada.

Bryce, R. (2010). *Power hungry: The myths of "green" energy and the real fuels of the future*. New York: Public Affairs.

Crutzen, P. J., Mosier, A. R., Smith, K. A., & Winiwarter, W. (2008). N_2O release from agro-biofuel production negates global warming reduction by replacing fossil fuels. *Atmospheric Chemistry and Physics, 8*(2), 389–395.

Ehrlich, P. R., & Ehrlich, A. H. (1972). *Population, resources, environment: Issues in human ecology* (Vol. 2). San Francisco: W.H. Freeman.

Ehrlich, P. R., & Ehrlich, A. H. (1990). *The population explosion*. New York: Random House.

Ehrlich, P. R., & Ehrlich, A. H. (1991). *Healing the planet*. New York: Addison Wesley.

Forrester, J. W. (1970). *World dynamics*. Portland, OR: Productivity Press.

Humphries, M. (2010). *Rare earth elements: The global supply chain* (Congressional Research Service Report # R41347/7-5700, pp. 17). Washington, DC: U.S. Congress.

Lindblom, C. E. (1959). The science of "muddling through". *Public Administration Review, 19*(2), 79–88.

Lomborg, B. (2001). *The skeptical environmentalist: Measuring the real state of the world.* Cambridge: Cambridge University Press.

Lomborg, B. (2007). *Cool it: The skeptical environmentalist's guide to global warming.* New York: Knopf.

Meadows, D. H., Meadows, D. L., Randers, J., & Behrens, W. W., III. (1972). *The limits to growth: A report for the Club of Rome's project on the predicament of mankind.* New York: Universe Books.

Meadows, D. H., Meadows, D. L., & Randers, J. (1992). *Beyond the limits: Confronting global collapse, envisioning a sustainable future.* Post Mills: Chelsea Green.

Morriss, A. P., Bogart, W. T., Dorchak, A., & Meiners, R. E. (2009). *Green jobs myth* (University of Illinois Law and Economics Research Paper Series No. LE09-001). Retrieved August 11, 2009, from http://ssrn.com/abstract=1358423

Moyo, D. (2009). *Dead aid. Why aid is not working and why there is a better way for Africa.* Vancouver: Douglas & McIntyre.

Prins, G., Galiana, I., Green, C., Grundmann, R., Hulme, M., Korhola, A., Laird, F., Nordhaus, T., Pielke, R. A., Jr., Rayner, S., Sarewitz, D., Shellenberger, M., Stehr, N., & Tezuka, H. (2010, May). *The Hartwell paper: A new direction for climate policy after the crash of 2009* (pp. 42). London, UK: Institute for Science, Innovation & Society, University of Oxford; LSE Mackinder Programme, London School of Economics and Political Science.

Searchinger, T. D., Heimlich, R., Houghton, R. A., Dong, F., Elobeid, A., Fabiosa, J., Tokgoz, S., Hayes, D., & Yu, T. (2008, February 29). Use of U.S. croplands for biofuels increases greenhouse gases through emissions from land-use change. *Science, 319*(5867), 1238–1240.

Simon, J. L. (1996). *The ultimate resource 2.* Princeton: Princeton University Press.

United Nations. (2011a). *The millennium development goals report 2011* (pp. 72). New York: United Nations.

United Nations. (2011b). *World economic and social survey 2011: The great green technological transformation* (E/2011/50/Rev.1.ST/ESA/333. Dept. of Social and Economic Affairs, 251 pp.). New York: United Nations.

van Kooten, G. C. (2004). *Climate change economics: Why international accords fail.* Cheltenham/Northampton: Edward Elgar.

Index

G.C. van Kooten, *Climate Change, Climate Science and Economics: Prospects for an Alternative Energy Future*, DOI 10.1007/978-94-007-4988-7,

D